PURINE AND PYRIMIDINE METABOLISM IN MAN VI

Part B: Basic Research and Experimental Biology

ADVANCES IN EXPERIMENTAL MEDICINE AND BIOLOGY

Recent Volumes in this Series

Volume 247A
KININS V, Part A
Edited by Keishi Abe, Hiroshi Moriya, and Setsuro Fujii

Volume 247B
KININS V, Part B
Edited by Keishi Abe, Hiroshi Moriya, and Setsuro Fujii

Volume 248
OXYGEN TRANSPORT TO TISSUE XI
Edited by Karel Rakusan, George P. Biro, Thomas K. Goldstick, and Zdenek Turek

Volume 249
MINERAL ABSORPTION IN THE MONOGASTRIC GI TRACT:
Chemical, Nutritional, and Physiological Aspects
Edited by Frederick R. Dintzis and Joseph A. Laszlo

Volume 250
PROGRESS IN POLYAMINE RESEARCH:
Novel Biochemical, Pharmacological, and Clinical Aspects
Edited by Vincenzo Zappia and Anthony E. Pegg

Volume 251
IMMUNOBIOLOGY OF PROTEINS AND PEPTIDES V: Vaccines—Mechanisms, Design, and Applications
Edited by M. Zouhair Atassi

Volume 252
DRUGS, SYSTEMIC DISEASES, AND THE KIDNEY
Edited by Alberto Amerio, Pasquale Coratelli, Vito M. Campese, and Shaul G. Massry

Volume 253A
PURINE AND PYRIMIDINE METABOLISM IN MAN VI, Part A:
Clinical and Molecular Biology
Edited by Kiyonobu Mikanagi, Kusuki Nishioka, and William N. Kelley

Volume 253B
PURINE AND PYRIMIDINE METABOLISM IN MAN VI, Part B:
Basic Research and Experimental Biology
Edited by Kiyonobu Mikanagi, Kusuki Nishioka, and William N. Kelley

---

PURINE AND PYRIMIDINE METABOLISM IN MAN VI

Part B: Basic Research and Experimental Biology

Edited by

Kiyonobu Mikanagi
Mikanagi Gotanda Hospital
Tokyo, Japan

Kusuki Nishioka
Tokyo Women's Medical College
Tokyo, Japan

and

William N. Kelley
University of Michigan Medical Center
Ann Arbor, Michigan

PLENUM PRESS • NEW YORK AND LONDON

Library of Congress Cataloging in Publication Data

International Symposium on Human Purine and Pyrimidine Metabolism (6th: 1988: Hakone-machi Japan)
Purine and pyrimidine metabolism in man VI.

(Advances in experimental medicine and biology; v. 253)
"Proceedings of the Sixth International Symposium on Human Purine and Pyrimidine Metabolism, held July 17–21, 1988, in Hakone, Japan"—T.p. verso.
Includes bibliographies and index.
Contents: pt. A. Clinical and molecular biology—pt. B. Basic research and experimental biology.
1. Purines—Metabolism—Congresses. 2. Pyrimidines—Metabolism—Congresses. I. Mikanagi, Kiyonobu. II. Nishioka, Kusuki. III. Kelley, William N., 1939- . IV. Title. V. Series. [DNLM: 1. Purine-Pyrimidine Metabolism, Inborn Errors—congresses. 2. Purines—metabolism—congresses. 3. Pyrimidines—metabolism—congresses. W1 AD559 v. 253 WD 205.5.P8 I605p 1988]
QP801.P8I55 1988 612.3'9 89-8546

ISBN 978-1-4684-5678-3

ISBN 978-1-4684-5678-3 ISBN 978-1-4684-5676-9 (eBook)
DOI 10.1007/978-1-4684-5676-9

Proceedings of the Sixth International Symposium on Human Purine and Pyrimidine Metabolism, held July 17–21, 1988, in Hakone, Japan

Softcover reprint of the hardcover 1st edition 1989

A Division of Plenum Publishing Corporation
233 Spring Street, New York, N.Y. 10013

PREFACE

These two volumes contain articles presented at the VIth International Symposium on Human Purine and Pyrimidine Metabolism held in Hakone, Japan, July 17 trough 21, 1988. The first meeting of this series of symposia convened in Tel Aviv, Israel, and since then meetings have taken place every three years in various parts of the world. The second meeting was held in Baden, Austria, the third in Madrid, Spain, the fourth in Maastricht, the Netherlands, and fifth in San Diego, California. The VIth meeting in Hakone marked the first such symposium held in Asia.

On occasion of publishing these books, I would like to describe how research in this field has evolved in Japan. Early in the 1950s, I was engaged in clinical practice treating various rheumatic diseases as an orthopedicist, and found that a substantial percentage of our patients had symptoms apparently compatible with gout. During the 1960s, the number of these gouty patients increased, and in the 1970s, research on the pathogenesis of gout was performed on the basis of approximately 2,000 cases of this disease, together with precise epidemiological studies concerning gouty and hyperuricemic individuals. Data derived from the two kinds of study had greatly changed the notion that gout was a rare disease among Japanese. My clinical studies have been succeeded by research at the molecular level on various purine metabolic abnormalities, including not only gout but also other diseases with various symptoms of wide clinical spectra. The latter studies were performed by various young researchers who had returned from such institutions as the University of California at San Diego, and the Scripps Clinics and Research Foundation.

The first milestone in the progress of this research field was the introduction by Rundles, Wyngaaden, Hitchings and others of such purine analogs as 6-mercaptopurine and, allopurinol. At that time cancer was introduced to the field of purine metabolism. To our great pleasure, we were informed while editing these books that Dr. Hitchings had won the 1988 Nobel Prize for medicine and physiology. Furthermore, Dr. Elion, who shares the prize with Dr. Hitchings and the other, is one of the researchers in our field who revealed a new, powerful aspect of purine and pyrimidine metabolism by introducing such analogs as acycloguanosine. Considering the fact that drugs effective for the treatment of AIDS and other viral diseases are virtually limited to purine and pyrimidine analogs, the future of this new field of purine and pyrimidine metabolism seems promising.

Another breakthrough in the research in this field was the discovery of an enzyme deficiency in Lesch-Nyhan syndrome by Seegmiller, Rosenbloom and Kelley. Later, when Giblett et al found two enzyme deficiencies causing immunodeficiencies, immunology entered the field of purine metabolism. The purine field continues to include other areas in biology and medicine owing to the discoveries of other abnormalities such as xanthine oxidase deficiency, adenine phosphoribosyltransferase deficiency and adenylosuccinase deficiency.

One of the remarkable features of the VIth International Symposium on Human Purine and Pyrimidine Metabolism was that the great power of the recombinant DNA technique was sufficiently demonstrated. For the diagnosis of various diseases and the analysis of disease-causing genes, the gene cloning and DNA sequencing techniques enjoyed adequate application. Moreover, the same techniques will also be applied to the therapeutic aspect of genetic diseases. Thus, gene therapy wherein normal genes are introduced to patients with genetic diseases, most likely will experience practical use in the near future.

Looking back at the progress of research in this field reminds us that the field of purine and pyrimidine metabolism has continuously expanded. Such prosperity probably has some relation to the fact that these compounds are the fundamental genetic codes in all biological materials. Taking all of this into account, the future development of research in this field seems guaranteed.

I would like to express my sincere gratitude to organizing and scientific committee members for their help in organizing and managing the symposium. In addition, I particularly thank two doctors, Naoyuki Kamatani and Hisashi Yamanaka, both at the Tokyo Women's Medical College, for their tremendous contributions. Without their devotion, neither the successful administration of the symposium nor the publication of these books would have been possible. I also thank the leading members of the Tokyo Women's Medical College and Japan Gout Research Foundation for their ongoing support of our endeavors.

The next symposium will meet in England during 1992, as organized by Dr. Simmonds and others. In closing, let us all hope for the further development of research in the field of purine and pyrimidine metabolism and the continuous success of this series of symposia as it unfolds through future decades.

Kiyonobu Mikanagi, M.D.
Chairman
Organizing Committee
VIth International Symposium on Human
Purine and Pyrimidine Metabolism

CONTENTS

ANTI-VIRAL PURINE AND PYRIMIDINE ANALOGS

ADENOSINE AND ADENOSINE RECEPTORS

PURINE METABOLISM IN VARIOUS CELLS

NUCLEOSIDE TRANSPORT

RAT LIVER PHOSPHORIBOSYLPYROPHOSPHATE SYNTHETASE : EXISTENCE AS HETEROGENEOUS AGGREGATES AND IDENTIFICATION OF THE CATALYTIC SUBUNIT

Kazuko Kita, Toshio Otsuki, Toshiharu Ishizuka
Sumio Ishijima, and Masamiti Tatibana

Department of Biochemistry
Chiba University School of Medicine
Inohana, Chiba 280, Japan

INTRODUCTION

5-Phosphoribosyl 1-pyrophosphate (PRPP) synthetase serves as the primary precursor of ribose monophosphate moiety of all nucleotides [1-3] and is a critical control factor for de novo synthesis of purines [3,4] and pyrimidines [5,6]. PRPP synthetase has been purified from *Salmonella typhimurium* [7], human erythrocytes [8], rat liver [9], and *Escherichia coli* [10]. The enzyme exists in many aggregated forms of various molecular mass [8,11], reportedly composed of 31-kDa to 40.5-kDa subunit [7-11]. Findings of the regulatory properties heretofore reported are not always in agreement and the control factors would not fully account for the *in situ* regulation of the enzyme.

We now report purification of rat liver PRPP synthetase to a specific activity, the highest value as a mammalian enzyme in the literature. The most purified enzyme is not homogeneous and together with the available evidence, we propose that this enzyme exists as complex aggregates of heterogeneous components. We identified the 34-kDa species as the catalytic subunit and determined amino acid sequence of the N-terminal region. In parallel with the present studies, we cloned cDNA for the enzyme [12]. More detailed description of the present results will be published elsewhere.

METHODS

Enzyme assay

PRPP synthetase was assayed essentially as described by Fox and Kelley [8]. Silicone-coated glassware was used for the purified enzyme to minimize the adsorption to the glass surface, in association with inactivation. One unit of the enzyme activity was defined as the amount catalyzing formation of 1 μmol of PRPP per min.

Enzyme purification

All buffers used contained enzyme-stabilizing agents [8,9], 0.3 mM ATP, 6 mM $MgCl_2$, 0.1 mM EDTA, and 2.5 mM 2-mercaptoethanol. The pH values indicated refer to those at 4°C and all steps were carried out at 4°C.

Livers from male Sprague-Dawley rats were homogenized with 9 volumes of homogenizing buffer containing 0.25 M sucrose, 20 mM potassium phosphate, pH 7.4, and 0.1 mM PMSF. The homogenate was centrifuged at 10000 x g for 20 min and the supernatant recovered. PRPP synthetase in the supernatant was precipitated between 3.8% and 6.8% (w/v) of polyethylene glycol 6000. The precipitate was dissolved in 50 mM potassium phosphate, pH 7.4, containing 0.1 mM PMSF. Protein concentration was adjusted to 10 mg/ml and the enzyme in the solution was precipitated between pH 6.1 and 5.36. After centrifugation, the precipitate was taken up in 20 mM potassium phosphate, pH 7.4, containing 0.1 mM PMSF. The acid precipitate fraction (60 ml from 600 g of liver) was adjusted to a final Pi concentration of 80 mM, and to a final volume of 240 ml, and mixed with hydroxyapatite (100 ml as pellet) equilibrated with 80 mM potassium phosphate. After centrifugation, the supernatant was recovered, and the hydroxyapatite was washed with 92 mM potassium phosphate. The supernatants were combined and the enzyme was precipitated at pH 5.8 with 10% polyethylene glycol.

The pellet was dissolved in 30 mM potassium phosphate, pH 7.4, and fractionated on a Toyopearl HW-65F column (2.7 x 60 cm), using the same buffer, plus 0.05% Triton X-100, at a flow rate of 85 ml/h. Active fractions were pooled and applied to a DEAE-Toyopearl 650S column (1.8 x 8.0 cm) equilibrated with 30 mM potassium phosphate, pH 7.0. Elution was performed first with 140 mM potassium phosphate, pH 6.3, and then with a linear gradient of 0 to 0.17 M KCl at a flow rate of 85 ml/h. The enzyme fractions (at about 0.08 M KCl) were dialyzed against 5 mM potassium phosphate, pH 6.8, and concentrated by adsorption and elution from a small hydroxyapatite column. The eluted enzyme in 250 mM potassium phosphate, pH 7.4, was dialyzed against 50 mM potassium phosphate, pH 7.4. A portion of the preparation was fractionated on a TSK G4000SW column (0.75 x 60 cm), using 50 mM potassium phosphate, pH 7.4, at a flow rate of 0.8 ml/min. Active fractions were pooled and stored at -80° C.

Dissociation of aggregated enzyme by gel filtration on TSK G2000SW in the presence of 1 M $MgCl_2$

The enzyme solution was diluted with an equal volume of 2 M $MgCl_2$, adjusted to pH 6.0, and then fractionated on a TSK G2000SW column (0.75 x 60 cm), using 50 mM MES, pH 6.0, containing 1 M $MgCl_2$, and 5 mM potassium phosphate, at a flow rate of 0.8 ml/min.

RESULTS

Enzyme purification

A typical result of purification is summarized in Table 1. The total recovery of the activity was 17% and the enzyme was purified 9800-fold relative to the tissue homogenate, to a specific activity of 7280 milliunits/mg, the highest value for a mammalian PRPP synthetase so far described. The enzyme emerged near the void volume in gel filtration on Toyopearl HW-65F, thereby indicating that the enzyme has molecular mass of over 1000 kDa, as was reported for the human erythrocyte enzyme [8]. As the enzyme appeared to be readily adsorbed to the glass surface and other materials (data not shown), nitrocellulose membrane products were not used in the steps of purification.

Heterogeneous components of purified enzyme

The enzyme at the final step showed, upon SDS-PAGE, three major protein bands of 34 kDa, 38 kDa, and 40 kDa, and their relative amounts were

Table 1. Purification of rat liver PRPP synthetase
Starting materıals was 600 g of rat liver.

Purification step	Total protein	Total activity	Specific activity
	mg	milliunits	milliunits/mg
Homogenate	119,000	87,800	0.738
Postmitochondrial fraction	54,200	69,200	1.28
Polyethylene glycol	5,120	64,200	12.5
Acid precipitation	1,020	61,000	59.8
Hydroxyapatite	157	43,000	274
Toyopearl HW-65F	21.8	31,000	1,420
DEAE-Toyopearl 650S	4.96	20,000	4,030
TSK G4000SW	2.02	14,700	7,280

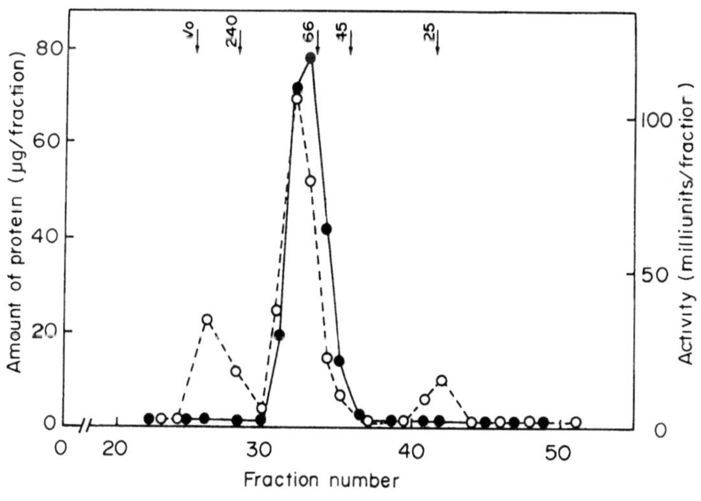

Fig. 1. Gel filtration on TSK G2000SW of purified enzyme in the presence of 1 M $MgCl_2$. The enzyme preparation at DEAE-Toyopearl step (0.6 ml, 380 µg) was treated with 1 M $MgCl_2$ and fractionated as described in Methods. The arrows marked with Vo, 240, 66, 45, and 25 denote the void volume, the elution positions of catalase (240 kDa), albumin (66 kDa), ovalbumin (45 kDa), and chymotrypsinogen (25 kDa) in the presence of 1 M $MgCl_2$, respectively. (●—●) PRPP synthetase activity; (O--O) Protein amount

approximately 70%, 24%, and 3%, respectively. In further attempts at purification we used electrofocusing on ampholytes, chromatography on activated thiol-Sepharose 4B, phospho-cellulose, and phenyl-Sepharose, and gel filtration on TSK G4000SW in the presence of 1 M KCl [8,11]. None of the procedures could separate the three major components and there was no indication of a different behavior (data not shown).

A;

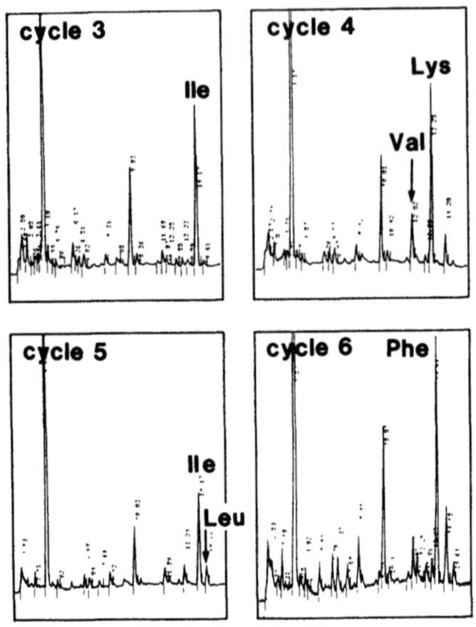


```
                    1                5                     10
B;  NH2-Pro-Asn-Ile-Lys-Ile-Phe-Ser-Gly-Ser-Ser-
                            Val Leu

        11                 15                    20
        His-Gln-Asp-Leu-Ser-Gln-Lys-Ile-Ala-Asp
                                Arg Val
```

Fig. 2. N-terminal sequence of the 34-kDa subunit. The 34-kDa component of the enzyme (catalytic subunit) was isolated by electroelution from SDS/polyacrylamide gel and sequenced. A; HPLC analysis data for determination of amino acids of cycles 3 to 6. B; The sequence determined up to 20 amino acids.

Identification of the catalytic subunit

A high concentration of $MgCl_2$ can dissociate the antigen-antibody complex [13], and has a physicochemical effect on protein molecules [14]. Therefore, 1 M $MgCl_2$ was used to disaggregate the PRPP synthetase. Upon succeeding gel filtration on TSK G2000SW in the presence of 1 M $MgCl_2$, the enzyme behaved as a peak of 68 kDa (Fig. 1). Four separate fractions (Fr. 31-34) within the peak were collected and analyzed by SDS-PAGE. The last fraction (Fr. 34) contained only the 34-kDa component, and showed a definite catalytic activity, with a specific activity higher than those in the other three fractions. These results show that the 34-kDa component carries the catalytic activity.

Partial amino acid sequence of the catalytic subunit

N-terminal sequence of the catalytic subunit was analyzed using an Applied Biosystems 470 sequencer with Model 120A on-line HPLC system. Nos. 4, and 5 amino acids (Lys and Ile) contained minor components of Val and Leu, respectively (Fig. 2A) and Nos. 17 and 18 (Lys and Ile) contained Arg and Val, respectively. The result for the sequence up to 20 amino acids was shown in Fig. 2B. In addition, two tryptic peptides of the catalytic subunit were obtained and sequenced (data not shown).

Our cloning experiments using synthetic oligonucleotide probes based on the above sequences, gave two distinct, but quite similar cDNAs [12]. The deduced amino acid sequences contained regions coinciding with the protein sequences. These results taken together indicate that the 34-kDa component from rat liver is actually a mixture of two highly homologous isoforms, PRS I and PRS II [12].

DISCUSSION

Despite the number of studies on mammalian PRPP synthetase, the nature of this enzyme is not completely understood. We purified rat liver PRPP synthetase to a specific activity, the highest for a mammalian enzyme reported in the literature. However, the final preparation did contain heterogeneous protein components. Further purification by conventional procedures gave only co-purification of these components. These findings suggest that rat liver PRPP synthetase molecule composed of heterologous subunits. The 34-kDa subunit was identified as the catalytic subunit. The value for the subunit molecular mass coincides with that for the subunit of human erythrocyte PRPP synthetase (34.5 kDa)[8], but differed from that (40.5 kDa) reported for the rat liver enzyme [9]. The physical properties of other components are unknown. The results of cDNA analysis [12] and peptide mapping (Ishizuka et al., unpublished) indicated that the 38-kDa or 40-kDa component is unlikely to be a precursor of the catalytic subunit.

Treatment of the purified enzyme with a high concentration of $MgCl_2$ facilitated separation of a part of the 34-kDa subunit from other components (Fig. 1). Component proteins of the four fractions within the peak were separated by SDS-PAGE, and their relative amounts were determined by densitometry. Specific enzyme activity in terms of 34-kDa component was calculated (data not shown). The last fraction that was composed exclusively of the 34-kDa species showed the highest specific activity, followed by 3rd, then 2nd or 1st fraction. Apparently the 38-kDa and/or other minor heterogeneous components exerted a suppressive effect on the activity of the 34-kDa subunit. This also implies that there is some molecular contact between 34-kDa subunit and other components. Therefore, we would like to propose that the 34-kDa and other components (the 38-kDa species at least) together form an integral molecule of rat liver PRPP synthetase. Further studies on this protein-protein interaction may reveal another aspects of the regulation of this enzyme. Consideration also should be given to the findings that the 34-kDa component is a mixture of two isoforms (this paper and ref. 12).

REFERENCES

1. A. Kornberg, I. Lieberman, and E. S. Simms, J. Biol. Chem. 215:417 (1955).
2. J. G. Flaks, M. J. Erwin, and J. M. Buchanan, J. Biol. Chem. 228:201 (1957).
3. I. H. Fox, and W. N. Kelley, Ann. Intern. Med. 74:424 (1971).

4. E. W. Holmes, in: "Handbook of Experimental Pharmacology," W. N. Kelley, and I. M. Weiner, eds., 51:21, Spriger-Verlag, Heidelberg, (1978).
5. M. Tatibana, and K. Shigesada, J. Biochem. (Tokyo) 72:549 (1972).
6. D. Keppler, and A. Holstege, in: "Metabolic Compartmentation," H. Sies, ed., 147, Academic Press Inc. (London) Ltd. (1982).
7. K. R. Schubert, and R. L. Switzer, J. Biol. Chem. 250:7492 (1975).
8. I. H. Fox, and W. N. Kelley, J. Biol. Chem. 246:5739 (1971).
9. D. G. Roth, E. Shelton, and T. F. Deuel, J. Biol. Chem. 249:291 (1974).
10. B. Hove-Jensen, K. W. Harlow, C. J. King, and R. L. Switzer, J. Biol. Chem. 261:6765 (1986).
11. M. A. Becker, L. J. Meyer, W. H. Huisman, C. Lazar, and W. B. Adams, J. Biol. Chem. 252:3911 (1977).
12. M. Taira, S. Ishijima, K. Kita, K. Yamada, T. Iizasa, and M. Tatibana, J. Biol. Chem. 262:14867 (1987).
13. S. W. Kessler, J. Immunol. 115:1617 (1975).
14. T. Arakawa, and S. N. Timasheff, Biochemistry 21:6545 (1982).

POTENT AND SPECIFIC INHIBITORS OF MAMMALIAN PHOSPHORIBOSYLPYROPHOSPHATE (PRPP) SYNTHETASE

Randall C. Willis, L.D. Nord, James M. Fujitaki, Timothy S. Breen, and Steven S. Matsumoto

Departments of Biochemistry and Cell Biology
The Nucleic Acid Research Institute
Costa Mesa, CA 92626 USA

INTRODUCTION

Phosphoribosylpyrophosphate (PRPP) synthetase catalyzes the formation of cellular PRPP, a central precursor for syntheses of purine and pyrimidine nucleotides, pyridine nucleotides, and the amino acids tryptophan and histidine (1). PRPP is a required substrate for phosphoribosylamine (PRA) synthetase and orotate phosphoribosyltransferase (OPRTase) in de novo purine and pyrimidine synthesis. The 'salvage' enzymes, APRTase and HPRTase, also require PRPP as does nicotinamide PRTase (2), Figure 1. The normal function and regulation of PRPP synthesis is critical as suggested by the absence of reported PRPP synthetase-deficient eukaryotic cell lines and by the clinically observed problems associated with overproduction of PRPP in man (1). Despite the importance of PRPP to the cell, the regulation of PRPP synthetase is not well understood (3); this is due in part to the lack of specific inhibitors of this critical enzyme.

The nucleoside, ARPP, 4-amino-8-(D-ribofuranosylamino)pyrimido-[5,4-d]pyrimidine has demonstrated antitumor activity against L1210 murine leukemia in vivo. ARPP accumulates intracellularly as the 5'-monophosphate (5), higher phosphate derivatives have not been observed. Studies with salvage enzyme-deficient cell lines show that adenosine kinase is required for phosphorylation. The mechanism of action of ARPP and a 4-methoxy cogener (MRPP) have been investigated in the B lymphoblast cell line, WI-L2. The results presented here with ARPP are applicable to MRPP and consistent with the monophosphates of these compounds being AMP analogs which provide specific inhibition of PRPP synthetase. In cytotoxicity assays with WI-L2, ARPP and MRPP displayed ID_{50} values of 0.3 and 0.03 uM respectively. As shown in Table I, the compounds appear to cause a block in the de novo pathway of purine nucleotide synthesis proximal to FGAR formation. As expected from cytotoxicity assays with adenosine kinase resistant cell lines, adenosine kinase deficient lymphoblast are resistant to the effects of ARPP on de novo purine nucleotide synthesis (Table IB), a further demonstration of the requirement for phosphorylation for the activity of these compounds.

The effects of ARPP were compared to 6-methylmercaptopurine riboside (MMPR), which is known to inhibit amidophosphoribosyl transferase, the first step in purine nucleotide de novo synthesis. As shown in Table II, incubation of cells with MMPR had no significant effect on the rate of hypoxanthine uptake (PRPP-dependent phosphoribosylation of hypoxanthine to IMP). In contrast, the rate of hypoxanthine uptake is severely diminished when cells are incubated with ARPP. In similar studies of adenosine uptake in the presence of EHNA and uridine uptake, each dependent on nucleoside kinase activity for conversion to the monophosphate, ARPP had little effect beyond that expected from the competition of the nucleoside for adenosine kinase and no effect on uridine uptake, respectively.

Table I. Effect of ARPP on de novo purine biosynthesis.

The incorporation of [^{14}C]formate into purine de novo intermediates was observed in separate experiments by treating WI-L2 cells with (in addition to ARPP): **A.** no inhibitor (6), **B.** hadacidin, (500 μM), alanosine (20 μM) and mycophenolic acid (20 μM)(3), or **C.** azaserine (50 μM)(7).

ARPP(μM)	[^{14}C]formate incorporation (% of control)	
A. into purine nucleotides		
	% synthesized	% excreted
10.0	17.0	64.0
1.0	50.0	12.0
0.1	95.0	1.0
B. into inosine 5′-monophosphate		
	WI-L2	AKase deficient WI-L2
3.2	12.0	99.4
1.0	10.1	96.6
0.3	16.2	97.1
0.1	47.3	94.6
C. into formylglycinamide ribotide (FGAR)		
	area units	% of control
10.0	0.30	4.5
0.0	6.73	100.0

The results thus far suggested that ARPP could be interfering with the availability of PRPP. De novo pyrimidine nucleotide synthesis requires PRPP in the fifth and penultimate step in the formation of UMP, ie orotate phosphoribosyltransferase. As shown in Table III and consistent with a PRPP restriction, ARPP inhibits de novo pyrimidine nucleotide biosynthesis and causes the accumulation of the carbamyl-L-aspartate and orotate (data not presented), products of the second and fourth steps.

Table II. ARPP inhibition of [^{14}C]hypoxanthine labeling of purine nucleotides.

WI-L2 cells (2 X 10^6 cells/mL) and compound (5 μM) were incubated for 5 min prior to the addition of [^{14}C]hypoxanthine (50 Ci/mol, 2.5 μM). Incorporation of [^{14}C]Hx into purine nucleotides was determined by DE-81 filter binding assay after 13 min.

Compound	CPM ± S.D.	% control
control	2949 ± 38.5	100.0
ARPP	430 ± 68.0	14.6
MMPR	2804 ± 46.5	95.1

Table III. ARPP inhibition of de novo pyrimidine and purine pathways.

WI-L2 cells were treated with ARPP (5 μM) for 3 h before resuspension in bicarbonate free media containing ARPP (7). [^{14}C]Bicarbonate was added (57 Ci/mol, 10 μL/mL of cell suspension) and incubation continued 1 h. The acid-soluble extracts were analyzed by SAX-HPLC (8).

Compound	CPM/10^6 cells			% control	
	PyrTP[a]	PurTP[b]	CAA[c]	PyrTP	PurTP
control	21692	59520	0	100	100
ARPP	0	0	1608	0	0

[a] Pyrimidine 5′-triphosphates (CTP + UTP).

[b] Purine 5′-triphosphates (ATP + GTP).

[c] Carbamyl aspartic acid is an intermediate in de novo pyrimidine biosynthesis which accumulates upon inhibition of the UMP synthetase complex.

Inosine-driven, adenine uptake was used to study the effects of ARPP on activity of PRPP synthetase in whole cells. The scheme used and the results of these studies are shown in Table IV. Under these assay conditions, ARPP reduced the availability of PRPP by greater than 50%. When the effect of ARPP on PRPP content was examined, Table V, the PRPP content of ARPP-treated cells was 45% of the control cells. The PRPP content of MMPR-treated cells was 3.5 times greater than that of the control cells.

Table IV. Inhibition of the inosine-driven synthesis of PRPP by ARPP.

HPRTase-deficient WI-L2 cells were incubated with inosine to drive the synthesis of PRPP (9).

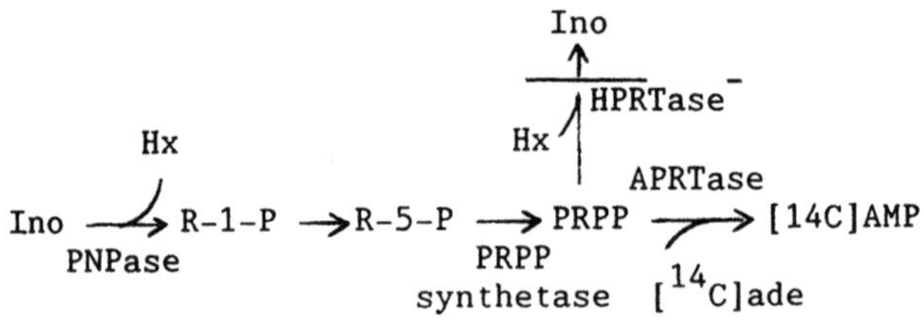

PRPP synthesis was measured by incorporation of [^{14}C]ade into adenine nucleotides over 20 min. Cells were pre-incubated with ARPP for 10 min prior to addition of inosine and 12 min prior to the addition of [^{14}C]ade.

Compound, μM		CPM	% control
control		53,000	100.0
ARPP	10	23,500	44.3

Table V. PRPP levels of WI-L2 cells treated with ARPP or MMPR.

WI-L2 cells were incubated for 3 h with drug (10 μM) before determination of PRPP levels (3).

Compound	PRPP pmol/10 cells	% control
control	170 ± 26	100
ARPP	76 ± 8	45
MMPR	593 ± 26	349

ARPP-5′P was synthesized chemically as well as isolated from human erythrocytes incubated with the nucleoside. The nucleoside monophosphate inhibited partially purified erythrocyte PRPP synthetase in a dose dependent manner (Table VI). MMPR-5′P synthesized and studied in parallel displayed no inhibition. Preliminary results suggest that ARPP-5′P may alter the distribution of subunit aggregation profiles (10) of PRPP synthetase and thereby maintain the enzyme in a less active state.

Table VI. Effect of ARPP-5′P on purified erythrocyte PRPP synthetase.

PRPP synthetase activity was determined as described by Becker and co-workers (10) at physiological phosphate concentrations (1-2 mM). PRPP was measured by the conversion of [^{14}C]adenine to [^{14}C]AMP in the presence of purified APRTase.

Compound, μM		[^{14}C]AMP formed, CPM	% control
control		17,700	100.0
ARPP	18	11,700	66.1
	36	10,300	58.2
	72	7,200	40.7

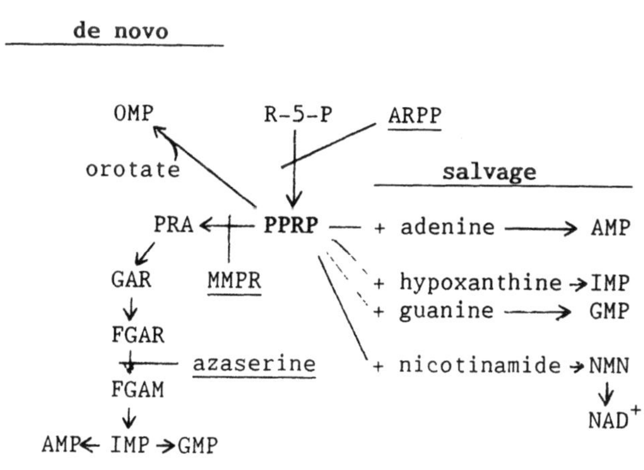

Figure 1. Schematic diagram of nucleotide biosynthesis by de novo and salvage pathways. PRPP serves a central role in de novo pyrimidine and purine biosynthesis and in the salvage of preformed nucleobases to form nucleoside 5′-monophosphates. Formation of pyridine nucleotides is also dependent upon PRPP availability.

REFERENCES

1. M.A. Becker, L.J. Meyer, and J.E. Seegmiller. Am. J. Med. 55, 232 (1973).

2. W.D.L. Musick. Critical Rev. Biochem. 11, 1 (1981).

3. R.B. Pilz, R.C. Willis and G.R. Boss. J. Biol. Chem. 259, 2927 (1984).

4. P.C. Srivastava, G.R. Revankar and R.K. Robins. J. Med. Chem. 24, 393 (1981).

5. R.J. Moss, C.R. Petrie, R.B. Meyer, Jr., L.D. Nord, R.C. Willis, R.A. Smith, S.B. Larson, G.D. Kini and R.K. Robins. J. Med. Chem. 31, 786 (1988).

6. M.S. Hershfield and J.E. Seegmiller. J. Biol. Chem. 252, 6002 (1977).

7. W.H. Huisman, K.O. Raivio and M.A. Becker. J. Biol. Chem. 254, 12595 (1979).

8. R.C. Willis, R.K. Robins and J.E. Seegmiller. Mol. Pharmacol. 18, 287 (1980).

9. S.D. Skaper, R.C. Willis and J.E. Seegmiller. Science, 193, 587 (1976).

10. M.J. Losman, S. Hecker, S. Woo and M.A. Becker. J. Clin. Lab. Med. 103, 932 (1984).

PRPP AND PURINE NUCLEOTIDE METABOLISM IN HUMAN LYMPHOBLASTS WITH BOTH PRPP SYNTHETASE SUPERACTIVITY AND HGPRT DEFICIENCY

Michael A. Becker , Mitchel Kim, and Kishwar Husain

Department of Medicine
The University of Chicago
Chicago, Illinois

INTRODUCTION

Much evidence supports the view that control of rates of purine synthesis *de novo* is effected primarily in the regulatory domain defined by the PRPP synthetase and amidoPRT reactions (figure 1)(1-5). Within this domain, PRPP and purine nucleotides act as antagonistic effectors (1,2,6), and a more precise concept of the relative contributions and sites of action of these determinants has emerged, in large part from the study of cells cultured from individuals with inherited defects leading to relaxation in the control of the pathway (1,4,5,7,8). In particular, comparison of normal fibroblasts, cells with severe HGPRT-deficiency and cells with several distinct types of kinetic defects leading to PRPP synthetase superactivity (1,7) has permitted reconciliation of several previously proposed regulatory schemes (2-4) (Figure 1).

Purine nucleotide products of the pathway inhibit both PRPP synthetase and amidoPRT, with the latter enzyme considerably more sensitive to inhibition (1,7). PRPP activates the pathway through antagonistic interaction with purine nucleotides at the level of the amidoPRT reaction (1,2,6) an interaction altering the quaternary structure of the enzyme (2). In addition, the amidoPRT reaction appears to be the rate-limiting reaction in the pathway (6). These studies indicate that while the amidoPRT reaction is the primary site of regulation of the pathway, dual sites of control permit flexibility in balancing the production of purine nucleotides with the need for PRPP in other pathways of its utilization.

Both HGPRT-deficiency and PRPP synthetase superactivity are X-chromosome-linked defects (9-12) associated with increased rates of purine synthesis *de novo* (4,5,7) uric acid overproduction (12,13) and increased intracellular PRPP concentrations (5,7). Important differences in PRPP and purine nucleotide synthesis exist, however, in these enzyme defects (Table 1). PRPP generation and purine nucleotide pool sizes are not prominently altered in HGPRT-deficient fibroblasts but salvage synthesis of purine nucleotides from hypoxanthine and guanine is severely impaired or absent (5,7). In contrast, PRPP generation and purine nucleotide pools are increased in all PRPP synthetase-superactive fibroblasts, and purine base salvage, a PRPP-dependent function, is accelerated (7). Moreover, in cells with catalytic defects in PRPP synthetase, purine excretion is normal and all of the PRPP and purine nucleotide synthetic abnormalities are attenuated in comparison to cells with

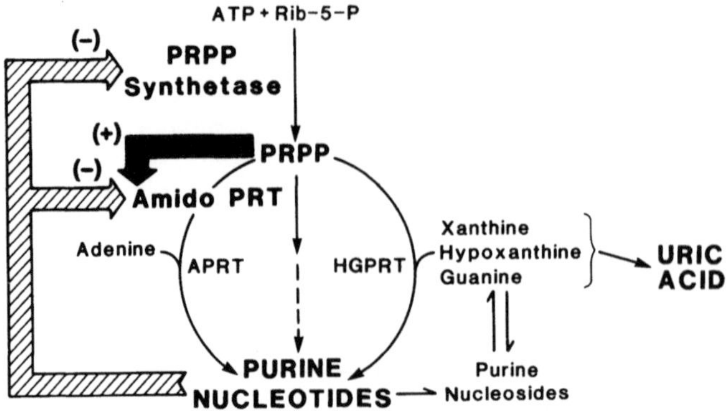

Figure 1. Schematic representation of the pathways of purine nucleotide synthesis and degradation to uric acid. Single step salvage synthesis of purine nucleotides is depicted by curved arrows, and purine synthesis de novo is shown by a solid arrow to denote the rate-limiting amidoPRT reaction and a dashed arrow to indicate the 9 subsequent steps to IMP synthesis. The proposed regulation of purine synthesis de novo is also shown. Purine nucleotide inhibition (-) of PRPP synthetase and amidoPRT is indicated by the heavy hatched arrow, and activation (+) of amidoPRT by PRPP, resulting from antagonistic interaction of PRPP and purine nucleotides on this enzyme, is shown by the heavy dark arrow.

purine nucleotide feedback-resistant forms of the enzyme. The latter class of PRPP synthetase superactive cells, like HGPRT deficient cells, excretes significant proportions of newly synthesized purines into the culture medium.

Table 1. PRPP and Purine Nucleotide Synthesis in HGPRT-deficient and PRPP Superactive Fibroblasts*

Defect	Purine synthesis de novo	Purine excretion	Purine base salvage +	Purine nucleotide conc.	PRPP conc.	PRPP gen.
HGPRT deficiency	↑↑	↑	Ade only	--	↑↑	--
PRPP synthetase-superactive						
Catalytic	↑	--	All ↑	↑	↑	↑
Feedback resistant	↑↑↑	↑	All ↑↑	↑↑	↑↑↑	↑↑

*--, normal; ↑, mild increase; ↑↑, moderate increase; ↑↑↑, marked increase.
+ Only adenine is salvaged in severe HGPRT-deficiency, but all purine bases are salvaged at increased rates in cells with PRPP synthetase superactivity.

Despite the insights gained to date, understanding of the control and integration of pathways of purine nucleotide synthesis is incomplete. Since the study of cells with single genetic defects has been productive (1,4,5,7,8) and some additional studies have demonstrated the utility of extending analyses to cells with multiple well-defined genetic defects (14,15) we have sought to answer some additional questions about regulation of PRPP and purine synthesis by selecting and analyzing cell lines bearing defects in both HGPRT and PRPP synthetase. We report here studies comparing the biochemical and growth characteristics of lymphoblasts defective in both of these enzymes with normal cells and cells with either HGPRT-deficiency or PRPP synthetase superactivity.

METHODS

A normal human B lymphoblast line (designated Li) and a cell line (SM) bearing a PRPP synthetase with diminished to responsiveness purine nucleotide inhibitors (8) served as the parental lines from which HGPRT-deficient clones were selected (16) among cells surviving growth in 6-thioguanine (6-TG) (17). After 60μM 6-TG was established as the minimal growth inhibitory concentration of the purine base analog in both parental cell lines, log phase cultures of each cell line were exposed for 48 hours to this concentration of 6-TG in minimal (RPMI-1640) medium. Cells were then resuspended in non-selective medium for 10 days, and the cycle of selection and expansion of the cultures was repeated twice more. At that point, substantial resistance of cultures to growth inhibition by 6-TG was achieved, and cells were then cloned twice at limiting dilution in selective medium using 96-well microtiter plates. Clonal lines were expanded in non-selective medium and tested for phenotypic expression of HGPRT and PRPP synthetase activities (as described below). Clones severely deficient in HGPRT were then further characterized with regard to PRPP and purine nucleotide metabolism, utilizing methods previously described (7,8).

RESULTS

Radiometric assay (18) showed that a high proportion of 6-TG-resistant clones had <0.2% of parental HGPRT activity. Clones with this degree of enzyme deficiency showed only background levels of incorporation of [^{3}H] hypoxanthine on autoradiography (11) and of [^{14}C] hypoxanthine into acid-soluble nucleotides and acid-precipitable nucleic acids (19). The following discussion will concentrate on HGPRT-deficient clones 31 and 71, derived, respectively from the Li and SM parental cell lines.

Figure 2 shows growth curves for the Li and SM cell lines and their HGPRT-deficient derivatives, clones 31 and 71, in minimal medium, medium with 60μM 6-TG and medium containing hypoxanthine, aminopterin and thymidine (HAT) (20). Of note are the following findings: clone 71 and, to a much lesser degree, clone 31 grow more slowly in minimal medium than the parental cell lines; growth of clones 31 and 71 is unaltered by the presence of 6-TG (and, in fact, LD_{50} for 6-TG are >200-fold higher for both clones than the corresponding LD_{50} values for the parental lines); clones 31 and 71 fail to grow in HAT medium, reflecting their inability to utilize hypoxanthine to bypass blockade of purine synthesis <u>de novo</u> in a manner comparable to that of the parental cell lines. The results of these kinetic and functional studies establish the presence of severe deficiency of HGPRT in clones 31 and 71.

We have previously shown that fibroblasts bearing feedback-resistant PRPP synthetases survive and grow in medium containing 10μM 6-methylthioinosine (MeSno), a concentration of this nucleoside analog preventing growth in normal cells and cells with catalytic defects in PRPP synthetase (21). MeSno is a potent inhibitor of purine synthesis <u>de novo</u> through inhibition of both amidoPRT and PRPP synthetase, the latter effect dependent on conversion of

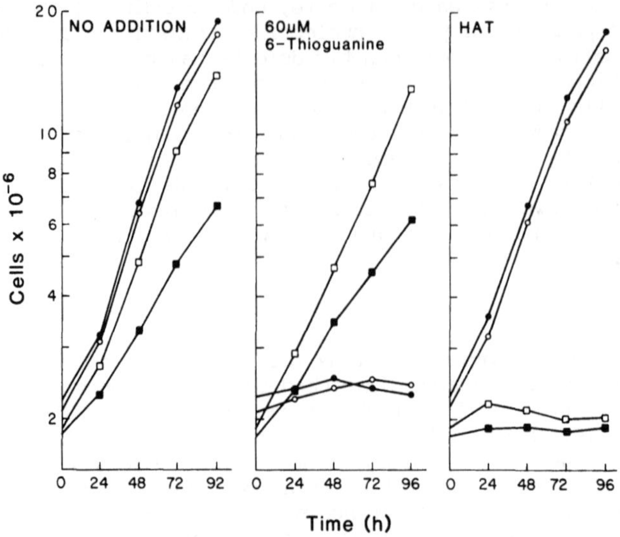

Figure 2. Growth of lymphoblast lines in minimal medium (RPMI-1640 with 10% fetal bovine serum and 2mM L-glutamine) (left panel) and in this medium with either 60μM 6-TG (middle panel) or HAT medium components (hypoxanthine, $1x10^{-5}M$; aminopterin, $1.6x10^{-5}M$; thymidine, $1x10^{-5}M$; and glycine, $4x10^{-4}M$) (right panel). (●), Li; (○), SM; (□), clone 31; (■), clone 71.

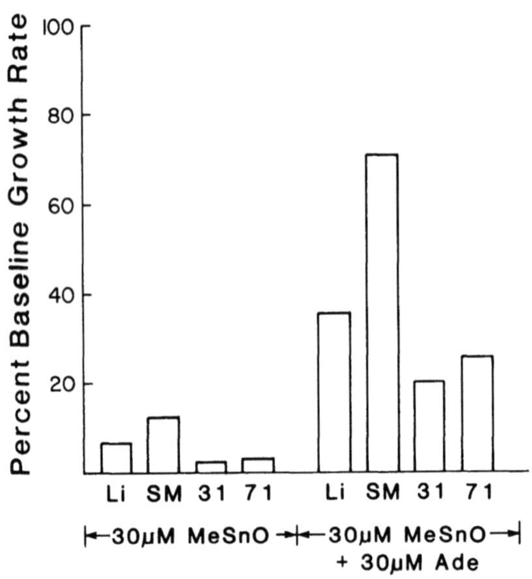

Figure 3. Inhibition of lymphoblast growth by MeSno and reversal of growth inhibition by adenine. Lymphoblasts were incubated in standard medium or this medium with 30μM MeSno or 30μM MeSno and 30μM adenine. Cells were enumerated after 72 hours of growth and values in experimental cultures are expressed as percent of growth in minimal medium.

the compound to a monophosphate noncompetitive inhibitor of the enzyme (21). growth of the 4 lymphoblast lines in medium containing 30μm MeSno is severely inhibited (Figure 3); addition of the purine base salvage substiate, adenine (30μM), to the medium ameliorates growth inhibition, especially in the SM cell line (Figure 3), reflecting resistance of PRPP synthesis in these cells to inhibition by MeSno (21). Both HGPRT-deficient clones 31 and 71 are, however, considerably more sensitive to MeSno than the parental lines and this effect persists even with the addition of adenine. This finding indicates that the enhanced sensitivity to growth inhibition by MeSno displayed by HGPRT-deficient lines is independent of rates of PRPP generation and more likely reflects a relative inability of HGPRT-deficient cells to utilize adenine derivatives to synthesize guanylates when IMP formation from purine synthesis de novo is blocked. This formulation suggests that in lymphoblasts, direct deamination of AMP to IMP is limited and that production of guanylates from adenine is therefore dependent on the sequence of reactions: Ade→AMP→adenosine→inosine→hypoxanthine→IMP→→GMP, a reaction sequence which is impaired in HGPRT-deficiency.

Phosphate activation of PRPP synthetase was compared in the lymphoblast lines (8) and shows that neither maximal enzyme activities nor patterns of the phosphate activation are altered by introduction of HGPRT-deficiency (Figure 4). This finding contrasts with the increase in PRPP synthetase activities previously reported for HGPRT-deficient lymphoblasts (22) and for HGPRT-deficient fibroblasts in subconfluent culture (23). Dissociation constants for substrates, phosphate and magnesium ion were comparable in all 4 lymphoblast lines but the previously documented resistance of SM PRPP synthetase to ADP (Figure 4) and GDP (8,24)as well as MeSno monophosphate (21) was readily demonstrable in clone 71. Selection for HGPRT-deficiency thus did not appreciably alter the kinetic characteristics of normal or superactive PRPP synthetases.

Table 2 shows PRPP concentrations and rates of PRPP production in the lymphoblast lines. PRPP concentrations in SM lymphoblasts are 3-fold greater than those of normal cells. The HGPRT-deficient derivatives of both of these cells have PRPP concentrations which are, respectively, 5-fold and 6-fold increased. PRPP concentration in clone 71, is thus nearly 20-fold greater than in normal cells. Despite the apparent major effect of HGPRT-deficiency on PRPP levels, PRPP generation is minimally increased (<8%) in clones 31 and 71 in comparison to the respective parental cell lines. Rates of purine synthesis de novo (determined by incorporation of [^{14}C] formate into the total of intracellular purines and purines excreted into the medium (7)) are, however, increased in HGPRT-deficient clone 31 compared to parental Li cell line, while the markedly increased rates of the process in the SM line are not exceeded in clone 71 (Table 2). The proportion of newly formed purines excreted into the medium by SM cells is considerably greater than that excreted by Li cells; the HGPRT-deficient derivative of each cell line excretes a much higher proportion of newly formed purines than its parental line and excretion approaches 50% of purines synthesized in the case of clone 71. Finally, ATP and GTP concentrations are comparable in SM and Li cells but are slightly lower in clone 31 than in Li lymphoblasts. In contrast, ATP and especially GTP pools are diminished in clone 71 compared with the other cells lines (Table 2).

DISCUSSION

The above findings support the concept that HGPRT-deficiency results in increased rates of purine synthesis de novo primarily as a consequence of redirection of PRPP utilization from purine base salvage to purine synthesis de novo (7,25). This is accomplished without detectable increase in overall

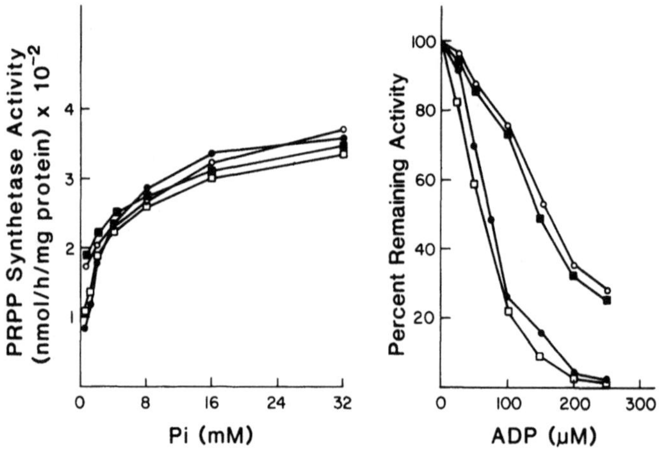

Figure 4. Pi activation of PRPP synthetase in dialyized lymphoblst extracts (left panel) and inhibition of PRPP synthetase activity by ADP in cell extracts chromatographed on Sephadex G-25. (●), Li; (○), SM; (□) Clone 31; (■) clone 71. Note hyperbolic Pi activation of SM and clone 71 PRPP synthetase and relative ADP resistance of enzyme from these cell lines.

Table 2. PRPP and Purine Nucleotide Synthesis in Lymphoblast Lines*

Cell line	PRPP Synthetase	HGPRT	PRPP conc.	PRPP gen.	Purine synthesis de novo	Purine excretion	Purine nucleotide conc.
Li	Normal	Normal	1.0	1.0	1.0	6%	1.00
Clone 31	Normal	Absent	4.7	1.1	2.3	26%	0.88
SM	Superactive	Normal	2.9	2.4	3.9	28%	0.93
Clone 71	Superactive	Absent	17.5	2.6	4.0	49%	0.77

*All values except for purine excretion are expressed relative to a value of 1.0 arbitrarily assigned to the Li (normal) cell line. Purine excretion is expressed as % of newly synthesized purine excreted. +Represents mean value of results of measurements of ATP and GTP concentrations.

PRPP production and implies a substantial normal rate of base reutilization catalyzed by HGPRT. In fact, in the absence of increased rates of PRPP generation and of PRPP utilization in purine synthesis de novo in clone 71, the increment in PRPP concentration in this cell line relative to the SM cell line should provide a minimal estimate of the rates of purine reutilization in the SM line. Our data do not exclude an accompanying contribution of slightly diminished nucleotide pools to the increased rate of purine synthesis de novo in clone 31 (relative to the Li cell line), although prior comparisons of nucleotide pools in normal lymphoblast lines and their HGPRT-deficient derivatives have not reported significant differences in nucleotide pools (26,27).

Increased rates of purine synthesis de novo in the SM cell line are not further stimulated in its HGPRT-deficient derivative clone 71 despite very high PRPP concentrations and reduced ATP and GTP concentrations in the latter clone. This finding most likely reflects saturation of amidoPRT by PRPP in SM lymphoblasts at baseline, a suggestion supported by relatively fixed rates of purine synthesis in fibroblasts with feedback-resistant PRPP synthetase under conditions promoting acceleration of the pathway in normal and HGPRT-deficient cells (1). We believe that the limited capacity of clone 71 to accelerate purine synthesis de novo in the face of a high rate of purine excretion due to impaired purine reutilization is the basis of nucleotide pool depletion in these cells. Although it seems paradoxical to propose that maximally accelerated rates of purine synthesis de novo in clone 71 are insufficient to maintain optimal purine nucleotide pools, a potential explanation may be provided by the finding that in cells with feedback-resistant PRPP synthetase (as opposed to HGPRT-deficient cells) an increased proportion of IMP synthesized by the pathway of purine synthesis de novo is directed toward adenylate synthesis rather than guanylate synthesis (25). If conversion of adenylates to guanylates by means of the multi-step sequence of reactions listed above is required to maintain guanylate pools in these cells, fulfillment of this requirement in cells also deficient in HGPRT would be unlikely. Cells with HGPRT-deficiency alone would, in fact, be better able to synthesize guanylates than clone 71 cells because of the more favorable distribution of IMP into guanylates even at the increased rates of purine synthesis de novo of these cells.

This formulation requires additional testing but the responses of the HGPRT-deficient cell lines to MeSno and adenine (Figure 3) are consistent with this view. The retarded growth of clone 71 with both enzyme defects may in turn be the consequence of nucleotide depletion resulting in nucleotide pools which are inadequate to support a growth rate comparable to that of parental or normal cell lines.

In summary, this study of lymphoblasts with both HGPRT-deficiency and PRPP synthetase superactivity has provided data pertinent to the following observations and interpretations: 1. introduction of HGPRT-deficiency does not alter activities or kinetic characteristics of normal or superactive PRPP synthetases and does not increase PRPP generation; 2. PRPP concentrations are substantially increased in HGPRT-deficient cells, apparently as a result of impaired PRPP utilization in purine base salvage rather than as a consequence of increased PRPP production; 3. increased PRPP availability (rather than nucleotide pool deficiency) is in the main responsible for accelerated purine synthesis in HGPRT-deficient cells; 4. in lymphoblasts with a feedback-resistant PRPP synthetase, amidoPRT may be saturated for PRPP, resulting in a fixed rate of purine synthesis de novo even when HGPRT-deficiency is introduced or when purine nucleotide pools are diminished; 5. depleted intracellular purine nucleoside triphosphate pools in lymphoblasts with dual defects in HGPRT and PRPP synthetase result from enhanced purine excretion, impaired purine salvage, and a fixed rate of purine synthesis de novo; 6. the reduced growth rate of lymphoblasts with dual enzyme defects may reflect reduced purine nucleotide pools; 7. HGPRT-deficient cells are more sensitive to blockade of purine synthesis *de novo* by MeSno even when adenine is provided, suggesting a limited direct conversion of AMP to IMP in lymphoblasts and consequent dependence on indirect conversion of adenylates to guanylates.

ACKNOWLEDGEMENTS

This work was supported in part by U.S. Public Health Service Grant DK-28554 and a grant from the Arthritis Foundation, Illinois Chapter. We thank Ms. Danette Shine and Ms. Tonja Hopkins for excellent assistance in manuscript preparation.

REFERENCES

1. M.A. Becker, and M. Kim, J. Biol. Chem. 262:14531 (1987).
2. E.W. Holmes, J.B. Wyngaarden, and W. N. Kelley, J. Biol. Chem. 248:6035 (1973).
3. A.S. Bagnara, A.A. Letter, and J.F. Henderson, Biochim. Biophys. Acta 373:259 (1974).
4. E. Zoref, A. deVries, and O. Sperling, J. Clin. Invest. 56:1093 (1975).
5. F.M. Rosenbloom, J.F. Henderson, I.C. Caldwell, W.N. Kelley, and J.E. Seegmiller, J. Biol. Chem. 243:1166 (1968).
6. E.W. Holmes, J.A. McDonald, J.M. McCord, J.B. Wyngaarden, and W.N. Kelley, J. Biol. Chem. 248:144 (1973).
7. M.A. Becker, M.J. Losman, and M. Kim, J. Biol. Chem. 262:5596 (1987).
8. M.J. Losman, D. Rimon, M. Kim, and M.A. Becker, J. Clin. Invest. 76:1657 (1985).
9. E. Zoref, A. deVries, and O. Sperling, Nature 260:786 (1976).
10. R.C.K. Yen, W.B. Adams, C. Lazar, and M.A. Becker, Proc. Natl Acad. Sci. USA 75:482 (1978).
11. F.M. Rosenbloom, W.N. Kelley, J.F. Henderson and J.E. Seegmiller, Lancet 2:305 (1967).
12. M. Lesch, and W.L. Nyhan, Am. J. Med. 36:561 (1964).
13. O. Sperling, P. Boer, S. Persky-Brosh, E. Kanarek, and A.deVries, Rev. Eur. Etud. Clin. Biol. 17:703 (1972).
14. M.S. Hershfield, F.F. Snyder, and J.E. Seegmiller, Science 197:1284 (1977).
15. B. Ullman, K. Kaur, and T. Watts, Mol. Cell. Biol. 3:1187 (1983).
16. J.E. Lever, G. Nuki, and J.E. Seegmiller, Proc. Natl. Acad. Sci. USA 71:2679 (1974).
17. K. Sato, R.S. Slesinski, and J.W. Littlefield, Proc. Natl. Acad. Sci. USA 69:1244 (1972).
18. A.W. Wood, M.A. Becker, and J. E. Seegmiller, Biochem. Genet 9:261 (1973).
19. G.W. Crabtree, And J.F. Henderson, Cancer Res. 31:985 (1971).
20. W. Szybalski, E.H. Szybalska, and G. Ragni, Natl Cancer Inst. Monogr. 7:75 (1962).
21. R.C.K. Yen, K.O. Raivio, and M.A. Becker, J. Biol. Chem. 256:1839 (1981).
22. G.H. Reem, Science 190:1098-1099 (1975).
23. D.W. Martin, Jr., and B.A. Maler, Science 193:408 (1976).
24. M.A. Becker, K.O. Raivio, B. Bakay, W. B. Adams, and W.L. Nyhan, J. Clin. Invest. 65;109 (1980).
25. E. Zoref-Shani, and O, Sperling, Biochim. Biophys. Acta. 607:503 (1980).
26. M.S. Hershfield, and J. E. Seegmiller, J. Biol. Chem. 252:6002 (1977).
27. D. Brenton, K. H. Astrim, M.K. Cruikshank, and J.E. Seegmiller, Biochem. Med. 17:231 (1977).

PURIFICATION OF RAT LIVER AMIDOPHOSPHORIBOSYLTRANSFERASE

Takashi Yamaoka, Fumiko Kagita, Hiroko Yoshikawa,
Mitsuo Itakura, and Kamejiro Yamashita

University of Tsukuba, Institute of Clinical Medicine
Tsukuba-city, Ibaraki 305 Japan

INTRODUCTION

Amidophosphoribosyltransferase (E.C.2.1.2.14) (ATase) is considered as a rate-limiting enzyme in de novo purine synthetic pathway under a skillful regulation system. High-grade purification of mammalian ATase is essential for understanding the regulatory mechanisms of de novo purine synthesis in mammals, examining its detailed properties, and supplying protein for molecular cloning and monoclonal antibody production. The present study describes a method for high-grade purification of rat liver ATase and some properties in its function and structure.

MATERIALS AND METHODS

MATERIALS

Seven-week-old male Wistar rats were purchased from Shizuoka Agricultural Cooperative Association for Laboratory Animals (Shizuoka, Japan). Radioactive [^{14}C]glutamine (258-292 μCi/μmol) was purchased from New England Nuclear. Anti-B. subtilis ATase antibody is a kind gift from Dr. R. Switzer, U.S.A. Anti-E. coli ATase antibody is a kind gift from Dr. Zalkin et al. (1).

METHODS

Enzyme Activity Assay of ATase

The methods for the assay of ATase are the same as previously reported (2). In principle PRPP-dependent glutaminase activity was regarded as ATase activity, which was determined by PRPP-dependent hydrolysis of [^{14}C]glutamine to [^{14}C]glutamate.

Purification

All procedures were performed at 4°C unless otherwise stated. The purification procedure included the following 8 steps. Step 1: Fresh livers obtained from seven-week-old Wistar rats were homogenized in cold 1.15% KCl solution. This homogenate was centrifuged at 1,200 x g for 20 min. The supernate was filtered through cotton gauzes and ultracentrifuged at 55,000 x g for 3 h after adding a spoonful each of leupeptin and pepstatin A. Then, concentrated potassium phosphate (KPi) and 2-mercaptoethanol were added to the supernate to make final concentrations of KPi 25 mM (pH 7.4) and 2-mercaptoethanol 70 mM. Step 2: After the sample was heated to 56°C in a water bath, iced again and centrifuged. Step 3: To adjust pH to 5.1, 5% acetic acid was added and the precipitate was resuspended with 25 mM KPi (pH 7.4) and 70 mM 2-mercaptoethanol (buffer A). Step 4: The precipitate by 30% ammonium sulfate was removed and that by 55% ammonium sulfate was resuspended with buffer A. The desalted enzyme preparation after this step was used as a partially purified enzyme preparation. Step 5: The sample solution was applied to gel filtration by Toyopearl HW55F. Step 6: The peak enzyme activity fraction was applied to an anion exchange column (TSK gel DEAE-5PW) and ATase was eluted with a linear gradient of 0 to 400 mM KCl in buffer A. Step 7: Enzyme was eluted with a linear gradient of 25 to 350 mM KPi in buffer A through a HPLC hydroxyapatite column. After this step, ATase activity in each fraction was compared with the amount of protein which was checked by sodium dodecyl sulfate polyacrylamide gel electrophoresis (SDS-PAGE) (3) and silver-staining (4). The peak fraction was also examined for the cross-reactivity with anti-B. subtilis and anti-E. coli ATase antibody. Step 8: Protein at 62 kD stained with 300 mM $CuCl_2$ (5) in SDS-PAGE was cut out. After destaining by incubation in 250 mM EDTA in 250 mM Tris-HCl pH 9.0, ATase was electoeluted in 25 mM Tris-tricine buffer with 0.1% SDS (6). ATase was collected into a dialysis membrane tube and extensively dialyzed against pure water. The purity of this preparation was checked by SDS-PAGE and silver-staining.

TABLE 1. PURIFICATION TABLE OF RAT LIVER ATASE

STEP	PROCEDURE	VOLUME (ml)	ACTIVITY (nmol/h/ml)	TOTAL PROTEIN (mg)	SPECIFIC ACTIVITY (nmol/h/mg)	PURIFI-CATION FOLD	YIELD (%)
1.	55,000xg Supernatant	1,410	113	22,100	7.19	1.0	100
2.	Heat Treatment	1,400	92.7	9,310	13.9	2.0	82
3.	Acid Precipitation	200	106	316	66.8	9.3	13
4.	Ammonium Sulfate Fractionation	300	1130	65.7	514	72.0	21
5.	Gel filtration (Toyopearl HW55F)	31.2	929	30.6	948	132	17
6.	Anion exchange (TSK gel DEAE-5PW)	90.0	157	13.5	1,050	146	8.7
7.	Hydroxyapatite (HA-1000)	24.0	57.2	0.199	6,890	958	4.1

Feedback Inhibition

ATase activity of partially purified preparation was assayed in the presence of 1 mM PRPP with 5 mM of each purine ribonucleotide. The percentage inhibition was calculated by comparing the activity with that with 1 mM PRPP and without purine ribonucleotide.

Oxygen and Orthophenanthroline Inactivation

Oxygen inactivation of partially purified ATase was examined at 37°C in 25 mM Tris pH 7.4 under 100% oxygen with argon as the control according to the published methods (2). The enzyme activity was assayed immediately

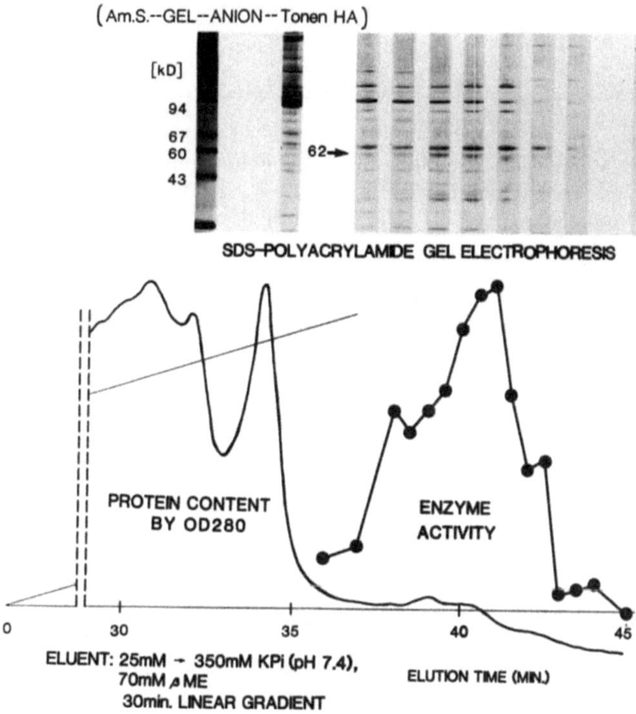

Fig. 1. ELUTION PROFILE OF RAT LIVER ATASE THROUGH A HPLC HYDROXYAPATITE COLUMN: The SDS-PAGE with silver staining of each fraction is shown in the upper half, and the enzyme activity and protein elution profile are shown in the lower half.

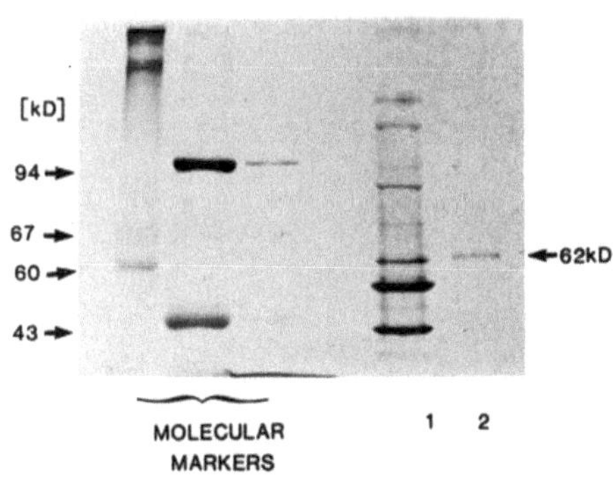

Fig. 2. HIGH-GRADE PURIFICATION OF RAT LIVER ATASE BY ELECTROELUTION: The results of SDS-PAGE analysis with silver-staining are shown. lane 1; The peak fraction after applying the post-ammonium sulfate enzyme preparation to a HPLC hydroxyapatite column, was applied to SDS-PAGE. lane 2; The recovered sample of 62 kD protein after electroelution is reapllied to SDS-PAGE.

after inactivation. A ferrous iron chelator, 1,10-orthophenanthroline and its isomer metaphenanthroline which lacks the chelating activity were included in the assay mixture and ATase activity was assayed according to the published methods (2).

Cross-Reactivity of Rat Liver ATase with anti-B. subtilis and anti-E. coli ATase Antibody

Rat liver ATase was separated by SDS-PAGE and transferred to a nylon membrane and its cross-reactivity to anti-B. subtilis or anti-E. coli ATase antibody was examined by the published methods (7).

Amino Acid Composition

After acid hydrolysis, the amino acid composition was examined by amino acid analyzer.

Result of Edman Degradation Analysis

The amino acid sequence of highly purified ATase was analyzed directly by an amino acid analyzer.

TABLE 2. FEEDBACK INHIBITION OF PARTIALLY PURIFIED A RAT LIVER ATASE: ATase activity was assayed in the presence of 1 mM PRPP and 5 mM of each purine ribonucleotide. The activity without purine nucleotides was regarded as 100% activity, and the percentage inhibition is shown in the table.

PURINE RIBONUCLEOTIDES	PERCENTAGE INHIBITION
AMP	39
ADP	16
ATP	6.6
GMP	43
GDP	26
GTP	23

RESULTS AND DISCUSSION

Enzyme Purification

As shown in Table 1, 1000-fold purification with a single component with silver-staining was possible by the method reported in this study. As shown in Fig. 1, the protein at 62 kD correlated with enzyme activity. As shown in Fig. 2, ATase at 62 kD was recovered after electroelution as a single protein. The amount of ATase protein in liver is estimated about 10 μg/g liver weight. The monomer size of rat liver ATase was thus estimated as about 62 kD.

Feedback Inhibition

As shown in Table 2, inhibition by purine ribonucleotides decreases in its intensity in the order of mono-, di- and tri-phosphate of adenine and guanine, which is consistent with previous reports.

Oxygen Inactivation and Orthophenanthrolin Inactivation

As shown in Fig. 3, inactivation of rat liver ATase by oxygen has shown a half life of about 15 min at 37°C in Tris buffer, while the activity is stable under argon. As shown in Fig. 4, ATase activity was strongly inhibited by orthophenanthroline but not by metaphenanthroline, suggesting the presence of ferrous iron essential for enzyme activity in rat liver ATase. These data are compatible with an iron-sulfur structure.

Amino Acid Composition

As shown in Table 3, the amino acid composition of rat ATase is more similar to that of B. subtilis ATase, which also contains iron-sulfur structure, than that of E. coli ATase.

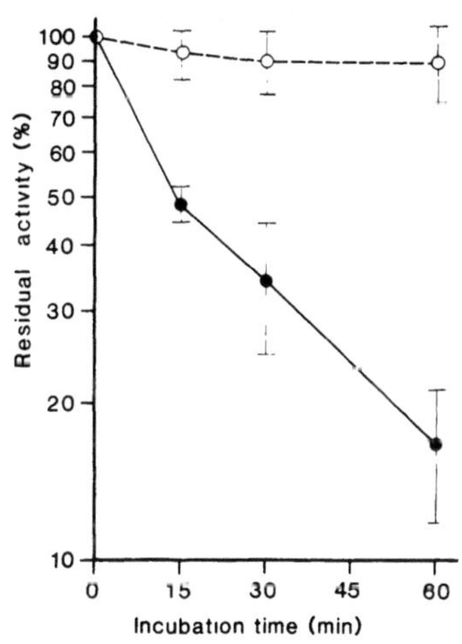

Fig. 3. OXYCEN INACTIVATION OF PARTIALLY PURIFIED RAT LIVER ATASE: Time in min is plotted on abscissa. The percent residual activity is plotted on ordinate. Enzyme activity under oxygen is shown by a solid line and that under argon by a broken line. Each point is the mean±SE of 4 experiments.

Cross-Reactivity of Rat Liver ATase with anti-B. subtilis and anti-E. coli ATase Antibody

As shown in Fig. 5, rat liver ATase shows cross-reactivity to anti-B. subtilis ATase antibody but not to anti-E. coli ATase antibody.

Result of Edman Degradation Analysis

The N-terminal amino acid of highly purified rat liver ATase is

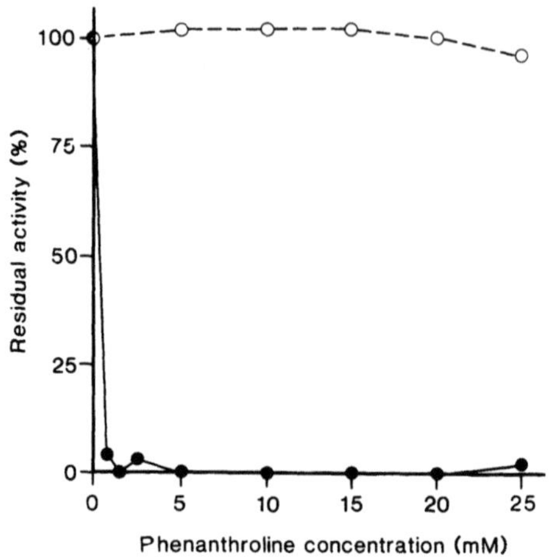

Fig. 4. INACTIVATION OF PARTIALLY PURIFIED RAT LIVER ATASE BY FERROUS IRON CHELATOR: The concentration in mM of ortho- and meta-phenanthroline is plotted on abscissa. The enzyme activity is plotted on ordinate. Each point is the mean±SE of 4 experiments.

TABLE 3. AMINO ACID COMPOSITION OF RAT LIVER ATASE COMPARED WITH E. COLI, B. SUBTILIS, AND YEAST ATASE: Data in this table were obtained from animo acid analysis (1), DNA sequence (2;J.Biol.Chem. 257:3532, 1982, 3;J.Biol.Chem. 258:10586, 1983, 4;J.Biol.Chem. 259:8478, 1984) and SDS-PAGE. (5)

	Rat[1]	E.coli[2]	B.subtilis[3]	Yeast[4]
Asp/Asn	36.0	63	36	62
Thr	21.6	20	27	24
Ser	34.6	28	36	26
Glu/Gln	39.2	52	53	48
Pro	20.9	18	14	22
Gly	61.2	33	49	42
Ala	27.9	48	35	37
Val	37.4	36	36	38
Cys	5.3	8	7	15
Met	5.5	12	14	10
Ile	23.0	44	33	32
Leu	57.0	35	38	41
Tyr	13.0	17	16	23
Phe	26.9	20	13	20
Trp	—	2	1	0
Lys	33.4	17	22	29
His	22.4	10	14	9
Arg	24.8	41	22	32
Total no.of amino acids	490.1	504	476	510
M.W.	ca. 62000[5]	56395	50397	56691

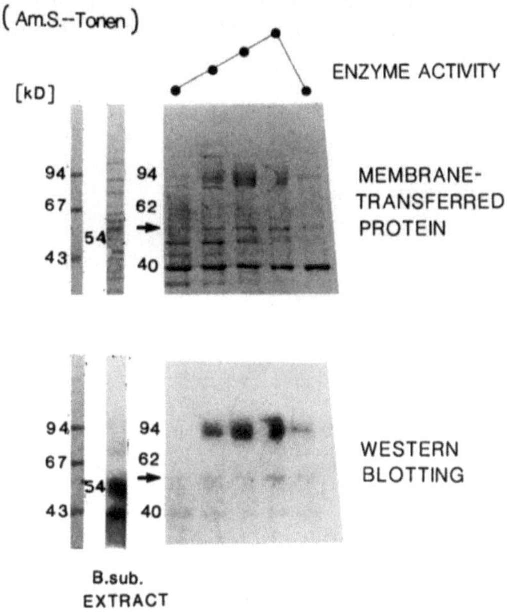

Fig. 5. CROSS-REACTIVITY OF RAT LIVER ATASE WITH ANTI-B. SUBTILIS ATASE ANTIBODY: The highly purified rat liver ATase by a HPLC hydroxyapatite column was examined for its cross-reactivity with anti-B. subtilis ATase antibody after SDS-PAGE and its transfer to a nylon membrane.

apparently modified, because no amino acid was released in Edman degradation analysis.

CONCLUSION

Based on these results we conclude firstly that a molecular weight of a rat liver ATase monomer is about 62 kD, secondly that our data on rat liver ATase are consistent with an iron-sulfur structure and a modified N-terminus, and thirdly that the amino acid composition and antigenicity of rat liver ATase are more similar to that of B. subtilis than that of E. coli.

REFERENCES

1. L. M. Messenger, and H. Zalkin, Glutamine phosphoribosylpyrophosphate amidotransferase from Escherichia coli, purification and porperties. J. Biol. Chem. 254:3382 (1978).
2. M. Itakura, and E. W. Holmes, Human amidophosphoribosyltransferase, an oxygen-sensitive iron-sulfur protein, J. Biol. Chem. 254:333 (1979).
3. U. K. Laemmli, Cleavage of structural proteins during the assembly of the head of bacteriophage T_4. Nature, 227:680 (1970).
4. B. R. Oakley, D. R. Kirsch, and N. R. Morris, A simplified ultrasensitive silver stain for detecting proteins in polyacrylamide gels. Anal. Bicohem. 105:361 (1980).
5. C. Lee, A. Levin, and D. Branton, Copper staining: A five-minute protein stain for sodium dodecyl sulfate polyacrylamide gel. Anal. Biochem. 166:308 (1987).
6. G. P. Tuszynski, C. H. Damsky, J. P. Fuhrer, and L. Warren, Recovery of concentrated protein samples from sodium dodecyl sulfate polyacrylamide gels. Anal. Biochem. 83:119 (1977).

7. H. Towbin, T. Staehelin, and J. Gordon, Electrophoretic transfer of proteins from polyacrylamide gels to nitrocellulose sheets: Procedure and some applications. Proc. Natl. Acad. Sci. USA. 76,9:4350 (1979).

PARADOXICAL EFFECTS OF GLUCAGON ON DE NOVO PURINE SYNTHESIS IN RAT LIVER: IN VITRO PHOSPHORYLATION OF AMIDOPHOSPHORIBOSYLTRANSFERASE

Mitsuo Itakura, Takashi Yamaoka, Hiroko Yoshikawa, and Kamejiro Yamashita

Division of Endocrinology and Metabolism, Institute of Clinical Medicine, University of Tsukuba, Tsukuba-city Ibaraki, 305 Japan

INTRODUCTION

The rate of de novo purine synthesis ("purine synthesis") may be regulated at amidophosphoribosyltransferase (E.C.2.1.2.14)(ATase). The regulatory mechanisms on ATase include activation by its substrate, 5-phosphoribosyl 1-pyrophosphate (PRPP), feedback inhibition by purine ribonucleotides and the amount of enzyme protein. (1-3) The parallel increases of "purine synthesis", glucagon, the second peak of cyclic AMP (cAMP), and PRPP in the prereplicative period in regenerating rat liver after a 70% hepatectomy (4) and increased "purine synthesis" by infusion of glucagon or dibutyryl cAMP (Bt_2cAMP) with increased hepatic cAMP and PRPP (5) suggested the cAMP-dependent regulatory mechanism on "purine synthesis". On the other hand a bolus of glucagon or Bt_2cAMP inhibited or did not increase "purine synthesis" in spite of increased cAMP and PRPP concentrations. (6) These paradoxical effects of glucagon made us to hypothesize that "purine synthesis" is regulated by cAMP-dependent phosphorylation of proteins including ATase. This possibility was tested in this study by examining in vitro phosphorylation of partially purified rat liver ATase by cAMP-dependent protein kinase.

MATERIALS AND METHODS

MATERIALS

Seven-week-old male Wistar rats were used. Radioactive $[^{14}C]$glycine, $[^{14}C]$glutamine, $[^{14}C]$adenine, and gamma$[^{32}P]$ATP were purchased from New England Nuclear (Boston). Glucagon assay kit was purchased from Daiichi, Tokyo, JAPAN (7), the Yamasa cAMP assay kit from Yamasa Company, Chyoshi, Japan (8). Glucagon was purchased from Novo Industry, Copenhagen. Bt_2cAMP, a catalytic subunit of cAMP-dependent protein kinase from a rabbit heart were purchased from Sigma Co (St. Louis).

METHODS

The infusion of glucagon or Bt_2cAMP was performed by an intravenous cannula at a dose of 0.005-1.0 mg or 1.0-10.0 μmol x 200 g body wt^{-1} x 12 h^{-1}. An intravenous bolus administration of glucagon or Bt_2cAMP was performed at the dose of 0.001-0.1 mg or 2 μmol x 200 g body wt^{-1}. The methods for [^{14}C]glycine incorporation were those previously described. (4-6) The cAMP concentrations were assayed with Yamasa cAMP assay kit. (8) PRPP concentrations were assayed by the method of Hisata et al (9) with the minor modifications. (4-6) Purine ribonucleotide concentrations were assayed by high performance liquid chromatography (HPLC). (4-6)

Enzyme purification was performed in purification buffer containing 25 mM potassium phosphate pH 7.4 with 70 mM 2-mercaptoethanol. ATase was detected by the enzyme activity assay. The 30,000 x g supernatant of liver homogenate was heated to 56°C, its supernatant was acidified by 5% acetic acid to pH 5.1. This pellet was resuspended in purification buffer, and fractionated between 30 and 55% ammonium sulfate. The pellet was resuspended and applied to a hydroxyapatite HPLC column. Phosphorylation experiment was performed by mixing the peak enzyme fraction and a catalytic subunit of cAMP-dependent protein kinase and 2 mM gamma[^{32}P]ATP. The identification of an ATase monomer as 62kD protein was made by the correlation of enzyme activity with the protein amount. Phosphorylation of ATase was detected by copper-staining and autoradiography of a sodium dodecyl sulfate polyacrylamide gel.

RESULTS

Time Course and Dose-dependency of Increase of "Purine Synthesis" after Glucagon Infusion

As shown in Fig. 1, glucagon infusion at a constant rate of 0.5 mg x 200 g body wt^{-1} significantly increased "purine synthsis" assayed by

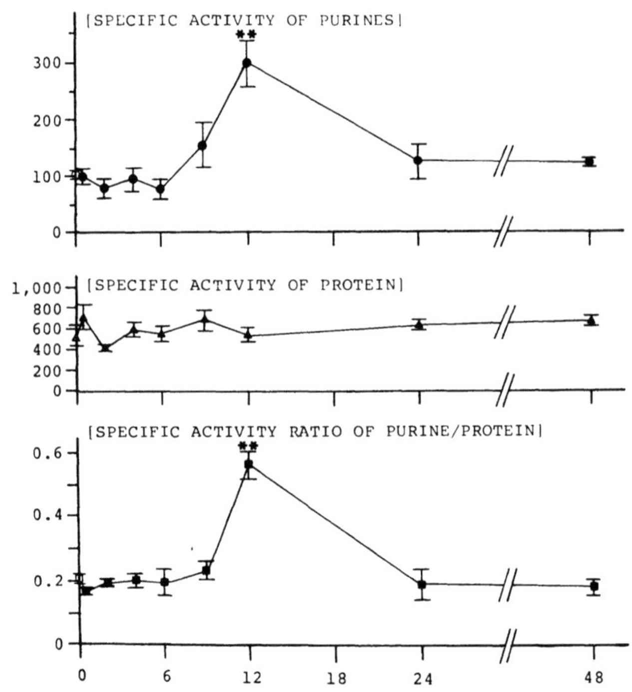

Fig. 1. TIME COURSE OF INCREASE OF "PURINE SYNTHESIS" AFTER GLUCAGON INFUSION: Glucagon was infused at a speed of 0.5 mg x 200 g body wt^{-1} x 12 h^{-1}. Duration in hour is plotted on abscissa. The top, middle and the bottom show incorporation to acid-soluble purines, acid-insoluble proteins, and specific activity ratio of purine to protein, respectively. Each point is a mean±SE of 4 animals.

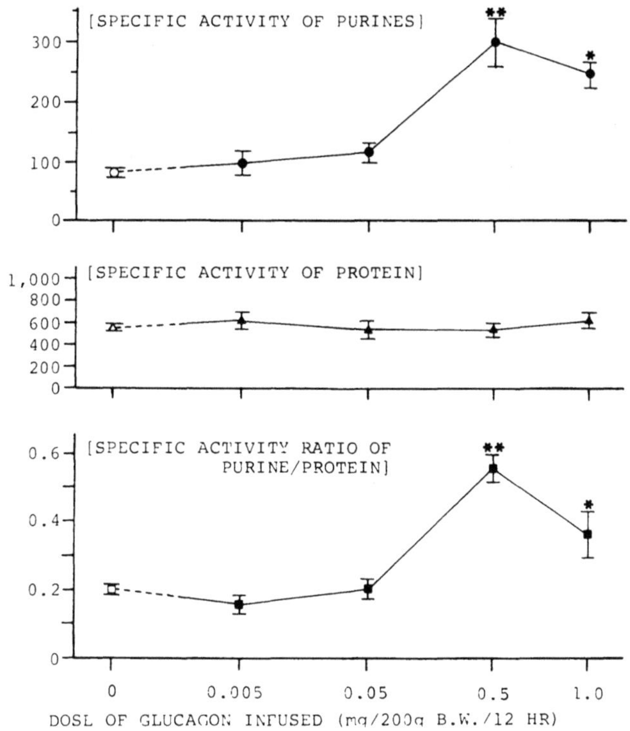

Fig. 2. DOSE-DEPENDENCY OF INCREASE OF "PURINE SYNTHESIS" AFTER GLUCAGON INFUSION: Dose of glucagon infusion as mg x 200 g body wt^{-1} x 12 h^{-1} is plotted on abscissa. The top, middle and the bottom show incorporation to acid-soluble purines as specific activity of purines, incorporation to acid-insoluble proteins and specific activity ratio of purine to protein, respectively. Each point is a mean±SE of 4 animals.

[^{14}C]glycine incorporation from 102±6 dpm/optic absorbance unit at 260 nm (n=4) at time 0 to a level of 304±39 (n=4) at 12 h. Specific activity of hepatic acid-insoluble proteins at time 0 was 520±110 dpm/mg protein (n=4), and it did not change after glucagon infusion. The specific activity ratio of purine to protein at time 0 was 0.22±0.03 (n=4), which increased to 0.56±0.03 (n=4) at 12 h.

As shown in Fig. 2, glucagon infusion at the dose of 0.5 or 1.0 mg x 200 g body wt^{-1} x 12 h^{-1} increased [^{14}C]glycine incorporation after 12 h of infusion to a level of 304±39 (n=4) or 214±31 (n=4) dpm/absorbance unit at 260 nm, respectively. On the other hand, the incorporation to acid-insoluble protein after various doses of glucagon stayed at the same level compared with a level of saline infusion of 585±57 dpm/mg protein (n=4). The specific activity ratio of purine to protein increased significantly to a level of 0.56±0.03 (n=4) or 0.36±0.07 (n=4), respectively, at the glucagon dose of 0.5 or 1.0 mg x 200 g body wt^{-1} x 12 h^{-1}. In the same experimental system, the infusion of Bt_2cAMP also increased "purine synthesis" at 12 h by 2.2- to 4.0-fold.

Specific Activity of ATase, Concentration of PRPP and Purine Ribonucleotides after Glucagon or Bt_2cAMP Infusion

The specific activity of ATase and concentrations of purine ribonucleotides were not significantly changed after glucagon or Bt_2cAMP

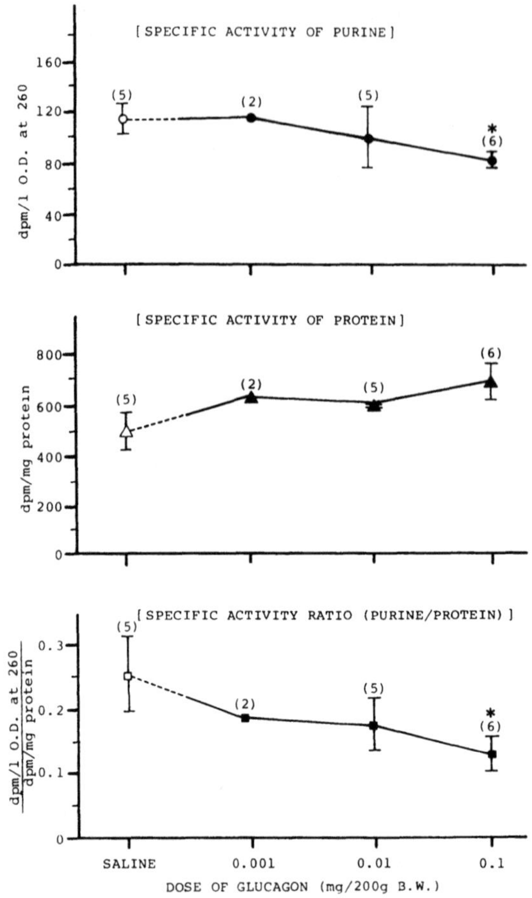

Fig. 3. DOSE-DEPENDENCY OF INHIBITION OF "PURINE SYNTHESIS" BY A BOLUS OF GLUCAGON: The dose of glucagon in mg x 200 g body wt^{-1} is shown on abscissa. The top shows the specific activity of hepatic purines as dpm/absorbance unit at 260 nm, the middle shows that of protein as dpm/mg protein, and the bottom shows the specific activity ratio. The number in parenthesis indicates the number of animals tested in the experiment. Each point is the mean±SE.

infusion, but PRPP concentrations increased by 1.5-fold.

Dose-dependency of Inhibition of "Purine Synthesis" by a Bolus of Glucagon

As shown in Fig. 3, the increasing amount of glucagon decreased the incorporation of [^{14}C]glycine into hepatic purines and the decrease became statistically significant at 0.1 mg x 200 g body wt^{-1} of glucagon. The changes in specific activity of proteins were not statistically significant. The specific activity ratio of purine to protein decreased in response to increasing amounts of glucagon.

The incorporation of [^{14}C]glycine into hepatic purines, protein, and purine to protein ratio after an intravenous bolus Bt_2cAMP at the dose of 2 μmol x 200 g body wt^{-1} were 118±15 dpm/absorbance unit at 260 nm, 540±91 dpm/mg protein and 0.23±0.04 (n=3), while those in saline administered control animals were 114±13, 489±74, and 0.26±0.06 (n=5), respectively. There were no statistically significant differences between the two groups.

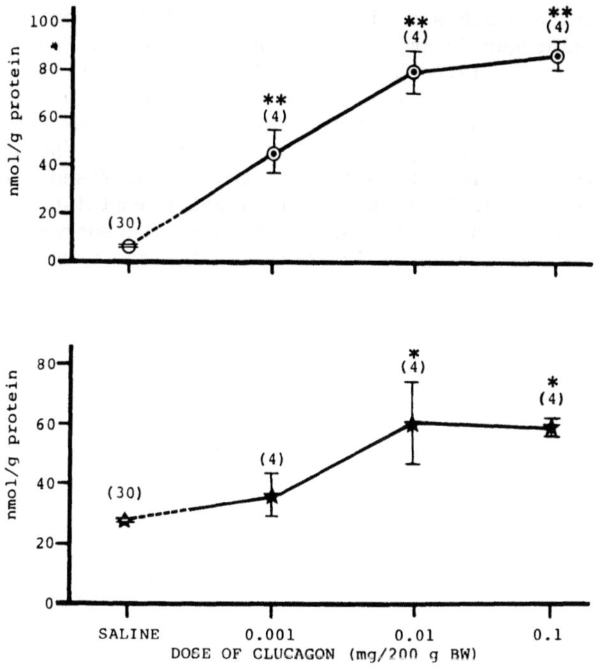

Fig. 4. DOSE-DEPENDENCY OF HEPATIC cAMP AND PRPP CONCENTRATIONS BY A BOLUS OF GLUCAGON: The dose of a bolus of glucagon in mg x 200 g body wt^{-1} is plotted on abscissa. The top and the bottom show, respectively, the hepatic concentrations of cAMP as nmol/g protein, and the concentrations of PRPP five minutes after administration as nmol/g protein. Each point is the mean±SE of 4 animals.

Specific Activity of ATase, Concentrations of RPRR, and Purine Ribonucleotides after a Bolus Glucagon or Bt_2cAMP

The specific activity of ATase and concentrations of purine ribonucleotides were not significantly changed after a bolus of glucagon or Bt_2cAMP. As shown in Fig.4, the cAMP and PRPP concentrations increased dose-dependently following administration of 0.001 to 0.1 mg x 200 g body wt^{-1} of glucagon. The increases in cAMP and PRPP concentrations were significant at doses higher than 0.001 and 0.01 mg x 200 g body wt^{-1}. Both a bolus of glucagon and Bt_2cAMP, respectively at 0.1 mg and 2 µmol x 200 g body wt^{-1} significantly increased hepatic PRPP concentrations, respectively, by 2.3-fold at 5 min and 2.1-fold at 15 min. There was no change in the concentrations of purine ribonucleotides after a bolus administration of glucagon or Bt_2cAMP.

Phosphorylation of Partially Purified ATase

As shown in Fig. 5, the enzyme preparation used for this study contained about 5 bands detectable by copper-staining. Partially purified rat liver ATase was phosphorylated only in the presence of a catalytic subunit of cAMP-dependent protein kinase.

DISCUSSION

The parallel increases of "purine synthesis", hepatic cAMP and plasma glucagon in regenerating rat liver after a 70% hepatectomy and the increased "purine synthesis" in liver after glucagon or Bt_2cAMP infusion support that glucagon is important as a regulator of "purine synthesis", and its signal is mediated by intracellular cAMP. On the other hand the inhibition or the absence of increase of "purine synthesis" by a bolus of glucagon or Bt_2cAMP in spite of increased hepatic cAMP and PRPP concentrations strongly suggested that the increased "purine synthesis" by glucagon infusion is not merely due to the increased PRPP concentrations because it is higher in a bolus administration of glucagon or Bt_2cAMP than that after infusion. Because concentrations of purine ribonucleotides and the specific activity of ATase were not changed, these paradoxical effects of glucagon on "purine synthesis" strongly suggest a new cAMP-dependent regulation mechanism.

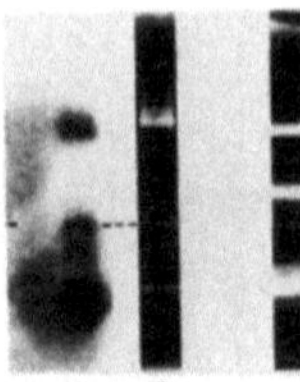

Fig. 5. PHOSPHORYLATION OF PARTIALLY PURIFIED ATASE: The first lane shows the copper-staing of sodium dodecyl sulfate polyacrylamide gel. The 2nd and 3rd lane show the autoradiogram with $[^{32}P]ATP$ in the absence and presence of a subunit of cAMP-dependent protein kinase.

The hypothesis of a new cAMP-dependent regulation mechanism of ATase was supported, although preliminary, by the presence of in vitro phosphorylation of partially purified rat liver ATase by cAMP-dependent protein kinase. Because the enzyme activity of carbamyl phosphate synthetase II, catalyzing the first enzymatic step for pyrimidine synthesis has been reported to be regulated by cAMP-dependent phosphorylation in vitro, (10) it is possible that ATase which catalyzes the first enzymatic step for de novo purine synthesis is regulated by the same mechanism. Based on these, we conclude that a cAMP-dependent phosphorylation of rat liver ATase is amenable to further study as a new regulation mechanism.

REFERENCES

1. Wyngaarden, J. B., and Kelly, W. N., 1983, Gout: The Metabolic Basis of Inherited disease, J. B. Stanbury, J. B. Wyngaarden, D. S. Fredrickson, J. L. Goldstein, and M. S. Brown ed., McGraw-Hill, New York
2. Holmes, E. W., Wyngaarden, J. B., and Kelley, W. N., 1973, Human glutamine phosphoribosylpyrophosphate amidotransferase: Two molecular forms interconvertible by purine ribonucleotides and phosphoribosylpyrophosphate, J. Biol. Chem., 248:6035.
3. Itakura, M., Sabina, R. L., Heald, P. W., and Holmes, E.W., 1981, Basis for the control of purine biosynthesis by purine ribonucleotides, J. Clin. Invest., 67:994.
4. Itakura, M., Maeda, N., Tsuchiya, M., and Yamashita, K., 1986, Increased rate of de novo purine synthesis and its mechanism in regenerating rat liver. Am. J. Physiol., 252:G373.

5. Itakura M., Maeda, N., Tsuchiya, M., and Yamashita, K., 1987, Glucagon infusion increases rate of purine synthesis de novo in rat liver. Am. J. Physiol., 253:E684.
6. Itakura, M., Maeda, N., Tsuchiya, M., and Yamashita, K., 1986, In vivo inhibition of the rate of de novo purine synthesis in rat liver by glucagon. Metabolism, 35:758.
7. Imagawa, K., Nishino, I., Shin, S., Uehata, S., Hashimura, E., Yanaihara, C., and Yanaihara, N., 1979, Production of anti-glucagon sera with a C-terminal fragment of pancreatic glucagon. Endocrinol. Jpn. 26:123.
8. Cailla, H. L., Racine-Weisbuch, M. S., and Delaage, M. A., 1973, Adenosine 3',5'-cyclic monophosphate assay at 10^{-15} mole level. Anal. Biochem. 56:394.
9. Hisata, T., 1975, An accurate method for estimating 5-phosphoribosyl-1-pyrophosphate in animal tissues with the use of acid extraction. Anal. Biochem. 68:448.
10. Carrey, E. A., Campbell, D. G., and Hardie, D. G., 1985, Phosphorylation and activation of hamster carbamyl phosphate synthetase II by cAMP-dependent protein kinase. A novel mechanism for regulation of pyrimidine nucleotide biosynthesis. EMBO Journal. 4:4735.

PROLIFERATIVE EFFECT OF INSULIN BY SELECTIVE GENE EXPRESSION IN PURINE METABOLIC PATHWAY IN PRIMARY CULTURED RAT HEPATOCYTES

Hiroko Yoshikawa, Masami Sato, Takashi Yamaoka
Mitsuo Itakura, and Kamejiro Yamashita

University of Tsukuba, Institute of Clinical Medicine
Tsukuba-city, Ibaraki, 305 Japan

INTRODUCTION

The growth of liver in vivo after a partial hepatectomy (1) and that of hepatocytes in vitro (2) are under the control of growth factors, but the mechanism by which the quiescent hepatocytes are brought to the proliferative stage is not well understood. Because "the rate of de novo purine synthesis" ("purine synthesis") increases in the prereplicative period in the regenerating rat liver after a partial hepatectomy, (3) de novo purine synthetic pathway can be regarded as one of the typical anabolic pathways essential for liver proliferation. In order to elucidate the mechanism of liver proliferation, the mechanism of the increased "purine synthesis" in primary cultured rat hepatocytes by insulin (4) was studied in regard to the induction of purine-metabolizing enzymes, including amidophosphoribosyltransferase (ATase), hypoxanthine-guanine-phosphoribosyltransferase (HGPRT), adenine-phosphoribosyltransferase (APRT), and xanthine dehydrogenase (XD).

MATERIALS AND METHODS

MATERIALS

Seven-week-old male Wistar rats were purchased from Shizuoka Agricultural Cooperative Association for Laboratory Animals (Shizuoka, Japan). Eagle's minimum essential medium was purchased from Nissui Pharmaceutical Co. Ltd., Tokyo, Japan, Williams' medium E and newborn calf serum were from Life Technologies, INF. New York, U.S.A. Dexamethasone was purchased from Towa Pharmaceutical Co. Ltd., Osaka Japan. Collagenase was purchased from Wako Pure Chemical Industries Ltd., Tokyo, Japan. Other chemicals were of the highest quality available. Radioactive [^{3}H]thymidine, [^{14}C]formate, [^{14}C]adenine, [^{14}C]glutamine, [^{14}C]hypoxanthine were purchased from New England Nuclear (Boston). Insulin was purchased from Novo Industry, Copenhagen.

METHODS

Isolation and Culture of Rat Hepatocytes

Hepatocytes were isolated from male Wistar rats, weighing 180 to

230 g by the two-step collagenase perfusion technique by Seglen. (5) Hepatocytes were initially cultured in Williams' medium E with 10% newborn calf serum, 10^{-7} M insulin, and 10^{-6} M dexamethasone. The initial cell density was 4-6 x 10^4 cells/cm^2. To determine the unmodified effect of insulin, insulin was administered to hepatocytes which were cultured for 20 h beforehand without serum and hormones following the initial plating culture for 4 h.

Preparation of Cell Extract

Hepatocytes were scraped by a straight edge of a thin plastic plate, and centrifuged at 12,000 x g for 10 seconds. After removing the supernatant, the pellet was resuspended in 25 mM KPi and 60 mM 2-mercaptoethanol for assaying HGPRT and APRT activity. This sample was sonicated twice each for 30 seconds and the 12,000 x g supernatant was used for enzyme assay.

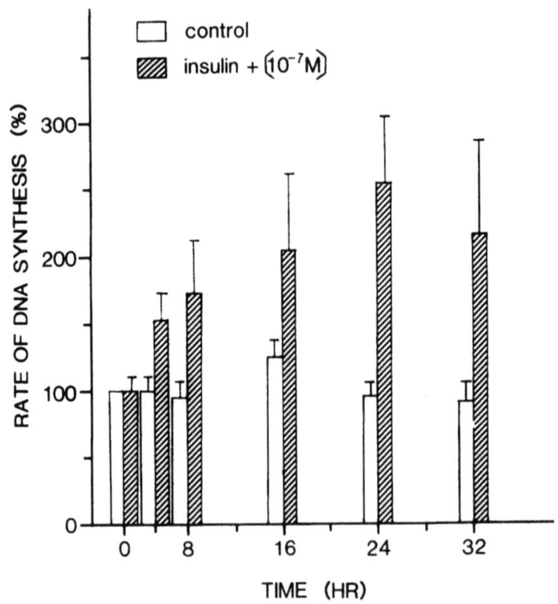

Fig. 1. TIME COURSE OF EFFECT OF INSULIN ON DNA SYNTHESIS: Time course of effect of insulin at 10^{-7} M on the rate of DNA synthesis is shown. Time after insulin administration in hour is shown on abscissa. Each bar represents the mean ± SE.

Assay of the Rate of DNA and de novo Purine Synthesis

The rate of DNA synthesis was determined by measuring the incorporation of [^{3}H]thymidine to DNA in hepatocytes. The rate of de novo purine synthesis was determined according to the published methods (6) by measuring the incorporation of [^{14}C]formate to acid-soluble adenine nucleotides in hepatocytes.

Assay of PRPP Availability

PRPP availability for nucleotide synthesis was determined according to the published methods (7) by measuring the incorporation of [^{14}C]adenine to acid-soluble adenine nucleotides in hepatocytes.

Assay of the Enzyme Activity of ATase, HGPRT, APRT and XD

PRPP-dependent glutaminase activity was regarded as ATase activity. It was assayed by radioactivity of [^{14}C]glutamate formed from [^{14}C]glutamine. (8) HGPRT activity was assayed by the radioactivity of [^{14}C]IMP formed from [^{14}C]hypoxanthine according to the published methods. (9) APRT activity was assayed by the radioactivity of [^{14}C]AMP formed from [^{14}C]adenine accoring to the published methods. (10) XD activity was assayed by the radioactivity of xanthine, uric acid and allantoin formed from [^{14}C]hypoxanthine according to the published methods. (11) The specific activity of these enzymes were determined by dividing the enzyme activity by the amount of protein.

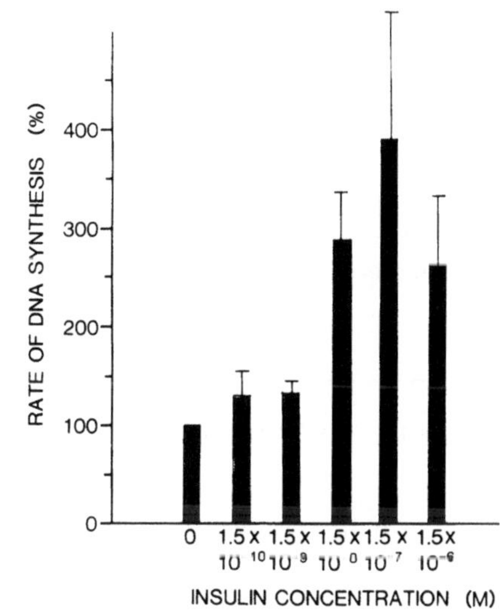

Fig. 2. DOSE-DEPENDENCY OF EFFECT OF INSULIN ON DNA SYNTHESIS: Dose-dependency of effect of insulin on rate of DNA synthesis at 24 h is shown.

RESULTS AND DISCUSSION

Time Course and Dose-dependency of Effect of Insulin on DNA Synthesis

Time course of effect of insulin at 10^{-7} M on the rate of DNA synthesis is shown in Fig. 1. Hepatocytes became competent by seeding them at the relatively low density, and they were introduced into S-phase by the addition of insulin. Dose-dependency of effect of insulin on the rate of DNA synthesis at 24 h is shown in Fig. 2. Insulin at the concentration of 10^{-8} to 10^{-7} M increased the rate of DNA synthesis at the plateau level with a 3- to 4-folds increase.

Time Course and Dose-dependency of Effect of Insulin on de novo Purine Synthesis and PRPP Availability

As reported previously (4), insulin at 10^{-7} M increased "purine synthesis" time-dependently and the increase became progressively small after 4 h and reached the maximal value of $24.5\pm1.8 \times 10^3$ dmp/μmol adenine

12 h after the administration which is 270-280% of a starting level. The increased rate of DNA synthesis and "purine synthesis" by insulin plateaued, respectively, at about 300 and 280% at 1.5 x 10^{-7} and 1.5 x 10^{-8} M. The comparable dose-dependency of DNA synthesis and "purine synthesis" suggests that the increased "purine synthesis" is an essential step for "cell proliferation". And the similar magnitute of increase of "purine synthesis" and the increase of DNA synthesis suggest that de novo synthetic pathway is, at least, as important as salvage pathway. This in vivo observation is in contrast with the hundred times higher Km for PRPP utilization by ATase than those of HGPRT and APRT in vitro.

Insulin also increased PRPP availability even at the dose of 1.5 x 10^{-9} M in the same time course, suggesting that the increased PRPP production is another important factor for increased "purine synthesis". The rate of de novo purine synthesis reached the near maximal level at 1.5 x 10^{-8} M. PRPP availability, on the other hand, reached the near maximal level at 1.5 x 10^{-9} M, both in a dose-dependent manner. This suggests that the increase of PRPP availability occurrs at one order of magnitude lower insulin concentration.

TABLE 1. EFFECT OF INSULIN ON THE SPECIFIC ACTIVITY OF ATASE, HGPRT, APRT AND XD: Effect of insulin at 10^{-7} M on the specific activity of purine metabolizing enzymes in primary cultured rat hepatocytes are summarized.

	INSULIN - (n)	INSULIN + (n)	%INCREASE
ATase (n mol glutamine deaminated/ hr/mg protein)	3.50±0.32 (9)	*5.39±0.75(10)	154
HGPRT (n mol hypoxanthine converted to IMP/hr/mg protein)	390±30 (10)	*503±38 (10)	129
APRT (n mol adenine converted to AMP/hr/mg protein)	112±28 (5)	*230±34 (5)	205
XD (n mol hypoxanthine converted to xanthine, uric acid, and allantoin/hr/mg protein)	30± 5 (5)	24± 1 (5)	80

(*$p<0.05$, **$p<0.01$)

Effect of Insulin on the Specific Activity of ATase, HGPRT, APRT, and XD

Effect of insulin at 10^{-7} M on the specific activity of purine metabolizing enzymes in primary cultured rat hepatocytes are summarized in TABLE 1. Insulin increased the specific activity of ATase to 150%, HGPRT to 130% and APRT to 200% with the increased Vmax and the unchanged Km of PRPP for ATase (4), whereas it did not increase the specific activity of XD. These data show that the activities of anabolic enzymes including ATase, HGPRT, and APRT were increased by insulin through induction of enzyme proteins, and that the activity of catabolic enzymes such as XD was unchanged by insulin. The increase of the specific activity of ATase was at most 154%, while that of DNA synthesis was about 300% in the presence of 10^{-7} M insulin. This may suggest that induction of not only ATase but also other anabolic enzymes such as PRPP synthetase is essentaial to increase "purine synthesis" by 250%.

Effect of Insulin with Protein Synthesis Inhibitors on the Specific Activity of ATase, HGPRT, and APRT

Effect of protein synthesis inhibitors on the increasing effect of insulin on the specific activity of ATase is shown in Fig. 3. When 1 μg/ml of actinomycin D or cycloheximide, as a protein synthesis inhibitor, was adminitered with insulin, the increasing effect of insulin on the specific activity of ATase was abolished and the specific activity of ATase remained at a level without insulin. Although not shown in this paper the increased specific activities of HGPRT and APRT were also completely suppressed by the addition of these protein synthesis inhibitors. These data suggest that the increased specific activities of ATase, HGPRT, and APRT by insulin are due to the induction of these enzyme proteins. Because there is the induction of APRT, the increased PRPP availability by insulin shown above can be now interpreted as firstly due to increased PRPP synthesis and secondly to the induction of APRT. In the latter case the apparent increase in PRPP availability by insulin should be the larger estimate than the real increase of PRPP-availability for purine ribonucleotide synthesis.

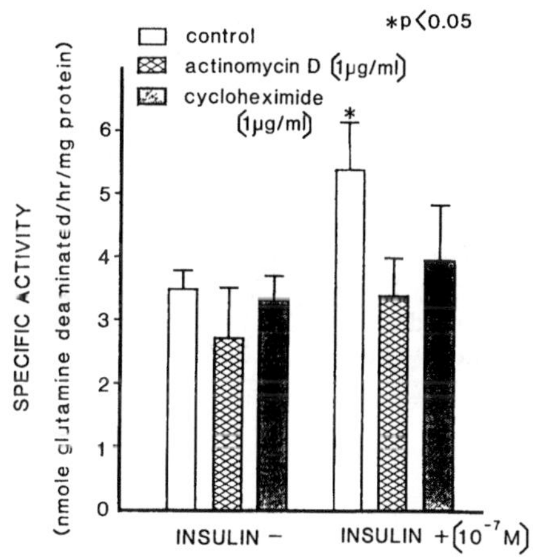

Fig. 3. EFFECT OF PROTEIN SYNTHESIS INHIBITORS ON THE INCREASE OF THE SPECIFIC ACTIVITY OF ATASE: Effect of Actinomycin D and cycloheximide both at 1 μg/ml on the specific activity of ATase is shown.

CONCLUSION

Because DNA synthesis in primary cultured rat hepatocytes is increased by insulin, this experimental system of primary cultured rat hepatocytes under insulin is regarded as a model for "cell proliferation". Proliferative effect of insulin on "purine synthesis" is a result of selective gene expression of a group of anabolic enzymes, associated with the unchanged specific activity of a catabolic enzyme. It is also concluded that de novo purine synthesis is as important as salvage pathway for "cell proliferation".

REFERENCES

1. N. L. R. Bucher, and R. A. Malt. Liver regeneration. In: Regeneration of Liver and Kidney. Boston, MA: Little Brown, p.17-178 (1971).
2. H. L. Leffert, K. S. Koch, P. J. Lad, B. DeHemptine, and H. Shelly. Glucagon and liver regeneration, In: handbook of Experimental Pharmacology. Glucagon I, edited by P. J. Lefebre. New York: Springer-Verlag, chapt. 21, p.453-484 (1983).
3. M. Itakura, N. Maeda, M. Tsuchiya, and K. Yamashita. Increased rate of de novo purine synthesis and its mechanism in regenerating rat liver. Am. J. Physiol. 251:G585 (1986).
4. M. Tsuchiya, M. Itakura, and K. Yamashita. The mechanism of insulin-induced increase of the rate of de novo purine synthesis in primary cultured rat hepatocytes, In: Purine and Pyrimidine Metabolism in Man V, Advances in experimental medicine and biology, vol 195B, ed. by W. L. Nyhan, L. F. Tompson, and R. W. W. Watts, Plenum Press, New York and London. p.623 (1985).
5. P. O. Seglen, Chapter 4. Preparation of isolated rat liver cells. Methods Cell Biol. 13:29 (1976).
6. G. R. Boss, Decreased phosphoribosylpyrophosphate as the basis for decreased purine synthesis during amino acid starvation of human lymphoblasts. J. Biol. Chem. 259:2936 (1984).
7. C. B. Thomas, J. C. Meade, and E. W. Holmes, Aminoimidazole carboxamide ribonucleoside toxicity: A model for study of pyrimidine starvation. J. Cell Physiol. 107:335 (1981).
8. M. Itakura, and E. W. Holmes, Human amidophosphoribosyltransferase, An oxygen-sensitive iron-sulfur protein. J. Biol. Chem. 254:333 (1979).
9. W. J. Arnold, and W. N. Kelley, Human hypoxanthine-guanine phosphoribosyltransferase, purification and subunit structure, J. Biol. Chem. 246:7398 (1971).
10. C. B. Thomas, W. J. Arnold, and W. N. Kelley, Human adenine phosphoribosyltransferase, purification, subunit structure, and substrate specificity, J. Biol. Chem. 248:2529 (1973).
11. T. Ikegami, Y. Natsumeda, and G. Weber, Decreased concentration of xanthine dehydrogenase (EC 1.1.1.204) in rat hepatomas, Cancer Res. 46:3838 (1986).

PURINE DE NOVO SYNTHESIS AND INOSINIC BRANCH POINT IN VIVO IN DIFFERENT TISSUES: A BIOMATHEMATICAL MODEL

Maria Pizzichini, Anna Di Stefano, Germano Resconi* and Enrico Marinello

Inst. of Biological Chemistry, University of Siena
* Inst. of Mathematics, University of Brescia - Italy

INTRODUCTION

Purine nucleotides can be synthetized trough the "de novo" pathway, a linear sequence of reactions, starting from PRPP and leading to IMP.

IMP is further converted to either AMP or GMP, through the "inosinic acid branch point", which is subjected to a fine regulation [1].

In this study, we present a comparative evaluation of the purine nucleotide synthesis rate in different rat tissues. Since nucleotides are precursors of coenzymes, of RNA and DNA, the rate of their synthesis should differ and relate to cell mitosis and to the different metabolic activities of the cell: however, such comparative evaluation is very difficult.

We studied the overall rate of the process by following the incorporation of labelled formate into nucleotides, elaborating a biomathematical model for a precise evaluation of experimental data, which may be calculated in terms of apparent rate constants.

MATERIALS AND METHODS

Male albino Wistar rats (200-250 g) were fasted 20 hours before killing and received 10 µCi/ 100 g b.w. of ^{14}C-formate (54.8 mCi/mmole) 1 h before decapitation. The organs were rapidly excised, homogenized with 0.5 N perchloric acid and centrifuged. The supernatant was utilized for the analysis of the acid soluble nucleotides, which were acid-hydrolyzed to free purines[2]. The bases were precipitated with $AgNO_3$, separated by chromatography on a Dowex-resin (AG 50W-X8, 200-400 mesh, hydrogen form) column and eluted with a linear gradient as previously described[3].

The radioactivity of the bases was determined on a Nuclear Chicago Scintillation Counter, using Instagel as a liquid scintillation cocktail.

Liver, kidney, spleen and heart were excised from one group of rats, whereas the muscles [levator ani (LAM), gastrocnemius (G), extensor digitorum longus (ELD)] were removed from another group.

We previously elaborated a biomathematical model[4] - which is not included here for sake of brevity - to evaluate purine biosynthesis. The sequence is considered a simplified series of monomolecular reactions:

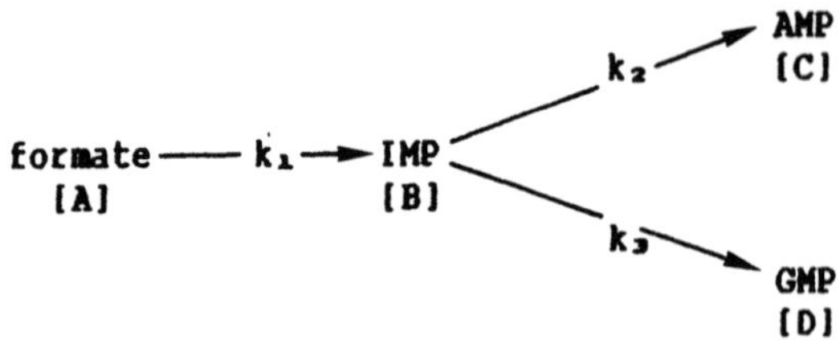

The three most important steps were evaluated in terms of apparent rate constants (k_1- k_2- k_3) through a system of differential equations.

It can be demonstrated that:

$$k_1 \, [A] = B^* + C^* + D^*$$

$$k_2 = C^* \, / \, [B]$$

$$k_3 = D^* \, / \, [B]$$

When we have the values B^*, C^*, D^* (dpm/g/sec) and the purine content (µmol/g), since [A] is constant, we can compute k_1, k_2, k_3, which are the apparent rate constants, proportional to the synthesis of three nucleotides IMP, AMP, GMP respectively. The k values are expressed as sec^{-1}.

The k_3/k_2 is an useful parameter to evaluate the regulation of the "inosinic branch point" and the relative formation rate of GMP and AMP.

RESULTS AND DISCUSSION

All apparent rate constants were very high in the spleen, in comparison to other organs. IMP and GMP formation (k_1- k_3) was rapid also in the kidney and heart, while they were lower in the liver and muscles. AMP formation (k_2) was very fast in the gastrocnemius and kidney.

k_3/k_2 ratio showed comparable values in the liver, kidney, spleen, in the LAM and ELD muscles (~0.1), indicating that the rate of AMP turnover was about ten times higher than GMP. In the heart, the k_3/k_2 ratio reached higher values (~0.5), because GMP formation was more rapid

Table 1 Apparent rate constants (k_1-k_2-k_3) in different rat organs and muscles. The values are the mean ± SE of three different rats.

	k_1	k_2	k_3
Liver	2.93 ± 0.24	4.58 ± 0.66	0.48 ± 0.07
Kidney	10.95 ± 1.02	14.80 ± 1.47	1.24 ± 0.16
Spleen	17.92 ± 0.34	118.20 ± 31.40	18.83 ± 5.53
Heart	5.40 ± 0.31	5.29 ± 0.93	2.58 ± 0.12
G	3.21 ± 0.67	18.22 ± 7.95	0.90 ± 0.34
LAM	1.78 ± 0.04	8.34 ± 2.53	0.76 ± 0.16
ELD	3.02 ± 0.25	8.08 ± 0.73	0.87 ± 0.05

than in other muscles, and about one half that of AMP. Guanine content in the heart was low (unpublished data), but present data show there was a surprising rate of GMP turnover; it would be interesting to demonstrate that such differences are in some way linked to cardiac function.

From our results it is evident that the rate of IMP, AMP and GMP synthesis differs in various organs and muscles.

As predicted, these differences were sometimes linked to the intensity of mitosis; for istance, kidney and spleen showed the highest values. However, the heart, (its cells do not undergo mitosis), showed higher values than the other muscles and the liver.

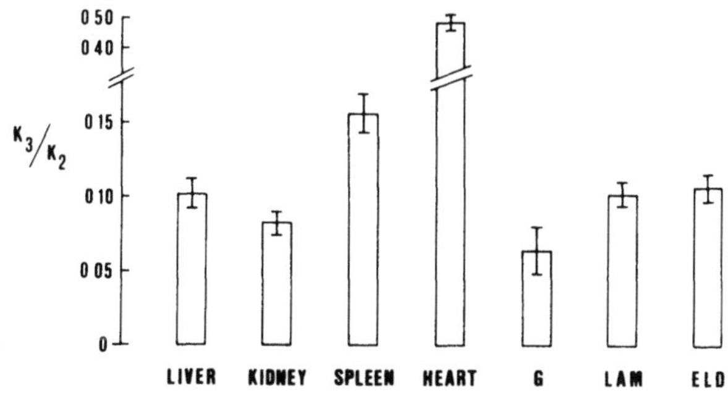

Figure1 k_3/k_2 ratio in different rat organs and muscles. The values are the mean ± SE of three animals.

The rate of AMP formation is related to ATP necessities, which explains the remarkable values found in the gastrocnemius muscle, while the exceptionally high rate in the spleen is difficult to be interpreted.

The k_3 values should represent GTP formation, which is related to protein synthesis, but its values are very low in the liver.

We conclude that there are differences in the rate of synthesis of nucleotides in different organs, which cannot always be explained on the basis of present knowledge, and further investigation must be undertaken. After many years of work on nucleotides, this problem still represents an undefined field of research.

REFERENCES

1. G. W. Crabtree and J. B. Handerson, Rate limiting steps in the interconversion of purine ribonucleotides in Ehrlich Ascites tumor cells in vitro, Cancer Research 31: 985 (1971).
2. G. Schmidt, Chemical and Enzymatic Methods for the identification and structural elucidation of nucleic acid and nucleotides, in: " Methods in Enzymology", S. P. Colowick and N. O. Kaplan, ed, Academic Press Inc., Publishers, New York 3 : 747 (1957).
3. M. Pizzichini, A. Di Stefano and E. Marinello, The regulations of purine ribonucleotide biosynthesis by glucocorticoid hormones, It.J.Biochem., 34 (5): 305 (1985).
4. A. Di Stefano, M. Pizzichini, G. Resconi and E. Marinello, submitted for publication on Hormone and Metabolic Research.

THE INOSINIC BRANCH POINT AND ITS HORMONAL REGULATION:

EVALUATION THROUGH A MATHEMATICAL MODEL

Maria Pizzichini, Anna Di Stefano, Germano Resconi*
and Enrico Marinello

Inst. of Biological Chemistry, University of Siena
* Inst. of Mathematics, University of Brescia, Italy

INTRODUCTION

Purine de novo synthesis consists of a "common segment", the sequence of the reactions leading from PRPP to IMP, and the "inosinic branch point", which converts IMP to either AMP or GMP.

We have evaluated nucleotide biosynthesis in different organs of normal, castrated and adrenalectomized rats, following the overall rate of the process through the incorporation of ^{14}C-formate into the acid-soluble bases, a method which has been widely described in recent literature [1,2].

We compute the overall rate of purine biosynthesis using a biomathematical model[3], (we do not describe it here for the sake of brevity), which permits calculation of the apparent rate constants of IMP, AMP and GMP formation. k_3/k_2 is used as an index of "IMP branch point" regulation.

MATERIALS AND METHODS

For our experiments we used 200-250 g male albino Wistar rats divided into different groups, of three animals each. In one group, the rats were adrenalectomized intraperitoneally under ether, given saline (1% NaCl) ad libitum and killed after 5 days. Another group was castrated under ether and kept at 25°C for 15 days. Normal rats, of the same age as treated animals, were used as controls.

10 µCi / 100 g b.w. of ^{14}C-formate (54.8 mCi / mmol) were injected intraperitoneally, 1 h before decapitation; liver, kidney and heart were rapidly removed from one group of rats, the muscles [gastrocnemius (G), levator ani (LAM), extensor digitorum longus (ELD)], were excised from another. The tissues were homogenized with 0.5 N perchloric acid, the nucleotides were submitted to acid hydrolysis and the purine bases were separated by ion exchange chromatography[4].

According to the above quoted mathematical model, the sequence of purine biosynthesis is visualized as a simplified series of reactions:

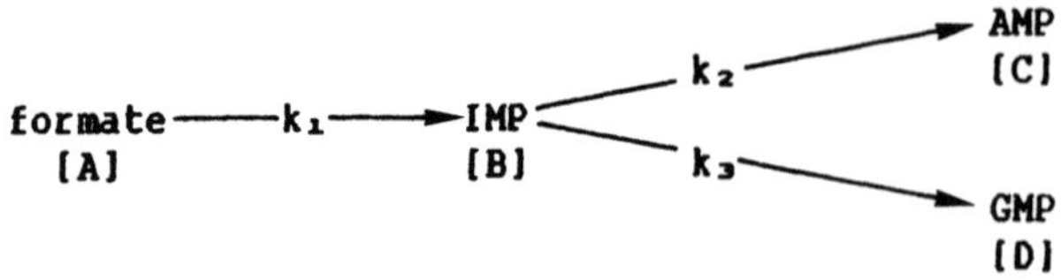

were k_1 represents IMP formation, k_2 the rate of AMP synthesis and k_3 that of GMP.

Once we have the values B^*, C^*, D^* (dpm/g/sec) and the purine content (µmol/g), since [A] is constant, we can calculate:

$$k_1 = B^* + C^* + D^*$$

$$k_2 = C^* / [B]$$

$$k_3 = D^* / [B]$$

which are the apparent rate constants, proportional to nucleotide synthesis.

RESULTS AND DISCUSSION

Table 1 illustrates the behavior of rate constants under the various experimental conditions.

In the liver and kidney all the k values increased after castration, showing that the reactions leading from formate to IMP, and from IMP to either AMP or GMP were accelerated.

Adrenalectomy determined a decrease in rate constants values which was more evident in the liver and heart, while in the kidney k_1 values remained unchanged.

Table 2 shows the behavior of k values in the muscles after castration; a decrease of k_2 and k_3 was evident only in LAM; there was a slight decrease of k_1 in ELD, of k_1 and k_2 in gastrocnemius.

Table 1. The behavior of k_1, k_2, k_3 in liver, kidney and heart, in normal, castrated and adrenalectomized rats. The values are the mean ± SE of three different rats. *$P<0.05$ **$P<0.01$

Organ	Treatment	k_1	k_2	k_3
Liver	normal	3.25 ± 0.19	5.71 ± 0.50	0.56 ± 0.08
	castrated	*6.54 ± 1.18	*12.90 ± 2.15	*1.65 ± 0.36
	adrenalectomized	*1.61 ± 0.51	*2.01 ± 0.09	*0.30 ± 0.11
Kidney	normal	9.38 ± 2.50	15.28 ± 3.18	1.44 ± 0.14
	castrated	*14.50 ± 1.68	**28.39 ± 4.96	*2.33 ± 0.40
	adrenalectomized	10.78 ± 3.74	*9.20 ± 1.60	*1.08 ± 0.15
Heart	normal	5.39 ± 0.63	6.60 ± 0.74	3.17 ± 0.37
	adrenalectomized	*3.68 ± 0.46	*2.14 ± 0.51	**0.82 ± 0.09

Table 2. The effect of castration on k_1, k_2, k_3 values in different rat muscles. The values are the mean ± SE of three rats. *P<0.05

Organ	Treatment	k_1	k_2	k_3
ELD	normal	3.04 ± 0.73	5.61 ± 0.47	0.66 ± 0.18
	castrated	2.45 ± 0.97	8.57 ± 0.51	0.83 ± 0.36
G	normal	3.24 ± 0.67	18.42 ± 2.58	0.90 ± 0.09
	castrated	2.28 ± 0.22	15.14 ± 3.03	1.02 ± 0.18
LAM	normal	1.12 ± 0.09	8.22 ± 0.98	0.89 ± 0.11
	castrated	1.10 ± 0.10	*2.24 ± 0.50	*0.25 ± 0.01

The k_3/k_2 ratio is reported in Figure 1. When the variations of k_2 and k_3 were about of the same order, the k_3/k_2 ratio remained constant, as was noted after castration. In liver, the situation was not identical after adrenalectomy, because of the differential decreases of k_2 and k_3. AMP formation was slower than GMP synthesis, while k_3/k_2 ratio was enhanced.

Our results demonstrate that nucleotide formation was substantially enhanced, after castration, in liver and kidney and decreased after adrenalectomy, which indicates that in normal rats, sexual and glucocorticoid hormones play an opposite regulatory role.

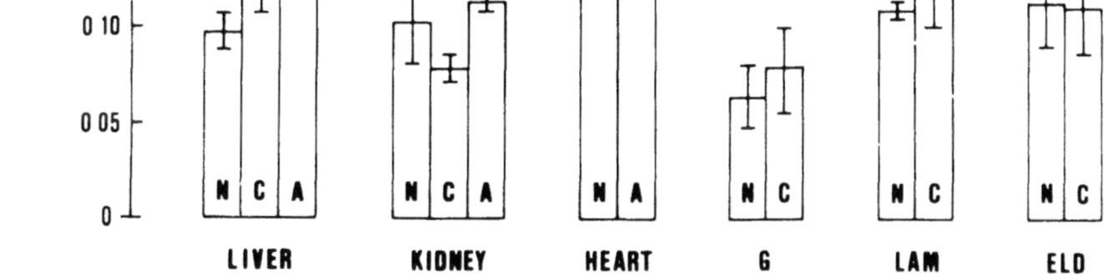

Figure 1. k_3/k_2 ratio in organs and muscles of normal (N), castrated (C) or adrenalectomized (A) rats. The values are the mean ± SE of three different animals. *P<0.05

The castration decreased AMP and GMP formation only in LAM, while in the other muscles there were no significant variations, suggesting that gastrocnemius and ELD were less influenced by testosterone.

After adrenalectomy, k_3/k_2 ratio showed a significant increase only in the liver, due to a slower formation of adenine nucleotide.

In the other situations, k_3/k_2 underwent no variations.

REFERENCES

1. A. Di Stefano, M. Pizzichini and E. Marinello, Effect of testosterone on purine synthesis de novo in rat perineal muscle in vivo, Biochim. Biophys. Acta 926: 1 (1987).
2. E. Zoref-Shani and O. Sperling, Dependance of the metabolic fate of IMP on the rate of total IMP synthesis, Biochim. Biophys. Acta 607: 503 (1980).
3. A. Di Stefano, M. Pizzichine, G. Resconi and E. Marinello Submitted for publication on Hormone and Metabolic Research.
4. M. Pizzichini, A. Di Stefano and E. Marinello, The regulation of purine ribonucleotide biosynthesis by glucocorticoid hormones, It. J. Biochem. 34 (5): 305 (1985).

KINETIC PROPERTIES OF IMP DEHYDROGENASE PURIFIED FROM RAT HEPATOMA 3924A

Yasukazu Yamada, Tadashi Ikegami, Yutaka Natsumeda and George Weber

Laboratory for Experimental Oncology, Indiana University School of Medicine, Indianapolis, IN 46223, U.S.A.

INTRODUCTION

IMP dehydrogenase (EC 1.1.1.205), the rate-limiting enzyme of de novo GTP biosynthesis, is a promising target in cancer chemotherapy.[1] Early studies in this laboratory showed that in all rat hepatomas examined, IMP dehydrogenase activity increased above that of normal liver and the rise correlated positively with the increase in tumor growth rates.[2] The oncolytic action of the antimetabolites targeted against IMP dehydrogenase has been shown to link with the depletion of GTP and dGTP pools.[3,4] Because of the strategic importance of the guanylate pathway and its key enzyme, IMP dehydrogenase, we purified the enzyme to homogeneity from rat rapidly growing hepatoma 3924A and elucidated the reaction mechanism by the observation of the kinetic properties of the purified enzyme with natural ligands (IMP, XMP, NAD, and NADH).

MATERIALS AND METHODS

[8-^{14}C]IMP (53 µCi/mmol) was purchased from Amersham. All other chemicals were of the highest purity available.

In enzyme purification, IMP dehydrogenase activity was determined by radiochemical and spectrophotometric assays.[5,6] In the kinetic studies with the purified enzyme, however, adenosine 5'-α,β-methylenediphosphate and allopurinol were omitted from the reaction mixture. The specific activity was expressed in µmol/h/mg protein. Protein concentration was determined by a Bio-Rad protein kit using crystalline bovine serum albumin as a standard.

IMP dehydrogenase was purified from 100,000 x g supernatant of transplantable rat hepatoma 3924A by ammonium sulfate fractionation, hydroxylapatite gel chromatography (Bio-Rad, HTP), gel filtration (Pharmacia, Sephacryl S-400) and an affinity chromatography of IMP Sepharose, as described previously.[7]

SDS-polyacrylamide gel electrophoresis was carried out according to a system of Laemmli,[8] and the polypeptide bands were detected by an ultrasensitive silver stain as reported by Oakley et al.[9]

RESULTS AND DISCUSSION

IMP dehydrogenase has been purified to homogeneity from rat hepatoma 3924A. The specific activity of the final enzyme preparation was 103μmol/h/mg protein and represented a 4980-fold purification over the original supernatant and a recovery of 30% of the initial activity. When either the 100,000 x g supernatant or the partially purified enzyme after hydroxylapatite was applied to a column of Sephacryl S-400 or S-300, the activity was eluted as a single peak with a molecular weight of 245,000. From a sucrose density (5-20%) gradient centrifugation of the purified enzyme at 100,000 x g for 12 h, a sedimentation coefficient value of 11.9 s was obtained and a molecular weight was calculated as about 250,000 by the method of Martin and Ames.[10] The molecular weight of the subunit was about 60,000 by using 10% polyacrylamide gel electrophoresis in the presence of SDS. These data suggest that the native IMP dehydrogenase of hepatoma 3924A consists of four subunits with a molecular weight of about 60,000 each.

The double reciprocal plots for dependence of velocity upon substrate concentration (Fig. 1) yielded K_m values for IMP and NAD^+ of 23 and 65 uM, respectively. NAD caused substrate inhibition in the purified enzyme as described previously.[11-13] When NAD^+ concentration was increased to 1.0 mM, the activity of IMP dehydrogenase decreased to 80% of the maximum rate observed at 0.25 mM NAD^+. An apparent K_m value for K^+ of 7.8 mM was obtained from the double reciprocal plots of various KCl concentration in the presence of 250 uM IMP and 250 μM NAD^+.

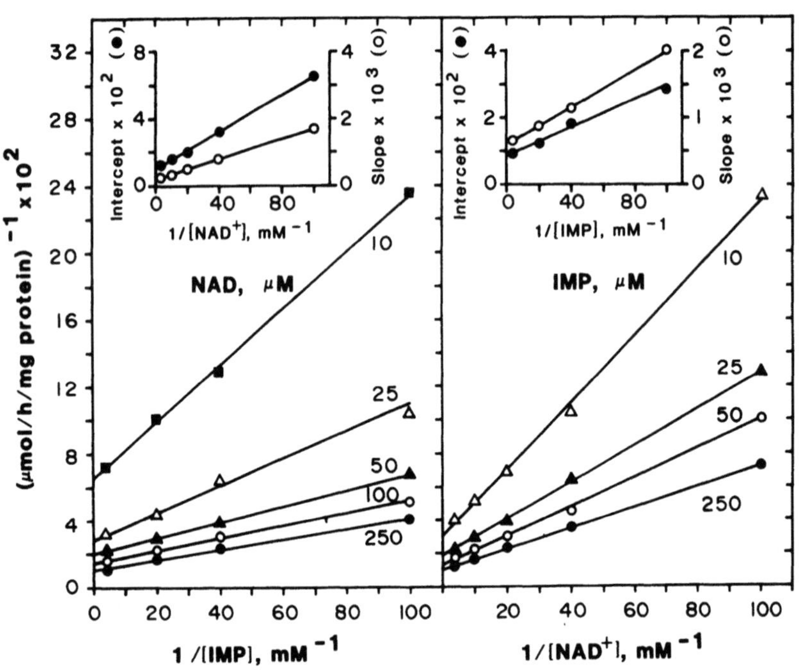

Fig. 1. Substrate dependence of IMP dehydrogenase activity. Double reciprocal plots are shown for various IMP (right) and NAD^+ (left) concentrations. (inset) The replots of slope and intercept versus 1/s.

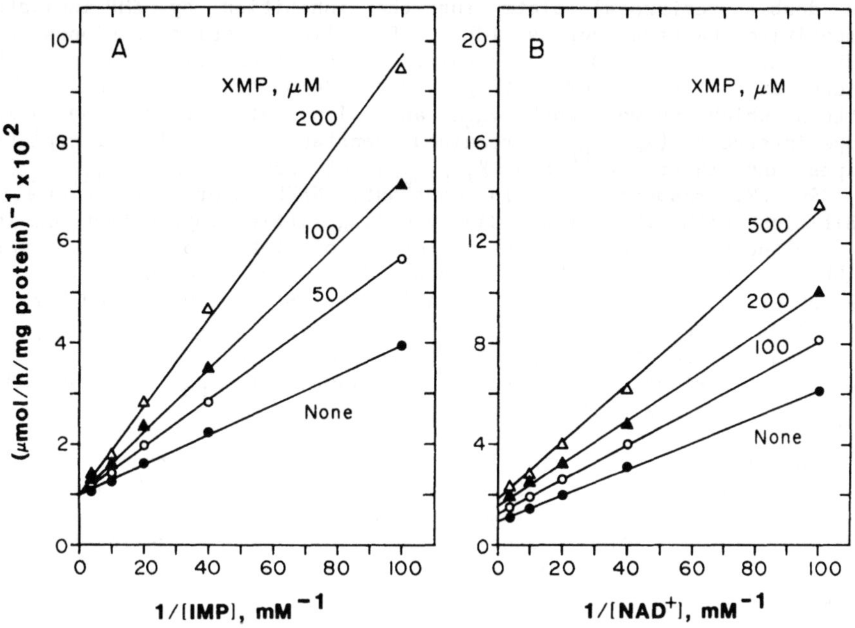

Fig. 2. Inhibition of IMP dehydrogenase by XMP. (A) Double reciprocal plots for various IMP concentrations. (B) Double reciprocal plots for various NAD^+ concentrations.

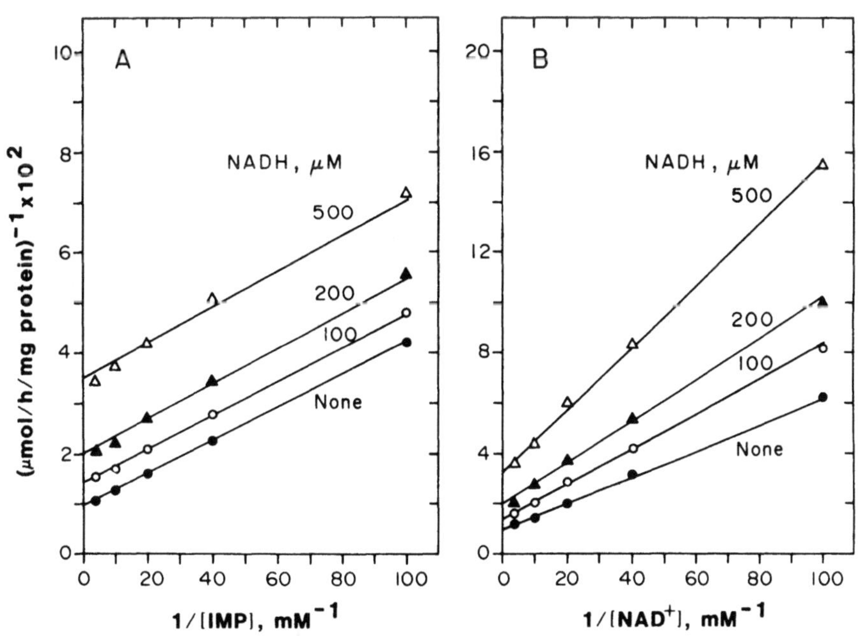

Fig. 3. Inhibition of IMP dehydrogenase by NADH. (A) Double reciprocal plots for various IMP concentrations. (B) Double reciprocal plots for various NAD^+ concentrations.

Product inhibition studies are shown in Figs. 2 and 3. In Fig. 2A the double reciprocal plots for the inhibition by XMP indicated a competitive pattern versus IMP. From linear replot of slopes versus [XMP], K_i of 136 μM was obtained. Inhibition by XMP at a fixed IMP concentration (250 μM) (Fig. 2A) demonstrated a noncompetitive pattern which showed both V_{max} and slope effects, but the ratio of slope/intercept ($K_{m\ app.}$) remained constant. The linear replots of slopes and intercepts gave $K_{i[slope]}$ of 320 μM and $K_{ii[intercept]}$ of 390 μM, respectively. In contrast, NADH exerted an uncompetitive inhibition with IMP (Fig. 3A), yielding little slope effect but large V_{max} effects. From a linear replot of intercept [$1/V_{max}$] versus [NADH], a K_i = 210 μM was obtained. A mixed type inhibition by NADH was observed with respect to NAD^+ (Fig. 3B). NADH exerted both V_{max} and slope effects, but the ratio of slope/intercept ($K_{m\ app.}$) was decreased, since the V_{max} effect was larger than the slope one. The replot of slope and intercept yielded $K_{i[slope]}$ of 290 μM and $K_{ii[intercept]}$ of 210 μM.

The initial rate kinetics and the results of product inhibition studies show a pattern of ordered Bi-Bi mechanism,[14] in which IMP binds to free enzyme followed by NAD^+. After rearrangement of the ternary complex, NADH dissociates and finally XMP is released from the enzyme as proposed in previous studies with partially purified enzyme.[11-13]

The role of IMP dehydrogenase during the neoplastic growth program has long been an important motive of investigation in this Laboratory.[1,2,5-7,12] Recent studies on purine metabolic fluxes in hepatoma 3924A cells indicated that relative utilization of IMP for guanylate production increased preferentially and that for adenylates decreased inversely during the display of their replicative program from resting to logarithmic growth phase.[15] The stepped-up channeling of IMP into guanylate biosynthesis is also reflected in elevated pools of GTP and dGTP in hepatoma cells in log phase.[16] These observations support the significance of de novo guanylate pathway as a potential target of chemotherapy. The idea that IMP dehydrogenase should be a sensitive target was taken up by R. K. Robins and his associates who synthesized a number of compounds as prospective inhibitors of IMP dehydrogenase.[17] Independently, Yoshida et al. synthesized a series of imidazole derivatives which possess a broad antitumor spectrum[18] and the mechanism of antitumor action was elucidated later to be due to inhibition of IMP dehydrogenase.[19] The inhibitory mechanisms of the active metabolites of these drugs are investigated in the purified IMP dehydrogenase and presented in a separate paper in this symposium.

ACKNOWLEDGEMENTS

This work was supported by an Outstanding Investigator Grant (OIG) USPH CA-42510 to G. Weber, and Am. Cancer Inst. Grant IN-161 to Y. Yamada.

REFERENCES

1. G. Weber, Biochemical strategy of cancer cells and design of chemotherapy: G.H.A. Clowes Memorial Lecture, Cancer Res. 43: 3466 (1983).
2. R. C. Jackson, G. Weber, and H. P. Morris, IMP dehydrogenase - an enzyme linked with proliferation and malignancy, Nature (London) 256: 331 (1975).

3. M. S. Lui, M. A. Federan, J. J. Liepnieks, Y. Natsumeda, E. Olah, H. N. Jayaram and G. Weber, Modulation of IMP dehydrogenase activity and guanylate metabolism by tiazofurin (2-β-D-ribofuranosylthiazole-4-carboxamide), J. Biol. Chem. 259: 5078 (1984).
4. M. Fukui, M. Inaba, S. Tsukagoshi and Y. Sakurai, Selective reduction of intracellular guanosine 5'-triphosphate pool by 4-carbamoylimidazolium 5-olate in murine tumor cells, Cancer Res. 46: 43 (1986).
5. T. Ikegami, Y. Natsumeda, and G. Weber, Direct assay method for inosine 5'-monophosphate dehydrogenase activity, Anal. Biochem. 150: 155 (1985).
6. T. Ikegami, Y. Natsumeda, and G. Weber, Purification of IMP dehydrogenase from rat hepatoma 3924A by a novel affinity chromatography, Life Sci. 40: 2277 (1987).
7. Y. Yamada, Y. Natsumeda, and G. Weber, Action of the active metabolites of tiazofurin and ribavirin on purified IMP dehydrogenase, Biochemistry 27: 2193 (1988).
8. U. K. Laemmli, Cleavage of structural protein during the assembly of the head of bacteriophage T4, Nature (London) 277: 680 (1970).
9. B. R. Oakley, D. R. Kirsch and N. R. Morris, A simplified ultra-sensitive silver stain for detecting proteins in polyacrylamide gels, Anal. Biochem. 105: 361 (1980).
10. R. G. Martin and B. N. Ames, A method for determining the sedimentation behavior of enzymes: Application to protein mixture, J. Biol. Chem. 236: 1372 (1961).
11. E. W. Holmes, D. M. Pehlke and W. N. Kelley, Human IMP dehydrogenase. Kinetics and regulatory properties, Biochim. Biophys. Acta 364: 209 (1972).
12. R. C. Jackson, H. P. Morris and G. Weber, Partial purification, properties and regulation of inosine 5'-phosphate dehydrogenase in normal and malignant rat tissues, Biochem. J. 166: 1 (1977).
13. D. T. Hupe, B. A. Azzolina and N. D. Behrens, IMP dehydrogenase from the intracellular parasitic protozoan Eineria tenella and its inhibition by mycophenolic acid, J. Biol. Chem. 261: 8363 (1986).
14. W. W. Cleland, The kinetics of enzyme-catalyzed reactions with two or more substrates or products. I. Nomenclature and rate equations, Biochim. Biophys. Acta 67: 104 (1963).
15. Y. Natsumeda, T. Ikegami, K. Murayama, and G. Weber, De novo guanylate synthesis in the commitment to replication in hepatoma 3924A cells, Cancer Res. 48: 507 (1988).
16. G. Weber, E. Olah, M. S. Lui, H. Kizaki, D. Y. Tzeng, and E. Takeda, Biochemical commitment to replication in cancer cells, Adv. Enzyme Regul. 18: 3 (1980).
17. R. K. Robins, P. C. Srivastava, V. L. Narayanan, J. Plowman, and K. D. Paull, 2-β-D-ribofuranosylthiazole-4-carboxamide, a novel potential antitumor agent for lung tumors and metastases, J. Med. Chem. 25: 107 (1982).
18. N. Yoshida, T. Kiyohara, M. Fukui, T. Atsumi, S. Ogino, M. Inaba, S. Tsukagoshi, and Y. Sakurai, Antitumor activities of newly synthesized 5-carbamoyl-1H-imidazol-4-yl 1-adamantane-carboxylate and 5-carbamoyl-1H-imidazol-4-yl piperonylate, Cancer Res. 40: 3810 (1980).
19. M. Fukui, M. Inaba, S. Tsukagoshi, and Y. Sakurai, New antitumor imidazole derivatives, 5-carbamoyl-1H-imidazole-4-yl piperonylate, as an inhibitor of purine synthesis and its activation by adenine phosphoribosyltransferase, Cancer Res. 42: 1098 (1982).

3. M. S. Lui, M. A. Faderan, J. J. Liepnieks, Y. Natsumeda, E. Olah, H. N. Jayaram and G. Weber, Modulation of IMP dehydrogenase activity and guanylate metabolism by tiazofurin (2-β-D-ribofuranosylthiazole-4-carboxamide), J. Biol. Chem. 259: 5078 (1984).
4. M. Fukui, M. Inaba, S. Tsukagoshi and Y. Sakurai, Selective reduction of intracellular guanine nucleotides [illegible] 6-[illegible] [illegible] Res. 42: 1098 (1982).
5. [illegible], T. Natsumeda, and [illegible] [illegible] 150: 155 (1985).
6. T. Ikegami, Y. Natsumeda and G. Weber, [illegible] IMP dehydrogenase from rat hepatoma 3924A by a novel affinity chromatography, Life Sci. 40: 2277 (1987).
7. [illegible], Y. Natsumeda, and G. Weber, [illegible] IMP dehydrogenase, Biochemistry [illegible] (1988).
8. [illegible] Nature (London) [illegible] (19[illegible]).
[illegible]
[illegible]

TWO DISTINCT TARGET SITES ON IMP DEHYDROGENASE IN CHEMOTHERAPY

Yasukazu Yamada, Yutaka Natsumeda, Yasufumi Yamaji and George Weber

Laboratory for Experimental Oncology, Indiana University School of Medicine, Indianapolis, IN 46223, U.S.A.

INTRODUCTION

IMP dehydrogenase (EC 1.1.1.205), the rate-limiting enzyme of _de novo_ GTP biosynthesis, is a sensitive target in anti-cancer and anti-viral chemotherapy.[1-4] Ribavirin is a potent anti-viral agent with a broad spectrum including hepatitis B virus.[5] Tiazofurin and SM-108 have wide murine anti-tumor activity including Lewis lung carcinoma which is refractory to most drugs.[2,6] The recent demonstration of the striking action of tiazofurin in the treatment of human leukemia has increased interest in drugs directed against IMP dehydrogenase.[7,8] In this study, the inhibitory mechanisms of ribavirin 5'-monophosphate (RMP), SM-108 nucleotide (bredinin 5'-monophosphate, BMP) and thiazole-4-carboxamide adenine dinucleotide (TAD), the chemotherapeutically active forms of ribavirin, SM-108 and tiazofurin, were investigated in IMP dehydrogenase purified to homogeneity from rat hepatoma 3924A. Combinations of the drugs with different modes of inhibitory action were tested in hepatoma 3924A cells in culture.

MATERIALS AND METHODS

Materials. [^{14}C]IMP and [^{14}C]formate were obtained from Amersham. Bactopeptone-free McCoy's 5A medium and tissue culture supplies were purchased from Grand Island Biochemical Co. Eagle's basal medium containing Earle's salt, L-glutamine and HEPES was from Sigma. Ribavirin, tiazofurin, RMP and TAD were kindly supplied by Dr. R. K. Robins, Nucleic Acid Research Institute (Costa Mesa, CA). SM-108 and BMP were kindly provided by Sumitomo Chemical Co., Ltd. (Takarazuka, Japan).

Enzyme assay and purification. IMP dehydrogenase activity was determined by a radiochemical assay and enzyme purification was according to the methods described previously.[9]

Cell Culture. Rat hepatoma 3924A cells in monolayer culture were maintained in log phase in McCoy's 5A medium supplemented with 10% fetal calf serum, penicillin, 100 units/ml and streptomycin, 100 ug/ml.[10] Clonogenic assays were performed in McCoy's 5A medium with dialyzed fetal calf serum without bactopeptone to be free from nucleosides and nucleobases. After 7 days of exposure to drugs, colonies of more than 50 cells were counted.[11] The initial rates of *de novo* guanylate synthesis were determined as outlined elsewhere.[10,11]

RESULTS AND DISCUSSION

Inhibitory mechanisms of RMP, BMP, and TAD, the active metabolites of ribavirin, SM-108 and tiazofurin, respectively, were investigated in rat hepatoma IMP dehydrogenase purified to homogeneity. RMP inhibited competitively with IMP and noncompetitively with NADH, and TAD inhibition was uncompetitive with respect to IMP and mixed type with NAD^+.[9] As shown in Fig. 1, the inhibition by BMP was competitive with IMP and noncompetitive with NAD^+ similar to that by RMP or XMP. K_i value of 0.11 uM with IMP was obtained from a linear replot of slope versus [BMP], and with respect to NAD^+, the replots of slope and intercept gave $K_{i[slope]}$ of 1.6 μM and $K_{ii[intercept]}$ of 0.6 uM, respectively.

The K_i values and the inhibition types of the products and antimetabolites are summarized in Table I. The results indicate that RMP and BMP exerted the same inhibitory mechanism as XMP, and the inhibition by TAD was similar to that by NADH. However, K_i values for RMP, BMP and TAD were 2 to 3 orders of magnitude lower than those for XMP and NADH. Thus, these drugs have different modes of action and interact at different ligand sites of the IMP dehydrogenase with higher affinities than the natural substrates and products; RMP and BMP with the IMP-XMP site and TAD with the NADH site.

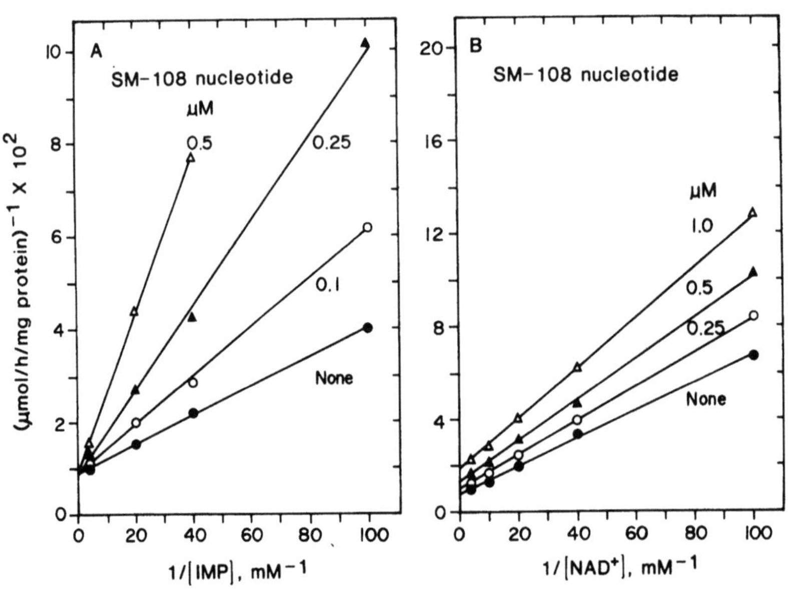

Fig. 1. Inhibition of IMP dehydrogenase by SM-108 nucleotide (BMP). (A) Double reciprocal plots for various IMP concentrations. (B) Double reciprocal plots for various NAD^+ concentrations.

Table 1. Inhibition of IMP dehydrogenase by Products and Antimetabolites

Compounds	Variable substrates	K_i values μM	Type of inhibition
Products			
XMP	IMP	136	Competitive
XMP	NAD^+	320	Noncompetitive
NADH	IMP	210	Uncompetitive
NADH	NAD^+	212	Mixed type
Antimetabolites			
RMP	IMP	0.8	Competitive
RMP	NAD^+	3.6	Noncompetitive
BMP	IMP	0.11	Competitive
BMP	NAD^+	0.6	Noncompetitive
TAD	IMP	0.13	Uncompetitive
TAD	NAD^+	0.13	Mixed type

Table 2. Effects of Preincubation of IMP Dehydrogenase with Various Ligands

Ligands	Concentration (μM)	Activity (% of control)		Activity recovered after preincubation
		Not preincubated	preincubated	
None		100*	100	100
RMP	2.5	89	62	70
BMP	2.5	71	70	99
TAD	0.5	73	73	100
XMP	1250	80	80	100
NADH	1250	63	62	98
IMP	1250	100	100	100
RMP +IMP	2.5 1250	89	91	102
RMP +XMP	2.5 1250	76	60	79
RMP +NADH	2.5 1250	53	37	70
RMP +TAD	2.5 0.5	65	41	63

*Control activity

The inhibitory effects were enhanced by preincubation of the purified IMP dehydrogenase with RMP in a temperature- and time-dependent manner.[9] Although BMP and XMP have the same inhibitory mechanism as RMP, preincubation of the enzyme with BMP or XMP did not increase the inhibition (Table 2). Incubation in the presence of TAD, XMP, NADH, or IMP did not exert any time-dependent change in the enzyme activity. However, IMP and XMP protected against the time-dependent action induced by RMP (Table 2). The results support the kinetic analysis that RMP shares the same binding site with IMP and XMP. In the presence of TAD, RMP further enhanced its inhibitory effect (Table 2).

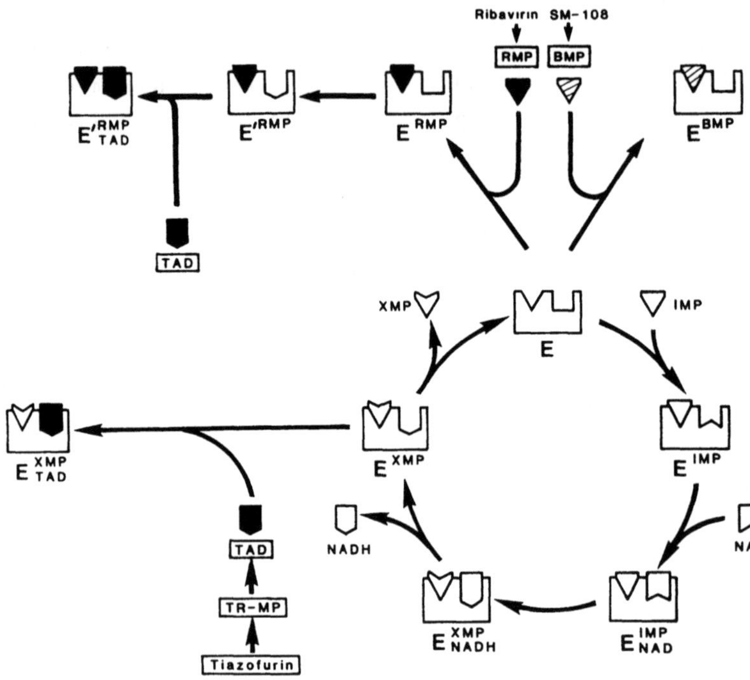

Fig. 2. Reaction mechanism of IMP dehydrogenase and the proposed mode of drug actions

The reaction mechanism of IMP dehydrogenase and the proposed mode of action of drugs are summarized in Fig. 2. TAD has the same inhibitory mechanism as NADH and is thought to interact with E·XMP and produce E·XMP·TAD complex. By contrast, RMP and BMP compete in binding to the free enzyme with IMP and XMP. Only RMP enhanced the inhibitory action in a time-dependent manner. The time dependent effects might be caused by the conversion from E·RMP to E'·RMP complex. If TAD would interact to the E'·RMP complex, TAD might shift the equilibrium to reduce E·IMP complex by producing E'·RMP·TAD complex.

Ribavirin, SM-108 and tiazofurin are converted to their respective active forms by different metabolic pathways. Ribavirin is phosphorylated to RMP by adenosine kinase,[12,13] whereas SM-108 is activated to BMP by adenine phosphoribosyltransferase.[14] Tiazofurin is metabolized in two steps, phosphorylation and NAD pyrophosphorylase-catalyzed, reaction, to TAD.[15] Thus, the three drugs targeted against IMP dehydrogenase possess different drug metabolism as well as different modes of inhibitory action. Combinations of these drugs with different mechanisms might potentiate inhibition of IMP dehydrogenase in cancer cells and provide good antitumor effects. This hypothesis was tested in hepatoma 3924A cells in clonogenic assays. Treatment of cells for 7 days with ribavirin, SM-108, tiazofurin and selenazofurin, a selenazole analogue of tiazofurin, yielded LD_{50} of 12, 3.5, 6.5 and 1.5, respectively. Treatment with 6 uM tiazofurin and 10 uM ribavirin resulted in a survival of 56 and 70%, whereas a combination of the two drugs yielded 18% survival, indicating synergistic action (Table 3). Selenazofurin plus ribavirin also exerted synergistic killing of the cells (Table 3). However, SM-108 in combination with the other drugs examined did not yield synergistic cytotoxic effects.

Table 3. Effects of Drugs on Survival of Hepatoma 3924A Cells

Drugs	uM	% of control survival	
		Observed	Predicted
None, control		100	
Tiazofurin	6	56 ± 6	
Selenazofurin	1	63 ± 3	
Ribavirin	10	70 ± 4	
SM-108	2	79 ± 4	
Tiazofurin + Ribavirin	6 10	18 ± 0*	39
Tiazofurin + SM-108	6 2	47 ± 6	44
Tiazofurin + Selenazofurin	6 1	38 ± 4	35
Selenazofurin + Ribavirin	1 10	15 ± 1*	44
Selenazofurin + SM-108	1 2	51 ± 4	50
Ribavirin + SM-108	10 2	61 ± 6	55

Means ± S. E. of 9 or more assays are given. Cells were treated for 7 days; survival measured by colony counts.
*Synergism

Table 4. Effects of Ribavirin and Tiazofurin on de Novo Guanylate Synthesis in Hepatoma 3924A Cells

Drugs, uM	Guanylate synthesis, % of control	
	Determined	Predicted
None, control	100	--
Ribavirin, 10	39	--
Tiazofurin, 8	52	--
Ribavirin, 10		
+ Tiazofurin, 8	6*	20

*Synergism

The effect of a combination of ribavirin and tiazofurin was further investigated in hepatoma 3924A cells by measuring the actual metabolic fluxes of de novo guanylate synthesis. When cells were treated for 48 h with 10 μM ribavirin or 8 uM tiazofurin, the rates of de novo guanylate synthesis decreased to 39 and 52%, respectively, of that in the control. A combination of the two drugs resulted in 6% of control value which indicated synergism, since a summation would have yielded 20% (Table 4).

Huggins et al. showed synergistic antiviral effects by a combination of ribavirin with tiazofurin or selenazofurin.[16] The present study demonstrated two distinct ligand sites of IMP dehydrogenase which provide promising drug targets; NADH site for tiazofurin through its active metabolite, TAD, and XMP-IMP site for ribavirin through its metabolite, RMP. A combination of ribavirin and tiazofurin exerted synergistic inhibitions of de novo guanylate synthesis and colony formation in rat hepatoma cells. The results offer a biochemical basis for combination chemotherapy directed to the two sites on IMP dehydrogenase.

ACKNOWLEDGEMEMTS

This work was supported by an Outstanding Investigator Grant (OIG) USPH CA-42510 to G. Weber, and Am. Cancer Inst. Grant IN-161 to Y. Yamada.

REFERENCES

1. G. Weber, Biochemical strategy of cancer cells and design of chemotherapy: G.H.A Clowes Memorial Lecture, <u>Cancer Res.</u> 43: 3466 (1983).
2. R. K. Robins, Nucleoside and nucleotide inhibitors of inosine monophosphate (IMP) dehydrogenase as potential antitumor inhibitor, <u>Nucleosides & Nucleotides</u> 1: 35 (1982).
3. M. S. Lui, M. A. Faderan, J. J. Liepnieks, Y. Natsumeda, E. Olah, H. N. Jayaram, and G. Weber, Modulation of IMP dehydrogenase activity and guanylate metabolism by tiazofurin (2-β-D-ribofuranosylthiazole-4-carboxamide), <u>J. Biol. Chem.</u> 259: 5078 (1984).

4. D. G. Streeter, J. T. Witkowski, G. P. Khare, R. W. Sidwell, R. J. Bauer, R. K. Robins, and L. N. Simon, Mechanism of action of 1-β-D-ribofuranosyl-1,2,4-triazole-3-carboxamide (Virazole), a new broad-spectrum antiviral agent, Proc. Natl. Acad. Sci. USA 70: 1174 (1973).
5. P. A. A. Galvao and I. O. Castro, The effect of 1-β-D-ribofuranosyl-1,2,4-triazole-3-carboxamide on acute viral hepatitis, Ann. N.Y. Acad. Sci. 284: 278 (1977).
6. N. Yoshida, T. Kiyohara, M. Fukui, T. Atsumi, S. Ogino, M. Inaba, S. Tsukagoshi, and Y. Sakurai, Antitumor activities of newly synthesized 5-carbamoyl-1H-imidazol-4-yl 1-adamantane-carboxylate and 5-carbamoyl-1H-imidazol-4-yl piperonylate, Cancer Res. 40: 3810 (1980).
7. G. J. Tricot, H. N. Jayaram, C. R. Nichols, K. Pennington, E. Lapis, G. Weber, and R. Hoffman, Hematological and biochemical action of tiazofurin (NSC 286193) in a case of refractory acute myeloid leukemia, Cancer Res. 47: 4988 (1987).
8. G. Weber, H. N. Jayaram, E. Lapis, Y. Natsumeda, Y. Yamada, Y. Yamaji, G. J. Tricot, and R. Hoffman, Enzyme-pattern-targeted chemotherapy with tiazofurin and allopurinol in human leukemia, Adv. Enzyme Regul. 27: 405 (1988).
9 . Y. Yamada, Y. Natsumeda, and G. Weber, Action of the active metabolites of tiazofurin and ribavirin on purified IMP dehydrogenase, Biochemistry 27: 2193 (1988).
10. Y. Natsumeda, T. Ikegami, K. Murayama, and G. Weber, De novo guanylate synthesis in the commitment to replication in hepatoma 3924A cells, Cancer Res. 48: 507 (1988).
11. Y. Natsumeda, Y. Yamada, Y. Yamaji, and G. Weber, Synergistic cytotoxic effect of tiazofurin and ribavirin in rat hepatoma 3924A cells, Biochem. Biophys. Res. Commun. 153: 321 (1988).
12. R. C. Willis, D. A. Carson, and J. E. Seegmiller, Adenosine kinase initiates the major route of ribavirin activation in a cultured human cell line, Proc. Natl. Acad. Sci., USA 75: 3042 (1978).
13. Y. Yamada, H. Goto, and N. Ogasawara, Adenosine kinase from human liver, Biochim. Biophys. Acta 660: 36 (1981).
14. M. Fukui, M. Inaba, S. Tsukagoshi, and Y. Sakurai, New antitumor Imidazole derivative, 5-carbamoyl-1H-imidazol-4-yl piperonylate, as an inhibitor of purine synthesis and its activation by adenine phosphoribosyltransferase, Cancer Res. 42: 1098 (1982).
15. D. A. Cooney, H. N. Jayaram, G. Gebeyehu, C. R. Bett, J. A. Kelley, V. E. Marquez, and D. G. Johns, The conversion of 2-β-D-ribofuranosylthiazole-4-carboxamide to an analogue of NAD with potent IMP dehydrogenase inhibitory properties, Biochem. Pharmacol. 31: 2132 (1982).
16. J. W. Huggins, R. K. Robins, and P. G. Canonico, Synergistic antiviral effects of ribavirin and the C-nucleoside analogs tiazofurin and selenazofurin against Togaviruses, Bunyaviruses, and Aranaviruses, Antimicrob. Agents Chemother. 26: 476 (9184).

A STUDY OF ADENOSINE DEAMINASE AND ITS CONVERSION FACTOR IN HUMAN SERUM

Jeffrey T. Starkey, Cynthia Ma and Pang Fai Ma

Center for Medical Education
Ball State University
Muncie, IN 47306 U.S.A.

INTRODUCTION

Adenosine deaminase (ADA) (EC 3.5.4.4) catalyzes the deamination of adenosine and deoxygadenosine to inosine and deoxyinosine respectively. Two molecular forms of ADA have been found in various human tissues (1,2). These enzyme forms differ distinctly in molecular sizes. The A form has a molecular weight of around 200,000 and the C form, 35,000. The two enzyme forms are found to have similar catalytic properties. It has been observed that the A form enzyme from most mammalian tissues tends to dissociate into the smaller C form to some extent (3). When two molecules of the C form of the enzyme aggregate with a large glycoprotein (the conversion factor), the complex becomes the A form (4,5).

A previous study (2) has shown that the distribution of the two molecular forms of ADA was tissue specific. The large form of the enzyme is predominant in human liver while the small form is predominant in human heart. It follows that the conversion factor is present in higher amounts within those tissues where the large form predominates. The purpose of this study was to examine the conversion process and determine if the serum level of the conversion factor was elevated in certain pathological conditions.

MATERIALS AND METHODS

Materials

Normal serum was obtained from healthy volunteers and from the Ball Memorial Hospital blood bank. Samples were obtained after routine screening of serum from daily blood donors. Pathological serum was obtained from the Ball Memorial Hospital chemistry lab after routine serum chemistries were performed. All samples were refrigerated until assayed, and used for the incubation experiments. Adenosine and adenosine deaminase from calf intestine were obtained from Sigma Chemical Co. Bio-Gel A-0.5 m was obtained from Bio-Rad.

Enzyme Assay

Enzyme activity was measured by following the decrease in absorbance at 265 nm (6). A unit of enzyme activity is defined as that amount of enzyme which catalyzes the hydrolysis of 1 umole of adenosine/min under standard conditions; i.e., with a 1-cm path-length cell containing 3.0 ml of 1.0 * 10^{-4} M adenosine in 0.2 M phosphate buffer (pH 7.0) at 38°.

Separation of ADA Molecular Forms

Bio-Gel A-0.5 m was suspended in 0.05 M phosphate buffer (pH 7.0) and kept in the refrigerator. Columns (0.9 cm * 30 cm) were poured in the cold and equilibrated overnight with flowing buffer. The void volume was determined with blue dextran. In all experiments a 0.5-1.0 ml samples was applied to the column and 0.5 ml fractions were collected. Enzyme assays were performed with the collected fractions. The ADA molecular forms were identified based on the elution profiles.

Assay for the Conversion Factor

An aliquot of 0.5 ml of a serum sample was mixed with an equal volume of a buffered solution containing a specified number of units of calf intestinal ADA (C form). The mixture was left at 22°C for 60 minutes. Immediately, the mixture was subjected to gel filtration colum chromatography to determine quantitatively the two molecular forms. For controls, the same volumes of a serum sample and the calf intestinal ADA solution were incubated with equal volumes of phosphate buffer, and the mixtures were subjected to the same procedures. When a sample to be tested contained an appreciable amount of the A form activity, the number of enzyme units determined was subtracted from that obtained with the mixture of the serum sample and the calf intestinal ADA solution. A unit of the conversion factor activity is arbitrarily expressed as the enzyme activity of the A form of ADA determined under specified conditions.

RESULTS

A fresh normal serum sample was divided into two equal portions. One portion was kept frozen in a freezer at -10°C immediately. Two 0.5 ml aliquots were taken from the other protion, and were mixed each with 100 ul of calf intestinal ADA (C form) containing 150 * 10^{-3} unit of enzyme activity. One mixed sample was left at 22°C for 30 minutes, and the other mixed sample, for 60 minutes. At the end of the incubation period, the sample was immediately passed through a gel filtration column. The eluent was collected as 0.5 ml fractions. The fractions were assayed for ADA activity.

Figure 1 represents a gel filtration column chromatographic profile of the mixture of normal serum and calf intestinal ADA, which was incubated at 22°C for 30 minutes. Three enzyme activity peaks were observed. The enzyme form in the middle peak was estimated to have a molecular weight of 142,000.

Figure 2 illustrates a gel filtration column chromatographic profile of the mixture of normal serum and calf intestinal ADA, which was incubated at 22°C for 60 minutes. The intermediate activity peak was not present.

After 24 hours, the frozen portion of the normal serum sample was thawed, and was used in the same experimental procedures. Results obtained were the same as shown in Figures 1 and 2. Increasing the incubation temperature to 37°C did not affect this conversion process.

As control experiments, the serum sample and the calf intestinal ADA

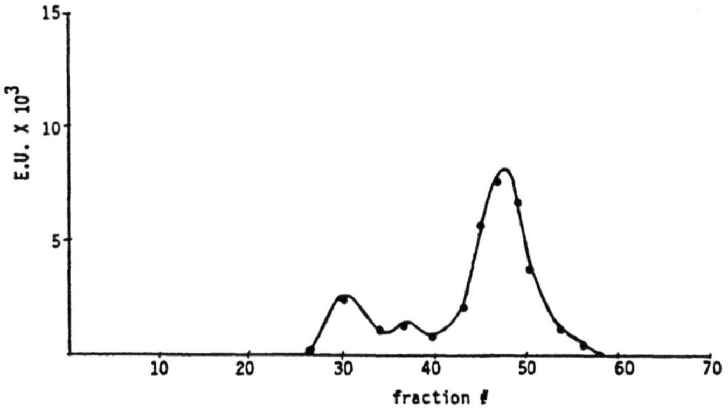

Fig. 1. A gel filtration column chromatographic pattern of the mixture of normal human serum and calf intestinal ADA, incubated at 22°C for 30 minutes.

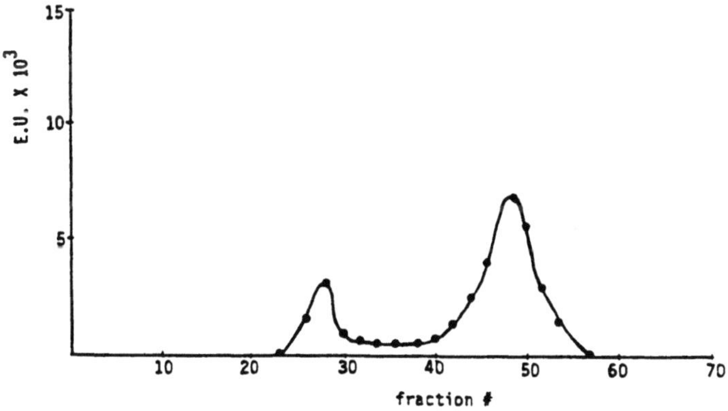

Fig. 2. A gel filtration column chromatographic pattern of the mixture of normal human serum and calf intestinal ADA, incubated at 22°C for 60 minutes.

solution were incubated separately, under the same conditions, with phosphate buffer. When the serum buffer mixture was passed through the gel filtration column, no enzyme activity was detectable in the fractions. As for the calf intestinal ADA buffer mixture, a single peak of enzyme activity was recorded at fraction 50, corresponding to the C form activity.

Increased conversion activity was observed when increasing amounts of serum were incubated with the same amount of calf intestinal ADA activity. Results are shown in Table 1.

The conversion factor activity of several pathological serum samples was examined and compared to that of normal serum. Results are summarized in Table 2.

DISCUSSION

When calf intestinal ADA was incubated with normal human sera, a high molecular weight ADA, corresponding to the A form, appeared as shown in Figure 2. This A form ADA was not observed in the calf intestinal or serum samples. Therefore, the results suggest that: 1. conversion factors in the free molecular form are present in normal human sera, and 2. a fraction of the calf intestinal ADA has been converted to the A form. An activity peak between those of the A and C forms, illustrated in Figure 1, represents an intermediate ADA form of molecular weight estimated to be 142,000. This observation implies that the conversion process of C form to A form is a sequential process between the conversion factor and the two molecules of the C form ADA, and that 30 minutes incubation time is not long enough for a complete conversion process. It is also shown in Table 1 that when the volume of serum sample increases, so does the conversion factor activity. Several pathological serum samples were tested for the level of the conversion factor activity. Results shown in Table 2 indicate that the conversion factor activity is greater than normal in serum samples from patients of renal failure, while it is lower in serum samples from lymphoma and peptic ulcer patients.

Table 1. Correlation of Conversion Factor Concentration and Increasing Volume of Normal Human Serum

Serum Volume (ul)	Conversion Factor (10^4 units)
100	25
200	43
400	77

Table 2. Conversion Factor Concentration in Normal and Pathological Human Sera

Serum Sample Type	Conversion Factor (10^4 units)
Normal	84
Renal failure	99
Breast cancer	92
Congestive heart failure	84
Leukemia	70
Peptic ulcer	47
Lymphoma	44

REFERENCES

1. H. Nishihara, H. Akedo, H., Okada, and S. Haltori, Clin. Chim. Acta, 30:251 (1966).
2. T. A. Magers and P. F. Ma, Int J. Biochem., 6:281 (1975).
3. P. F. Ma and J. R. Fisher, Comp. Biochem. Physiol., 31:771 (1969).
4. P. E. Daddona and W. N. Kelley, Biochim. Biophys. Acta, 580:302 (1979).
5. W. P. Schrader, F. J. Woodward, and B. Pollara, J. Biol. Chem., 254:11964 (1979).
6. H. M. Kalckar, J. Biol. Chem., 167:461 (1947).

CYTIDINE DEAMINASE FROM HUMAN SPLEEN

Alberto Vita, Tiziana Cacciamani, Paolo Natalini,
Silverio Ruggieri and Giulio Magni

Departimento di Biologia Cellulare, Universita di Camerino
and Scuola di Specializzazione in Biochimica e Chimica Clinica;*
Instituto di Biochimica, Universita di Ancona (Italy)

INTRODUCTION

Cytidine deaminase (EC3.5.4.5) catalyzes the irreversible hydrolytic deamination of cytidine, deoxycytidine and some nucleoside analogs including antineoplastic agents. The enzyme has been previously purified and characterized in our laboratory from Escherichia coli[1]. Studies "in vivo" have shown that the enzyme is inducible and involved in catabolic events[2]. In animal cells its role is less well defined. Cytidine deaminase has been detected and partially purified from a variety of animal sources[3-6]. Recently we have developed a convenient purification procedure that allowed us to obtain the enzyme from chicken liver in pure form[7]. The human enzyme has been studied, using partially purified preparations, in liver[4,8-10], in normal and leukaemic granulocytes[11-13] and in lymphoblasts[14]. Due to its ability to inactivate cytidine-derived drugs, such as cytosine arabinoside (Ara-C), 5-azacytidine (Aza-CR), 5-aza-2'-deoxycytidine (Aza-dCR) and many other antineoplastic agents[4,10,12], cytidine deaminase has been a target of considerable interest. The present work describes for the first time the purification and the characterization of the enzyme from human spleen.

MATERIALS AND METHODS

Enzyme assay. Cytidine deaminase activity was determined by a direct spectrophotometric assay based on the decrease of absorbance when the cytidine or cytidine analog is deaminated to uridine or uridine analog. Using cytidine, deoxycytidine, 5-methylcytidine and Ara-C as the substrates, ΔE_{mM} is 2.1, at 290nm; with Aza-CR and Aza-dCR, the optimum ΔE_{mM} is 1.3(at 290nm) and 3.2(at 260nm), respectively. The optimum changes in extinction coefficients (ΔE_{mM}) were determined enzymatically from the change in absorbance after prolonged incubation of the enzyme with the substrate. The standard reaction mixture was as previously reported[1].

In inhibition experiments the assay was performed at 290nm using cytidine as the substrate; 0.1M KCl was included as an "ionic strength buffer".

Purity. The purity of the final enzyme preparation was assessed by analytical polyacrylamide disc gel electrophoresis (PAGE) carried out as described by Ornstein[15] and Davis[16]. Protein bands were detected by Coomassie brilliant blue R-250 staining. To correlate the stained band with the enzyme activity, duplicates of non denaturating gel, loaded with 0.2 units of cytidine deaminase activity, were run simultaneously at 4°C. One gel was fixed and stained for proteins, while the second gel was sliced into 2-mm segments and placed in 200 µl of 100mM buffer A (see below, under "Purification") pH7.5 for 30min at 4°C. The enzyme activity was assayed on 150 µl of supernatant.

Molecular weight determination. The molecular weight of the purified native enzyme was determined by gel filtration chromatography on an FPLC system (LKB) using a Superose -12 HR 10/30 column (Pharmacia) equilibrated with 100mM Tris-HCl, pH7.5, containing 150mM KCl. 200 µl samples containing either cytidine deaminase or the appropriate protein markers were applied to the column. The elution was performed at a flow rate of 0.5 ml/min, and 0.5-ml fractions were collected. For the estimation of subunit molecular weight, pure cytidine deaminase was incubated in a denaturing mixture containing 2% SDS, 5% 2-mercaptoethanol and 0.125M Tris-HCl, pH6.8. Samples were loaded on 15% polyacrylamide SDS slab gel[17] and run at 30mA.

Fast Protein Liquid Chromatography (FPLC). The FPLC system (LKB) consisted of a single pump model 2150, a controller 2152, a detector Uvicord SD 2158 with a 280nm filter, a 2-channel recorder 2210 and an automatic fraction collector Helirac 2212. Gradients were formed by a mixing valve model 2040-203, between two solvents, A and B, previously degassed and filtered. All the samples were filtered on 0.45 µm filter units Millex MV (Millipore) prior to column injection.

Chromatofocusing. The isoelectric point of the purified enzyme was determined by FPLC (LKB) chromatofocusing, using a MONO-P HR 5/20 column) (Pharmacia). The column was equilibrated with 0.025M methylpiperazine-HCl, pH5.5 and eluted with Polybuffer 74 (Pharmacia) pH4.0, which generated a linear pH gradient from 4.5 to 3.5. The flow rate was kept constant at 0.5ml/min; 0.5-ml fractions were collected.

RESULTS AND DISCUSSION

Purification. The results of the purification procedure are summarized in Table I. All steps except FPLC analysis were performed at 0-4°C. One human spleen obtained from a normal individual following emergency splenectomy, (about 180g tissue) was homogenized in a Waring Blendor for 2min with 500ml of extraction buffer A (50mM Tris HCl pH8.0, 5mM 2-mercaptoethanol, 1mM EDTA, 0.1mM PMSF) containing 0.25M sucrose. The homogenate was centrifuged and the supernatant, containing 33.6 units of cytidine deaminase, with a specific activity of 0.0025 units/mg was mixed with 300ml of DE-52 cellulose equilibrated with buffer A without sucrose. After 30min at 4°C the mixture was transferred on a sintered glass filter funnel, washed with 600ml of 0.1M buffer A, and then packed on a column (5.0 x 15.0 cm). The column was washed until the absorbance at 280nm reached zero, then a linear gradient from 0.1M to 0.5M buffer A (2 liters total volume) was applied. The active fractions eluted at about 0.3M buffer A were collected and concentrated by ultrafiltration on an Amicon PM-10 membrane and dialyzed against 50mM buffer A, pH7.5. The active pool was brought to 30% ammonium sulphate (A.S.) saturation and applied on Pentyl-Agarose column (2.5 x 20 cm), equilibrated with 50mM buffer A, pH 7.5, 30% A.S.; the elution was performed with linear gradient from 30% to 0% A.S. saturation (1 liter total volume). The active fractions, eluted at about 15% A.S. saturation, were pooled, concentrated by ultrafiltration and dialyzed against 20mM buffer A, pH7.5, 25% A.S. saturation. All further steps were performed using an FPLC system. The pentyl-agarose pool was loaded on a TSK Phenyl-5PW column (0.8 x 7.0 cm) equilibrated with 20mM buffer A, pH7.5, 25% A.S. saturation (Solvent A) and eluted with 20mM buffer A (solvent B). The elution program was: 10min at 0%B, 20min at 20%B, linear gradient from 20%B to 47%B in 40min, then 10min at 47%B and finally 10min at 100%B, at a flow rate of 0.5ml/min. An asimmetrical peak of activity was eluted between 23%-35% solvent B. The active fractions were pooled, concentrated by ultrafiltration, dialyzed against 20mM buffer A, pH8.0 and applied on to a Mono-Q HR 5/5 column equilibrated with the same buffer A (Solvent A). The elution buffer was buffer A containing 0.5M KCl (Solvent B). The elution program was: 10min at 0%B, 10min at 12%B, linear gradient from 12%B to 38%B, for 30 min, at a flow rate of 0.5 ml/min. The cytidine deaminase activity peak eluted at 20%B. The specific activity of the enzyme at this step was 38.5 U/mg, with yield of 10% and a purification factor of 15,400. This preparation

Table I. Purification of Cytidine Deaminase from Human Spleen.

Step	Total Protein	Total Activity	Specific Activity	Purification	Yield
	(mg)	(Units[a])	(Units/mg)	(-fold)	(%)
Crude Extract	13,464	33.6	0.0025	-	100
DEAE-Cellulose	1,152	16.0	0.0138	5.5	48
Pentyl-Agarose	81	15.5	0.19	76.5	46
TSK Phenyl-5PW	2.09	8.2	3.9	1562.5	24
Mono Q	0.09	3.4	38.5	15400.0	10

[a]One unit is the amount of enzyme which catalyzes the deamination of 1 umol of cytidine per min at 37°C.

was subjected to 7.5% PAGE under non denaturating conditions. A diffuse band associated with the enzyme activity having a relative mobility (Rm) of 0.41 and a minor band with Rm 0.72, were shown. When the final enzyme preparation was applied on to a chromatofocusing column, the cytidine deaminase activity eluted at three different pH, corresponding to pI values of 4.05, 4.25 and 4.45, respectively. These results are consistent with the above described electrophoretic behaviour and suggest the existence of microheterogeneity in the enzyme preparation.

Molecular weght. The native molecular weight was estimated to be 52,000 by gel filtration as described in Materials and Methods. The native PAGE of the active fraction from gel filtration showed the same contaminant as for the Mono-Q pool. In an attempt to define the subunit composition of the enzyme in the absence of interfering contaminant, the active band from native PAGE was extracted and run on 15% SDS-PAGE.Two bands corresponding molecular weight 14,000 and 56,000 were obtained. These results appear to support the conclusion that the enzyme has a tetrameric structure which is not completely dissociated under our denaturating conditions, even though the presence of a contaminant cannot be discounted. We have previously reported an apparently tetrameric structure for cytidine deaminase from chicken liver[7]. The present results may confirm the hypothesis of Teng[13] assigning a tetrameric structure to the human reticulocyte cytidine deaminase, based on its electrophoretic multibanded pattern. The molecular weights reported for the enzyme from other sources are: 51,000 (human granulocytes[11]), 74,000 (mouse spleen[6]), 57,000 (yeast[18]), 66,000 (Escherichia coli[1], with two apparently identical subunits).

Kinetic properties. On the basis of the reaction catalyzed by cytidine deaminase, two kinetic mechanisms can be hypothesized according to Cleland's nomenclature: ordered UNI-BI and random UNI-BI, depending on the ordered or not release of the products, uridine and ammonia. This was further analyzed by performing product inhibition studies, using a pure preparation of the enzyme. Double-reciprocal plots of initial velocities at various concentrations of cytidine and at different fixed concentrations of uridine or ammonia both resulted in a family of lines intersecting on the ordinate axis. The competitive nature of both product inhibition patterns clearly indicated that the catalytic mechanism was of the rapid-equilibrium random UNI-BI type. The secondary plot of slope vs. product concentration showed that both inhibitions are linear, allowing the calculation of the inhibition constants (Ki) for uridine and ammonia, 0.75mM and 1.5M respectively. As shown in Table II, cytidine deaminase catalyzes the deamination of various cytidine analogs including antineoplastic agents active against leukemic cells. The Km values for Aza-CR and Aza-dCR were one order of magnitude higher than Km values for the other nucleosides tested. Since the Vmax/Km ratio for Ara-C is similar to the values for the physiological substrates, the presence of cytidine deaminase represents a serious obstacle to the use of this antimetabolite in chemotherapy. The Km value for Ara-C obtained in the present study, is in good agreement with the values reported for the enzyme from human liver[10], human myeloblastic leukemic cells[19], and human granulocytes[11]. The enzyme is competitively inhibited by various nucleosides and nucleo-

Table II. Kinetic Constants for Substrate Analogs of Cytidine Deaminase from Human Spleen.

Substrate	Km (mM)	Vmax (U[a]/mg)	Vmax/Km (x 10^3)
Cytidine	0.06	0.9	15
Deoxycytidine	0.03	1.2	40
5-Methylcytidine	0.03	1.4	47
Cytosine Arabinoside	0.10	1.1	11
5-Azacytidine	1.00	1.7	1.7
5-Aza-2'-Deoxycytidine	1.42	2.8	1.9

[a] One unit is the amount of enzyme which catalyzes the deamination of 1 umol of cytidine per min at 30°C.

Table III. Inhibition Kinetics of Cytidine Deaminase from Human Spleen.

Inhibitor	Ki (mM)	Type of inhibition
5,6-Dihydrouridine	0.04	Competitive
5,6-Dihydrothymidine	0.90	Competitive
Deoxyuridine	1.00	Competitive
Deoxythymidine	0.70	Competitive
Thymine Riboside	0.55	Competitive
Deoxyadenosine	1.05	Competitive
dTTP	0.60	Competitive
CMP	0.45	Competitive
dCMP	0.50	Competitive
Azidothymidine	1.12	Competitive

tides (Table III). All the inhibition constants (Ki), except for 5,6-dihydrouridine, were one order of magnitude higher than the affinity constants (Km) for the physiological substrates. The inhibition exerted by dTTP may suggest a possible feed back regulation of the enzyme by this metabolite, while the inhibition by CMP and dCMP might play a significant role in the catabolic pathway of these two nucleotides. However, the high Ki values for all these nucleotides appear not to be completely consistent with their physiological concentrations.

REFERENCES

1. A.Vita, A.Amici, T.Cacciamani, M.Lanciotti and G.Magni, Cytidine deaminase from Escherichia coli B. Purification and enzymatic and molecular properties, Biochemistry, 24: 6020 (1985).
2. G.Magni, A.Vita and A.Amici, Pyrimidine nucleoside-catabolizing enzymes in Escherichia coli B, in: "Current topics in cellular regulation", R.L.Levine and A.Ginsburg ed., Academic Press,NewYork (1985)
3. R.Tomchick, L.D.Saslaw and V.S.Waravdekar, Mouse kidney cytidine deaminase, J.Biol.Chem. 243: 2534 (1968).
4. M.P.Fanucchi, K.A.Watanabe, J.J.Fox and T-C.Chou, Kinetics and substrate specificity of human and canine cytidine deaminase, Biochem. Pharmac., 35: 1199 (1986).
5. G.B.Wisdom and B.A.Orsi, The purification and properties of cytidine aminohydrolase from sheep liver, Eur.J.Biochem., 7: 223 (1969).
6. V.G.Malathi and R.Silver, Effect of murine viral leukemia on spleen nucleoside deaminase: purification and properties of the enzyme from leukemic spleen, Biochim.Biophys.Acta, 238: 377 (1971).
7. A.Vita, T.Cacciamani, P.Natalini, S.Ruggieri, I.Santarelli and G.Magni, Purification of cytidine deaminase from chicken liver, Ital.J. Biochem., 36:275 (1988).
8. D.F.Wentworth and R.Wolfenden, On the interaction of 3,4,5,6-tetrahydrouridine with human liver cytidine deaminase, Biochemistry, 14: 5099 (1975).
9. G.W.Camiener, Studies of the enzymatic deamination of cytosine arabinoside-II. Properties of the deaminase of human liver, Biochem. Pharmac. 16: 1681 (1967).
10. G.G.Chabot, J.Bouchard, R.L.Monparler, Kinetics of deamination of 5-ara-2'-deoxycytidine and cytosine arabinoside by human liver cytidine deaminase and its inhibition by 3-deazauridine, thymidine or uracil arabinoside, Biochem.Pharmac., 32: 1327 (1983).
11. B.A.Chabner, D.G.Johns, C.N.Coleman, J.C.Drake and W.H.Evans, Purification and properties of cytidine deaminase from normal and leukemic granulocytes, J.Clin.Invest., 53: 922 (1974).
12. B.A.Chabner, J.C.Drake and D.G.Johns, Deamination of 5-azacytidine by a human leukemia cell cytidine deaminase, Biochem.Pharmac., 22: 2763 (1973).
13. Y-S.Teng, J.E.Anderson and E.R.Giblett, Cytidine deaminase: a new genetic polymorphism demonstrated in human granulocytes, Am.J.Human. Genet., 27: 492 (1975).

14. J-L.Pérignon, J.Chaléon, G.Leverger, A-M.Houllier, L.Thuillier and P.Cartier, Cytidine deaminase activity of human normal and malignant lymphoid cells, Clin.Chim.Acta, 147:67 (1985).
15. L.Ornstein, Disc electrophoresis. Part.I, Ann.N.Y.Acad.Sci., 121: 321 (1964).
16. B.J.Davis, Disc electrophoresis. Part.II, Ann.N.Y.Acad.Sci., 121: 404 (1964).
17. U.K.Laemmli, Cleavage of structural proteins during the assembly of the head of Bacteriophage T4, Nature (London), 277: 680 (1970).
18. P.L.Ipata, G.Cercignani and E.Balestrieri, Partial purification and properties of cytidine deaminase from baker's yeast,Biochemistry, 9: 3390 (1970).
19. Y-C.Cheng, R-S.Tan, J.L.Ruth and G.Dutschman, Cytotoxicity of 2'-fluor-5-iodo-1-ß-D-arabinofuranosylcytosine and its relationship to deoxycytidine deaminase, Biochem.Pharmac., 30: 213 (1982).

AMP DEAMINASE AND THYMIDINE KINASE DEFICIENCIES IN A MUTANT MOUSE S49 CELL CLONE

Sheri Hanson and Buddy Ullman

Department of Biochemistry
The Oregon Health Sciences University
Portland, Oregon 97201

ABSTRACT

From a mutagenized population of wild type S49 cells, a clone was isolated in a single step that possessed functional and biochemical deficiencies in both AMP deaminase and thymidine kinase activities. This mutant cell line, DTB6, was selected in semi-solid medium containing 1mM thymidine and 1mM dibutyryl cyclic AMP. In comparative growth rate experiments, DTB6 cells were considerably less sensitive than parental cells to the growth inhibitory effects of thymidine. In contrast, DTB6 cells were much more sensitive to the cytotoxic effects of adenine and adenosine. The supersensitivity of DTB6 cells toward adenine could be ameliorated by the addition of hypoxanthine to the culture medium. The growth phenotype of the mutant cells could be attributed to deficiencies in two enzyme activities. First, DTB6 cells possessed a 60-70% deficiency in AMP deaminase activity, although the residual activity appeared kinetically similar to the wild type enzyme. Second, DTB6 cells possessed a virtual complete deficiency in thymidine kinase activity. Both enzyme deficiencies behaved in a recessive fashion in intraspecies hybrids. Revertants of DTB6 cells possessed wild type levels of AMP deaminase activity but remained deficient in thymidine kinase activity, while another revertant of DTB6 cells expressed 11 % of the wild type thymidine kinase level but did not perceptibly change its AMP deaminase activity. The ability to isolate single step mutants with two seemingly independent biochemical abnormalities raises the speculation that there may be some link between cellular functions responsible for purine nucleotide and thymidine metabolism.

INTRODUCTION

Adenine and adenosine have a variety of physiological and pharmacological effects (1), and both can be growth inhibitory and cytotoxic toward mammalian cells (2). Most, but not all, of the effects of these purines require metabolism to the nucleotide level. Adenine and adenosine are converted to AMP by the enzymes adenine phosphoribosyltransferase and adenosine kinase, respectively. Subsequent metabolism of AMP in animal cells depends on the relative intracellular activities of two enzymes, AMP kinase and AMP deaminase. The latter is a ubiquitous mammalian enzyme and catalyzes the effectively irreversible deamination of AMP to IMP and ammonia. Several important regulatory functions for AMP deaminase have been proposed. These include ammonia production in muscle (3), the maintenance of adenylate pool sizes and overall energy charge (4), and the balance between adenylate and guanylate nucleotides in cells (5). It is also an important component of the purine nucleotide cycle (6). In addition, AMP is an important enzyme modulator, and therefore, the AMP deaminase plays an indirect regulatory role for many pathways of intermediary metabolism (7).

Alterations in AMP deaminase activity have been implicated in a variety of disease states. A marked reduction in AMP deaminase activity has been reported in certain genetically dystrophic mice (8), in patients with Duchenne's muscular dystrophy (9), in a patient with hypokalemic periodic paralysis (10), and in several patients with muscular weakness and cramping after excercise (11). Second, due to the highly regulated nature of the enzyme by nucleotides and by inorganic anions, Hers and van den Berghe have implicated a defect in the regulation of the enzyme as a basis for overproduction hyperuricemia and gout (12). Third, Bagnara and Hershfield have indicated that increased flux through AMP deaminase may be the basis for the depletion of red blood cell ATP observed in leukemic patients treated with the adenosine deaminase inhibitor, deoxycoformycin (13). Finally, Cohen and coworkers have attributed the unique deoxyribonucleotide abnormalities observed in adenosine deaminase deficiency to the substrate specificity of AMP deaminase (14).

We have exploited somatic cell genetic approaches to analyze the role of AMP deaminase in purine metabolism. In this investigation, we report the isolation of an S49 cell line with altered AMP deaminase activity and cellular growth resistance to thymidine. This cell line, DTB6, was isolated in a single step and expressed two apparent biochemical lesions; the first, a 60% deficiency in AMP deaminase activity, and the second, a complete deficiency in thymidine kinase activity.

MATERIALS AND METHODS

Chemicals and reagents. The sources of all chemicals, materials, and reagents have been described previously and were of the highest quality commercially available.

Cell culture and maintenance. The wild type S49 cell line is a T cell lymphocytic lymphoma originally derived from a Balb/c mouse injected with mineral oil. Wild type and mutant S49 cells were routinely propagated in Dulbecco's modified Eagle medium supplemented with 10% horse serum which had been previously heat-inactivated at 56° for 30 min.

Isolation of the DTB6 cell line. Wild type S49 cells were mutagenized by a 3 hr exposure to 3 ug/ml MNNG. Cells were separated from the mutagen by centrifugation and resuspended in fresh complete medium lacking MNNG. The mutagenized cells were allowed to proliferate for seven cell divisions to ensure expression of the mutant phenotype. The cells were then plated in semi-solid (0.3%) agarose (Seakem-ME) overlaying mouse embryo fibroblasts as described by Coffino et al. The DTB6 cell line was isolated as a single colony from ten million cells plated in selective medium containing 1mM thymidine and 1mM dibutyryl cyclic AMP.

Generation of intraspecies tetraploid hybrids. Intraspecies cell-cell hybrids between wild type and DTB6 cells (DTB6 X wild type) were generated by polyethylene glycol fusion as described by this laboratory (15). The wild type cell line possessed mutations in the cAMP-dependent protein kinase and hypoxanthine-guanine phosphoribosyltransferase activities, mutations which behaved in a dominant and recessive fashion, respectively. A wild type X wild type hybrid was isolated and served as a control (15).

Isolation of revertants of DTB6 cells. To isolate derivatives of DTB6 cells, which had reverted their AMP deaminase activity, a genetic deficiency of hypoxanthine-guanine phosphoribosyltransferase (HGPRTase) was first inserted into DTB6 cells by clonal selection in semi-solid agarose containing 20uM 6-thioguanine. This HGPRTase-deficient derivative of the DTB6 cell line, DTB6-Tg3, was then mutagenized with 3 ug/ml MNNG as described for the original selection of the DTB6 cell line, and plated on semi-solid agarose containing 200 uM adenine. Two adenine-resistant derivatives of DTB6-Tg cells, DTB6-Tg-Ade2 and DTB6-Tg-Ade3, were analyzed further. The HGPRTase deficiency was essential for isolating adenine-resistant revertants of DTB6 cells, since DTB6 cells had a survival frequency of about 0.1% in semi-solid medium containing concentrations of adenine as high as 1.5 mM.

To generate derivatives of DTB6 cells that had reverted their thymidine kinase activity, DTB6 cells were mutagenized as described above and plated selectively in medium containing 0.5 mM hypoxanthine, 0.4 uM amethopterin, and 10 uM thymidine (HAT medium). DTB6 cells could not thrive in this selective medium, since cell survival in HAT medium requires the ability of the cell to convert exogenous thymidine to cellular thymidylate nucleotides in order to circumvent the cellular depletion of tetrahydrofolate pools caused by amethopterin. One such HAT-resistant cell line, DTB6-HAT^r, was evaluated further for its functional and biochemical alterations in the AMP deaminase and thymidine kinase activities.

Growth rate experiments. The growth inhibitory and cytotoxic effects of various purines and pyrimidines and their analogs on wild type and DTB6 cells were ascertained in Costar multiwell (24 well) tissue culture plates as previously described (16).

AMP deaminase assay. AMP deaminase activity was measured using a colorimetric assay in which the ammonia generated from AMP forms a stable blue color in the catalyzed indophenol reaction (17). The reaction mixture contained 0.1M imidazole/HCl, pH 7.0, 0.1M KCl, 2mM ATP, 2mM AMP, and a small volume of cell extract to make the final volume 0.4ml and was incubated at 37°.

Thymidine kinase assay. Thymidine kinase activity was measured in 50mM Tris, pH 7.4, 5mM $MgCl_2$, 1mM dithiothreitol, 5mM ATP and either 12uM or 30uM [^{14}C]thymidine (56mCi/mmol) in a volume of 300ul at 37°. The reaction was initiated by the addition of a small volume of extract, and 50ul portions were removed at 10 min intervals and applied to DEAE filter papers. The filters were pooled in 2 l of H_2O and washed four times with 2 l of 1mM ammonium formate. The filter papers were dried and counted by liquid scintillation.

RESULTS

Growth rate experiments. DTB6 cells were isolated from mutagenized wild type cells by virtue of their ability to grow in selective medium containing 1mM concentrations of both thymidine and dibutyryl cyclic AMP. Initially, the sensitivity of the mutant cells to growth inhibition by each agent alone was compared to that of wild type parental cells. Although the DTB6 cells were much less sensitive to the cytotoxic effects of thymidine than wild type cells, they were equally sensitive to dibutyryl cyclic AMP. Interestingly, however, DTB6 cells were much more sensitive to the toxicity of adenine (Figure 1) and adenosine in the presence of 10uM EHNA (data not shown). Whereas the effective concentration of adenine that inhibited the growth of wild type cells was 400 uM (E.C.$_{50}$ value), the E.C.$_{50}$ value of DTB6 cells for adenine was about 10 uM. The E.C.$_{50}$ value of mutant cells for adenine could be increased to that of wild type cells by the addition of 0.5 mM hypoxanthine to the culture medium (Figure 1). Hypoxanthine at that concentration did not alter the sensitivity of wild type cells to adenine.

The ability of hypoxanthine to ameliorate adenine toxicity in DTB6 cells is a property shared by a previously S49 cell line (18) in which adenine induces a cellular starvation for guanylate nucleotides.

Measurements of AMP deaminase activities. The increased sensitivity of DTB cells to growth inhibition by either adenosine-EHNA or adenine and the abil ty of hypoxanthine to ameliorate the effects of the latter were properties of a previously described mutant S49 cell line, 100-10, that possessed a genetic alteration in AMP deaminase activity (18). Therefore, the levels o AMP deaminase in wild type and DTB6 cell extracts were compared. The data presented in Figure 2 indicated that the DTB6 cells possessed only 30-40% o the levels of the AMP deaminase activity of wild type cells. However, unlike the enzyme from the 100-10 cell line (18), the residual enzyme activity in DTB6 cells had a similar apparent K_m value for AMP (1mM) and similar sensitivities to the AMP deaminase inhibitors, inorganic phosphate and GTP, as the wild type enzyme (data not shown). This decreased AMP deaminase activity of DTB6 cells could account for their greater sensitivit to inhibition by adenine and adenosine.

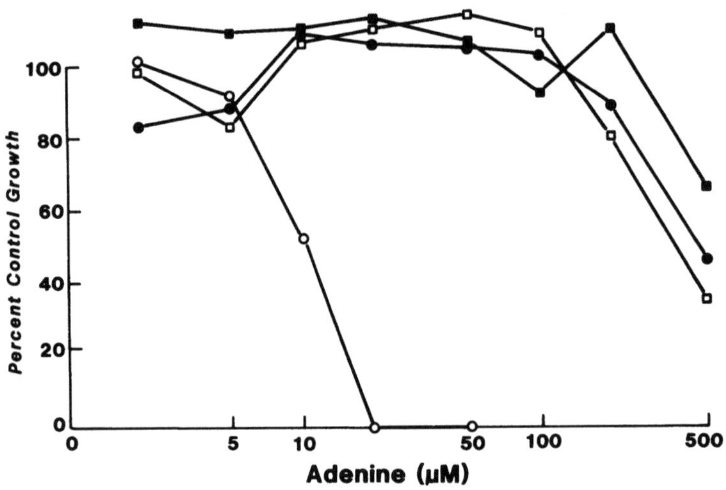

Figure 1. Adenine toxicity in wild type and DTB6 cells in the absence and presence of hypoxanthine. The effects of increasing concentrations of adenine on the growth rate of wild type (●,■) and DTB6 (○,□) cells in the absence (●,○) and presence of 0.5mM hypoxanthine were determined as described. 0.5mM hypoxanthine alone did not influence the growth rate of either cell line.

Measurements of thymidine kinase activities. The resistance of DTB6 cells to thymidine and its analogs and their inability to proliferate in HAT medium suggested that this mutant cell line lacked thymidine kinase activity. Comparison of the levels of thymidine kinase in wild type and mutant cell extracts indicated that the latter were virtually completely deficient in thymidine kinase activity (data not shown).

Characterization of intraspecies tetraploid hybrids. In order to assess whether some unusual regulatory mutation was responsible for both the deficiency in AMP deaminase and and that in thymidine kinase, a DTB6 X wil type hybrid cell line was constructed. As indicated by the similar growth sensitivities of the DTB6 X wild type and the wild type X wild type cell lines to adenine and 5-fluorodeoxyuridine, both enzyme deficiencies behaved in recessive fashions (data not shown).

Characterization of derivatives of DTB6 cells selected for reversion of either AMP deaminase or thymidine kinase activities - The isolation of a mutant cell line in a single step that possessed two biochemically defined enzyme deficiencies contributing to its genetically altered phenotype raised the possibility that the two deficiencies were controlled by a single genetic function. To analyze the genetic relationship between the AMP deaminase and thymidine kinase deficiencies, the DTB6 cells were selected for reversion of one enzyme defect in order to determine whether the other enzyme deficiency also reverted to the parental phenotype. As demonstrated in Table I, restoration of the AMP deaminase activity in the DTB6-Tg-Ade2 and DTB6-Tg-Ade3 cell lines to wild type levels did not result in any increase in thymidine kinase activity as measured by direct enzyme assay and by their insensitivity to growth inhibition by 5-fluorodeoxyuridine.

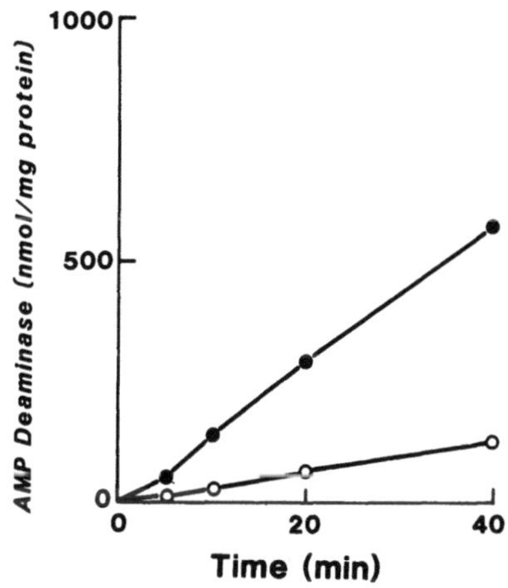

Figure 2. AMP deaminase assays. The levels of AMP deaminase activity in soluble extracts of wild type (0) and DTB6 (0) were quantitated as a function of time. This is one of four completely independent determinations. The specific activity of the AMP deaminase activity in mutant cells was 30-40% of that in wild type parental cells in all four determinations.

Similarly, a HAT-resistant derivative of DTB6 cells, DTB6-HAT^r, was also analyzed for its biochemical and growth phenotypes (Table I). DTB6-HAT^r cells possessed about 11 % of the wild type thymidine kinase activity, which could account for its survival in HAT medium. However, this 11 % restoration of wild type thymidine kinase activity did not quite confer wild type growth sensitivity to 5-fluorodeoxyuridine. DTB6-HAT^r cells were almost as sensitive to growth inhibition by adenine (Table I) and contained as much AMP deaminase activity as the DTB6 parental cells (Table I).

DISCUSSION

The DTB6 cell line is the second mutant S49 cell line that has been isolated that possesses a genetically-induced alteration in the AMP deaminase. The biochemical alteration in the previously described S49 cell line, 100-10 (18), is dissimilar to that of the DTB6 clone. The AMP deaminase

TABLE I

Growth sensitivities and enzyme levels of revertants of DTB6 cells. The sensitivities of wild type, DTB6, DTB6-Tg-ade2, DTB6-Tg-ade3, and DTB6-HAT^r cells to growth inhibition by adenine and 5-fluorodeoxyuridine and the intracellular levels of AMP deaminase and thymidine kinase activity were measured as described in Materials and Methods. The thymidine kinase values reported in parentheses were carried out at a substrate concentration of 12 uM thymidine, while all others were carried out at 30 uM thymidine. Units for the AMP deaminase assay are nmol/min, and units for the thymidine kinase assay are pmol/min.

Cell Line	E.C.$_{50}$ Values		Enzyme Levels		
	Adenine (uM)	5-fdUrd (nM)	AMP deaminase (units/mg protein)	Thymidine kinase	
Wild Type	400	1	33.7	183	(61)
DTB6	10	800	14.5	1.6	(1)
DTB6-Tg-ade2	400	1000	37.6	2.2	
DTB6-Tg-ade3	400	1000	33.8	3.6	
DTB6-HAT^r	20	10	17.3		(6.6)

activity of 100-10 cells possesses a higher K_m value for AMP, increased sensitivity to inhibition by inorganic phosphate, and an increased K_i for GTP compared to the wild type enzyme. The AMP deaminase of DTB6 cells possesses none of these kinetic aberrations but rather, a 60% deficiency in enzyme activity. The in situ phenotypes of the 100-10 and DTB6 cells are similar, however, in that both cell lines are supersensitive to adenine and adenosine-EHNA toxicity and can be rescued by the provision of an exogenous source of guanylate nucleotides, i.e., hypoxanthine, to the culture medium.

The phenotype of the 100-10 cells is also complicated by the presence of a second mutation, one that encodes a genetically altered ribonucleotide reductase activity that is insensitive to inhibition by deoxyATP (18). This mutation confers a growth resistance toward thymidine to the 100-10 cells. The resistance of the DTB6 cells to thymidine cannot be attributed to a genetic alteration in a gene for ribonucleotide reductase, since unlike S49 cells possessing genetic lesions in the activity or substrate specificity site of ribonucleotide reductase (19,20), DTB6 cells are just as sensitive to growth inhibition by deoxyadenosine-EHNA and deoxyguanosine and possess no deoxyribonucleotide pool derangements (data not shown). That the two independent S49 clones, 100-10 and DTB6, isolated by different methods contain different genetic alterations in their AMP deaminase activities as well as a second mutation in a pyrimidine metabolizing function raises the intriguing possibility of some sort of genetic linkage or genetic regulatory mechanism between elements of the purine and pyrimidine nucleotide pathways.

The complex phenotype of the DTB6 cell line can be attributed to biochemical and functional deficiencies in two enzymes, AMP deaminase and thymidine kinase. The former enzyme catalyzes the deamination of AMP, while the latter phosphorylates thymidine and its analogs. The 60-70% deficiency in AMP deaminase caused the DTB6 cells to be extraordinarily sensitive to the toxicity of adenine and adenosine. Although adenine and adenosine are converted to AMP by separate enzymes, AMP deaminase deficiency results in a more efficient conversion of both to adenylate nucleotides. The levels of the purine nucleoside triphosphates in wild type cells are not diminished by exposure to adenosine, since the nucleoside is a source of adenylate nucleotides via direct phosphorylation and a source of guanylate nucleotides, a process initiated by the deamination of AMP to IMP. In situations of AMP deaminase deficiency, however, AMP cannot be efficiently converted to IMP in the presence of either adenine or adenosine. Therefore DTB6 cells, unlike wild type cells, become starved for guanylate nucleotides after exposure to adenine or adenosine (data not shown), which accounts for the observation that hypoxanthine, a source of guanylate nucleotides, protected mutant cells, but not parental cells, from adenine toxicity (Figure 1). Therefore these studies corroborate the hypothesis that AMP deaminase plays a critical role in the regulation and balance of the ratio between adenylate and guanylate nucleotide levels in mammalian cells.

ACKNOWLEDGEMENTS

This work received financial support from grant No. DK38809 from the National Institutes of Health. B.U. is a recipient of a National Institutes of Health Research Career Development Award.

REFERENCES

1. Fox, I.H. & Kelley, W.N. Ann. Rev. Biochem. 47, 655-686 (1978).

2. Aronow, L. Biochim. Biophys. Acta 47, 184-185 (1961).

3. Embden, G. & Wassermeyer, H. Hoppe-Seyler's Z. Physiol. Chem. 179, 226-237 (1928).

4. Chapman, A.G. & Atkinson, D.E. J. Biol. Chem. 248, 8309-8312 (1973).

5. Setlow, B., Burger, R., & Lowenstein, J.M. J. Biol. Chem. 241, 1244-1245 (1966).

6. Lowenstein, J.M. & Tornheim, K. Science 171, 397-400 (1971).

7. Lee, Y.-P. & Wang, M.H. J. Biol. Chem. 243, 2260-2265 (1968).

8. Pennington, R.J. Nature 192, 884-885 (1961).

9. Pennington, R.J. Proc. Assoc. Clin. Biochem. 2, 17-18 (1962).

10. Engle, A.G., Potter, C.S., & Rosevear, J.W. Nature 202, 670-672 (1964).

11. Fishbein, W.N., Armbrustmacher, V.W., & Griffin, J.L. Science 200, 545-546 (1978).

12. Hers, H.-G. & van den Berghe, G. Lancet 1, 585-586 (1979).

13. Bagnara, A.S. & Hershfield, M.S. Proc. Natl. Acad. Sci. USA 79, 2673-2677 (1982).

14. Barankiewicz, J. & Cohen, A. J. Biol. Chem. 259, 15178-15181 (1984).

15. Aronow, B., Allen, K., Patrick, J., & Ullman, B. J. Biol. Chem. 260, 6226-6233 (1985).

16. Ullman, B., Clift, S.M., Gudas, L.J., Levinson, B.B., Wormsted, M.A., & Martin, D.W. Jr. J. Biol. Chem 255, 8308-8314 (1979).

17. Chaney, A.L. & Marbach, E.P. Clin. Chem. 8, 130-132 (1962).

18. Buchwald, M., Ullman, B., & Martin, D.W. Jr. J. Biol. Chem. 256, 10346-10353 (1981).

19. Ullman, B., Gudas, L.J., Clift, S.M., & Martin, D.W. Jr. Proc. Natl. Acad. Sci. USA 76, 1074-1078 (1979).

20. Eriksson, S., Gudas, L.J., Ullman, B., Clift, S.M., & Martin, D.W. Jr. J. Biol. Chem. 256, 10184-10188 (1979).

PROPERTIES OF HIGHLY PURIFIED HUMAN T LYMPHOBLAST (MOLT-4) DEOXYCYTIDINE KINASE

Nabanita S. Datta, Donna S. Shewach, Mary C. Hurley
Beverly S. Mitchell, and Irving H. Fox

Departments of Internal Medicine
and Biological Chemistry
Ann Arbor, Michigan, 48109-0108

INTRODUCTION

Lymphoid cells are characterized by substantial quantities of nucleoside phosphorylating activities[1-3]. In S-49 lymphoma cells deoxyguanosine kinase (dGK) activity is associated with deoxycytidine kinase (dCK) and deoxyadenosine kinase (dAK) is associated with adenosine kinase (AK) and deoxycytidine kinase[4,5]. In cultured human lymphoblasts, thymus and L1210 cells deoxyadenosine is phosphorylated by adenosine kinase and deoxycytidine kinase[6-13]. These conclusions concerning substrate specificity have been derived largely from studies with mutant cell lines and direct biochemical verification of the results has been limited by the use of partially purified or crude cell lysates which could contain more than one enzyme activity.

To clarify the properties and the physiologic role of deoxycytidine kinase we have undertaken the purification of this enzyme from cultured human lymphoblasts. Our studies show that highly purified deoxycytidine kinase is a very labile protein with multiple nucleoside phosphorylating activities.

MATERIALS AND METHODS

Materials and methods are described elsewhere in detail[14]. The enzyme was assayed by a radiochemical method which quantitated the formation of radioactively labeled nucleotide from radioactively labeled nucleoside[15,16]. Deoxyguanosine, deoxyadenosine, adenine arabinoside and cytidine phosphorylating activities were measured under the same conditions.

The enzyme was purified from 50 g of frozen MOLT 4 cells by ammonium sulfate precipitation between 35 and 60 percent, gel filtration chromatography on a Pharmacia superose 12 column (50 X 1.6 cm), ion exchange chromatography on DE52, AMP sepharose 4B affinity chromatography, and finally by dCTP sepharose 4B chromatography.

RESULTS

Deoxycytidine kinase was purified 50,000 fold to a specific activity of 8 μmol/min/mg with a 4% recovery (Figure 1).

This highly purified enzyme was found to have activity for deoxyguanosine, deoxyadenosine, cytidine and cytosine arabinoside (Table I). The apparent Km for deoxycytidine was

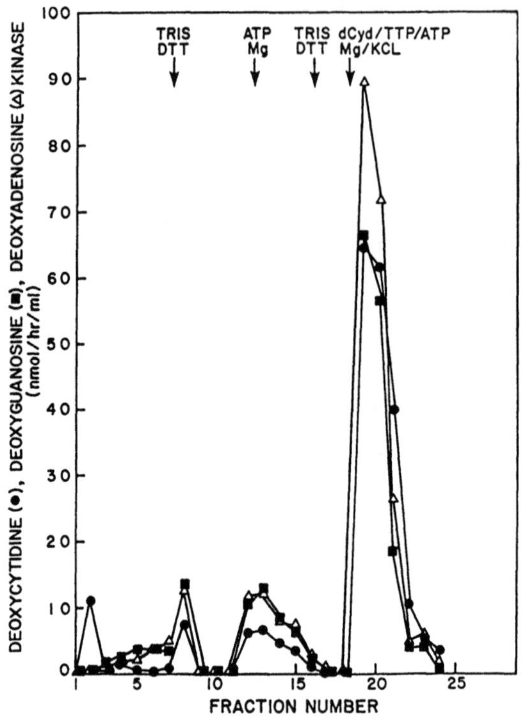

Fig 1. Final Purification Step of Deoxycytidine Kinase on dCTP Sepharose-4B. Two hundred milliliters of diluted AMP-Sepharose pool were applied to a dCTP Sepharose column (8.5 cm x 7 cm). Fractions 1 through 17 contained 10 ml each and fractions 18 to 24 contained 2 ml each. The enzyme was eluted with 50 mM Tris-HCl pH 7.4, 25 mM dithiothreitol and 5% glycerol, (Peak I, fraction number 8). Peak II and III (Fraction numbers 12-16 and 19-24, respectively) were eluted with 2 mM ATP, 2.4 mM $MgCl_2$ and 100 μM deoxycytidine, 2 mM ATP, 2.4 mM $MgCl_2$, 2 mM TTP, respectively. Peak III fractions were desalted prior to asaying for nucleoside kinase activity. Protein was undetectable using a commassie blue assay.DTT, dithiothreitol.

0.8 μM and the apparent Km for the other substrates were 1 to 2 orders of magnitude higher (Table I). Dideoxycytidine was phosphorylated at 5% of the rate of deoxycytidine and inhibited the enzyme competitively with a Ki of 0.9 mM. Dideoxyadenosine was a competitive inhibitor with a Ki of 2 mM.

TABLE I
Nucleoside Kinetic Constants for
Phosphorylation by Deoxycytidine Kinase

Compounds	Apparent Km (μM)	Apparent Vmax (μmol/min/mg)
Deoxycytidine *	0.8	8.0
Deoxyguanosine	430	12
Deoxyadenosine	500	22
Cytidine	450	7.8
Cytosine Arabinoside	40	0.5

*True Km and Vmax values

TABLE II
Properties of Human T-Lymphoblast (MOLT-4)
Deoxycytidine Kinase

- 50,000 fold purification to a specific activity of 8 μmol/min/mg
- Molecular weight of 60,000
- Subunit molecular weight 30,500
- Substrate specificity includes deoxyguanosine, deoxyadenosine, cytidine, cytosine arabinoside
- Absolute requirement for Mg^{2+}
- pH optimum 6.5 to 9.0
- ADP and dCTP are important regulators.

Using gel filtration chromatography the enzyme had an apparent molecular weight of 60,000 and a stokes radius of 32 Å. Sodium dodecylsulfate polyacrylamide gel electrophoresis demonstrated one major band of protein that migrated with an estimated molecular weight of 30,500 (Table II).

The rate of deoxycytidine phosphorylation is greatest at pH 7 but the enzyme had a broad pH optimum ranging from 6.5 to 9. Deoxycytidine kinase has an absolute requirement for magnesium (Figure 2). Deoxycytidine is phosphorylated most efficiently at equal concentrations of ATP and magnesium.

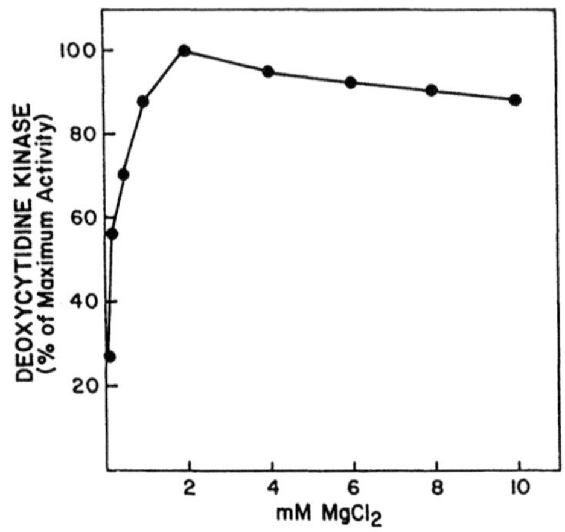

Fig 2. Effect of Magnesium Concentration on Deoxycytidine Kinase Activity. The enzyme was incubated with 2 mM ATP, 10 μM deoxycytidine and varying concentrations of magnesium under standard assay conditions. Results have been expressed as the percentage of maximum deoxycytidine kinase activity which is estimated to be 7.5 μmol/min/mg protein. There was no enzyme activity with 0.2 mM EDTA added.

DISCUSSION

The substantial variation in the apparent substrate specificity of deoxycytidine kinases is probably related to the study of partially purified enzymes which may contain more than one active protein and tissue specific differences. In our study deoxycytidine, deoxyguanosine and deoxyadenosine phosphorylating activities coeluted during gel filtration, ion exchange chromatography, and two affinity chromatography columns. Deoxycytidine kinases in continuous cell lines and human leukemic T-lymphoblasts and myeloblasts phosphorylate deoxycytidine, deoxyguanosine and deoxyadenosine[1-13]. Phosphorylation of cytidine by partially purified deoxycytidine kinase has been a controversial matter. In partially purified L1210 cells deoxycytidine, deoxyguanosine, deoxyadenosine and cytidine phosphorylation are associated with the same deoxynucleoside kinase[8]. Our studies revealed definite cytidine phosphorylating activity asociated with the highly purified deoxycytidine kinase.

The substrate specificity for deoxycytidine kinase includes a variety of nucleoside analogs with therapeutic efficacy in humans. This enzyme is the rate-limiting step in the activation of the chemotherapeutic agent cytosine arabinoside to its 5'-triphosphate[17,18]. Consistent with that finding our studies demonstrate that cytosine arabinoside is a substrate for highly purified deoxycytidine kinase. Adenine arabinoside, an antiviral drug, is phosphorylated by adenosine kinase and deoxycytidine kinase[19-23]. In our studies this compound has minor if any activity with MOLT-4 deoxycytidine kinase. 2',3'-Dideoxycytidine, which is active against human immunodeficiency virus[24], has a very low Vmax. This latter compound is a poor substrate for partially purified human T-lymphoblastic cytoplasmic and mitochondrial deoxycytidine kinases[24]. Studies with radioactive derivatives will be necessary to clarify this matter.

The existence of highly purified human T-lymphoblastic deoxycytidine kinase will facilitate our delineation of the role for this enzyme in nucleoside and nucleoside analog phosphorylation. With this method it should be possible to purify similar activities from other tissues to compare their properties. This is important for the development of nucleoside analogs with specific anticancer or antiviral activities.

SUMMARY

Nucleoside kinases phosphorylate endogenous nucleosides and nucleoside analogs which are anticancer and antiviral drugs. Deoxycytidine kinase was purified from cultured human T-lymphoblasts to a specific activity of 8 µmol/min/mg protein and 92% purity. The molecular weight was 60 kD and the stokes radius was 32 Å. The subunit molecular weight was 30.5 kD. Deoxyguanosine, deoxyadenosine and cytidine phosphorylating activites copurified with deoxycytidine kinase to final specific activities of 7.2, 13.5 and 4 µmol/min/mg protein, respectively. This enzyme had apparent Km values of 1.5, 430, 500, 450 and 40 µM for deoxycytidine, deoxyguanosine, deoxyadenosine, cytidine and cytosine arabinoside, respectively. The pH optimum ranged from 6.5-9.0. Mg^{2+} is absolutely required for activity. This highly purified enzyme will facilitate our delineation of the role for this enzyme in nucleoside or nucleoside analog phoshorylation and the structural basis for the regulation of this enzyme.

ACKNOWLEDGMENT

This work was supported by United States Public Health Service Grants 1R01GM32837, 5M01RR42, 2R01CA34085. The authors wish to thank Dan Link for technical support and Paige Bigelow for typing the manuscript.

REFERENCES

1. W.R. Gower, M.C. Carr, and D.H. Ives, J. Biol. Chem. 254:2181 (1979).
2. T.A. Krenitsky, J.V. Tuttle, G.W. Koszalka, I.S. Chen, L.M. Beacham III, J.L. Rideout, and G.B. Elion, J. Biol. Chem. 251:4055 (1976).

3. W.R.A. Osborne and C.R. Scott, Biochem. 12:3294 (1983).
4. L.J. Gudas, B. Ullman, A. Cohen, and D.W. Martin, Jr., Cell 14:531 (1978).
5. B. Ullman, L.J. Gudas, A. Cohen, and D.W. Martin, Jr., Cell 14:365 (1978).
6. A. Baxter, L.M. Currie, and J.P. Durham, Biochem. J. 173:1005 (1978).
7. D.A. Carson, D.B. Wasson, J. Kaye, B. Ullman, D.W. Martin, Jr., R. Robins, and J.A. Montgomery, Proc. Natl. Acad. Sci. USA. 77:6865 (1980).
8. C.H. Chang, R.W. Brockman, and L.L. Bennett, Jr. Cancer Res. 42:3033.(1982).
9. J.P. Durham and D.H. Ives, J. Biol. Chem. 245:2276 (1970).
10. J.P. Durham and D.H. Ives, Biochim. Biophys. Acta. 228:9 (1971).
11. M.S. Hershfield, J.E. Fetter, W.C. Small, A.S. Bagnara, S.R. Williams, B. Ullman, D.W. Martin, Jr., D.B. Wasson, and D.A. Carson, J. Biol. Chem. 257:6380 (1982).
12. B. Ullman, B.B. Levinson, M.S. Hershfield, and D.W. Martin, Jr., J. Biol. Chem. 256:848 (1981).
13. V. Verhoef, J. Sarup, and A. Fridland, Cancer Res. 41:4478 (1981).
14. N.S. Datta, D.S. Shewach, M.C. Hurley, B.S. Mitchell, and I.H. Fox (Submitted for publication).
15. M.C. Hurley, and I.H. Fox, Biochem. Med. 30:89 (1983).
16. M.C. Hurley, T.D. Palella, and I.H. Fox, J. Biol. Chem. 258:15021 (1983).
17. S.S. Cohen, in: "Progress in Nucleic Acid Research and Molecular Biology," J.H. Davidson and W.E. Dohn, eds., Academic Press, New York (1966).
18. J. Mejer, Scand. J. Clin. Lab. Invest. 42:401 (1982).
19. N.R. Cozzarelli, Ann. Rev. Biochem. 46:641 (1977).
20. J.F. Henderson, F.W. Scott, and J.K. Lowe, Pharm. Ther. Part A Chemother. Toxicol. Metab. Inhibitors 8:573 (1980).
21. W. Plunkett, S. Chubb,L. Alexander, and J.A. Montgomery, Cancer Res. 40:2349 (1980).
22. J.C. Sarup, and A. Fridland, Biochemistry 26:590 (1987).
23. Y. Yamada, H. Goto, and N. Ogasawara, Int. J. Biochem. 17:425 (1985).
24. M. Cate, and Y. Cheng, J. Biol. Chem. 262:988 (1987).

CELL CYCLE REGULATION OF RIBONUCLEOTIDE REDUCTASE M2 SUBUNIT SPECIFIC RNA IN WILD TYPE AND MUTANT S49 CELLS

Daniel A. Albert, Edwardine Nodzenski, and Gloria Yim

Department of Medicine
The University of Chicago
Chicago, Illinois

The enzyme ribonucleotide diphosphate reductase (EC 1.17.4.1) reduces ribose diphosphates to deoxyribose diphosphates which are subsequently phosphorylated and incorporated into DNA(1). This pathway is the only <u>de novo</u> pathway for the in vivo synthesis of all four deoxynucleoside triphosphates(2). The enzyme activity varies with the cell cycle and appears to result primarily from an increase in the activity of M2 subunit of the enzyme in S phase and G2/M phase(3,4). The enzyme is composed of two subunits M1 and M2. The M1 subunit contains the allosteric binding sites for the complex regulation of enzyme activity. The M2 subunit contains the tyrosyl radicals necessary for the reduction reaction. Hydroxyurea is a specific inhibitor of ribonucleotide reductase that binds to the tyrosyl radicals on M2 and inactivates them(5). Both subunits are necessary for catalytic activity. Exposure of cells to increasing concentrations of hydroxyurea results in resistance that is conferred by increasing ribonucleotide reductase activity. This increase in enzyme activity is a consequence, in part, of amplifying the gene coding for the M2 subunit resulting in an increase in M2 protein(6). These and other data suggests that under normal conditions the availability of the M2 subunit is rate limiting. However, hydroxyurea resistant cells have normal cell cycle regulation of ribonucleotide reductase indicating that gene amplification per se does not normally alter transcriptional regulation(7).

We observed that some hydroxyurea resistant cells were also cyclic AMP resistant(7). While wild type S49 cells arrest reversibly in the G1 phase of the cell cycle when cyclic AMP is increased, mutant cells having cyclic AMP dependent protein kinase do not cell cycle arrest. We subsequently determined that hydroxyurea resistant cell lines that are cyclic AMP resistant have mutated to cyclic AMP dependent protein kinase deficient phenotype(8). More recent evidence suggests that the M2 subunit of ribonucleotide reductase may be phosphorylated and inactivated by cAMP dependent protein kinase(8). However, since only some hydroxyurea resistant cell lines are cyclic AMP resistant and all hydroxyurea resistant cell lines appear to have normal cell cycle regulation of ribonucleotide reductase activity we hypothesized that the major effect of cyclic AMP in generating cell cycle arrest is most likely transcriptional rather than post translational. Thus, in preparation for analyzing the effect of cyclic AMP on transcriptional regulation of the M2 subunit of ribonucleotide reductase, we first examined the normal cell cycle regulation of M2 transcription using centrifugal elutriation to obtain cell cycle specific populations of wild type, cyclic AMP dependent protein kinase deficient, and hydroxyurea resistant cell lines both cyclic AMP sensitive an resistant.

METHODS

Cell Culture

Wild type and mutant S49 cells were grown as previously described. All cell lines used have been characterized and described previously(7,8,9,10).

Centrifugal Elutriation

The elutriation system consists of a Beckman J2-21 Centrifuge a JE-6 elutriation rotor with standard separation chambers and a Masterflex pump. sample of $4x10^8$ cells in 1.5L of Dulbeccos modified Eagles media with 10% horse serum is loaded at 9.5ml/min rotor speed 2000 RPM. Subsequently 50ml fractions are removed as the flow rate is increased from 11.5ml/min to 32ml/min(7).

RNA Extraction

Cells are washed in PBS and resuspended in 10 x cell pellet volume of a Guanidine mix that consists of 3.5M Guanidine Hydrochloride .02M potassium acetate 10mM EDTA. RNA is precipitated with 1/2 volume absolute ETOH at -20°C. RNA is reprecipitated in guanidine then resuspended in TE prior to phenol chloroform isoamylalcohol extraction(11).

Dot Blot

RNA was denatured in 50% deionized formamide and 6% formaldehyde at 50° for 1 hour(12). Samples were applied to Gene Screen Plus (New England Nuclear) nylon membranes with a dot blot manifold (BioRad)(13). The filter was dried and baked for 2 hours at 80°C.

Nick Translation and Probing

1-2μg of cDNA(14)was nick translated (Bethesda Research Laboratories). Hybridization was performed at 65°C overnight with salmon sperm DNA to reduc backround. Blots were washed at high stringency which included .1xSSC at 25 and 2xSSC with 1% SDS at 65°C then exposed to autoradiographic film.

RESULTS

The amount of mRNA coding for the M2 subunit of ribonucleotide reductas varies in the cell cycle. In Fig.1 wild type cells fractionated into five cell cycle populations show variations with a large amount of M2 specific mR in the G1 phase, a low amount in S phase, and a high concentration in G2/M. In contrast CDP reductase activity increases in S phase and continues to be high in G2/M. Thus one might infer that message transcribed in G1 is translated in S phase and that it may be augmented by additional transcripti in G2/M.

Cell lines deficient in cyclic AMP dependent protein kinase exemplified by KIN^-(Fig.2) have the same cell cycle variation as wild type. In other experiments hydroxyurea resistant cell lines both with and without cyclic AM dependent protein kinase activity have similar cell cycle variation albeit with increased amount of message for M2 per ug of total cellular RNA. In

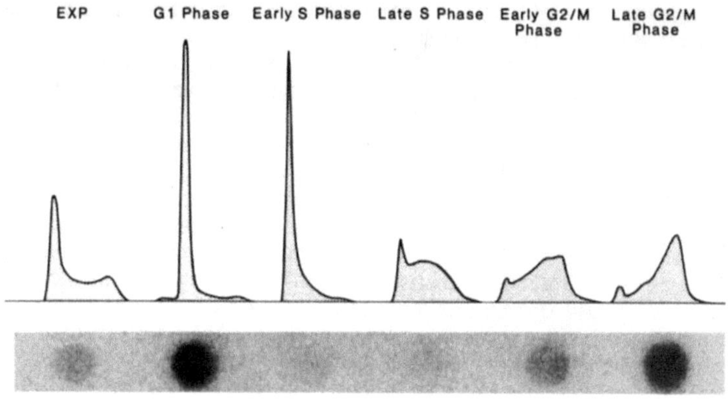

Fig. 1 RNA dot blot quantitation of M2 specific RNA. RNA from wild type cells (20μg total cellular RNA per dot) in exponential growth phase, G1, early S, late S and G2/M phase was extracted and bound to nylon membranes. Cell cycle specific populations were obtained by centrifugal elutriation as detailed in the Methods section. Dot blot was probed with nick translated M2 specific washed at high stringency and exposed for a variable length of time.

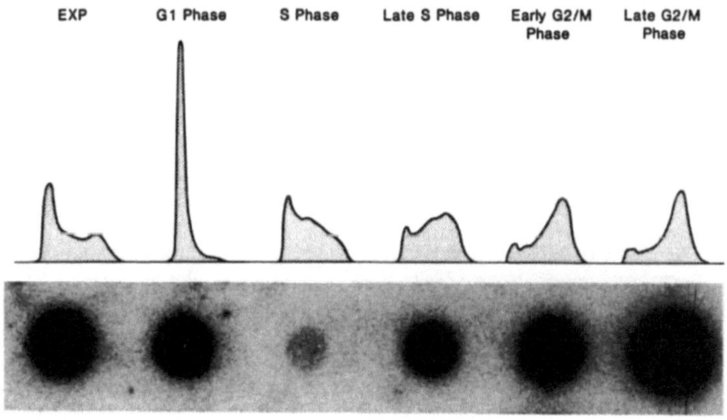

Fig. 2 RNA dot blot quantitation of M2 specific RNA from KIN$^-$(10μg per dot) cells. Dot blot and hybridization preferred as above.

preliminary experiments subfractionating the G1 phase cells appear to indicate a narrow period of increased message in late G1 phase but low concentrations of M2 specific mRNA in early G1 phase.

DISCUSSION

The enzyme ribonucleotide diphosphate reductase is essential for DNA synthesis. The enzyme is cell cycle regulated and this regulation appears to be due to increased M2 subunit activity.

In turn enzyme activity appears to be due to new protein synthesis. Although previously we were unable to demonstrate increased spot density during S phase(9)we now believe the spot tentatively thought to be M2 is another protein that is also amplified with exposure to hydroxyurea(8). Newer analysis revealed that the M2 protein comigrated with actin which may have obscured its cell cycle variation(8). Based on data presented above, we now believe that the increased M2 activity is probably due to increased synthesis and that this is regulated at the transcriptional level. Of interest KIN^- cells that lack cyclic AMP dependent protein kinase have normal cell cycle regulation. Thus the effect of cyclic AMP on cell cycle regulation may be mediated through a post translational mechanism. However, this does not preclude transcriptional regulation by a phosphorylated protein.

Further experiments will delineate the mechanism of cyclic AMP effect on cell cycle regulation of the M2 subunit of ribonucleotide reductase.

ACKNOWLEDGEMENTS

Support for these investigations was provided by the Athritis Foundation National, Arthritis Foundation of Illinois and the USPHS. cDNA probes for M2 were kindly provided by L. Thelander. Expert secretarial assistance was provided by D. Shine and T. Hopkins.

REFERENCES

1. E.C. Moore and R.B. Hurlbert. Regulation of Mammalian Deoxyribonucleotide Biosynthesis by Nucleotide Activators and Inhibitors. J Biol Chem. 241:4802-4809, 1966
2. L. Thelander and P. Reichard. Reduction of Ribonucleotides. Ann Rev Biochem 48:133-158, 1979
3. Y. Engstrom, S. Eriksson, I. Jildevik, S. Skog, L. Thelander and B. Tribukait. Cell Cycle-Dependent Expression of Mammalian Ribonucleotide Reductase: Differential Regulation of the Two Subunits. J Biol Chem 260:9114-9116, 1985
4. S. Eriksson, A. Graslund, S. Skog, L.Thelander, B. Tribukait. Cell Cycle - Dependent Regulation of Mammalian Ribonucleotide Reductase: The S Phase Correlated Increase in Subunit M2 is Regulated by De Novo Protein Synthesis. J Biol Chem 259:11695-11700, 1984
5. L. Akerblom, A. Ehrenberg, A. Graslund, J. Lankinen, P. Reichard and L. Thelander. Overproduction of the Free Radical of Ribonucleotide Reductase in Hydroxyurea-Resistant Mouse Fibroblast 3T6 Cells. Proc Natl Acad Sci USA 78:2159-2163,1981
6. D.A. Albert and E. Nodzenski: unpublished results
7. D.A. Albert, L.J. Gudas, and E.Nodzenski. Deoxyribonucleotide Metabolism and Cyclic AMP Resistance in Hydroxyurea-Resistant S49 T-Lymphoma Cells. J Cell Physio 130:262-269, 1987
8. D.A. Albert and E. Nodzenski. The M2 Subunit of Ribonucleotide Reductase is a Target of Cyclic AMP Dependent Protein Kinase (submitted)

9. D.A. Albert and L.J. Gudas. Deoxyribonucleoside Triphosphate Metabolism During the Cell Cycle. J Biol Chem 260:679-684, 1985
10. Wetter, T.V.D., M.P. Murtaugh, and P. Coffino. Revertants of a Trans-Dominant S49 Mouse Lymphoma Mutant that Affects Expression of cAMP-Dependent Protein Kinase. Cell 35:311-320, 1983
11. T. Maniatis, E.F. Fritsch, J. Sambrook. Molecule Cloning: A Laboratory Manual. Cold Spring Harbor Laboratory, 1982
12. H. Lehrach, D. Diamond, J.M. Wozney, and H. Boedtker. RNA Molecular Weight Determinations by Gel Electrophoresis under Denaturing Conditions, A Critical Re-examination. Biochemistry 16:4743-4751, 1977
13. F.C. Kafatos, C.W. Jones, and A. Efstratiadis. Determination of Nucleic Acid Sequence Homologies and Relative Concentrations by a Dot Blot Hybridization Procedure. Nucleic Acids Res 7:1541-1552, 1979
14. L. Thelander and P. Berg. Isolation and Characterization of Expressible cDNA Clones Encoding the M1 and M2 Subunits of Mouse Ribonucleotide Reductase. Mol and Cell Biol 6:3433-3442, 1986.
15. M.W. McDonell, M.N. Simon, F.W. Studier. Analysis of Restriction Fragments of T7 DNA and Determination of Molecular Weights by Electrophoresis in Neutral and Alkaline Gel. J Mol Biol 110:119-146, 1977

PHOSPHOCREATINE AS A POSSIBLE MODULATOR OF THE ADENYLATE POOL

Giovanni Ronca, Angela Pezzini, Angela Conte,
Paola Galbani, Riccardo Zucchi, and Simonetta Ronca-Testoni

Institute of Biological Chemistry, Faculty of Medicine
via Roma 55, 56100 Pisa, Italy

INTRODUCTION

Creatine (Cr) is mainly synthesized in liver and pancreas, then transported by the blood into the other tissues. Muscle, heart and brain which show high creatine kinase activity are very efficient in taking up creatine from the blood and in accumulating it as phosphocreatine (PCr) and creatine (Walker, 1979). These tissues also have a high level of adenine nucleotides. The adenine nucleotide pool is mainly regulated through a feedback inhibition on the de novo synthesis and on the salvage pathways. However different levels of adenine nucleotides are observed in different tissues suggesting that other factors may influence feedback regulation.

From the data of Fitch et al. (1974) it appears that the depletion of creatine and phosphocreatine in skeletal muscle of rats fed β-guanidinopropionic acid results in a decrease of the adenine nucleotide pool and of the ATP/ADP ratio. We observed that the intravenous injection of phosphocreatine increased the energy charge and the adenine nucletide pool as well as the creatine-phosphocreatine pool in the rat heart (Ronca-Testoni et al., 1985).

In this paper we report the effect of oral administration of phosphocreatine on the adenine nucleotide pool of the rat heart, muscle, and brain. We also report the effect of the increased creatine-phosphocreatine pool on the adenine nucleotide pool in cell cultures. The results obtained suggest that the creatine-phosphocreatine pool may modulate the adenine nucleotide pool in some tissues.

MATERIALS AND METHODS

Phosphocreatine disodium salt . 5 H_2O was dissolved in 15% ethanol (v/v) at a concentration of 0.73 mmoles/ml immediately before use. The solution was then brought to 4°C and the pH was carefully lowered to 3.0 with 6 N HCl on ice and under vigorous stirring. Female Wistar rats (200 to 250 g) were anaesthetized by ethyl ether vapours and oxygen and the phosphocreatine solution was administered intragastrically by intubation with a rubber catheter (2.9 mmoles/Kg body wt.). After 4-6 hours the rats were anaesthetized and the hearts were quickly excised and perfused by the retrograde Langendorff technique with Krebs-Henseleit buffer, pH 7.4, containing 11 mM glucose and equilibrated with O_2:CO_2 (95%:5%). After a washout of 90 sec the hearts were rapidly frozen in liquid nitrogen with precooled Wollenberger clamps. Intercostal (red) muscles, brain, liver, kidney and blood were removed immediately after heart excision and put into liquid nitrogen.

The frozen tissues were freeze-dried and kept at -80°C until used. Samples of the dried tissues were powdered with a mechanical homogenizer (Ultraturrax) at setting of 1/4 maximal speed for 1 min. About 5 mg of the dried tissue powder were dissolved in 1 ml of 0.42 M perchloric acid and kept on ice for 10 min with intermittent shaking. After neutralization with 2 N KOH the extracts were centrifuged at 3000 rpm for 5 min. Supernatants kept on ice were analysed for adenine nucleotides, IMP, creatine and phosphocreatine by HPLC according to Sellevold et al. (1986). Supernatant aliquots were injected on a Waters Bondapak C-18 column (3.9 mm x 30 cm) and eluted isocratically with 215 mM potassium dihydrogen phosphate, 2.3 mM tetrabutylammonium hydrogen sulphate and 3% acetonitrile, pH 6.2, at a flow rate of 1 ml/min. The eluent was monitored at 206 nm. The levels of creatine and phosphocreatine in liver, kidney and blood were also determined on the neutralized perchloric extracts according to Eggleton et al. (1943) and Opie et al. (1971), respectively.

Human lymphoblastoid Raji cells cultured in RPMI 1640 medium containing 10% heat-inactivated fetal calf serum at 37°C for 2 days and mouse hybridoma ascites cells harvested 7-8 days after inoculation in Balb/c mice were washed three times and resuspended in RPMI 1640 medium supplemented with 20 mM Hepes and 10% heat-inactivated fetal calf serum or in the same solution containing 20 mM creatine. The cells at a concentration of 9-12 x 10^6 for hybridoma cells and 12-20 x 10^6 for Raji cells were incubated for 4 hours at 37°C, washed three times and extracted with perchloric acid according to Annesley and Walker (1978). After neutralization with KOH and centrifugation supernatant aliquots were analysed for nucleotides, creatine and phosphocreatine as above described.

RESULTS AND DISCUSSION

The levels of creatine and phosphocreatine in blood at different times between 4 and 6 hours after oral administration of phosphocreatine are reported in Table 1 and Fig. 1. During all the interval creatine was about twice that found in the blood of control animals and did not vary significantly throughout the experiment. Phosphocreatine which was absent or present in trace amounts in controls was markedly increased at the 4th hour (620 ± 80 nmoles/g dry wt.), then it progressively decreased. A 2.6-fold increase of the creatine level which remained constant during the

Table 1. Levels of creatine in liver, kidney and blood and of phosphocreatine in blood after oral phosphocreatine administration to rats.

		0 hours	4 hours
		(µmoles/g dry wt.)	
Liver	creatine	8.0 ± 1.8	26.8 ± 0.4 ***
Kidney	creatine	15.7 ± 4.1	40.8 ± 2.2 ***
Blood	creatine	1.4 ± 0.1	3.3 ± 0.2 ***
	phosphocreatine	traces	0.6 ± 0.1

Data are mean ± S.D. of 4 experiments
*** $p<0.001$

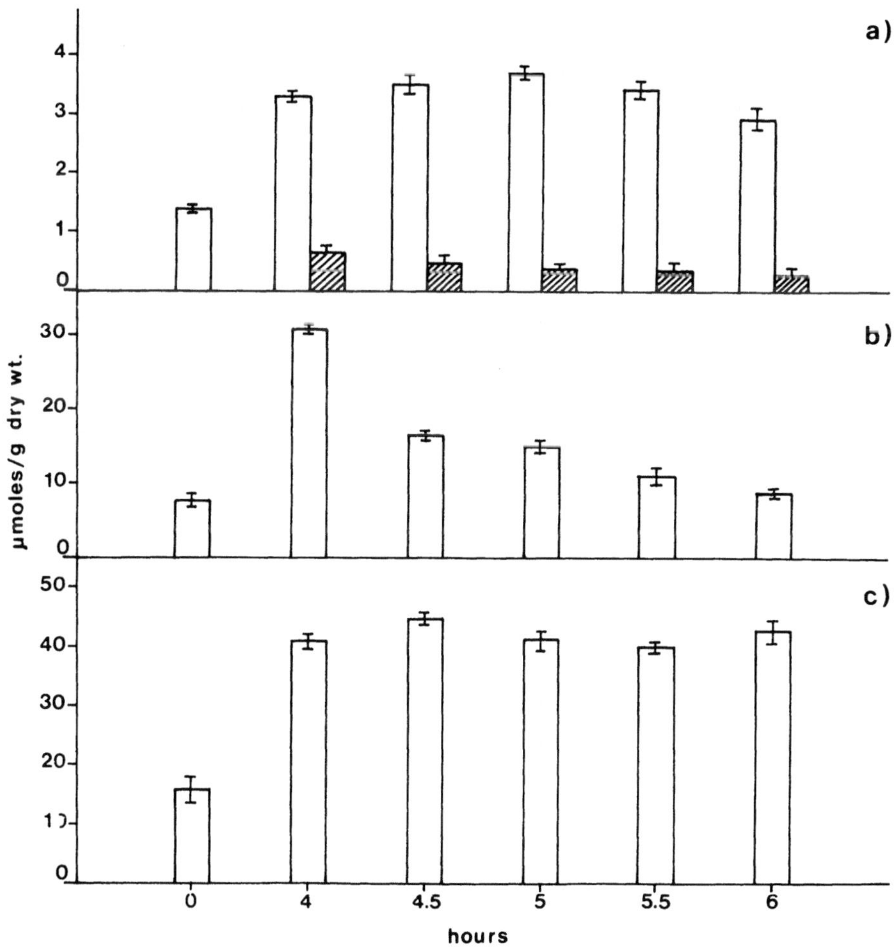

Fig. 1. Levels of creatine (blank bars) in blood (a), liver (b) and kidney (c) and of phosphocreatine in blood (a, hatched bars) at intervals after oral administration of phosphocreatine.

interval examined was found in kidney. In liver a high level of creatine (3.3 times higher than the basal value) was observed at the 4th hour, then it rapidly decreased. Trace amounts of phosphocreatine were detected in the liver and kidney only of the treated animals, probably due to residual blood in the tissues.

These data show that orally administered phosphocreatine may be absorbed not only as creatine but also as phosphocreatine which was still found in blood between the 4th and the 6th hour. The phosphocreatine level decreased in this interval and it is known that the compound may be hydrolyzed by blood phosphatases. The absorption of phosphocreatine was favoured by the presence of ethanol in the aqueous solution through which it was administered and by lowering the solution pH to avoid buffering gastric juice. It is likely that gastric and intestinal absorption both occurred, anyhow it was rapid since a decrease of phosphocreatine in blood and of creatine in liver was already observed between the 4th and the 6th hour.

As far as phosphocreatine rich tissues are concerned (Table 2) the level of the creatine-phosphocreatine pool was not modified in brain during the interval examined; on the other hand a significant increase (20%) of the adenine nucleotide pool was observed. Red skeletal muscle and heart showed a significant increase of the creatine-phosphocreatine pool at the 4th hour and at the 4th and 6th hour, respectively. Adenine nucleotide pool also increased by 18% in the heart (6th hour) and by 34% in skeletal muscle (4th hour).

The levels of creatine, phosphocreatine and adenine nucleotides in the rat heart are reported in Table 3. The increase of the creatine-

Table 2. Effect of oral administration of phosphocreatine on the creatine-phosphocreatine pool and on the adenine nucleotide pool in rat heart, skeletal muscle and brain.

	Cr + PCr			ΣAXP		
	hours			hours		
	0	4	6	0	4	6
Heart	74.9 ±2.3	79.2* ±1.4	94.8*** ±4.4	24.6 ±1.6	26.2 ±1.3	29.1* ±2.9
Red skeletal muscle	105.8 ±3.6	126.7** ±7.2	115.8 ±8.6	25.4 ±3.2	34.1* ±4.9	31.5* ±2.6
Brain	64.3 ±6.2	62.9 ±1.8	63.7 ±6.3	13.9 ±1.0	16.7* ±1.4	15.0 ±1.8

Data are mean ± S.D. of 4 experiments and are expressed as µmoles/g dry wt.

* $p<0.05$ ** $p<0.01$ *** $p<0.001$

Table 3. Effect of oral administration of phosphocreatine on nucleotide, creatine and phosphocreatine levels in the rat heart.

	hours		
	0	4	6
Creatine	36.3 ± 1.4	37.3 ± 3.2	44.6 ± 3.8**
Phosphocreatine	38.6 ± 3.7	42.0 ± 4.2	50.2 ± 7.2*
Cr + PCr	74.9 ± 2.3	79.3 ± 1.4	94.8 ± 4.4***
PCr/Cr	1.06	1.13	1.13
ATP	19.1 ± 1.7	21.3 ± 0.6	23.6 ± 2.7*
ADP	4.3 ± 0.6	4.0 ± 0.5	4.5 ± 1.8
AMP	1.2 ± 0.4	0.9 ± 0.2	1.0 ± 0.3
Σ AXP	24.6 ± 1.6	26.2 ± 1.3	29.1 ± 2.9*
E.C.	0.86	0.87	0.86

Data are mean ± S.D. of 4 experiments and are expressed as µmoles/g dry wt.

* $p<0.05$ ** $p<0.01$ *** $p<0.001$

phosphocreatine pool resulted from a rise of both compounds, however a small increase of the PCr/Cr ratio was observed. ATP too increased by 11% and 24% at the 4th and at the 6th hour, respectively, while ADP and AMP were not modified. The E.C. value (ATP + ½ADP/ Σ AXP) as also the PCr/Cr ratio were slightly increased at the 4th and 6th hour.

Annesley and Walker (1978) reported that Erlich ascites tumor cells accumulated creatine and phosphocreatine when incubated in vitro with creatine and that phosphocreatine was fully functional as a reservoir of high energy phosphate. However they did not determine adenine nucleotides. We used the same experimental conditions to increase the creatine-phosphocreatine pool in hybridoma cells and in Raji cells and examined the levels of adenine nucleotides. Both cell lines had low levels of creatine and phosphocreatine as reported for ascites tumor cells. The nucleotide concentration was comparable to that found in lymphoid cells (Matsumoto et al. 1979) or generally in cells with a low content of creatine and phospho creatine (Table 4 and 5). When creatine was added to the culture medium the concentrations of creatine and phosphocreatine markedly increased. The adenylate pool also was higher. The increase was more evident for ATP so that the energy charge value shifted from 0.85 to 0.89. Hybridoma and Raji cells had creatine kinase activity of about 52.3 mU and 44.1 mU/10^6 cells, respectively, which corresponded to 3.1% of the activity found in the pig heart by the same method and was comparable to that of ascites tumor cells.

Although other studies are in progress on the effects of the orally administered phosphocreatine the results so far obtained and those found for cell cultures confirm and extend to other tissues our previous data on

Table 4. Effect of creatine on the levels of creatine, phosphocreatine and adenine nucleotides in mouse hybridoma ascites cells.

	Control	Creatine treated
	(nmoles/10^6 cells)	
Creatine	1.05 ± 0.60	5.46 ± 0.84***
Phosphocreatine	1.30 ± 0.81	16.44 ± 3.43***
Cr + PCr	2.35 ± 0.93	21.90 ± 3.92***
PCr/Cr	1.24	3.01
ATP	2.49 ± 0.67	4.62 ± 0.83***
ADP	0.79 ± 0.10	1.00 ± 0.14**
AMP	0.13 ± 0.02	0.09 ± 0.01**
ΣAXP	3.41 ± 0.80	5.71 ± 1.02**
E.C.	0.85	0.89

Data are mean ± S.D. of 4 experiments.
** $p < 0.01$ *** $p < 0.001$

Table 5. Effect of creatine on the levels of creatine, phosphocreatine and adenine nucleotides in human lymphoblastoid Raji cells.

	Control	Creatine treated
	(nmoles/10^6 cells)	
Creatine	0.81 ± 0.52	6.81 ± 1.18***
Phosphocreatine	0.93 ± 0.60	14.10 ± 2.25***
Cr + PCr	1.74 ± 0.83	20.91 ± 4.62***
PCr/Cr	1.15	2.07
ATP	2.25 ± 0.43	4.38 ± 1.20***
ADP	0.50 ± 0.13	0.76 ± 0.18**
AMP	0.16 ± 0.04	0.11 ± 0.02**
ΣAXP	2.91 ± 0.69	5.25 ± 1.42**
E.C.	0.86	0.90

Data are mean ± S.D. of 4 experiments.
** $p < 0.01$ *** $p < 0.001$

Table 6. Effect of increased ATP, energy charge and ΣAXP on adenine nucleotide metabolism

	ATP	EC	ΣAXP
CATABOLISM			
5'-nucleotidase	-	-	+
AMP deaminase	+	-	+
DE NOVO SYNTHESIS			
PRPP synthetase	+	+	-
amidophosphoribosyl transferase	-	+	-
adenylosuccinate synthetase	+ (GTP)	+	
SALVAGE PATHWAYS			
PRPP synthetase	+	+	-
adenosine kinase	+	+	+

+ = increased enzyme activity
- = decreased enzyme activity

the rat heart after intravenous injection of phosphocreatine (Ronca-Testoni et al., 1985). It appears that there is a relationship between the creatine-phosphocreatine pool and the adenylate pool and along with the well known functions of creatine and phosphocreatine as energy reservoir and energy shuttle we suggest that the creatine-phosphocreatine pool may modulate the adenylate pool. As far as this modulation mechanism is concerned an increased concentration of the creatine-phosphocreatine pool may determine a rise of the energy charge that allows the adenylate pool to reach a higher level. The effects of ATP, energy charge and adenylate pool on the enzymes of adenine nucleotide metabolism are known and are summarized in Table 6. Phosphocreatine may also directly inhibit 5'-nucleotidase (Preobrazhenskii et al., 1986) and AMP deaminase (Ronca-Testoni et al., 1970) with a decrease of adenylate catabolism.

REFERENCES

Annesley, T.M. and Walker, J.B., 1978, Formation and utilization of novel high energy phosphate reservoirs in Erlich ascites tumor cells, J. Biol. Chem., 53:8120

Eggleton, P., Eldsen, S.R., and Gough, N., 1943, The estimation of creatine and of diacetyl, Biochem. J., 37:526.

Fitch, C.D., Jellinek, M., and Mueller, E.J., 1974, Experimental depletion of creatine and phosphocreatine from skeletal muscle, J. Biol.Chem. 249:1060.

Matsumoto, S.S., Raivo, K.O., and Seegmiller, J.E., 1979, Adenine nucleotide degradation during energy depletion in human lymphoblasts, J. Biol Chem., 254:8956.

Opie, L.H., Mansford, K.R.L., and Owen, P., 1971, Effects of increased heart work on glycolisis and adenine nucleotides in the perfused heart of normal and diabetic rats, Biochem. J., 124:475.

Preobrazhenskii, A.N., Dzhavadov, S.A., and Saks, V.A., 1986, Study on the hypothetical mechanism of the protective effect of phosphocreatine on ischemic myocardium, Biokhimiya, 51:675.

Ronca-Testoni S., Raggi, A., and Ronca G., 1970, Muscle AMP aminohydrolase III. A comparative study on the regulatory properties of skeletal muscle enzyme from various species, Biochim.Biophys. Acta, 198:101.

Ronca-Testoni, S., Galbani, P., and Ronca, G., 1985, Effect of creatine phosphate administration on the rat heart adenylate pool, J. Mol. Cell. Cardiol., 17:1185.

Sellevold, O.F.M., Jynge, P., and Aarstad, K., 1986, High performance liquid chromatography: a rapid isocratic method for determination of creatine compounds and adenine nucleotides in myocardial tissue, J. Mol. Cell. Cardiol., 18:517.

Walker, J.B., 1979, Creatine: biosynthesis, regulation, and function, in: "Advances in Enzymology", A. Meister, ed., Wiley & Sons, New York.

REGULATION OF SOLUBLE 5'-NUCLEOTIDASE I FROM RABBIT HEART

Y. Yamazaki,[1,2] A. R. Collinson,[1] V. L. Truong,[1] and J. M. Lowenstein[1]

[1]Graduate Department of Biochemistry, Brandeis University Waltham, Massachusetts 02254, U.S.A.
[2]Department of Molecular Genetics, National Institute of Genetics, Mishima, Shizuoka 411 Japan

SUMMARY

Rabbit heart contains two soluble 5'-nucleotidases, termed N-I and N-II, which can be separated using phosphocellulose chromatography. N-I prefers AMP over IMP as substrate, in contrast to N-II which prefers IMP over AMP. Both enzymes require Mg^{2+}, but the optimum Mg^{2+} concentrations for the two enzymes are different. High concentrations of NaCl inhibit N-I and activate N-II. Purified N-I is activated by ADP but not by ATP. According to Itoh et al. (1986), purified N-II is activated by both ADP and ATP. N-I has been purified approximately 1000-fold to a specific activity of approximately 100 μmol/mg protein/min. The properties of N-I suggest that it is the enzyme responsible for the release of adenosine from AMP under conditions of hypoxia or increased work load.

INTRODUCTION

Adenosine is a coronary vasodilator (Berne, 1963). During ischaemia the adenosine content of myocardial cells increases (Rubio et al., 1973; Jennings et al., 1981; Swain et al., 1982). ADP and AMP levels also increase, but relatively little IMP accumulates (Swain et al., 1982; Williamson, 1966). Most mammalian tissues contain membrane-bound and soluble 5'-nucleotidases. Experiments with isolated guinea pig heart (Schutz et al., 1981) and neonatal rat heart cells in culture (Meghji et al., 1985) show that a soluble 5'-nucleotidase is most likely to be responsible for the release of adenosine from AMP under conditions of hypoxia or increased work load. Cytosolic 5'-nucleotidases have been reported in heart of chicken (Itoh and Oka, 1985), rat (Lowenstein et al., 1983; Itoh, 1986), and pigeon (Gibson and Drummond, 1972). The cytosolic enzyme purified from rat heart by Itoh et al. (1986) shows a substrate preference for IMP over AMP; it is activated by both ADP and ATP. The partially purified, soluble enzyme from pigeon heart shows a preference for AMP over IMP at high substrate concentrations. At AMP concentrations < 6 mM this enzyme is activated by ATP; at higher AMP concentrations it is inhibited by ATP (Gibson and Drummond, 1972).

We have found that rabbit heart possesses two distinct types of solubl 5'-nucleotidase. We refer to these as N-I and N-II on the basis of their order of elution from phosphocellulose columns (Truong et al, 1988). N-II is similar to the 5'-nucleotidase isolated from rat and chicken heart (Itoh et al., 1986; Itoh and Oka, 1985). N-I clearly differs from the N-II enzym described by Itoh's group. Purified N-I also differs from the pigeon enzym with respect to activator specificity. Activation of purified N-I by ADP but not ATP, and its greater activity with AMP as compared to IMP, is consi tent with a role for N-I in adenosine production during ischaemia and increas work load.

METHODS

Purification of N-I. The method up to and including column chromatography on phosphocellulose follows the steps described by Truong et al.(198 Frozen rabbit hearts are pulverized in a percussion mortar cooled with liqu nitrogen. The powder is suspended in 4 ml of buffer A (40 mM Na Hepes buffe pH 7.0, 25% glycerol, 1 mM EDTA, 1 mM DTT, and 0.2 mM PMSF) per gram of tissu The suspension is homogenized in a Brinkman Polytron at maximum speed using 3 bursts of 45 seconds duration, with a 2 minute pause between each burst. The containing vessel is submerged in crushed ice during this procedure. The homogenate is centrifuged at 14,000 x g for 45 min and the supernatant is filtered through gauze and centrifuged at 150,000 x g for 60 min. Solid ammonium sulfate (351 g/L) is added to the supernatant with stirring. The suspension is centrifuged at 14,000 x g for 30 min. The resulting precipitat is resuspended in a minimum volume of buffer A and dialyzed against buffer The dialyzed sample is applied to a phosphocellulose column (2.6 x 23 cm) equilibrated with buffer A. The column is washed with 400 ml of buffer A and eluted with a linear gradient of NaCl from 0 to 0.5 M in 800 ml buffer A. Fractions containing enzyme activity are pooled. Subsequent steps involv chromatography on Mono-S FPLC, and affinity chromatography on AMP-Agarose.

Assays for 5'-Nucleotidases I and II. A. The reaction mixture contair 50 mM Na MOPS buffer, pH 7.0, 5 mM AMP, 3 mM $MgCl_2$, and 1 mM ADP in a final volume of 0.5 ml. B. The reaction mixture contains 50 mM Na MOPS buffer, pH 7.0, 6 mM AMP, 50 mM $MgCl_2$, 500 mM NaCl in a final volume of 0.5 ml. Unless otherwise indicated, the reactions are started by adding enzyme and are run at 37 ^{o}C. The reactions are stopped by adding 2 ml of ice-cold 10% trichloracetic acid. Precipitated protein is removed by centrifugation and 2 ml of the supernatant fluid is used for orthophosphate determination by the method of Sanui (1974). One unit of activity is the amount of enzyme that hydrolyzes 1 μmol AMP per min.

RESULTS

Two Soluble 5'-Nucleotidases. When the ammonium sulfate fraction made from the soluble fraction of rabbit heart is subjected to phosphocellulose chromatography, two 5'-nucleotidases are observed (Fig. 1). N-I is eluted by 0.25 M NaCl and N-II by 0.8 M NaCl.

Some Properties of N-I and N-II. N-I is completely inhibited by 200 mM NaCl, whereas N-II is activated by NaCl (Fig. 2). Both N-I and N-II requir Mg^{2+} for activity; however, they show different responses to increasing MgCl concentrations (Fig. 2). Under the conditions used, N-I is activated maximal by 2-3 mM $MgCl_2$ and is strongly inhibited by higher concentrations. N-II is activated at all concentrations of $MgCl_2$ up to 50 mM.

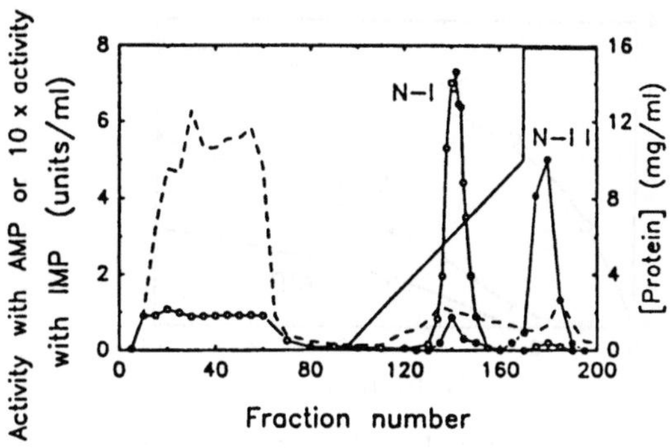

Fig. 1. Phosphocellulose column chromatography of rabbit heart soluble fraction. 5'-Nucleotidase was assayed as described under Methods. o, assay A; •, assay B; broken line without symbols, protein; solid line without symbols, NaCl concentration in eluent (0-0.5 M between tubes 94 and 170, then 0.8 M to tube 200).

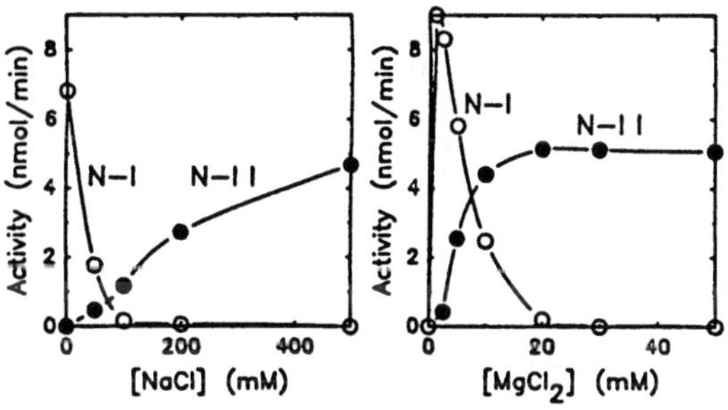

Fig. 2. Effect of NaCl and $MgCl_2$ on N-I and N-II. 5'-Nucleotidase activity was measured with assay A, except that NaCl was added (left panel) or $MgCl_2$ was varied as indicated (right panel).

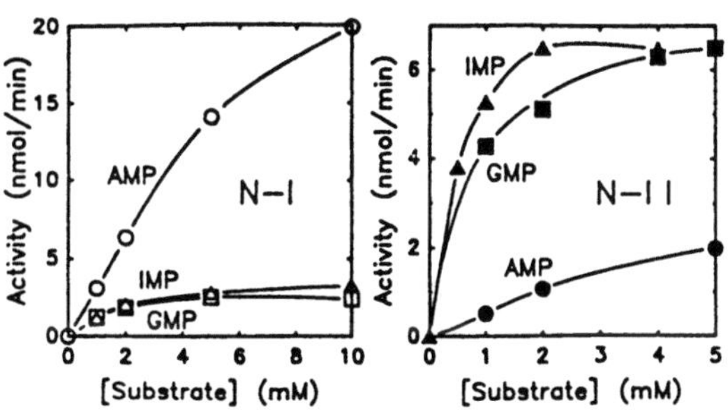

Fig. 3. Comparison of substrate specificity of N-I and N-II. 5'-Nucleotidase activity was measured with assay A. Where indicated, AMP was replaced with IMP or GMP.

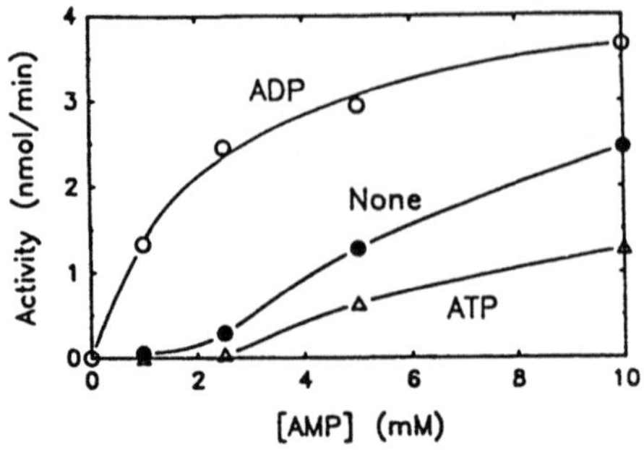

Fig. 4. Effect of ADP and ATP on N-I. The reaction mixture contained 50 m Na MOPS buffer, pH 7.0, 3 mM $MgCl_2$, AMP as indicated, and N-I from the DEAE Sepharose column. Where indicated, the reaction also contained 1 mM ADP or ATP.

The substrate specificities of N-I and N-II also differ (Fig. 3). N-I prefers AMP to IMP and GMP in contrast to N-II which prefers IMP and GMP to AMP.
N-I obtained by phosphocellulose chromatography is activated by ADP and ATP However, when the enzyme is subjected to further chromatography on DEAE-Sep rose, activation by ATP disappears, while activation by ADP persists (Fig. 4). More highly purified N-I is also activated by ADP, but not by ATP. A possible explanation is that the ATP activation of the phosphocellulose fra tion occurs because this fraction is contaminated by myokinase, and that myokinase is removed by further purification.

DISCUSSION

N-I is detected readily with assay A in which AMP is used as substrate but N-II shows little or no activity with this assay. Conversely, N-II is detected readily with assay B in which IMP is used as substrate, but N-I shows relatively little activity with this assay (Fig. 1). N-I is more uns ble than N-II, especially at enzyme concentrations less than 10 µg/ml and i the absence of glycerol. It is therefore easily missed unless care is take to stabilize the enzyme. The two enzymes differ substantially with respect to substrate specificity, and the effects of Mg^{2+} and NaCl. ADP, which increases during ischaemia, activates N-I. ATP, which decreases during ischaemia, does not stimulate the activity of N-I. According to Itoh et al (1986), both ATP and ADP activate N-II. N-I dephosphorylates AMP preferentially to IMP. As has already been stated, these properties are consistent with a role for N-I in adenosine production in response to ischaemia and increased work load.

Soluble 5'-nucleotidases have been purified from heart of rat (Itoh et al., 1986), chicken (Itoh & Oka, 1985), and pigeon (Gibson & Drummond, 1972). The rat and chicken enzymes show kinetic properties similar to N-II The enzyme from pigeon heart was not purified extensively, but it resembles N-I in some respects. An enzyme purified highly from erythrocytes is also of the N-II type, because it prefers IMP over AMP and is activated by ATP and to a lesser extent by ADP (Bontemps et al., 1988). Zekri et al. (1988) purified a soluble 5'-nucleotidase from bovine liver which is more active with pyrimidine than purine nucleotides. This enzyme is activated equally by ATP and ADP. The authors have proposed classifying soluble 5'-nucleotidases in terms of pyrimidine nucleotide-specific and purine nucleotide-specif. enzymes. A detailed comparison of the substrate and activator specificitie of the different forms of 5'-nucleotidase must await more complete studies.

REFERENCES

Berne, R. M., 1963, Am. J. Physiol. 204:317-322

Bontemps, F., Van den Berghe, G., and Hers, H.-G., 1988, Biochem. J. 250, 687-696

Collinson, A. R., Peuhkurinen, K. J., and Lowenstein, J. M., 1987, Regulation and Function of 5'-Nucleotidases, in: "Topics and Perspectives in Adenosine Research," E. Gerlach and B. F. Becker, eds., Springer Verlag, New York, 133-144

Gibson, W. B., and Drummond, G. I., 1972, Biochemistry 11:223-229

Itoh, R., and Oka, J., 1985, Comp. Biochem. Physiol. 81B:159-163

Itoh, R., Oka, J., and Ozasa, H., 1986, Biochem. J. 235:847-851

Jennings, R. B., Reimer, K. A., Hills, M. L., and Mayer, S. E., 1981, Circ. Res. 49:829-900

Lowenstein, J. M., Yu, M.-K., and Naito, Y., 1983, Regulation of Adenosine Metabolism by 5'-Nucleotidases, in: Regulatory Function of Adenosine, R. M. Berne, T. W. Rall, & B. Rubio, eds., Martinus Nijhoff, Boston, 117-131

Meghji, P., Holmquist, C. A., and Newby, A. C., 1985, Biochem. J. 229:799-805

Rubio, R., Berne, R. M., and Dobson, J. G., 1973, Am. J. Physiol. 225:938-953

Schutz, W., Schrader, J., and Gerlach, E., 1981, Am. J. Physiol. 240:H963-H970

Swain, J. L., Sabina, R. L., McHale, P. A., Greenfield, J. C., Jr., and Holmes, E. W., 1982, Am. J. Physiol. 242:H818-H826

Truong, V. L., Collinson, A. R., and Lowenstein, J. M., 1988, Biochem. J., 253:117-121

Williamson, J. R., 1966, J. Biol. Chem. 241:5026-5036

Zekri, M., Harb, J., Bernard, S., and Meflah, K., 1988, Eur. J. Biochem. 172:93-99

PURINE 5'-NUCLEOTIDASE--ITS REESTIMATED SUBUNIT MOLECULAR MASS AND IMMUNOCYTOCHEMICAL LOCALIZATION IN CHICKEN LIVER

Jun Oka,[a] Hisashi Ozasa,[a*] Roichi Itoh,[a] and Sadaki Yokota[b]

[a]Division of Adult Nutrition, The National Institute of Nutrition, Tokyo 162, Japan, [b]Department of Anatomy Yamanashi Medical School, Yamanashi 409-38, Japan

INTRODUCTION

Purine 5'-nucleotidase, formerly termed cytosol 5'-nucleotidase (1), is one of soluble nucleotidases including pyrimidine 5'-nucleotidase (2) and deoxyribonucleotidase (3), and preferentially hydrolyzes IMP, GMP, and AMP in the presence of magnesium ion. The enzyme has been investigated to have allosteric properties (1) characterized by activation by ATP, ADP (4), diadenosine tetraphosphate (5), and 2,3-diphosphoglycerate (6). Purine 5'-nucleotidase has been purified from various sources (7-11), but structural studies remain incomplete.

In this study, the subunit molecular mass of purine 5'-nucleotidase was reestimated by use of immunoblotting (12) and immunoprecipitation of <u>in vitro</u> translation products. The localization of purine 5'-nucleotidase in chicken liver was also investigated by means of an immunoenzyme technique (13). And subcellular distributions of various nucleotidases are discussed.

EXPERIMENTAL PROCEDURES

Chicken liver purine 5'-nucleotidase was purified as described previously (9). Chicken livers used in the purifications were obtained from a local meat market. An antibody against the purified enzyme was raised in a rabbit (12, 13). Fab' fragment of the antibody was conjugated with peroxidase by the maleimide method (I) of Ishikawa <u>et al</u>. (14).

Sodium dodecyl sulfate-polyacrylamide gel electrophoresis (SDS-PAGE) in 10% slab gels was carried out according to the method of Laemmli (15). A gel was calibrated with molecular mass standards as described previously (12). For immunoblot analysis, proteins were electrophoretically transferred to nitrocellulose paper after SDS-PAGE (16). Immunoblotting was carried out using the Fab'-peroxidase (12).

Chicken liver polysomes were prepared by the magnesium precipitation method of McGillivray <u>et al</u>. (17). Cell-free translation of chicken liver

*Present address: Kidney Center, Tokyo Women's Medical College, Tokyo 162, Japan

polysomes and immunoprecipitation using S. aureus cells (IgGsorb; The Enzyme Center, MA) were carried out as described by Mori et al. (18). Briefly, the mixture (50 µl) containing 20 µl of nuclease-treated rabbit reticulocyte lysate (N-90; Amersham International, UK), 10 µCi of [^{35}S] methionine, and 0.43 A_{260} unit of chicken liver polysomes was incubated for 90 min at 25°C, and then treated with a solution (1.0 ml) containing 10 mM Tris/HCl (pH 7.4), 0.1% SDS, 0.1% Triton X-100, 2 mM EDTA, and 40 µg of anti-5'-nucleotidase antibody with and without 35 µg of purified chicken liver purine 5'-nucleotidase as excess unlabeled antigen. Immunoprecipitated and then extracted proteins were subjected to SDS-PAGE as described above, and radioactive bands were visualized by fluorography (4-week exposure at -70°C) using Amplify (Amersham International) as intensifying reagent.

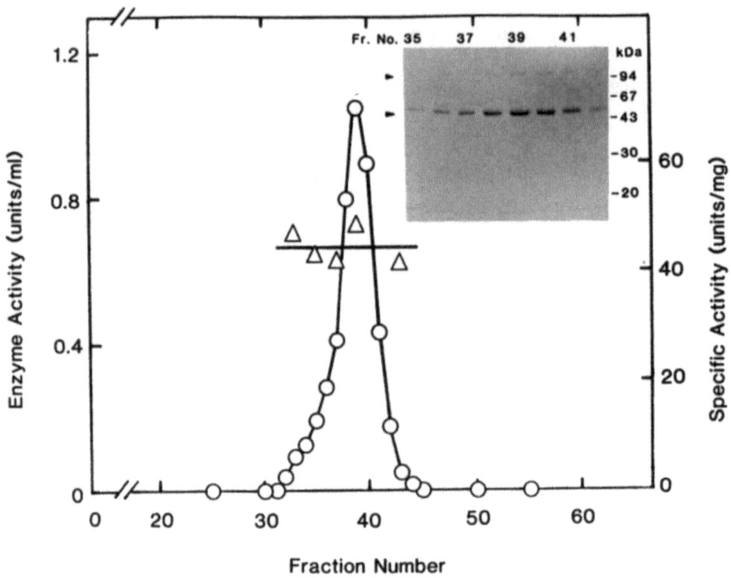

Fig. 1. Gel filtration chromatography of chicken liver purine 5'-nucleotidase on Sephacryl S-300. The purified enzyme was applied to a column of Sephacryl S-300 as described previously (12); the enzyme activity (○) and the specific activity (△). Inset, SDS-PAGE of the fractions 35-42. Arrowheads indicate 51 and 100 kDa-bands, respectively.

The localization of purine 5'-nucleotidase in chicken liver was investigated by use of a pre-embedding immunoenzyme technique as described elsewhere (13); the small tissue sections were incubated with the Fab' conjugated with peroxidase.

RESULTS AND DISCUSSION

By gel filtration on a column of Sephacryl S-300, a final preparation of chicken liver purine 5'-nucleotidase was eluted as a single peak with constant specific activity across the peak (Fig. 1). Upon SDS-PAGE, each fraction from the column showed a major peptide chain and also a minor one with molecular masses of 51 and 100 kDa, respectively (Fig. 1, inset); the staining intensities of both bands appeared parallel to the enzyme activity fraction by fraction.

A polyclonal antibody against the final preparation of the enzyme was raised in a rabbit, and the Fab' was conjugated with peroxidase (12, 13). By immunoblot analysis using the Fab'-peroxidase after SDS-PAGE, the two peptide bands were detected in the purified preparation of the enzyme (lane 1 in Fig. 2A). The minor 100 kDa peptide had originally been assumed to be a dimer of the 51 kDa subunit (7). From a different viewpoint, however, the

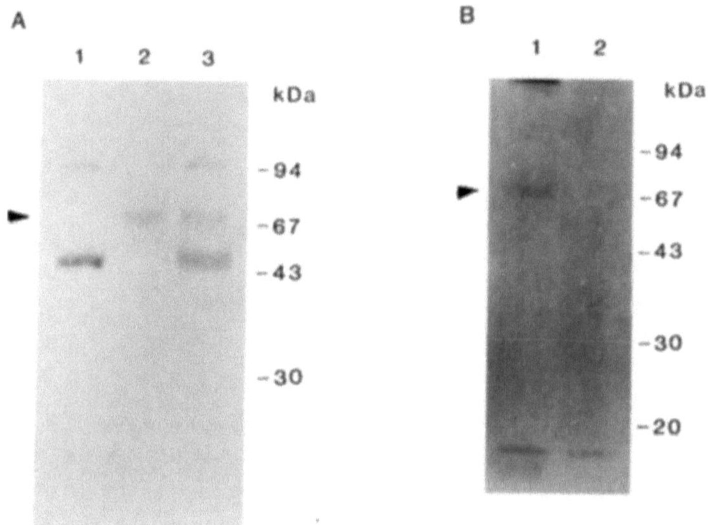

Fig. 2. Immunoblot analysis of chicken liver purine 5'-nucleotidase (A), and immunoprecipitation of cell-free translation products of the polysomes (B). A, the purified enzyme (lane 1), the SDS-extracted proteins (lane 2), and the purified enzyme plus the SDS-extracted proteins (lane 3) were immunostained after SDS-PAGE (12). B, cell-free translation products were immunoprecipitated with no addition (lane 1) or with excess unlabeled enzyme (lane 2). An arrowhead each indicates a 70 kDa band.

100 kDa peptide might be assumed to be a protomeric subunit producing the 51 kDa fragment by proteolysis during homogenization. In order to test this latter possibility under the conditions that the effects of tissue proteinases were minimized, freshly isolated chicken liver was solubilized directly in hot SDS sample buffer (15), and thus SDS-extracted proteins were analyzed; this resulted in the detection of a single 70 kDa band (lane 2 in Fig. 2A).

In order to further elucidate the molecular mass of the protomeric subunit, chicken liver polysomes were prepared, and their in vitro translation products immunoprecipitated. A radioactive 70 kDa band was observed fluorographically (lane 1 in Fig. 2B), and proved to be immunologically specific because of its disappearance when immunoprecipitated with excess unlabeled purine 5'-nucleotidase (lane 2 in Fig. 2B). These results of this study indicated that the molecular mass of the protomeric subunit of chicken liver purine 5'-nucleotidase appeared to be 70 kDa.

Proteolytic processing occurs in posttranslational maturation of proteins, as required for their transport to membrane organelles, secretion, or control of activity (19), though it is generally accepted that proteins in the cytosol are almost always devoid of this processing. In this context, immunocytochemical study of chicken liver purine 5'-nucleotidase was conducted (13), revealing that the antigen was located exclusively in the cytoplasmic matrix of liver sinusoidal and parenchymal cells (Fig. 3); this suggests that physiological processing might not take place with purine 5'-nucleotidase, and that the 51 kDa peptide in the purified enzyme preparation might be a proteolytic artifact. And purine 5'-nucleotidase is the first to be immunocytochemically located in the cytosol among the known cytosolic nucleotidases (see below).

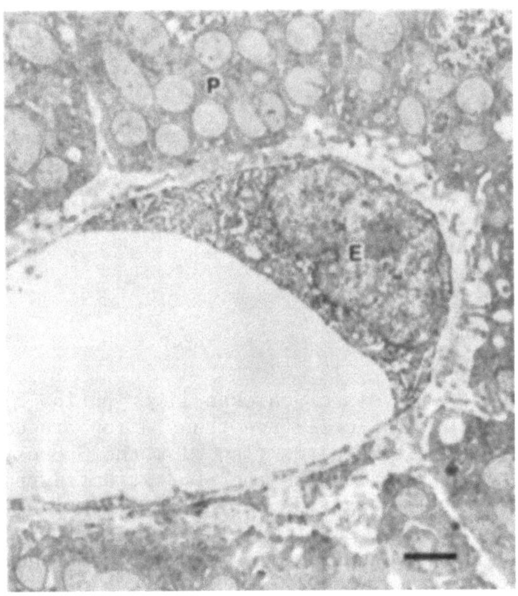

Fig. 3. Electron micrography of chicken liver stained for purine 5'-nucleotidase by immunoenzyme technique (13). The cytoplasmic matrix of an endothelial cell (E) is heavily stained but that of parenchymal cells (P) only weakly stained. x 8,700. Bar = 1 μm.

The dephosphorylation activity of nucleoside 5'-monophosphates has been found in different subcellular fractions including the plasma membrane (20), the cytoplasm (1-3), and the lysosome (21, 22). Plasma membrane ecto-5'-nucleotidase hydrolyzes extracellular purine and pyrimidine 5'-ribo- and deoxyribonucleotides with micromolar Km values, while several distinct cytosolic enzymes degrade intracellular nucleotides with millimolar Km values (6, 23); purine 5'-nucleotidase preferentially hydrolyzes purine 5'-ribo- and deoxyribonucleotides, pyrimidine 5'-nucleotidase specifically cleaves pyrimidine 5'-ribo- and deoxyribonucleotides (2), and deoxyribonucleotidase prefers purine and pyrimidine deoxyribonucleotides (3). Comparative analyses of active center structures of these membrane and cytosolic nucleotidases would be of interest in future. Intracellular nucleotides might also be degraded by Type A acid phosphatase (24), and 5'-nucleotidases which prefer AMP as substrate (6, 25). The latters seem to necessiate further elucidation of molecular properties.

Some investigators (26-29) reported cytoplasmic presence of another 5'-nucleotidase whose molecular and enzymological properties resemble those of ecto-5'-nucleotidase. The provenance of the enzyme is uncertain, but it is possible that it derives from ecto-5'-nucleotidase attached to the plasma membrane by a phospholipid tail (30) as a result of phospholipase activation (31).

ACKNOWLEDGMENTS

This work was supported in part by a Grant-in-aid for Atomic Research at National Research Institutes from Science and Technology Agency of Japan, and a Research Grant for the Intractable Diseases from the Ministry of Health and Welfare of Japan.

REFERENCES

1. Tsushima, K. (1986) Adv. Enzyme Regul. 25, 181-200
2. Paglia, D. E., and Valentine, W. N. (1975) J. Biol. Chem. 250, 7973-7979
3. Fritzson, P. (1978) Adv. Enzyme Regul. 16, 43-61
4. Van den Berghe, G., Van Pottelsberghe, C., and Hers, H. -G. (1977) Biochem. J. 162, 611-616
5. Pinto, R. M., Canales, J., Sillero, M. A. G., and Sillero, A. (1986) Biochem. Biophys. Res. Commun. 138, 261-267
6. Bontemps, F., Van den Berghe, G., and Hers, H. -G. (1988) Biochem. J. 250, 687-696
7. Naito, Y., and Tsushima, K. (1976) Biochim. Biophys. Acta 438, 159-168
8. Itoh, R. (1981) Biochim. Biophys. Acta 657, 402-410
9. Itoh, R., and Oka, J. (1985) Comp. Biochem. Physiol. 81B, 159-163
10. Itoh, R., Oka, J., and Ozasa, H. (1986) Biochem. J. 235, 847-851
11. Pinto, R. M., Canales, J., Faraldo, A., Sillero, A., and Sillero, M. A. G. (1987) Comp. Biochem. Physiol. 86B, 49-53
12. Oka, J., Ozasa, H., and Itoh, R. (1988) Biochim. Biophys. Acta 953, 114-118
13. Yokota, S., Oka, J., Ozasa, H., and Itoh, R. (1988) J. Histochem. Cytochem. 36, in press
14. Ishikawa, E., Yoshitake, S., Imagawa, M., and Sumiyoshi, A. (1983) Ann. N. Y. Acad. Sci. 420, 74-89
15. Laemmli, U. K. (1970) Nature 227, 680-685
16. Towbin, H., Staehelin, T., and Gordon, J. (1979) Proc. Natl. Acad. Sci. U. S. A. 76, 4350-4354
17. MacGillivray, R. T. A., Chung, D. W., and Davie, E. W. (1979) Eur. J. Biochem. 98, 477-485
18. Mori, M., Miura, S., Tatibana, M., and Cohen, P. P. (1981) J. Biol. Chem. 256, 4127-4132
19. Holzer, H., and Heinrich, P. C. (1980) Annu. Rev. Biochem. 49, 63-91
20. DePierre, J. W., and Karnovsky, M. L. (1974) J. Biol. Chem. 249, 7111-7120
21. Arsenis, C., and Touster, O. (1968) J. Biol. Chem. 243, 5702-5708
22. Wada, I., Eto, S., Himeno, M., and Kato, K. (1987) J. Biochem. (Tokyo) 101, 1077-1085

23. Iizasa, T., Takeuchi, F., Honda, Z., Nishida, Y., Kamatani, N., and Miyamoto, T. (1986) Biochim. Biophys. Acta 882, 228-233
24. Chen, C. -H., and Chen, S. C. (1988) Arch. Biochem. Biophys. 262, 427-438
25. Truong, V. L., Collinson, A. R., and Lowenstein, J. M. (1988) Biochem. J. 253, 117-121
26. Madrid-Marina, V., and Fox, I. H. (1986) J. Biol. Chem. 261, 444-452
27. Fritzson, P., Haugen, T. B., and Tjernshaugen, H. (1986) Biochem. J. 239, 185-190
28. Zekri, M., Harb, J., Bernard, S., and Meflah, K. (1988) Eur. J. Biochem. 172, 93-99
29. Le Hir, M., and Dubach, U. C. (1988) Am. J. Physiol. 254, F192-F195
30. Low, M. G. (1987) Biochem. J. 244, 1-13
31. Thompson, L. F., Ruedi, J. M., and Low, M. G. (1987) Biochem. Biophys. Res. Commun. 145, 118-125

THE APPLICATION OF AFFINITY CHROMATOGRAPHY FOR THE SEPARATION OF "HIGH Km" AND "LOW Km" 5'-NUCLEOTIDASE AND OTHER AMP METABOLIZING ENZYMES

Jozef Spychala and Irving H.Fox

From the Human Purine Center
Internal Medicine and Biological Chemistry
Clinical Research Center, University of Michigan
Ann Arbor, Michigan 49109-0108

ABSTRACT

AMP-sepharose 4B has been widely used as a general ligand affinity chromatography for purification of AMP deaminase, 5'-nucleotidase, adenosine kinase and other adenine nucleotide metabolizing enzymes. Since these enzymes generally differ in their kinetic properties related to the values of Km for AMP and analogous compounds, it was assumed that there may be a specific elution pattern of some of the enzymes which would enable sequential elution from the column during a single run. Using 0.5 M NaCl, 10 mM ATP and 5 mM adenosine as eluting agents, it was possible to separate on AMP-sepharose column AMP deaminase "high Km" and "low Km" 5'-nucleotidase and adenosine kinase. Adenylate kinase, adenosine deaminase and nonspecific phosphatase did not bind to the column. Using human placental extract, AMP deaminase, "high Km" and "low Km" 5'-nucleotidase and adenosine kinase were purified 2.8, 2.9, 105 and 1240 fold, respectively. AMP deaminase and "high Km" 5'-nucleotidase were further separated using phosphocellulose column chromatography and the final purification was 227 and 143 fold, respectively. The specific activities of purified enzyme preparations were 9.1, 1.0, 0.4 and 0.5 µmols/min/mg protein of AMP deaminase, "high Km" 5'-nucleotidase and adenosine kinase, respectively. This approach provides a rapid method for initial purification of these enzymes from crude soluble extracts.

INTRODUCTION

AMP-sepharose and AMP-agarose has been used as a general ligand affinity chromatography for purification of AMP deaminase [1], 5'-nucleotidase [2,3], adenosine kinase [4-6] and other adenine nucleotide metabolizing enzymes [7,8]. These enzymes differ in their kinetic properties related to Km values for AMP and related compounds. Preliminary studies have suggested that enzymes which have lower affinity for AMP may be eluted from the affinity column with only high salt concentrations. On the other hand, enzymes which have high affinity to AMP such as adenosine kinase [4] are to be eluted using specific affinity ligands.

The chromatographic properties of AMP deaminase, adenylate deaminase, adenosine deaminase, 5'-nucleotidases and adenosine kinase on AMP-sepharose affinity column indicate that there may be selective eluting agents which would sequentially separate these enzymes from a single extract in one column. This is especially useful for the separation of soluble specific 5'-nucleotidases. It also may be useful for quantitating of various soluble 5'-nucleotidases. The substantial error in measuring activities directly in crude tissue extracts is related to the existence of nonspecific phosphatases and other interfering activities.

In this paper we describe partial purification and separation of AMP deaminase, adenosine deaminase, high Km 5'-nucleotidase, low Km 5'-nucleotidase, and adenosine kinases using a single run on AMP Sepharose 4B.

MATERIALS AND METHODS

Chemicals

AMP-sepharose (8-(6-aminohexyl)-amino-5'-AMP-sepharose 4B) was purchased either from Sigma Chemical Co., St. Louis, MO or from Pharmacia Fine Chemicals, Piscataway, NJ. (8-^{14}C)AMP (50 mCi/mmol), (8-^{14}C)IMP (53 mCi/mmol) and (U-^{14}C)adenosine were purchased from Amersham Corporation, Chicago, Il. Nucleoside monophosphates, adenosine, inosine, imidazole, magnesium chloride, dithiothreitol, ethyleneglycol-bis-(-aminoethyl ether) N,N,N',N'-tetraacetic acid (EGTA), phenylmethylsulfonylfluoride were from Sigma Chemical Co. St. Louis, MO. Phosphocellulose P-11 and 3 MM chromatography paper were from Whatman, Clifton, NJ. Bradford protein reagent was purchased from Bio-Rad, Rockville Center, NY.

Preparation of Human Placental Extract

Amnion and chorion were removed from fresh human placenta which was than washed with 0.9 % NaCl at 4°C. The tissue (20 g) was chopped and homogenized with 4 volumes of the homogenization buffer containing 50 mM Tris-HCl pH 7.5, 250 mM sucrose, 1 mM EDTA, 0.1 mM EGTA, 0.1 mM phenanthroline and 0.1 mM phenylmethylsulphonylfluoride. The homogenate (100 ml) was filtered twice through a single layer of gauze and was spun at 2000 x g for 10 min and then again at 49 000 x g for 2 hours. The supernatant was dialyzed overnight against 6 volumes of 20 mM imidazole-HCl pH 70, 20 mM $MgCl_2$, 0.1 mM EGTA and 0.1 mM phenylmethylsulfonylfluoride (buffer A). After dialysis, the extract was spun again at 49 000 x g for 10 min.

Affinity Chromatography on AMP-Sepharose 4B

The clear supernatant (20 ml, 9 mg protein per ml) was applied to an AMP-sepharose column (10 x 2 cm), prepared as recommended by the manufacturer. The column volume was 20 ml, the flow rate was 60 ml/h and 10 ml fractions were collected (Fig.1). After application of the extract the column was washed with 50 ml buffer A. The following elution sequence was then applied: 1) 50 ml of 0.5 M NaCl in buffer A, 2) 50 ml of

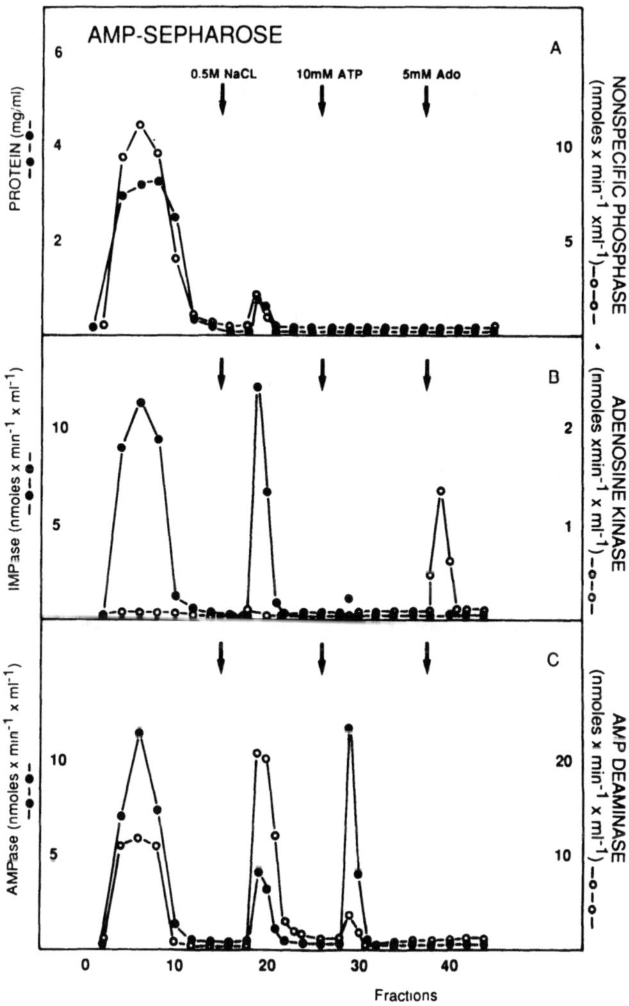

Figure 1. AMP-sepharose chromatography of human placental extract.

buffer A, 3) 50 ml of 10 mM ATP in buffer A, 4) 50 ml of buffer A, and 5) 5 mM adenosine and 5 mM ß-mercaptoethanol in buffer A (Fig. 1).

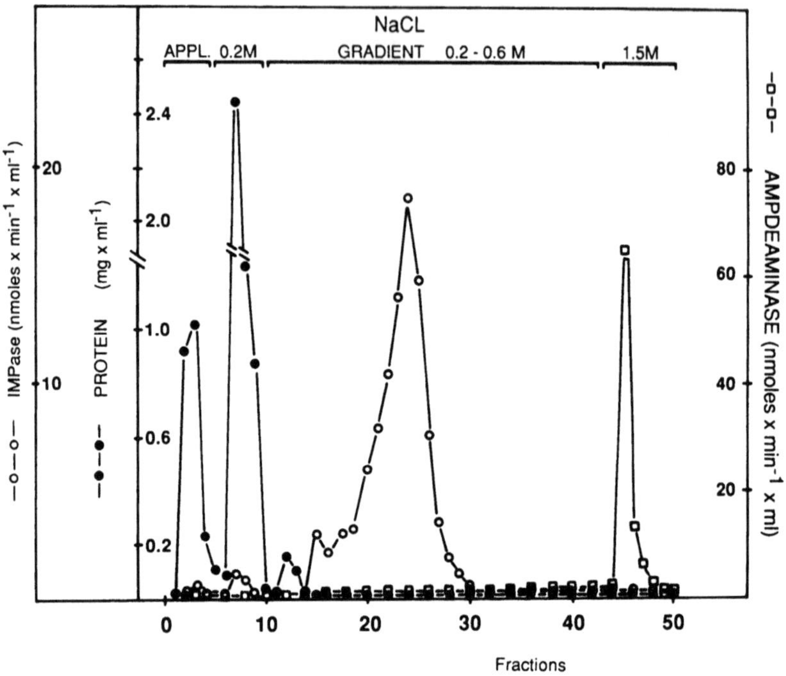

Figure 2. Phosphocellulose column chromatography of AMP deaminase and "high Km" 5'-nucleotidase from human placenta.

Phosphocellulose Chromatography

To further separate "high Km" 5'-nucleotidase and AMP deaminase, activities which coelute on AMP-sepharose column, the active fractions after 0.5 M NaCl from 5 purifications were pooled, precipitated with 50% ammonia sulfate dialyzed against 50 mM imidazole-HCl pH 7.0 and 5 mM mercaptoethanol (buffer B), and applied to a P-11 column (8 x 1.6 cm). The column was washed with 30 ml buffer B at a flow rate of 15 ml/h and 6 ml fractions were collected. The column was then washed with 0.2 M NaCl in buffer B. "High Km" 5'-nucleotidase was eluted with a gradient from 0.2 to 0.6 M NaCl in buffer B (180 ml of total volume). AMP deaminase was eluted with 1.5 M NaCl in buffer B (Fig. 2).

Enzyme Assays

"High Km" 5'-nucleotidase was assayed by a colorimetric method with 5 mM IMP in imidazole-HCl pH 6.5, 20 mM $MgCl_2$, 0.5 M NaCl, 5 mM beta-mercaptoethanol and 1 mg/ml bovine serum albumen in 200 ul total volume. The reaction was stopped by adding 50 ul of 10 % trichloroacetic acid and spun at 3000 rpm. To 200 ul of the supernatant, 1 ml of Chen reagent was added and the absorbance was measured at 820 nm [9]. Fractions containing ATP were dialyzed with added bovine serum albumen (to final concentration of 1 mg/ml) before being assayed for the 5'-nucleotidase and nonspecific phosphatase activities.

"Low Km" 5'-nucleotidase was assayed radiochemically with 100 uM (8-^{14}C)AMP in 60 mM imidazole-HCl pH 7.4, 20 mM $MgCl_2$ and 1 mg/ml bovine serum albumin in 100 ul of total volume. The reaction was stopped by heating at 90°C for 1.5 min. Twenty microliters of the supernatant, clarified by centrifugation, was spotted on 3 MM Whatman chromatography paper and the reaction products were separated using high voltage electrophoresis with 50 mM sodium borate pH 8.8 as previously described [3].

Non-specific phosphatase activity was measured in the incubation mixture containing 60 mM imidazole-HCl pH 6.5, 20 mM $MgCl_2$, 1 mg/ml bovine serum albumin, 0.5 M NaCl, 5 mM mercaptoethanol and 5 mM beta-glycerophosphate. Aliquots were processed as described for "high Km" 5'-nucleotidase.

AMP deaminase was assayed in 50 mM potassium succinate pH 6.5, 150 mM KCl, 1 mM ß-mercaptoethanol, 20 mM AMP and 1 mg/ml bovine serum albumin in a total volume of 200 ul. The reaction was stopped by adding ice-cold phenol and hypochlorite reagents and the absorbance was read at 625 nm as described before [10].

Adenosine kinase was assayed in 60 mM imidazole buffer pH 7.4, 150 mM KCl, 1 mM ATP, 1 mM $MgCl_2$, 5 mM DTT and 1 uM (U-^{14}C)adenosine in a total volume of 100 ul. The assays were stopped by heat inactivation at 90°C for 1.5 min. Separation of adenosine from AMP was on high voltage paper electrophoresis as previously described [3].

Adenylate kinase activity was measured in 60 mM imidazole buffer pH 7.4, 150 mM KCl, 2 mM ATP, 20 mM $MgCl_2$, 1 mM dithiothreitol and 0.5 mM (8^{14}C)AMP in a total volume of 200 ul. The reaction was stopped by the addition of 50 ul 10 % TCA and the reaction products were separated using high voltage electrophoresis in citrate buffer pH 3.4. The nucleotide spots were visualized under ultraviolet light, cut out and counted in toluene based scintilation fluid.

Adenosine deaminase was assayed in 60 mM cacodylate pH 7.4, 150 mM KCl, 10 mM inorganic phosphate and 5 mM adenosine in a total volume of 200 µl by ammonia determination. The reaction was stopped by adding phenol and hypochlorite reagents and the absorbance was read at 625 nm [10].

Protein assay

Protein concentration was measured with Bradford assay using bovine serum albumin as a standard [11].

RESULTS AND DISCUSSION

AMP-sepharose 4B column chromatography was used as a single step for the separation of enzymes related to AMP metabolism in human placental extract including AMP deaminase, "low Km" and "high Km" 5'-nucleotidase and adenosine kinase (Fig. 1). The terms "high Km" and "low Km" 5'-nucleotidase refer to kinetic properties of the separated enzymes: "high Km" 5'-nucleotidases is specific for IMP and has Km in the range of 0.3 to 0.5 mM and "low Km" 5'-nucleotidase is specific for CMP
and AMP and has Km in the range of 15 to 20 μM [12]. These enzymes were separated from nonspecific phosphatase, adenosine deaminase and adenylate kinase which do not bind to the column under the conditions described. "High Km" 5'-nucleotidase and AMP deaminase were eluted using 0.5 M NaCl. "Low Km 5'-nucleotidase was eluted using 10 mM ATP and adenosine kinase using 5 mM adenosine (Fig. 1). With "low Km" 5'-nucleotidase 10 mM ADP was as efficient as ATP (data not shown).

The degree of purification of AMP deaminase and "high Km" 5'-nucleotidase was low probably because of the ion-exchange properties of the column at low ionic strength (Table I). The introduction of 0.1 M NaCl wash before 0.5 M NaCl minimized the nonspecific binding of other proteins and gave better purification of 5'-nucleotidase. However, a significant amount of AMP deaminase activity was lost (data not shown). Significant purification was achieved with "low Km" 5'-nucleotidase and especially adenosine kinase (105 and 1240-fold, respectively).

Further separation of "high Km" 5'-nucleotidase and AMP deaminase was achieved using phosphocellulose column chromatography. "High Km" 5'-nucleotidase was eluted during 0.2 to 0.6 M NaCl gradient and AMP deaminase with 1.5 M NaCl (Fig. 2). The degree of purification of both enzymes was significantly improved by this step (Table I).

Using these chromatographic methods, purified AMP deaminase, "high Km" 5'-nucleotidase,"low Km" 5'-nucleotidase and adenosine kinase preparations had specific activities of 9.1, 1.0, 0.4 and 0.5 umol/min/mg, respectively (Table 1). Therefore, this approach seems to be an efficient separation method for these enzymes in human placenta or other tissues prior to further purification. Furthermore, this separation allows for more accurate measurement of these enzymes in different tissues. In fact, these activities are difficult to quantitate accurately in crude tissue extracts due to interfering activities. This is especially true for nucleotide metabolizing activities as a result of nonspecific phosphatases.

The separated enzymes specifically interact with the AMP immobilized at the N^6 position on Sepharose through a 6 carbon aliphatic side chain. The basis for the binding of the enzymes to the matrix-bound AMP appears to be the affinity of the catalytic site for this ligand. The evidence for this is based on the elution sequence of all four separated enzyme activities. The descending Km values of AMP deaminase, "high Km" 5'-nucleotidase, "low Km" 5'-nucleotidase and adenosine kinase directly corresponds with the elution sequence from the affinity column (Table 2). In the case of adenosine kinase the Ki value for AMP ranges from 0.14 to 0.5 mM and AMP is a competitive inhibitor with respect to adenosine.[13] However, since AMP-sepharose may be subjected to dephosphorylation, it is not certain whether adenosine kinase binds to AMP-sepharose or adenosine-sepharose. The fact that adenosine elutes the enzyme suggests that the enzyme recognizes the adenosine moiety of AMP.

TABLE I. Purification of AMP deaminase, "High Km" and "Low Km" 5'-nucleotidases and adenosine kinase using AMP sepharose and phosphocellulose

Enzyme	Purification step	Total activity nmol/min	Specific activity µmol/min/mg	Yield %	Fold x
AMP deaminase					
	Extract	7200	0.040	100	1.0
	AMP-sepharose	1792	0.112	25	2.8
	Phosphocellulose	1400	9.080	19	227
"High Km" 5'-nucleotidase					
	Extract	1260	0.007	100	1.0
	AMP-sepharose	320	0.02	25	2.9
	Phosphocellulose	220	1.002	16	143
"Low Km" 5'-nucleotidase					
	Extract	720	0.004	100	1.0
	AMP-sepharose	240	0.420	33	105
Adenosine Kinase					
	Extract	72	0.0004	100	1.0
	AMP-sepharose	60	0.4960	83	1240

Thus, the application of AMP-sepharose chromatography for the separation of AMP metabolizing activities in human placenta may serve as an practical example for using the affinity resin with selective eluting agents for the sequential separation of enzymes which differ in their affinities for the immobilized ligand.

TABLE II. Elution sequence of AMP-metabolizing enzymes from AMP-sepharose and their Km values

Enzyme	Elution sequence	Km (mM)	Elution mode
Nonspecific phosphatase	not bound	3.5 (AMP)	-
Adenylate kinase	not bound	1.0 (AMP)	-
AMP deaminase	1	0.6 (AMP)	NaCl
"High Km" 5'-nucleotidase	1	0.3 (IMP)	NaCl
"Low Km" 5'-nucleotidase	2	0.02 (AMP)	ADP or ATP
Adenosine kinase	3	0.0004 (Ado)	Adenosine

ACKNOWLEDGMENTS

The authors wish to thank the nurses in the Women's Hospital for providing fresh placenta and Jeanne Schmaltz and Paige Bigelow for typing the manuscript. This research was supported by United States Public Health Service Grants 5M01RR42, R01AM19674 and R01 GM32837.

REFERENCES

1. W. Makarewicz, and A. Stankiewicz, Purification of AMP-deaminase from human skeletal muscle on 5'-AMP sepharose 4B, Int. J. Biochem. 7:245 (1976).
2. V. Madrid-Marina and I.H. Fox, Human placental cytoplasmic 5'-nucleotidase, J. Biol. Chem. 261:244 (1986).
3. C. Peeters, M. deWolf, G. Van Dessel, A. Lagrou, H. Hilderson, and W. Dierick, Topography, purification and characterization of thyroidal 5'-nucleotidase, Int. J. Biochem. 20:409 (1988).
4. C.H. Andres and I.H. Fox, Purification and properties of human placental adenosine kinase, J. Biol. Chem. 254:11388 (1979).
5. Y. Yamada, H. Goto, and N. Ogasawara, Adenosine kinase from human liver, Biochim. Biophys. Acta. 60:36 (1981).
6. R.L. Miller, D.L. Adamczyk, and W.H. Miller, Adenosine kinase from rabbit liver I. Purification by affinity chromatography and properties, J. Biol. Chem. 254:2339 (1979).
7. L.H. Lazarus, C.-Y. Lee, and B. Vermuth, Application of general ligand affinity chromatography for the mutual separation of deoxyribonuclease and ribonuclease free of protease contamination, Anal. Biochem. 74:138 (1976).
8. H.V. Hershey, and M.W. Taylor, Purification of adenine phosphoribosyltransferase by affinity chromatography, Prep. Biochem. 8:453 (1978).
9. P.S. Chen, T.Y. Toribara, and H. Warner, Microdetermination of phosphorus, Anal. Chem. 28:1756 (1956).

10. J. Spychala, J. Marszalek, and E. Kucharczyk, AMP deaminases from rat small intestine, Biochim. Biophys. Acta. 880:123 (1986).
11. M. Bradford, A rapid and sensitive method for the quantitation of microgram quantities of protein utilizing the principle of protein-dye binding, Anal. Biochem. 72:248 (1976).
12. J. Spychala, V. Madrid-Marina, and I.H. Fox, Evidence for "low Km" and "high Km" soluble 5'-nucleotidases in human tissues and rat liver, Adv. Exp. Med. In Press (1989).
13. T.D. Palella, C.M. Andres, and I.H. Fox, Human placental adenosine kinase, J. Biol. Chem. 255:5264 (1980).

EVIDENCE FOR "LOW Km" and "HIGH Km" SOLUBLE

5'-NUCLEOTIDASES IN HUMAN TISSUES AND RAT LIVER

Jozef Spychala, Vicente Madrid-Marina, and
Irving H. Fox

From the Human Purine Research Center
Internal Medicine and Biological Chemistry
Clinical Research Center, University of Michigan
Ann Arbor, Michigan 48109-0108

ABSTRACT

Two kinetically distinct purine 5'-phosphomono-esterase activities were isolated from soluble fractions of human placenta, cultured human T- and B-lymphoblasts and rat liver using AMP-sepharose chromatography. We have defined these activities as "high Km" and "low Km" 5'-nucleotidase. The relative content of "high Km" and "low Km" activities in the tissues studied ranged from 2 to 264. The optimum pH of "low Km" 5'-nucleotidases ranged from 7.4 to 9.0, Km for AMP from 7 to 15 uM and for IMP from 10 to 26 uM. ATP and ADP were inhibitors of "low Km" enzymes with the apparent Ki values of 55 to 20 uM and 8 to 20 uM for ATP and ADP, respectively. "High Km" 5'-nucleotidases had an optimum pH at 6.5, Km for IMP of 0.3 to 0.5 mM and Km for AMP of 1.0 to 9.4 mM. "High Km" enzymes were activated by ATP with $A_{0.5}$ values, of 1.7 to 2.3 mM at 100 µM IMP. The data indicate that soluble "low Km" and "high Km" 5'-nucleotidases coexist in mammalian cells and fulfill different functions. These observations suggest a complex system for the regulation of AMP and IMP dephosphorylation

INTRODUCTION

At least two different soluble specific nucleoside 5'-monophosphate dephosphorylating activities have been described. One 5'-nucleotidase in rat and chicken liver[1,2], rat heart[3], human lymphoblastoid cells[4,5], and human erythrocytes[6] has maximum activity with IMP and GMP. It is characterized by millimolar Km values for nucleoside 5'-monophosphate substrates, activation by ADP and ATP, inhibition by inorganic phosphate at physiological concentrations, and an optimum activity at pH 6.3 to 6.5. We have used the term "high Km" 5'-nucleotidase to describe this enzyme. Another soluble 5'-nucleotidase is present in human placenta[7,8], sheep, bovine and mouse brain[9-11], and bovine liver[12]. This activity is characterized by maximum activity with CMP, UMP and AMP, inhibition by ATP and ADP, inhibition by inorganic phosphate only at high nonphysiologic concentrations, and optimum activity at pH from 7.4 to 9.0. We have used the term "low Km" 5'-nucleotidase to define this enzyme. We have used the ability to separate soluble 5'-nucleotidases on a AMP-sepharose 4B affinity column[13] to investigate several tissues for the coexistence of "high Km" and "low Km" enzymes.

EXPERIMENTAL PROCEDURES

Preparation of Tissue Extracts

Human placental extracts were obtained as previously described[8]. The frozen MOLT-4 and MGL-8 cells were thawed, suspended (1:10, w/v) in 20 mM imidazole-HCl pH 7.0, 20 mM $MgCl_2$, 100 uM phenylmethylsulfonylfluoride and 1 mM EGTA (buffer A) and were lysed by nitrogen bomb cavitation at 750 lb/in2 for 60 min at 4°C and by Dounce tissue homogenizer with 20 to 30 strokes. The lysate was spun for 10 min at 10,000 x g to remove cellular debris. The supernatant was spun at 49,000 x g for two hours to eliminate the remaining cellular membranes. Rat liver extract was obtained by homogenizing frozen tissue in buffer A (1:10, w/v) using a Sorvall Omnimixer. The homogenate was processed by two spins as described for T- and B-lymphoblasts. All extracts were dialyzed overnight at 4°C against buffer A. After additional spinning at 49,000 x g for 10 min, the clarified supernatants were used immediately for affinity chromatography.

Enzyme Separation and Purification

A. 5'-AMP Sepharose 4 B Affinity Chromatography. The gel was prepared according to manufacturer's directions and equilibrated with buffer A. From 40 to 80 ml of each extract (depending on protein concentration and on initial binding studies) were applied to a 20 ml column of AMP-Sepharose 4B. Fractions of 10 ml were collected. The column was washed with 100 ml of buffer A, 45 ml of 0.5 M sodium chloride in buffer A, 45 ml of buffer A, and then 60 ml of 10 mM ATP in buffer A [13].

B. Further Enzyme Purification. "High Km" 5'-nucleotidases obtained after elution with 0.5 M NaCl in buffer A were concentrated, then diluted 10-fold with 20 mM imidazole-HCl pH 7.0 and 5 mM dithiothreitol (buffer B) and 0.5 mg/ml of bovine serum albumin, and applied to a phosphocellulose P-11 column, 1.6 cm X 6 cm. After washing with 20 ml buffer B, 5'-nucleotidase was eluted using a linear gradient of NaCl (from 0 to 0.6 M) in 180 ml buffer B. Fractions corresponding to the major peak of activity (30 ml) were pooled, concentrated, and used for kinetic studies or for molecular weight determination.

"Low Km" 5'-nucleotidases, obtained from AMP-Sepharose column after elution with 10 mM ATP, were further purified using DE-52 ion exchange chromatography (0 to 0.5 M NaCl gradient elution), gel filtration chromatography on Sephacryl S-300 and Con A-Sepharose chromatography[7,8]. Rat liver was purified only through the gel filtration step.

TABLE 1. CHARACTERISTICS OF SOLUBLE 5'-NUCLEOTIDASES FROM RAT LIVER, HUMAN T-LYMPHOBLASTS (MOLT-4), B-LYMPHOBLASTS (MGL-8) AND PLACENTA

	T-Lymphoblasts		B-Lymphoblasts		Rat Liver		Human placenta	
	"High Km"	"Low Km"	"High Km"	"Low Km"	"High Km"	"Low Km"	"High Km"	"Low Km"
pH Optimum	6.5	9.0	6.5	7.4-9.0	6.5	7.5-8 0	6 5	7 4-9.0
Km for AMP	9.4 mM	7 μM	9.4 mM	12 μM	3.0 mM	15 μM	1.0 mM	14 μM
Km for IMP								
Control	0.4 mM	7 μM	0.4 mM	25 μM	0.5 mM	26 μM	0.3 mM	30 μM
+ATP	0.07 mM	-	0.06 mM	-	-	-	0.09 mM	-
+ATP,Pi*	0.57 mM	-	0.31 mM	-	-	-	0.13	-
Ki for								
ATP	-	55 μM	-	110 μM	-	70 μM	-	100 μM
ADP	-	8 μM	-	20 μM	-	20 μM	-	15 μM
Ka for ATP**	1.7 mM	-	0.9 mM	-	1.1 mM	-	1.4 mM	-

Kinetic constants were estimated with substrate concentrations from 2.5 to 5000 uM IMP and AMP(using radiochemical assay [8, 13]) and 0.05 to 30.0 mM of AMP (using colorimetric assay [8, 13]). Incubation mixture contained 60 mM imidazole pH 7.4, 150 mM KCl, 1 mM $MgCl_2$ in excess of ATP and ADP and 1 mg/ml of BSA. * Measured with 3 mM ATP and 4 mM Pi, and ** with 100 uM IMP.

RESULTS AND DISCUSSION

We used AMP-sepharose 4B affinity chromatography to assess whether the two distinct 5'-nucleotidase activities coexist in human placenta B and T-lymphoblasts and rat liver. The data show that the "low Km" and "high Km" 5'-nucleotidases were found in all four extracts. The example of AMP-sepharose 4B affinity chromatography, which separates these two activities in human placenta, is shown in another chapter of this book [13].

The separation of two specific 5'-nucleotidases and nonspecific phosphatase, which does not bind to AMP-sepharose column, allows a direct comparison of the content of each specific enzyme (assayed at their optimal pH) in the starting extract. There is a large variation in the "high Km" to "low Km" ratio in the four extracts studied ranging from 2.0, 5.5 for human placenta and rat liver to 27 and 266 for B- and T-lymphoblasts respectively. Rat liver and Molt-4 cells have the largest quantity of "high Km" 5'-nucleotidase (56.1 and 52.7 nmoles/min/g tissue respectively, while human placenta and rat liver have the largest quantity of the "low Km" enzyme. Lymphoblasts have the highest relative quantity of "high Km" 5'-nucleotidase. The soluble 5'-nucleotidases eluted with 0.5 M NaCl and 10 mM ATP most probably each represent a single species, since extensive further purification of both enzymes from human placenta reveal no other activities (data not shown).

The two soluble purine 5'-phosphomonoesterase activities have unique properties (Table 1). "High Km" nucleotidase has a pH optimum of 6.5 while the "low Km" nucleotidase a pH plateau from 7.4 to 9.0. The Km values for AMP and IMP are in the micromolar range for "low Km" nucleotidase and in the milimolar range for "high Km" nucleotidase. All the "high Km" enzymes are activated by ATP and inhibited by inorganic phosphate and all the "low Km" enzymes are inhibited by ATP and ADP.

The properties of "high Km" soluble 5'-nucleotidases from the tissues and cells examined in this study resemble those described previously in chicken liver, rat liver, rat heart and human lymphoblasts[1-3,4-6,14]. Our previous work has defined the properties of soluble "low Km" 5'-nucleotidase in human placenta[7,8] In the current study rat liver and human T- and B-lymphoblasts have this activity in addition to the previously described "high Km" enzyme. The Km values and the inhibition by ATP and ADP resemble other "low Km" enzymes that we have described. However, it is not yet clear whether there is a structural relationship of the soluble enzyme to the membrane 5'-nucleotidase which faces the external cell suface and can be released by phospholipase[15].

The distinctive kinetic and regulatory properties of both soluble 5'-nucleotidases and the variable relative distribution in different tissues and cells indicate that these two activities may fulfill different functions. The "high Km" enzyme may primarily dephosphorylate IMP[1-3,6]. The data to support this idea include its preference for IMP (Table 1) and its lack of activity for AMP with a physiological mixture of effectors[6]. The "low Km" soluble 5'-nucleotidase may dephosphorylate AMP to release adenosine, a biologically active compound[16,17] which is produced intracellularly[18]. The data to support this proposal include the kinetic properties of this soluble enzyme with a Km value in the range of physiological concentrations for AMP (Table 1). The relatively high levels of "low Km" 5'-nucleotidase in placenta and liver in our study corresponds to the known ability of these tissues to produce a substantial quantity of adenosine[18-20]. In addition, B-lymphoblasts have twice the quantity of "low Km" 5'-nucleotidase and dephosphorylate AMP and dAMP at a higher rate than T-lymphoblasts [21].

ACKNOWLEDGMENTS

The authors wish to thank the nurses in the Women's Hospital for providing fresh placenta, Sandy Ostrowsky for providing cultured cells and Jeanne Schmaltz and Paige Bigelow for typing the manuscript. This research was supported by United States Public Health Service grants 5M01RR42, R01AM19674, and R01GM32837.

REFERENCES

1. R. Itoh, A. Mitsui, and K. Tsushima, 5'-Nucleotidase of chicken liver, Biochim. Biophys. Acta. 146:151 (1967).
2. R. Itoh, Purification and some properties of cytosol 5'-nucleotidase from rat liver, Biochim. Biophys. Acta. 657:401 (1981).
3. R. Itoh, J. Oka, and H. Ozasa, Regulation of net heart cytosol 5'-nucleotidase by adenylate energy charge, Biochem. J. 235:847 (1986).
4. D.A. Carson, J. Kate, and D.B. Wasson, The potential importance of soluble dexoynucleotidase activity in mediating deoxyadenosine toxicity in human lymphoblasts, J. Immunol. 126:348 (1981).
5. D.A. Carson and D.B. Wasson, Characterization of an adenosine 5'-triphosphate and deoxyadenosine 5'-triphosphate-activated nucleotidase from human malignant lymphocytes, Cancer Res., 42:4321 (1982).
6. F. Bontemps, G. Van den Berghe, and H.-G. Hers, 5'-nucleotidase activities in human erythrocytes, Biochim. J. 250:687 (1988).
7. V. Madrid-Marina and I.H. Fox, Human placental cytoplasmic 5'-nucleotidase, J. Biol. Chem. 261:244 (1986).
8. J. Kaminska, V. Madrid-Marina, and I.H. Fox, Purification and properties of human placental cytoplasmic 5'-nucleotidase, J. Biol. Chem. 261:249 (1986).
9. P.L. Ipata, Sheep brain 5'-nucleotidase: some enzymic properties and allosteric inhibition by nucleoside triphosphates, Biochem. J. 7:507 (1968).
10. J. Mallol and J. Bozal, Modification of 5'-nucleotidase activity by divalent cations and nucleotides, J. Neurochem. 40:1205 (1983).
11. J.M. Montero and J.B. Fes, Purification and characterization of bovine brain 5'-nucleotidase, J. Neurochem. 39:982 (1982).
12. R. Itoh and K. Tsushima, 5'-Nucleotidase of bovine liver, Int. J. Biochem. 2:652, 1971.

13. J. Spychala and I.H. Fox, Affinity chromatography on AMP-sepharose 4B for the separation of "high Km" and "low Km" 5'-nucleotidases and other AMP metabolizing enzymes, Adv. Exp. Med. In Press (1989).
14. G. Van den Berghe, C. Van Pottelsberghe, and H.G. Hers, A kinetic study of the soluble 5'-nucleotidase of rat liver, Biochim. J. 162:611 (1977).
15. M.G. Low, Biochemistry of the glycosyl-phosphatidylinositol membrane protein anchors, Biochem J. 244:1 (1987).
16. J.R.S. Arch and E.A. Newsholm, The control of the metabolism and the hormonal role of adenosine, Essays Biochem., 14:82 (1978).
17. I.H. Fox and W.N. Kelley, The role of adenosine and 2'-deoxyadenosine in mammalian cells, Ann. Rev. Biochem. 47:655 (1978).
18. A.C. Newby, Y. Worku, and P. Meghji, Critical evaluation of the role of ecto and cytosolic 5'-nucleotidase in adenosine formation, in: "Topics and Perspectives in Adenosine Research," E. Gerlach and B.F. Becker, eds., Springer-Verlag, Berlin (1987).
19. F. Bontemps, G. Van den Berge, and H.G. Hers, Evidence for substrate cycle between ATP and adenosine in isolated hepatocytes, Proc. Natl. Acad. Sci., USA 80:2829 (1983).
20. M.K. Sim and M.H. Maguire, Presence of adenosine in the human term placenta: determination of adenosine content and pathways of adenosine metabolism, Circulation Res. 31:779 (1972).
21. N.L. Edwards, D. Recker, J. Manfredi, R. Rembecki, and I.H. Fox, Regulation of purine metabolism by plasma membrane and cytoplasmic 5'-nucleotidases, Am. J. Physiol. 243:C270 (1982).

LOW 5´NUCLEOTIDASE ACTIVITY IN MONONUCLEAR CELLS OF PATIENTS WITH DEFECT T-CELL FUNCTION

Lisa Dalh Christensen*, Viggo Faber*,
Johannes Mejer** and Per Nygaard***

Department of Infectious Diseases*, Rigshospitalet
Blood Serology Department**, Bispebjerg Hospital
and Institute of Biological Chemistry B***
Copenhagen, Denmark

INTRODUCTION

Purine enzymes are of special importance for the functioning of T- and B-cells. Inheritable deficiencies of adenosine deaminase (ADA) and purine nucleoside phosphorylase (PNP) are known to cause severe immunodeficiency syndromes. Decreased ecto 5´nucleotidase (5´N) activity has been observed in peripheral blood lymphocytes from patients with congenital X-linked and common variable hypogammaglobulinaemia[1-7], and acquired immune deficiency syndrome[8,9]. Abnormal levels of ADA, PNP and 5´N are observed in different types of leukemic cells[3,10,11] in mononuclear cells from patients with infectious mononucleosis[12] and with typhoid fever[13]. Immature cells exhibit high ADA and low 5´N activity. The level of 5´N is somewhat higher in B-cells than in T-cells[1,3,4,6]. But the level is not directly correlated the proportion of B-cells[3,4]. In this study the coherence between tests of immune function and purine enzymes in immuno compromised patients has been evaluated.

MATERIALS AND METHODS

Patients

A study of 24 patients admitted to the Department of Infectious Diseases and suspected for immune deficiency has been carried out. The group included all patients within a 40 months period in which both 5´N and LT had been determined. Their main diagnoses are listed in Table 1. Classification of the primary immunodeficency diseases were performed according to Rosen et al.[14].

Lymphocyte Transformation and Lymphocyte Subpopulations

The lymphocyte transformation tests (LT) and lymphocyte subpopulations (LS) determinations were performed at the

Tissue Typing Laboratory, Rigshospitalet[15]. The proliferative response of mononuclear cells to phytohemaglutinin, concanavalin A, pokeweed mitogen and different antigens: Candida albicans, Staphylococcus aureus, Escherichia coli and pooled allogenic cells was measured by radiolabeled thymidine incorporation. The response to mitogens was registered as decreased when the response to 2 or more mitogens was below half the mean value of 2 normal persons investigated on the same day. The response to antigens was registered as low when the response was regarded severe and generally depressed. The LS were evaluated in mononuclear cells by immunofluorescence using antibodies against CD5 and human immunoglobulins.

Purine Enzymes

Enzyme activities of mononuclear cells isolated from heparinized peripheral blood, were determined in cell-free extracts as described before[11,12,16]. Twenty-three normal persons served as controls. Purine nucleoside phosphorylase (PNP) activity was measured spectrophotometrically[11]. Adenosine deaminase (ADA) activity and 5´-nucleotidase (5´N) activity were determined by radiochemical methods[12,16]. Enzyme activity (μ/mg) is given as nmoles of products formed per hour per mg protein at 37°C.

RESULTS

The diagnoses of the patients, together with laboratory findings are given in Table 1. The patients have been divided into two major groups according to their diagnoses, one with and one without a known immune disease. Fifteen of the patients had low 5´N activity and among them were all the patients with decreased LT (Table 2). Low 5´N was not correlated to the response to specific mitogens or antigens, to LS or to plasma immunoglobulin levels.

Three patients had reduced ADA activity. One of them, a one month old boy, had very low ADA and 5´N activity and suffered from a severe combined immunodeficiency syndrome. When admitted to the department he was seriously ill with sepsis and he died shortly after. The level of ADA in granulocytes was below 10% of the normal level. The patient was the first affected child in the second Danish family with ADA deficiency. Nine patients had ADA activities above the normal upper range and were distributed equally in the two patient groups. In six of the patients from the group without known immune deficiency low PNP levels were found. Neither the level of ADA nor that of PNP appear to be correlated with any of the markers used for the determination of the immune function.

DISCUSSION

This study was primarily designed to enable a comparison between purine enzyme levels in peripheral blood mononuclear cells and some frequently used methods for the determination of the immune function in patients with/or suspected for immune deficiency. In the 24 patients, representing 19 different diagnoses, a good correlation between low 5´N activity and decreased LT was observed (Table 2). In several

Table 1. The Main Diagnosis and some Laboratory Data of the Patients Included in this Study

Sex	Age	Main Diagnosis	ADA	PNP	5'N	LT	LS	Ig
			Laboratory Findings					
				u/mg				
I. Patients with Immune Diseases:								
M	15	X-linked agammaglobulinaemia	644	10613	15	+	-	-
F	10	Autosomal recessive hypogamma-	611	3692	51	+	-	-
F	16	-do-	684	3489	34	-	-	-
F	31	Common variable immunodeficiency with predominant B-cell defect	1467	6309	31	-	-	-
F	77	-do-	1027	6163	26	-	-	-
F	61	-do-	2171	2701	15	-	-	-
M	62	Hypogammaglobulinaemia without	642	6047	13	+	+	-
M	25	-do-	1096	5440	156	+	+	-
M	8	-do-	2052	5098	160	+	-	-
M	16	IgA deficiency	653	5384	27	+	+	-
M	14	Severe combined immunodeficiency	750	3466	15	-	-	-
M	0.1	Severe combined immunodeficiency with ADA deficiency	36	5216	8	-	-	-
M	10	Ataxia telangiectasia, IgA deficiency	767	6663	8	-	+	-
F	8	DiGeorge syndrome	1526	3275	8	-	-	-
M	29	Alimentary intolerance	512	6540	131	+	+	+
II. Patients without Known Immune Diseases:								
M	0.3	Pneumonia	1925	2313	21	-	+	+
F	46	Recurrent cutaneous abcesses	279	1592	48	+	+	+
M	6	Susp.for lymphoproliferative disease or chronic Epstein-Barr virus infection	506	2496	159	+	+	-
F	17	Myelitis transversalis antae, Meningitis purolentae antae	268	1416	81	+	+	+
M	42	Recurrent fever (reason unknown)	1253	7244	6	+	+	+
M	1.3	Ectodermal dysplasia	2038	4444	213	+	+	+
M	6	Multiple congenital ectodermal lesions	907	3367	150	+	+	+
M	2	Aplastic anaemia, sepsis	456	1536	26	-	-	-
M	40	AIDS obs. pro. not found	1797	<500	30	+	+	+

Laboratory findings: (-) = abnormal,values outside the normal range limits; (+) = normal. Ig level is marked as abnormal, if one or more of the subclasses are outside the normal range limits. ADA (normal range 337-1040 u/mg, median 552 u/mg); PNP (normal range 2592-8940 u/mg, median 5198 u/mg); 5'N (see Table 2).

investigations decreased 5´N activity has been observed in one or a few immune deficiency diseases [1,2,4,7]. But, attempts to correlate low 5´N levels with diminished T-lymphocyte function were not made. It is not known whether a low 5´N level reflects a population of cells with low 5´N activity or a general decrease in the cellular level. Variations in 5´N activity caused by alterations in the composition of T- and B-lymphocytes can not account for the differences observed (Table 1). In conclusion: We have found a good correlation between low 5´N activity and defect T-cell function and we have identified a new family in Denmark with severe combined immunodeficiency syndrome and ADA deficiency.

Table 2. Correlation between Decreased Lymphocyte Transformation (LT) and low 5´Nucleotidase (5´N) Activity

	5´N Activity	
LT Response	Normal	Low
Normal	9 patients	5 patients
Decreased	0 patients	10 patients

Normal 5´N (>35 μ/mg), low 5´N (<35 μ/mg). For normal persons, range 35-346 μ/mg, median 103 μ/mg.

ACKNOWLEDGMENTS

The expert technical assistance of I. Carlsen and I. Skibshøj is appreciated. We thank Dr. C. Koch, Department of Pediatrics TG, Rigshospitalet, for allowing us to include patients, under his care, in this study. The work was supported by the Danish Medical Research Council (Nos. 12-2421, 512-20630).

REFERENCES

1. A. Cohen, A. Mansour, H. M. Dosch, and E.W. Gelfand: Association of lymphocyte purine enzyme deficiency (5´- nucleotidase) with combined immunodeficiency. Clin. Immunol. Immunopathol. 15:245 (1980).
2. N. L. Edwards, D. B. Magilavy, J. T. Cassidy, and I. H. Fox: Lymphocyte ecto-5´-nucleotidase deficiency in agammaglobulinemia. Science 201:628 (1978).
3. E. El-Mohandes and F. G. S. Hayhoe: 5´-Nucleotidase in lymphocytes from various clinical disorders. Leuk. Res. 7:57 (1983).

4. L. A. Freeman and H. G. Herrod: Lack of correlation between two markers of lymphocyte differentiation: 5′nucleotidase activity and T-lymphocyte colony formation. Blut 49:389 (1984).
5. L. F. Thompson, G. R. Boss, H. L. Spiegelberg, J. V. Jansen, R. D. O′Conner, T. A. Waldmann, R. N. Hamburger, and J. E. Seegmiller: Ecto-5′Nucleotidase activity in T and B lymphocytes from normal subjects and patients with cogenital X-linked agammaglobulinemia. J. Immunol. 123:2475 (1979).
6. A. D. B. Webster, M. North, J. Allsop, G.L. Asherson, and R. W. E. Watts: Purine metabolism in lymphocytes from patients with primary hypogammaglobulinemia. Clin. Exp. Immunol. 31:456 (1978).
7. P. A. Østergaard, A. Deding, J. Eriksen, and J. Mejer: Common variable immunodeficiency and purine nucleotidase and nucleoside phosphorylase deficiency. Acta Path. Microbiol. Scand. Sect. C. 88:299 (1980).
8. J. L. Murray, D. W. Bywaters, J. M., Reuben, P. W. Mansell, and E.M. Hersh: Decreased 5′ nucleotidase activity in suppressor (OKT8) T lymphocytes from homosexuals with AIDS- related complex: Nonassociation with enhanced deoxynucleoside toxicity. Clin. Immunol. Immunopathol. 42:10 (1987).
9. J. F. Salazar-Gonzales, D. J. Moody, J. V. Giorgi, O. M. Maza, R. T. Mitsuyasu, and J. L. Fahey: Reduced ecto-5′-nucleotidase activity and enhanced OKT10 and HLA-DR expression on CD8 (T suppressor/cytotoxic) lymphocytes in the acquired immune deficiency syndrome: Evidence of CD8 cell immaturity. J. Immunol. 135:1778 (1985).
10. T. Eng Gan, L. Hallam, and M.B. van der Weyden: Purine and pyrimidine activities in acute and chronic lymphocytic leukaemia: Relation to cellular proliferation status. Leuk. Res. 6:839 (1982).
11. J. Mejer and P. Nygaard: Adenosine deaminase and purine nucleoside phosphorylase levels in acute myeloblastic leukemia cells. Relationship to diagnosis and clinical course. Leuk. Res. 3:211 (1979).
12. O. Gadeberg, V. Andersen, J. Cohn, J. Hesse, J. Mejer, P. Nygaard, J. Petersen, and V. Faber, V.: Clinical symptoms, immunoglobulin levels, immunoglobulin secreting cell numbers and mononuclear cell purine enzyme activities during the course of infectious mononucleosis. Dan. Med. Bull. 31:507 (1984).
13. B. Galanti, S. Nardiello, M. Russo, and F. Fiorentino: Increased lymphocyte adenosine deaminase in typhoid fever. Scand. J. Infect. Dis. 13:47 (1981).
14. F. S. Rosen, M. D. Cooper, and R. J. P. Wedgwood: The primary immunodeficiencies I. New Eng. J. Med. 311:235 (1984).
15. P. Platz, T. Fog, N. Morling, A. Svejgaard, G. Sønderstrup, L.P. Ryder, M. Thomsen, and C. Jersild: Immunological in vitro parameters in patients with multiple sclerosis and in normal individuals. Acta Path. Microbiol. Scand. Sect. C. 84:501 (1976).
16. J. Mejer, S. Hørbov, and P. Nygaard: Purine metabolizing enzymes in lymphocytes from patients with solid tumors. Acta Med. Scand. 215:5 (1984).

STIMULATION BY GLYCERATE 2,3-BISPHOSPHATE : A COMMON PROPERTY OF CYTOSOLIC PURINE 5'-NUCLEOTIDASE IN VARIOUS TISSUES

F. Bontemps, M.F. Vincent, F. Van den Bergh, G. van Waeg, and G. Van den Berghe

Laboratory of Physiological Chemistry, International Institute of Cellular and Molecular Pathology, Brussels, Belgium

INTRODUCTION

As other tissues [1-6], human erythrocytes [7,8] contain a cytosolic purine 5'-nucleotidase which has a higher affinity for IMP and GMP than for AMP and other 5'-nucleotides. This enzyme is stimulated by ATP and other nucleoside triphosphates and inhibited by Pi. In addition, its activity was found to be potently enhanced by glycerate 2,3-bisphosphate (2,3-BPG) [7,8], the most abundant phosphate ester in human erythrocytes. The present study shows that other cell types and tissues, in man and in rat, also contain a cytosolic purine 5'-nucleotidase which is stimulated by 2,3-BPG.

METHODS

Adult fed male Wistar rats were killed by decapitation. Liver, kidney cortex, spleen, cardiac ventricular muscle, hind leg skeletal muscle and cerebral hemispheres were excised rapidly, rinsed with physiologic saline, blotted on filter paper, and immediately homogenized in 4 vol. of 0.25 M sucrose. Cytosol fractions were prepared by centrifugation at 8,000 *g* for 15 min, followed by high-speed centrifugation at 100,000 *g* for 120 min. The supernatants were made 20 mM in Tris-Cl pH 7.4, 1 mM in EDTA, and 1 mM in dithiothreitol, and dialyzed at 4 °C against 500 vol. of the same buffer; a first dialysis of 2 h duration was followed by a second one, performed overnight. The pellets were washed twice in the same buffer, followed by centrifugation at 10,000 *g* for 5 min. Thereafter they were resuspended for 15 min in the same buffer, containing 0.2 % Triton X-100, and recentrifuged at 10,000 *g* for 5 min. The resulting supernatants were termed the membrane fractions.

Human haemolysates were prepared as described in [8], except that haemolysis was followed by high-speed centrifugation as given above, and that 0.2 M NaCl was added to the dialysis buffer. Human polymorphonuclear neutrophil leucocytes (PMNL) were isolated from citrated venous blood by dextran sedimentation, and mixed peripheral blood lymphocytes (MPBL) by

Ficoll-Hypaque (Pharmacia, Uppsala, Sweden) density gradient centrifugation [9]. Both preparations were sonicated for 3 x 10 sec. Platelet-rich plasma was prepared by the method of Holmsen et al. [10], and frozen and thawed thrice. Human cultured skin fibroblasts were grown in Minimal Essential Medium, harvested by scraping with a rubber policeman, washed 3 times in phosphate-buffered saline, and sonicated for 3 x 10 sec. All these preparations were subjected to high-speed centrifugation and dialysis as described above.

5'-Nucleotidase activities were measured with [8-^{14}C]IMP, [U-^{14}C]GMP or [8-^{14}C]AMP as substrates [8]. Incubations were performed at 37 °C in a medium containing 0.2 mM substrate, 10 mM $MgCl_2$, 50 mM HEPES buffer pH 7.2, and additions as indicated, in a total volume of 100 µl. When ATP was added, or when the dephosphorylation of [8-^{14}C]AMP was measured, the cytosol fractions were preincubated for 15 min with 100 µM coformycin prior to the addition of substrate, in order to inhibit both adenosine deaminase and AMP deaminase [11,12]. At appropriate time intervals, 10 µl of the incubation medium were spotted on polyethyleneimine cellulose t.l.c. plates, on which carrier solutions (50 nmol) of the 5'-nucleotide substrate, its corresponding nucleoside and base, xanthine, uric acid and allantoin had been applied. After development in butan-1-ol/acetic acid/water (2:1:1, by vol.), the enzyme activity was calculated from the sum of the radioactive nucleoside and further catabolites formed. It was verified that coformycin did not influence the activity of the cytosolic 5'-nucleotidase.

Protein was measured by the method of Bradford [13], with bovine gamma globulin as the standard, and haemoglobin by the method of de Duve [14]. ATP and the release of Pi from 2,3-BPG were measured, respectively, by HPLC [15] and colorimetrically [16], in neutralized $HClO_4$ extracts prepared from the incubates as described in [12].

RESULTS

As shown in Fig. 1, the cytosol fractions of all human blood cell types and of cultured fibroblasts, displayed an IMP-dephosphorylating activity which was stimulated by 5 mM 2,3-BPG. Stimulation was 3- to 6-fold in blood cells and approx. 2-fold in fibroblasts. In blood cells the effect of 5 mM ATP was comparable or slightly less pronounced than that of 5 mM 2,3-BPG. In fibroblasts, however, there was nearly no effect of ATP. In order to verify that the cytosol fractions were not contaminated by membrane 5'-nucleotidase, the effect of adenosine 5'-methylene diphosphonate (AOPCP), a powerful inhibitor of this enzyme [17], was also investigated. Unexpectedly, AOPCP, at 2.5 mM concentration, stimulated the dephosphorylation of IMP approx. 1.5- to 3-fold, except in fibroblasts, where it had a slight inhibitory effect.

In cytosol fractions from rat tissues (Fig. 2), stimulation of the dephosphorylation of IMP by 2,3-BPG was most pronounced in heart muscle, where it reached 12-fold, followed by, in order, liver, brain, kidney and spleen. In the latter tissue, as well as in skeletal muscle (results not shown), stimulation was only 1.5- to 2-fold. As in human blood cells, stimulation by ATP was comparable or slightly lower than by 2,3-BPG, with the exception of muscle, in which stimulation by ATP was not recorded. AOPCP also stimulated, by 60 to 300 %, except in kidney and skeletal muscle, where it had no effect.

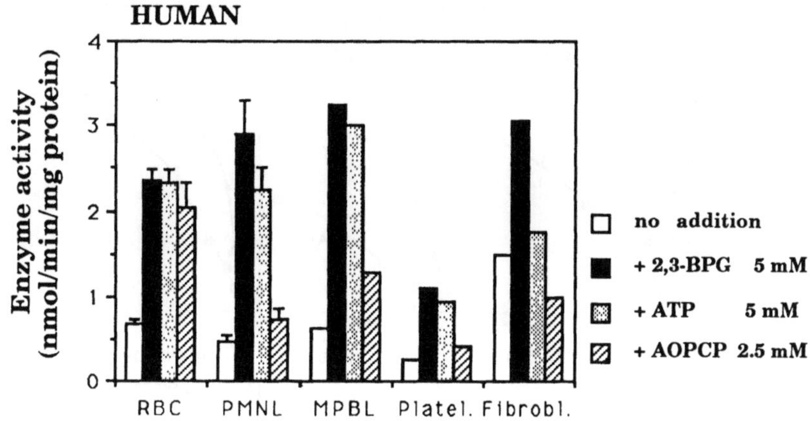

Figure 1. Effect of 2,3-BPG, ATP and AOPCP on the dephosphorylation of IMP in human red blood cells (RBC), polymorphonuclear leucocytes (PMNL), mixed peripheral blood lymphocytes (MPBL), platelets and fibroblasts. Activities in RBC are expressed as nmol/min per 100 mg of Hb. Values shown are either means of two different experiments, or means ± S.E.M. (bars) of three different experiments.

Kinetic studies performed with cytosol fractions of rat liver, heart and brain, and of human PMNL, platelets and MPBL (Fig. 3) showed that the dephosphorylation of IMP was more sensitive to 2,3-BPG than to ATP. Half-maximal stimulation was reached at 0.5 - 1 mM with 2,3-BPG, and at approx. 5-fold higher concentrations with ATP. HPLC analysis and measurement of Pi in the incubates showed that in all tissues and cell types, the degradation of ATP and 2,3-BPG never exceeded 10 % over the duration of the incubations.

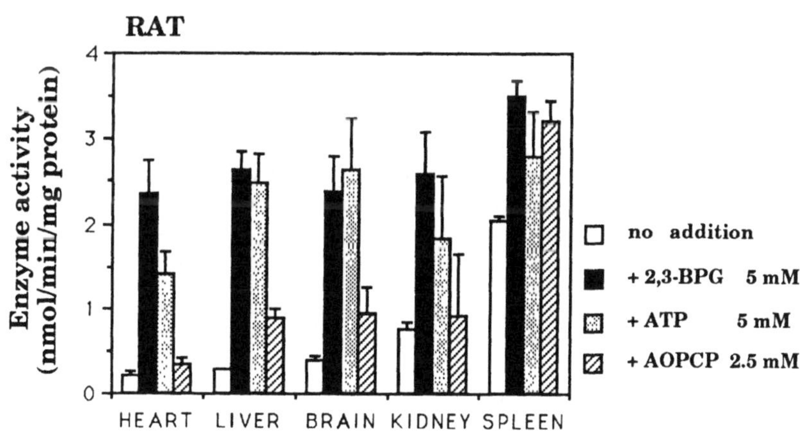

Figure 2. Effect of 2,3-BPG, ATP and AOPCP on the dephosphorylation of IMP in rat tissues. Values shown are means ± S.E.M. of three different experiments.

The effect of 2,3-BPG was also investigated on the dephosphorylation of GMP. As shown in Fig. 4, depicting an experiment performed with a cytosol

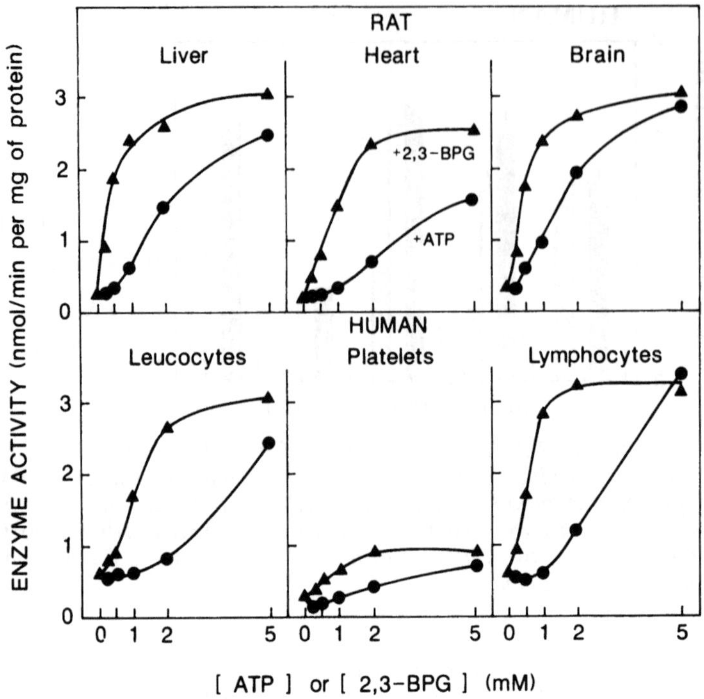

Figure 3. Effect of 2,3-BPG (▲) and of ATP (●) on the dephosphorylation of IMP in cytosol fractions of various rat tissues and human blood cells. Activities were measured at 0.2 mM IMP.

fraction of human platelets, the dephosphorylation of 0.2 mM GMP was stimulated 17-fold by the addition of 5 mM 2,3-BPG. During the first 10 min of incubation, the stimulatory effect of 5 mM ATP was exactly comparable to that of 2,3-BPG; at longer time intervals, however, it nearly completely disappeared. HPLC analysis revealed that the addition of ATP induced conversion of GMP into GTP, accompanied by production of ADP. Similar results were obtained with rat liver cytosol. In human PMNL's, however, no nucleotide interconversions occurred upon addition of ATP, and the dephosphorylation of 0.2 mM GMP was stimulated approx. 6-fold, both by 5 mM ATP and 2,3-BPG.

Table 1. Comparison of the effect of 2,3-BPG on the dephosphorylation of AMP and of IMP in cytosol fractions of rat tissues.

	AMP (0.2 mM)		IMP (0.2 mM)	
	no addition	+2,3-BPG	no addition	+2,3-BPG
heart	0.50 ± 0.01	0.40 ± 0.06	0.41 ± 0.10	3.79 ± 0.37
liver	0.17 ± 0.02	0.30 ± 0.03	0.31 ± 0.01	2.26 ± 0.23
brain	0.31	0.34	0.47	4.16

Rates of dephosphorylation are expressed as nmol/min per mg of protein, and are means of two or three (± S.E.M.) different experiments. 2,3-BPG was added at 5 mM concentration.

In cytosol fractions of rat heart, liver and brain, the dephosphorylation of AMP, in contrast to that of IMP, was not or little stimulated by 2,3-BPG (Table 1). In the presence of ATP, no stimulation of the dephosphorylation of AMP could be recorded, because of an extensive conversion of AMP into ADP.

In membrane fractions prepared from rat liver, heart and brain, the dephosphorylation of IMP was not influenced by the addition of 5 mM 2,3-BPG, and inhibited by 60 to 95 % by 2.5 mM AOPCP. Membrane fractions of human erythrocytes and PMNL's displayed no detectable IMP-dephosphorylating activity.

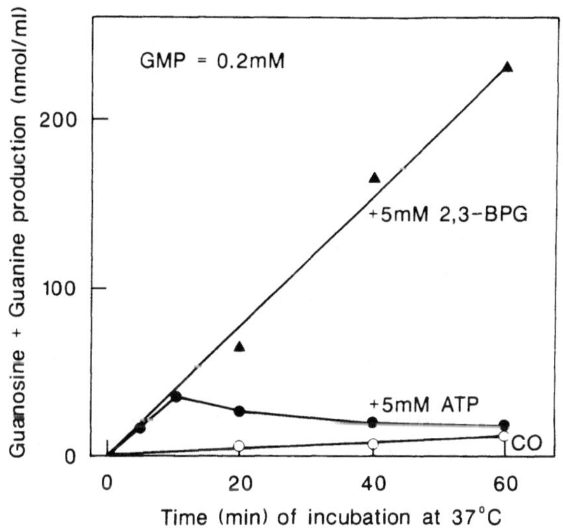

Figure 4. Effect of 2,3-BPG and ATP on the dephosphorylation of GMP by cytosol from human platelets.

DISCUSSION

This study indicates that the stimulatory effect of 2,3-BPG, discovered while investigating a cytosolic purine 5'-nucleotidase with maximal affinity for IMP and GMP, purified from human erythrocytes [7,8], is not restricted to this cell type, but may be a common property of the enzyme in all tissues. Indeed, in the high-speed supernatants of all cell types studied so far, a stimulatory effect of 2,3-BPG on the dephosphorylation of IMP and GMP was found. A small stimulation by 2,3-BPG of the hydrolysis of AMP, for which this cytosolic 5'-nucleotidase has much less affinity, was only recorded in liver, out of three tissues in which it was investigated. That the stimulatory effect of 2,3-BPG was markedly dependent on tissue and substrate, may be explained by the presence, in the cytosol, of different amounts of other 5'-nucleotidases, which are not stimulated by 2,3-BPG. In human erythrocytes for instance, the stimulatory effect of 2,3-BPG on the dephosphorylation of AMP, observed with the purified purine cytosolic 5'-nucleotidase, is much lower in the haemolysate, owing to the presence of a predominant AMP-dephosphorylating activity, carried by another enzyme [8]. It is therefore proposed that the 2,3-BPG-stimulated cytosolic purine 5'-nucleotidase with high affinity for IMP and GMP, should be referred to as the IMP-GMP 5'-nucleotidase. Our results also show that 2,3-BPG does not influence the activity of the membrane 5'-nucleotidase. The absence of membrane 5'-nucleotidase in human RBC and PMNL is in accordance with the observations of others [18].

In most preparations, the effect of 5 mM 2,3-BPG was of similar magnitude as that of 5 mM ATP, the first reported stimulator of the cytosolic IMP-GMP 5'-nucleotidase [1]. In the cell types in which it was investigated, the enzyme was more sensitive to 2,3-BPG than to ATP (Fig. 3), as also observed with the cytosolic purine 5'-nucleotidase of human erythrocytes [7,8]. The greater sensitivity of the enzyme to 2,3-BPG does not reside in its better resistance to degradation, since ATP was not more hydrolyzed than 2,3-BPG in the high-speed supernatants.

We have also found that AOPCP, added to the high-speed supernatants to verify that they were devoid of membrane 5'-nucleotidase, exerts an up to 3-fold, hitherto unreported stimulatory effect on the cytosolic IMP-GMP 5'-nucleotidase. The use of AOPCP to measure cytosolic 5'-nucleotidase activity by substraction from total 5'-nucleotidase activity in unfractionated cell homogenates, should thus be carefully evaluated. The inhibitory effect of AOPCP in high-speed supernatants of human fibroblasts might be due to a contamination of this preparation by membrane fragments, since the ADP analogue is a powerful inhibitor of the membrane 5'-nucleotidase [17].

As discussed before [7,8], the approx. 3 mM concentration of free 2,3-BPG, prevailing in human erythrocytes, is maximally stimulatory for their cytosolic IMP-GMP 5'-nucleotidase, and also renders it nearly insensitive to physiological concentrations of ATP and of its inhibitor, Pi. In other mammalian tissues, the stimulatory effect of 2,3-BPG most likely lacks physiological significance, because they contain concentrations of the bisphosphate which are 2 to 3 orders of magnitude lower [reviewed in 19], than the concentration of 2,3-BPG providing half-maximal stimulation of the enzyme. Nevertheless, the present study suggests that the cytosolic IMP-GMP 5'-nucleotidase of different tissues may have a common origin. Stimulation by 2,3-BPG is also a useful tool to study this enzyme in crude preparations, particularly with GMP or AMP, since the use of ATP as stimulator is excluded by its reaction with these substrates. These reactions result from the presence of adenylate kinase and of other nucleotide kinases.

ACKNOWLEDGEMENTS

We thank Mrs A. Delacauw and T. Timmerman for expert technical assistance. This work was supported by grant n° 3.4539.87 of the Fund for Medical Scientific Research (Belgium). G. Van den Berghe is Research Director of the Belgian National Fund for Scientific Research.

REFERENCES

1. G. Van den Berghe, C. van Pottelsberghe and H.G. Hers, A kinetic study of the soluble 5'-nucleotidase of rat liver, *Biochem. J.* 162 : 611 (1977).
2. R. Itoh, C. Usami, T. Nishino and K. Tsushima, Kinetic properties of cytosol 5'-nucleotidase from chicken liver, *Biochim. Biophys. Acta* 526 : 154 (1978).
3. R. Itoh, Purification and some properties of cytosol 5'-nucleotidase from rat liver, *Biochim. Biophys. Acta* 657 : 402 (1981).
4. D.A. Carson and D.B. Wasson, Characterization of an adenosine 5'-triphosphate- and deoxyadenosine 5'-triphosphate-activated nucleotidase from human malignant lymphocytes, *Cancer Res.* 42 : 4321 (1982).
5. R. Itoh and J. Oka, Evidence for existence of a cytosol 5'-nucleotidase in chicken heart : comparison of some properties of heart and liver enzymes, *Comp. Biochem. Physiol.* 81 B : 159 (1985).
6. R. Itoh, J. Oka and H. Ozasa, Regulation of rat heart cytosol 5'-nucleotidase by adenylate energy charge, *Biochem. J.* 235 : 847 (1986).
7. F. Bontemps, G. Van den Berghe and H.G. Hers, Identification of a purine 5'-nucleotidase in human erythrocytes, *Adv. Exp. Med. Biol.* 195 B : 283 (1986).
8. F. Bontemps, G. Van den Berghe and H.G. Hers, 5'-Nucleotidase activities in human erythrocytes. Identification of a purine 5'-nucleotidase stimulated by ATP and glycerate 2,3-bisphosphate, *Biochem. J.* 250 : 687 (1988).
9. A. Böyum, Isolation of mononuclear cells and granulocytes from human blood, *Scand. J. clin. Lab. Invest.* 21, suppl. 97 : 77 (1968).
10. H. Holmsen, H.J. Day and E. Storm, Adenine nucleotide metabolism of blood platelets. VI. Subcellular localization of nucleotide pools with different functions in the platelet release reaction, *Biochim. Biophys. Acta* 186 : 254 (1969).
11. R.P. Agarwal and R.E. Parks, Potent inhibition of muscle 5'-AMP deaminase by the nucleoside antibiotics coformycin and deoxycoformycin, *Biochem. Pharmacol.* 26 : 663 (1977).
12. G. Van den Berghe, F. Bontemps and H.G. Hers, Purine catabolism in isolated rat hepatocytes. Influence of coformycin, *Biochem. J.* 188 : 913 (1980).
13. M.M. Bradford, A rapid and sensitive method for the quantitation of microgram quantities of protein utilizing the principle of protein-dye binding, *Anal. Biochem.* 72 : 248 (1976).
14. C. de Duve, A spectrophotometric method for the simultaneous determination of myoglobin and hemoglobin in extracts of human muscle, *Acta Chem. Scand.* 2 : 264 (1948).
15. R.A. Hartwick and P.R. Brown, The performance of microparticle chemically-bonded anion-exchange resins in the analysis of nucleotides, *J. Chromatogr.* 112 : 651 (1975).
16. K. Itaya and M.Ui, A new micromethod for the colorimetric determination of inorganic phosphate, *Clin. Chim. Acta* 14 : 361 (1966).
17. R.M. Burger and J.M. Lowenstein, Preparation and properties of 5'-nucleotidase from smooth muscle of small intestine, *J. Biol. Chem.* 245 : 6274 (1970).
18. P.M. Shirley, P. Wang, L.R. DeChatelet and M. Waite, Absence of the membrane marker enzyme 5'-nucleotidase in human polymorphonuclear leukocytes, *Biochem. Med.* 15 : 289 (1976).
19. J. Carreras, R. Bartrons, F. Climent, and R. Cusso, Bisphosphorylated metabolites of glycerate, glucose and fructose : functions, metabolism and molecular pathology, *Clin. Biochem.* 19 : 348 (1986).

FUNCTIONAL CHARACTERIZATION OF ECTO-5'-NUCLEOTIDASE (ECTO-5'-NT) POSITIVE AND NEGATIVE HUMAN LYMPHOCYTES

Linda F. Thompson and Julie M. Ruedi

Department of Immunology, Scripps Clinic and Research Foundation, La Jolla, California

Ecto-5'-nucleotidase (ecto-5'-NT) is located on the external surface of subpopulations of human T and B cells (1-7). This enzyme has been postulated to be a lymphocyte maturation marker because its activity increases during lymphocyte ontogeny (2,8-10). Patients with a variety of immunodeficiency diseases show reduced ecto-5'-NT activity in either their T cells, B cells, or both (1,5,11-16). Especially in the case of B cells, this reduced activity has been proposed to be caused by a block in lymphocyte maturation (10). This may also be the case in some patients with T cell ecto-5'-NT deficiency. However, since ecto-5'-NT is differentially expressed on subsets of T lymphocytes (1,3-5), an imbalance in the proportions of regulatory T cells could also lead secondarily to T cell ecto-5'-NT deficiency.

We recently described the development of goat antibodies specific for ecto-5'-NT which can be used to detect cell surface 5'-NT by indirect immunofluorescence (3,17). Here we report the use of these antibodies to prepare highly-enriched populations of ecto-5'-NT positive and negative T and B cells using the fluorescence-activated cell-sorter. These populations were evaluated in several assays to determine whether ecto-5'-NT expression defines functional subsets of lymphocytes and to gain more insight into the significance of lymphocyte ecto-5'-NT deficiency in immunodeficient patients.

METHODS

Separation of T cells, B cells, and monocytes. Peripheral blood lymphocytes were isolated by dextran sedimentation, carbonyl iron treatment to remove monocytes, and Ficoll-Hypaque density gradient centrifugation (1). T lymphocytes were isolated by one cycle of rosetting with neuraminidase-treated sheep erythrocytes for 15 min at 37°C and 1 h at 4°C followed by Ficoll-Hypaque density gradient centrifugation. Non-T (B-enriched) cells were isolated at the interface of the second Ficoll-Hypaque gradient.

To isolate monocytes, peripheral blood mononuclear cells were first isolated from blood anticoagulated with 0.2 M EDTA by Ficoll-Hypaque density gradient centrifugation. They were washed in phosphate buffered saline + 0.25% bovine serum albumin and the monocytes were then isolated at the interface between 0 and 50% Percoll on a discontinuous gradient. The monocyte-enriched populations were >80% pure as judged by nonspecific esterase staining.

Preparation of antiserum, immunofluorescent staining and use of the FACSTAR. The anti-5'-NT antiserum was prepared by immunizing a goat with highly-purified human placental 5'-NT and absorbed as previously described (17). Lymphocytes were incubated with $F(ab')_2$ anti-5'-NT IgG at 250 µg/ml, followed by affinity purified $F(ab')_2$ rabbit anti-goat IgG, which had been absorbed on a Cohn II-Sepharose column and then biotinylated (17), and fluorescein isothiocyanate (FITC)-avidin at 4°C. Non-T cells were also stained with Leu 12 (anti-CD19, Becton Dickinson) plus phycoerythrin (PE)-anti-mouse κ and two-color sorting was used to separate pure B cells into ecto-5'-NT positive and negative populations.The cells were resuspended in RPMI + 2.5% fetal bovine serum (FBS) + 25 mM NaHepes, pH 7.4 and kept on ice until they were sorted into Fl^+ and Fl^- populations using the FACSTAR fluorescence-activated cell sorter (Becton-Dickinson) as previously described (3). The positive and negative populations were >95% pure when reanalyzed. Sorted ecto-5'-NT^+ and ecto-5'-NT^- T cells were analyzed for the expression of CD4 and CD8 using Leu 3a and Leu 2a, respectively, plus PE-anti-mouse κ (Becton-Dickinson).

Phytohemagglutinin (PHA) and phorbol myristate acetate (PMA) stimulation. One hundred thousand T cells plus 5 x 10^3 monocytes were incubated in the presence of PHA (Sigma, PHA-P) at 10 µg/ml in a total volume of 0.1 ml of HB101 (serum-free, New England Nuclear) medium in flat-bottomed "half-area" microtiter plates. Serum-free medium was used in these experiments because FBS contains 5'-NT activity. All cultures were set up in triplicate and incubated for 3 da at 37°C in an humidified atmosphere of 95% air and 5% CO_2. After 68 h, the cultures were pulsed with 1.0 µCi of ^3H-TdR/well for 4 h and then harvested with a multiple automated sample harvester (MASH). The amount of radioactivity taken up was quantitated by liquid scintillation counting. PMA stimulation experiments were performed in a similar manner with PMA at concentrations from 0.3 to 30 ng/ml.

Mixed lymphocyte reaction (MLR). Fifty thousand T cells plus 5 x 10^3 monocytes were incubated in the presence of irradiated (3000 rads) allogeneic lymphocytes in 0.2 ml of HB101 medium in round-bottom microtiter plates. All cultures were set up in triplicate and incubated for 7 da at 37°C as for the PHA experiments. During the last 8 h, the cultures were pulsed with 1.0 µCi of ^3H-TdR/well and then harvested with a MASH. The amount of radioactivity taken up was quantitated by liquid scintillation counting.

Pokeweed mitogen (PWM) stimulation. One half million B cells, one half million autologous irradiated (3000 rads) T cells, and 1 x 10^4 monocytes were incubated in the presence of PWM (Gibco, 1:50 final dilution) in round-bottomed tubes in a total volume of 1.0 ml of RPMI 1640 + 10% FBS. After 7 da at 37°C in an humidified atmosphere of 95% air and 5% CO_2, the cells were centrifuged and the supernatants collected for quantitation of IgM and IgG secreted. In some experiments, the volume of the cultures was reduced to 0.5 ml, but the cell concentrations remained constant.

Epstein Barr virus (EBV) stimulation. One half million B cells and 10^4 monocytes were incubated in the presence of 20% (v/v) EBV-containing supernatant from B-95-8 cells in round-bottomed tubes in a total volume of 1.0 ml of RPMI 1640 + 10% FBS. After 10 da at 37°C in an humidified atmosphere of 95% air and 5% CO_2, the cells were centrifuged and the supernatants collected for quantitation of IgM and IgG secreted. In some experiments, the volume of the cultures was reduced to 0.5 ml, but the cell concentrations remained constant.

Quantitation of IgM and IgG in culture supernatants. The quantities of IgM and IgG in culture supernatants were measured using a solid-phase enzyme immunoassay. All supernatants were tested at three serial dilutions in duplicate. The standards were the Hyland International Reference Serum for IgM and a Cohn II fraction of human serum for IgG (gift of Dr. Hans Spiegelberg). The range of concentrations in the standard curve was 250-32,000 pg/ml for IgM and 1000-125,000 pg/ml for IgG.

RESULTS

Cell surface phenotype of ecto-5'-NT positive and negative T cells. The expression of CD4 and CD8 on ecto-5'-NT$^+$ and ecto-5'-NT$^-$ T cells isolated in four different experiments is shown in Table 1. Ecto-5'-NT$^+$ T cells contained an average of 30% CD4$^+$ and 62% CD8$^+$ cells while ecto-5'-NT$^-$ T cells contained an average of 71% CD4$^+$ and 19% CD8$^+$ cells.

PHA response. The ability of ecto-5'-NT$^+$ and ecto-5'-NT$^-$ T cells to proliferate in response to PHA is shown in Figure 1. Untreated T cells incorporated 117,000 cpm of ^{3}H-TdR in a 4-h pulse. The purified populations of ecto-5'-NT positive and negative T cells showed an equal ability to incorporate ^{3}H-TdR in response to PHA (56,800 vs 55,300 cpm). Similar results were obtained in a second experiment. In a third experiment, ecto-5'-NT$^-$ T cells incorporated four-times as much ^{3}H-TdR as ecto-5'-NT$^+$ T cells (72,000 cpm vs 18,000 cpm). Thus, after PHA stimulation, ecto-5'-NT$^-$ T cells incorporate as much, or more, ^{3}H-TdR as ecto-5'-NT$^+$ T cells.

Mixed lymphocyte reaction. The ability of ecto-5'-NT$^+$ and ecto-5'-NT$^-$ T cells to proliferate in response to irradiated allogeneic lymphocytes (i.e., in a MLR) is shown in Figure 1. Untreated T cells incorporated 3600 cpm of ^{3}H-TdR in an 8-h pulse. After sorting, ecto-5'-NT$^-$ T cells gave a 3-fold higher ^{3}H-TdR incorporation after MLR than ecto-5'-NT$^+$ T cells (6100 cpm vs 2100 cpm). In two other experiments, ecto-5'-NT$^-$ T cells incorporated twice as much ^{3}H-TdR as ecto-5'-NT$^+$ cells and in a fourth experiment they incorporated 75% as much ^{3}H-TdR as the ecto-5'-NT$^+$ T cells. Therefore, as in the case with PHA stimulation, ecto-5'-NT$^-$ T cells can respond at least as well as ecto-5'-NT$^+$ T cells in a MLR.

Table 1. CELL SURFACE PHENOTYPE OF ECTO-5'-NT POSITIVE AND NEGATIVE T CELLS

	Untreated T cells		Ecto-5'-NT$^+$ T cells		Ecto-5'-NT$^-$ T cells	
	% CD4$^+$	% CD8$^+$	% CD4$^+$	% CD8$^+$	% CD4$^+$	% CD8$^+$
Expt. #1	52	35	16	78	79	13
Expt. #2	56	34	37	57	64	21
Expt. #3	60	27	42	53	64	26
Expt. #4	56	33	23	60	77	16
Mean	56	32	30	62	71	19
± 1 S.D.	± 3	± 4	±12	±11	± 8	± 6

Untreated T cells and sorted ecto-5'-NT positive and negative T cells were analyzed for CD4 and CD8 expression by indirect immunofluorescence.

PMA response. The ability of ecto-5'-NT positive and negative T cells to proliferate in response to PMA was compared in two experiments (Table 2). In the first experiment, the [PMA] varied from 0.3 to 1.0 ng/ml. Ecto-5'-NT$^+$ T cells showed proliferation at 0.6 ng/ml with a stimulation index of 13 and at 1.0 ng/ml with a stimulation index of 29. In contrast, ecto-5'-NT$^-$ T cells failed to proliferate at all three concentrations tested. In the second experiment, the [PMA] varied from 3.3 to 30 ng/ml. Ecto-5'-NT$^+$ T cells showed a similar degree of proliferation at all three concentrations with stimulation indices of approximately 100. For the ecto-5'-NT$^-$ T cells, the stimulation index increased with increasing [PMA] and ranged from 13 at 3.3 ng/ml up to 75 at 30 ng/ml. Thus, the two populations show markedly different dose response curves, with ecto-5'-NT$^+$ cells being responsive at ten-fold lower [PMA] than ecto-5'-NT$^-$ cells.

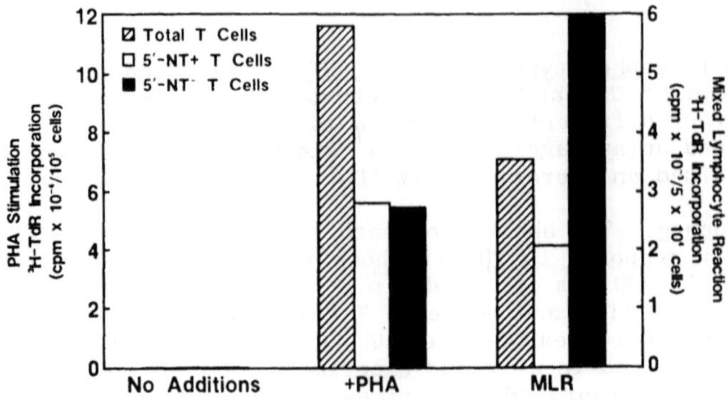

Figure 1. Ability of ecto-5'-NT positive and negative T cells to proliferate in response to PHA and in a mixed lymphocyte reaction. Untreated T cells and ecto-5'-NT positive and negative T cells were cultured with PHA or irradiated allogeneic cells. After 3 da with PHA, proliferation was assessed by ^{3}H-TdR incorporation in a 4-h pulse. After 7 da of MLR, proliferation was assessed by ^{3}H-TdR incorporation in an 8-h pulse.

T cell help for PWM-stimulated immunoglobulin (Ig) biosynthesis. Irradiated ecto-5'-NT positive and negative T cells were also compared for their ability to provide help to autologous B cells for PWM-driven Ig biosynthesis. Data from a representative experiment are given in Figure 2. After sorting, ecto-5'-NT$^-$ T cells performed nearly as well as the original untreated T cells and 4- to 5-fold better than ecto-5'-NT$^+$ cells.

Table 2. ABILITY OF ECTO-5'-NT$^+$ AND ECTO-5'-NT$^-$ T CELLS TO PROLIFERATE IN RESPONSE TO PHORBOL MYRISTATE ACETATE

[PMA]	Untreated T cells	Ecto-5'-NT$^+$ T cells	Ecto-5'-NT$^-$ T cells
ng/ml	cpm ^{3}H-TdR incorporation/5 x 10^4 cells		
Experiment #1			
0	1574	276	1736
0.3	1169	519	625
0.6	1133	3691	1598
1.0	4722	8019	294
Experiment #2			
0	515	489	494
3.3	7083	52425	6443
10.0	14427	48200	17630
30.0	20734	54127	37022

Untreated T cells and sorted ecto-5'-NT positive and negative T cells were examined for the ability to proliferate in response to various concentrations of PMA.

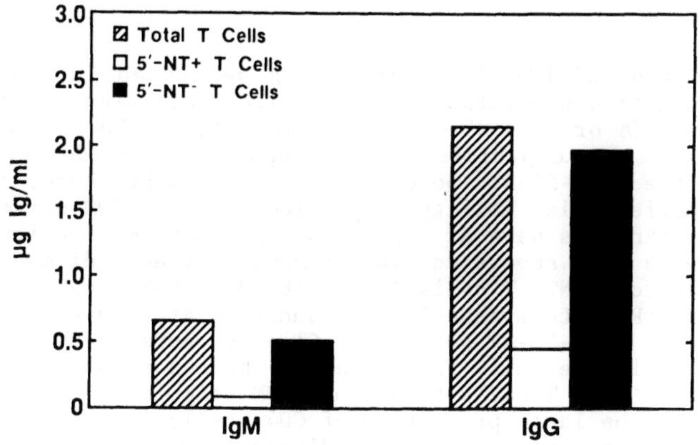

Figure 2. Ability of ecto-5'-nucleotidase positive and negative T cells to provide help for PWM-stimulated Ig biosynthesis. Untreated T cells and ecto-5'-NT positive and negative T cells were irradiated and cultured for 7 da with B cells in the presence of PWM. T cell help was assessed by quantitating the IgM and IgG secreted into the medium.

Immunoglobulin synthesis in response to PWM and EBV. Ecto-5'-NT positive and negative B cells were compared for their ability to synthesize IgM and IgG in response to either a T cell-dependent (PWM) or T cell-independent stimulator (EBV) of polyclonal Ig biosynthesis. As shown in Figure 3, in PWM-stimulated cultures, ecto-5'-NT^{+} B cells synthesized 2-fold more IgM and 26-fold more IgG per cell than the ecto-5'-NT^{-} B cells. In EBV-stimulated cultures, ecto-5'-NT^{+} B cells synthesized 1.6-fold more IgM and 8-fold more IgG per cell than ecto-5'-NT^{-} B cells.

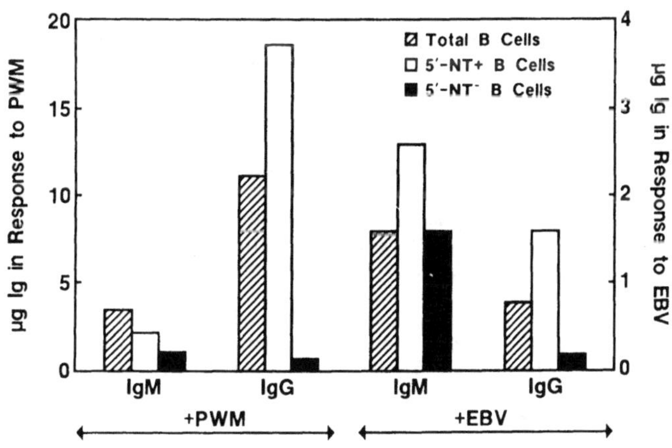

Figure 3. Ability of ecto-5'-NT positive and negative B cells to synthesize polyclonal Ig in response to PWM and EBV. Untreated B cells and ecto-5'-NT positive and negative B cells were cultured for 7 da with irradiated T cells and PWM or for 10 da with EBV. B cell function was assessed by quantitating the IgM and IgG secreted into the medium.

DISCUSSION

Human peripheral blood T lymphocytes separated into ecto-5'-NT positive and negative populations, show similar abilities to proliferate in response to PHA or in a MLR as determined by ^{3}H-TdR incorporation. However, the ability to proliferate in response to low doses of PMA is restricted to the ecto-5'-NT$^+$ population. On the other hand, ecto-5'-NT$^-$ T cells have a preferential ability to provide help for PWM-driven Ig biosynthesis. These findings have at least two important implications. First, although patients with severe combined immunodeficiency disease often have predominantly ecto-5'-NT$^-$ T cells (11,12,16,18), their inability to respond to PHA or in a MLR or to synthesize Ig cannot be attributed solely to an absence of T cell ecto-5'-NT activity. Clearly, normal adult peripheral ecto-5'-NT$^-$ T cells are fully functional in these assays. The preferential ability of ecto-5'-NT$^-$ T cells to support PWM-stimulated Ig synthesis may be due merely to the large proportion of CD4$^+$ cells in this population. It is true that early stages of thymocyte differentiation are characterized by low ecto-5'-NT activity (8,9) and the lymphocytes of SCID patients may be blocked in maturation at this stage. However, it is not possible to ascertain the maturation state of T cells merely by measuring their ecto-5'-NT activity.

Second, ecto-5'-NT positive and negative T cells have markedly different dose response curves for activation by PMA. At doses of 1.0 ng/ml, or less, proliferation is restricted to the ecto-5'-NT$^+$ population. Ecto-5'-NT$^+$ cells become maximally responsive at 3.3 ng/ml, while the response of ecto-5'-NT$^-$ T cells is still increasing at 30 ng/ml. To our knowledge, this is the first functional assay in which ecto-5'-NT$^-$ T cells have been found to be impaired compared to ecto-5'-NT$^+$ T cells. An understanding of the mechanism responsible for this impairment may lead to an understanding of the significance of reduced numbers of ecto-5'-NT$^+$ T cells in immunodeficient patients.

The mechanism of action of PMA is through the activation of protein kinase C (PKC, 19). PHA-induced mitogenesis is also mediated through PKC activation, in this case by diacylglycerol (DAG), derived from the hydrolysis of phosphatidylinositol 4,5-bisphosphate (20). Since ecto-5'-NT$^-$ T cells are fully responsive to PHA, it is not immediately obvious why they should be unresponsive to low [PMA]. Perhaps the PKC in ecto-5'-NT$^-$ T cells has a more stringent requirement for activation than that in ecto-5'-NT$^+$ T cells and low [PMA] provide a signal which is not as "strong" as that provided by DAG. Alternatively, low [PMA] might initiate an alternate pathway of T cell activation which is restricted to ecto-5'-NT$^+$ cells. Recent experiments in our lab demonstrate that ecto-5'-NT$^+$ T cells can be stimulated to proliferate by anti-5'-NT antibodies and submitogenic doses of PMA (1.0 ng/ml or less) (21), suggesting that ecto-5'-NT itself may be involved in transmembrane signal transduction. Experiments are in progress to determine whether ecto-5'-NT$^+$ T cells can utilize unique biochemical pathways of cellular activation.

Both ecto-5'-NT$^+$ and ecto-5'-NT$^-$ B cells can synthesize IgM in response to either PWM or EBV, but the synthesis of IgG *in vitro* is largely restricted to the ecto-5'-NT$^+$ subpopulation. Our data, therefore, provide the first direct evidence that ecto-5'-NT is a marker for the functional maturation of human B cells. In fact, since approximately 80% of adult peripheral blood B cells are ecto-5'-NT$^+$, we calculate that in bulk cultures, 97-99% of the IgG synthesized in response to PWM or EBV is made by ecto-5'-NT$^+$ B cells. Similar calculations predict that 50-85% of the IgM in such cultures is made from ecto-5'-NT$^+$ B cells. Clearly, not all ecto-5'-NT$^+$ B cells respond to PWM, however. Approximately 80% of adult peripheral blood B cells are ecto-5'-NT$^+$ by either indirect immunofluorescence or histochemical staining (2,3), whereas limiting dilution analysis suggests that <1% of B cells synthesize Ig in response to stimulation with PWM (22).

Our data also clearly show that ecto-5'-NT$^-$ B cells synthesize as much IgM as ecto-5'-NT$^+$ B cells, on a per cell basis. These results are consistent with our earlier studies of both patients with common variable immunodeficiency (CVI) and normal infants. Although ecto-5'-NT deficient B cells from most CVI patients have a markedly reduced capacity for Ig synthesis *in vitro*, B cells from some patients can synthesize IgM (2). Similarly, normal newborn infants' B cells have low ecto-5'-NT activity and can synthesize IgM, but not IgG in response to EBV *in vitro* (23,24). Ecto-5'-NT expression increases in the first six months of life (24) and is then maintained in the subset of B cells that is PWM-responsive. Thus, ecto-5'-NT differs from other B cell maturation markers such as IgD, HB-4, Leu 8, and the mouse erythrocyte rosette receptor, which are lost as B cells mature and gain the ability to respond to PWM (25-27). Therefore, analysis of ecto-5'-NT expression, especially in combination with other B cell differentiation antigens, should prove a valuable tool in delineating stages in human B cell maturation.

ACKNOWLEDGMENTS

This work was supported by grants AI18220 and GM39699 and the Biomedical Research Support Program grant RR05514 from the National Institutes of Health. This is publication no. 5406-IMM from the Research Institute of Scripps Clinic. The authors acknowledge the excellent secretarial support of Margaret Stone, and the technical expertise of the members of the Cell Sorting Facility.

REFERENCES

1. Thompson, L.F., Saxon, A., O'Connor, R.D., Fox, R.I.Ecto-5'-nucleotidase activity in human T cell subsets. J. Clin. Invest. 71:892-899, 1983.
2. Thompson, L.F., Ruedi, J.M., O'Connor, R.D., Bastian, J.F. Ecto-5'-nucleotidase expression during human B cell development. An explanation for the heterogeneity in B lymphocyte ecto-5'-nucleotidase activity in patients with hypogammaglobulinemia.J. Immunol. 137:2496-2500, 1986.
3. Thompson, L.F., Ruedi, J.M., Low, M.G., Clement, L.T. Distribution of ecto-5'-nucleotidase on subsets of human T and B lymphocytes as detected by indirect immunofluorescence using goat antibodies. J. Immunol. 139:4042-4048, 1987.
4. Dianzani, U., Massaia, M., Pileri, A., Grossi, C.E., Clement, L.T. Differential expression of ecto-5'-nucleotidase activity by functionally and phenotypically distinct subpopulations of human Leu-2+/T8+ lymphocytes. J. Immunol. 137:484-489, 1986.
5. Salazar-Gonzalez, J.F., Moody, D.J., Giorgi, J.V., Martinez-Maza, O., Mitsuyasu, R.T., Fahey, J.L. Reduced ecto-5'-nucleotidase activity and enhanced OKT10 and HLA-DR expression on CD8 (T suppressor/cytotoxic) lymphocytes in the acquired immune deficiency syndrome: Evidence of CD8 cell immaturity. J. Immunol. 135:1778-1785, 1985.
6. Rowe, M., De Gast, C.G., Platts-Mills, T.A.E., Asherson, G.L., Webster, A.D.B., Johnson, S.M. 5'-nucleotidase of B and T lymphocytes isolated from human peripheral blood. Clin. exp. Immunol. 36:97-101, 1979.
7. Recker, D.P., Edwards, N.L., Fox, I.H. Histochemical evaluation of lymphocytes in hypogammaglobulinemia. Decreased number of 5'-nucleotidase-positive cells. J. Lab. Clin. Med. 95:179, 1980.
8. Edwards, N.L., Gelfand, E.W., Burk, L., Dosch, H.-M., Fox, I.H. Distribution of 5'-nucleotidase in human lymphoid tissues. Proc. Natl. Acad. Sci. 76:3474-3476, 1979.
9. Cohen, A., Lee, J.W.W., Dosch, H.-M., Gelfand, E.W. The expression of deoxyguanosine toxicity in T lymphocytes at different stages of maturation. J. Immunol. 125:1578-1582, 1980.
10. Rowe, M., DeGast, G.C., Platts-Mills, T.A.E., Asherson, G.L., Webster, A.D.B., Johnson, S.M. Lymphocyte 5'-nucleotidase in primary hypogammaglobulinemia and cord blood. Clin. exp. Immunol. 39:337-343, 1980.

11. Boss, G.R., Thompson, L.F., O'Connor, R.D., Ziering, R.W., Seegmiller, J.E. Ecto-5'-nucleotidase deficiency: Association with adenosine deaminase deficiency and non-association with deoxyadenosine toxicity. Clin. Immunol. Immunopathol. 19:1-7, 1981.
12. Thompson, L.F., O'Connor, R.D., Bastian, J.F. Phenotype and function of engrafted maternal T cells in patients with severe combined immunodeficiency. J. Immunol. 133:2513-2517, 1984.
13. Webster, A.D.B., North, M., Allsop, J., Asherson, G.L., Watts, R.W.E. Purine metabolism in lymphocytes from patients with primary hypogammaglobulinemia. Clin. exp. Immunol. 31:456-463, 1978.
14. Edwards, N.L., Magilavy, D.B., Cassidy, J.T., Fox, I.H. Lymphocyte ecto-5'-nucleotidase deficiency in agammaglobulinemia. Science 201:628-630, 1978.
15. Thompson, L.F., Boss, G.R., Spiegelberg, H.L., Jansen, I.V., O'Connor, R.D., Waldmann, T.A., Hamburger, R.N., Seegmiller, J.E. Ecto-5'-nucleotidase activity in T and B lymphocytes from normal subjects and patients with congenital X-linked agammaglobulinemia. J. Immunol. 123:2475-2478, 1979.
16. Gelfand, E.W., McCurdy, D., Rao, C.P., Cohen, A. Absence of lymphocyte ecto-5'-nucleotidase in infants with reticuloendotheliosis and eosinophilia (Omenn's Syndrome). Blood 63:1475-1480, 1984.
17. Thompson, L.F., Ruedi, J.M., Low, M.G. Purification of 5'-nucleotidase from human placenta after release from plasma membranes by phosphatidylinositol-specific phospholipase C. Biochem. Biophys. Res. Comm. 145:118-125, 1987.
18. Cohen, A., Mansour, A., Dosch, H.-M., Gelfand, E.W. Association of a lymphocyte purine enzyme deficiency (5'-nucleotidase) with combined immunodeficiency. Clin. Immunol.& Immunopathol. 15:245-250, 1980.
19. Isakov, N., Mally, M.I., Scholz, W., Altman, A. T-lymphocyte activation: The role of protein kinase C and the bifurcating inositol phospholipid signal transduction pathway. In: Immunological Reviews. Edited by Moller,G. 95th Edition. pp. 89-111. Munksgaard, Copenhagen, 1987.
20. Isakov, N., Scholz, W., Altman, A. Signal transduction and intracellular events in T-lymphocyte activation. Immunology Today 7:271-277, 1986.
21. Thompson, L.F., Ruedi, J.M., Low, M.G. Anti-5'-nucleotidase antibodies cause human peripheral blood T cells to proliferate. In: Adv. Exp. Med. Biol., this volume.
22. Thiele, C.J., Stevens, R.H. Antibody potential of human peripheral blood lymphocytes differentially expressing surface membrane IgM. J. Immunol. 124:1898-1904, 1980.
23. Tosato, G., Magrath, I.T., Koski, I.R., Dooley, N.J., Blaese, R.M. B cell differentiation and immunoregulatory T cell function in human cord blood lymphocytes. J. Clin. Invest. 66:383-388, 1980.
24. Bastian, J.F., Ruedi, J.M., MacPherson, G.A., Golembesky, H.E., O'Connor, R.D., Thompson, L.F. Lymphocyte ecto-5'-nucleotidase activity in infancy: increasing activity in peripheral blood B cells precedes their ability to synthesis IgG in vitro. J. Immunol. 132:1767-1772, 1984.
25. Lucivero, G., Lawton, A.R., Cooper, M.D. Rosette formation with mouse erythrocytes defines a population of human B lymphocytes unresponsive to pokeweed mitogen. Clin. exp. Immunol. 45:185-190, 1981.
26. Tedder, T.F., Clement, L.T., Cooper, M.D. Development and distribution of a human B cell subpopulation identified by the HB-4 monoclonal antibody. J. Immunol. 134:1539-1544, 1985.
27. Kansas, G.S., Wood, G.S., Engleman, E.G. Maturational and functional diversity of human B lymphocytes delineated with anti-Leu-8. J. Immunol. 134:3003-3006, 1985.

ANTI-5'-NUCLEOTIDASE ANTIBODIES CAUSE HUMAN PERIPHERAL BLOOD T CELLS TO PROLIFERATE

Linda F. Thompson, Julie M. Ruedi, and Martin G. Low

Department of Immunology, Scripps Clinic and Research Foundation, La Jolla, CA and Department of Physiology New York, NY.

Ecto-5'-nucleotidase (ecto-5'-NT) is an enzyme found on the external plasma membrane (1) of subpopulations of human peripheral blood lymphocytes. By indirect immunofluorescence with goat anti-5'-NT antibodies (2), approximately 32% of CD3$^+$, 19% of CD4$^+$, and 50% of CD8$^+$ T cells express 5'-NT on their surfaces. Ecto-5'-NT activity increases during T cell ontogeny, as adult peripheral blood T cells have at least 10 times the ecto-5'-NT activity of thymocytes (3). Reduced or absent ecto-5'-NT activity is found on the T cells of patients with a variety of immunodeficiency diseases (2). Nevertheless, the function of ecto-5'-NT on human T cells is unknown. Human leukemic T cell lines are uniformly deficient in ecto-5'-NT activity (4) and yet they grow in tissue culture with doubling times comparable to ecto-5'-NT$^+$ B cell lines. Purified populations of ecto-5'-NT$^-$ T cells proliferate in response to phytohemagglutinin (PHA) or allogeneic cells and provide help for pokeweed mitogen-driven immunoglobulin synthesis in a manner similar to ecto-5'-NT$^+$ T cells (5). Although ecto-5'-NT activity can be demonstrated on human T cells resuspended in optimal concentrations of Tris, $MgCl_2$ and bovine serum albumin, Cohen et al. (7) have shown that when T cells are incubated in RPMI 1640, they are unable to degrade extracellular AMP, raising the possibility that the enzyme may not be active under some physiologic conditions.

Here we report that antibodies to ecto-5'-NT cause human peripheral blood T cells to proliferate in the presence of submitogenic doses of phorbol myristate acetate (PMA). Our results are very similar to those obtained with antibodies to other T cell surface structures believed to be involved in lymphocyte activation and provide the first evidence that ecto-5'-NT may be involved in lymphocyte signal transduction.

METHODS

<u>Lymphocyte isolation.</u> Peripheral blood mononuclear cells were isolated from freshly-drawn heparinized blood by Ficoll-Hypaque density gradient centrifugation. T lymphocytes were isolated by one cycle of rosetting with neuraminidase-treated sheep erythrocytes, E_N, for 15 min at 37°C and 1 h at 4°C followed by Ficoll-Hypaque density gradient centrifugation. E_N were lysed with 0.83% NH_4Cl. These preparations contained greater than 95% T cells as determined by immunofluorescent staining with Leu 4 (anti-CD3) and 1-3% monocytes as determined by esterase staining.

<u>Preparation of antiserum.</u> The anti-5'-NT antiserum was prepared by immunizing a goat with highly-purified human placental 5'-NT and absorbed as described (8).

Proliferation assay. T cells at 1 x 10^6/ml were cultured in flat-bottomed "half-area" microtiter plates (0.1 ml volumes) in RPMI 1640 supplemented with 10% fetal bovine serum (FBS), 2 mM glutamine, penicillin (100 units/ml), streptomycin (100 μg/ml), and amphotericin B (0.25 μg/ml) in the presence of anti-5'-NT antibodies and varying doses of PMA. After three days, the cultures were pulsed for 4 h with 1.0 μCi ^{3}H-thymidine (TdR) per well and then harvested with a multiple automated sample harvester (MASH). The amount of radioactivity taken up was quantitated by liquid scintillation counting. All cultures were performed in triplicate. The results are expressed as mean cpm ^{3}H-TdR incorporated/10^5 cells; triplicate values usually agreed within 15%. The response to PHA is given as a positive control.

IL-2 receptor expression and IL-2 secretion. Human peripheral blood T cells were cultured with goat anti-5'-NT IgG at 125 μg/ml and 1.0 ng/ml PMA as described above. At various intervals, the cells were examined for IL-2 receptor (IL-2R, CD25) expression by immunofluorescent staining with a fluorescein isothiocyanate (FITC)-conjugated murine monoclonal anti-CD25 antibody (Becton-Dickinson). The concentrations of IL-2 secreted into the media after 22 hrs were determined in a proliferative assay using the IL-2 dependent murine cell line, CTLL. Human IL-2 secreted by Jurkat cells was used as a standard.

RESULTS

Stimulation of T cell proliferation by anti-5'-NT antibodies and PMA. Human peripheral blood T cells were cultured with goat anti-5'-NT antibodies and PMA at 0.3, 0.6, or 1.0 ng/ml for three da. At 72 h, the degree of proliferation was assed by ^{3}H-TdR incorporation in a 4-h pulse (Table 1). Both intact IgG and $F(ab')_2$ fragments isolated from the goat anti-5'-NT antiserum caused T cells to proliferate when co-cultured with submitogenic doses of PMA. The proliferative response increased with increasing doses of PMA. Whole IgG and $F(ab')_2$ fragments were equally effective at the concentration employed (250 μg/ml). The anti-5'-NT antibodies alone were not mitogenic and they did not influence the ability of T cells to proliferate in response to PHA. Intact IgG or $F(ab')_2$ fragments from pre-immune serum had no effect. The proliferative response of human T cells to anti-5'-NT antibodies (either intact IgG or $F(ab')_2$ fragments) was dose-dependent and reached a maximum at 125 μg/ml of intact IgG or 60 μg/ml of $F(ab')_2$ fragments (data not shown).

Table 1. STIMULATION OF HUMAN T CELLS WITH ANTI-5'-NT ANTIBODIES AND PMA

	No additions	Pre-immune IgG	Anti-5'-NT IgG	Pre-immune $F(ab')_2$	Anti-5'-NT $F(ab')_2$
	Mean ^{3}H-Thymidine incorporation, cpm/100,000 cells				
No additions	389	965	826	225	1,525
PHA	62,310	77,209	66,786	73,324	74,629
PMA @ 0.3 ng/ml	1,841	918	17,154	1,077	40,881
PMA @ 0.6 ng/ml	3,758	3,085	53,376	4,490	50,143
PMA @ 1.0 ng/ml	7,593	7,154	54,328	7,844	51,288

Human peripheral blood T cells were cultured with pre-immune or anti-5'-NT IgG or $F(ab')_2$ fragments and the indicated concentrations of PMA. After 3 da, proliferation was assessed by ^{3}H-TdR incorporation in a 4-h pulse.

Table 2. PROLIFERATIVE RESPONSE OF ECTO-5'-NT POSITIVE AND NEGATIVE T CELLS TO ANTI-5'-NT ANTIBODIES AND PMA

		T cells		
		Untreated	Ecto-5'-NT$^+$	Ecto-5'-NT$^-$
		Mean ^{3}H-TdR incorporation, cpm/5 x 10^4 cells		
1.	No additions	1,574	276	1,736
2.	+ PHA	11,099	4,425	9,594
3.	+ PMA	4,722	8,019	294
4.	+ PMA + pre-immune IgG	1,976	5,573	466
5.	+ PMA + anti-5'-NT IgG	24,182	39,604	2,153

Untreated T cells and sorted ecto-5'-NT positive and negative T cells were tested for the ability to proliferate in response to anti-5'-NT IgG (125 μg/ml) and PMA (1.0 ng/ml).

Anti-5'-NT response is limited to the 5'-NT T cell population. To confirm that the proliferative response to our goat antibodies was caused by anti-5'-NT and not by antibodies to a contaminating protein in the enzyme preparation used to immunize our goat, we performed a proliferative assay on purified populations of ecto-5'-NT positive and negative T cells prepared by cell sorting. Only the ecto-5'-NT$^+$ T cells were able to proliferate in response to anti-5'-NT IgG and PMA (Table 2).

IL-2 receptor expression and IL-2 secretion. Human peripheral blood T cells were cultured with anti-5'-NT IgG at 125 μg/ml and PMA at 1.0 ng/ml. At various times after the initiation of the cultures, aliquots of cells were removed and stained with FITC-anti-CD25 to determine the extent of IL-2R induction (Figure 1). Goat anti 5'-NT antibodies + PMA induced IL-2R expression on human T cells. The degree of IL-2R expression was greater than that seen with PMA alone and increased with increasing time after the initiation of the cultures. Unlike PHA, which caused IL-2R expression on virtually all peripheral T cells by 96 h, anti-5'-NT antibodies caused IL-2R expression on only a subpopulation, consistent with the fact that ecto-5'-NT is expressed on only a subpopulation of adult peripheral blood T cells.

Supernatants from the above cultures were also analyzed for IL-2 content using the murine IL 2 dependent cell line, CTLL (Table 3). Goat anti-5'-NT IgG plus PMA at 1.0 ng/ml caused the secretion of 0.51 units/ml of IL-2, a ten-fold increase over the amount secreted in response to PMA alone. This concentration of IL-2 was approximately one-third that secreted in response to PHA.

Inhibition of the anti-5'-NT response by pre-treatment of T cells with phosphatidylinositol-specific phospholipase C (PI-PLC). T cells were pre-incubated with purified PI-PLC from *B. thuringiensis* at 0.2 units/ml for 1 h at 37°C and then tested for the ability to proliferate in response to PMA and anti-5'-NT IgG at 125 μg/ml. PI-PLC was present throughout the entire 3 da culture period. The results are shown in Table 4. At the highest dose of PMA, pre-treatment of T cells with PI-PLC inhibited anti-5'-NT induced thymidine uptake on da 3 by 84% (from 117,172 to 28,126 cpm). This amount of inhibition closely resembled the percent of ecto-5'-NT activity released (80%) by PI-PLC treatment for 1 h at 37°C from a separate aliquot of T cells. In two additional experiments, PI-PLC treatment of T cells inhibited their ability to respond to PHA by only 22 and 0%.

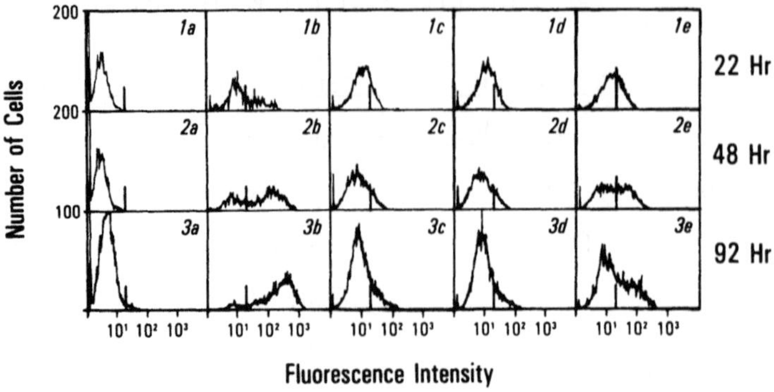

Figure 1. IL-2 receptor expression on T cells activated with PHA or goat anti-5'-NT antibodies + PMA. IL-2 receptor expression on human peripheral blood T cells was measured by immunofluorescence with FITC-anti-CD25 after 22, 48, and 92 h of culture. a. Media alone, b. + PHA (10 μg/ml), c. PMA at 1.0 ng/ml, d. PMA at 1.0 ng/ml + pre-immune IgG at 125 μg/ml, e. PMA at 1.0 ng/ml + anti-5'-NT IgG at 125 μg/ml.

Table 3. IL-2 SECRETION BY T CELLS STIMULATED WITH GOAT POLYCLONAL ANTI-5'-NT ANTIBODIES + PMA

Culture Conditions	Concentration of IL-2 Secreted, units/ml
1. Cells alone	<0.01
2. + PHA	1.82
3. + PMA + goat pre-immune IgG	0.05
4. + PMA + goat anti-5'-NT IgG	0.51

Human peripheral blood T cells were cultured with pre-immune or anti-5'-NT IgG and PMA at 1.0 ng/ml for 22 h. The concentrations of IL-2 secreted into the supernatants were determined in a proliferative assay with the murine IL-2 dependent cell line, CTLL.

DISCUSSION

Goat antibodies to ecto-5'-NT, when combined with submitogenic doses of PMA, cause human peripheral blood T cells to proliferate, express IL-2 receptors, and secrete IL-2. These responses are limited to ecto-5'-NT$^+$ T cells (demonstrated by cell sorting), which is consistent with the responses being mediated by anti-5'-NT antibodies rather than by antibodies to unknown antigens which may have contaminated the enzyme preparation used to immunize our goat. Furthermore, similar results are obtained with a murine monoclonal antibody to ecto-5'-NT recently developed in our laboratory (manuscript in preparation), leaving no doubt that the responses described are truly caused by anti-5'-NT antibodies. Our observations are very similar to those obtained with antibodies to other human and murine T

Table 4. INHIBITION OF ANTI-5'-NT INDUCED T CELL PROLIFERATION BY PRE-TREATMENT WITH PHOSPHATIDYLINOSITOL-SPECIFIC PHOSPHOLIPASE C

	No additions	PI-PLC	Pre-immune IgG	Anti-5'-NT IgG
	Mean ^{3}H-Thymidine incorporation, cpm/100,000 cells			
No additions	436	-	n.t.	n.t.
	1,750	+	n.t.	n.t.
PHA	101,982	-	n.t.	n.t.
PMA @ 0.3 ng/ml	3,938	-	4,009	5,192
		+	4,487	3,219
PMA @ 0.6 ng/ml	5,672	-	6,494	23,673
		+	8,165	12,409
PMA @ 1.0 ng/ml	11,446	-	10,108	117,172
		+	10,295	28,126

Human peripheral blood T cells were pre-treated with *B. thuringiensis* PI-PLC for 1 h at 37° and then cultured with pre-immune or anti-5'-NT IgG at 125 μg/ml and PMA at 1.0 ng/ml. After 3 da, proliferation was assessed by ^{3}H-TdR incorporation in a 4-h pulse.

cell surface antigens believed to be involved in lymphocyte activation, including CD3 (9), the T cell activating protein, TAP (10), Thy-1 (11), Ly-6 (12), decay accelerating factor (13), and CD28, the 44 kD antigen recognized by murine monoclonal antibody 9.3 (14). Antibodies to the T cell antigen receptor (TCR) (15), the sheep red blood cell receptor (CD2) (16), and the rat alloantigen, RT-6 (17), are also mitogenic, even in the absence of PMA. Antibodies to LFA-3, an antigen found on thymic epithelial cells, induce IL-1 secretion (18). Our data provide the first evidence that ecto-5'-NT may also play a role in lymphocyte signal transduction.

Most antibodies against T cell surface structures are not mitogenic and the majority of well-studied mitogenic antibodies have proven to be directed against physiologic triggering structures. CD3 is known to be intimately linked to the TCR (19). This is probably also true for TAP, since mutants of a murine T-T hybridoma specific for chicken albumin that are defective in TAP expression, can no longer be triggered by antigen (20). However, the role of the other antigens listed above in T cell triggering is less well-understood. These molecules, including ecto-5'-NT, may serve to amplify or modify signals originating from the CD3/TCR complex, or they may be components of distinct alternative antigen-independent pathways of T cell activation.

The best-studied mechanism of lymphocyte activation is the phosphoinositide signal transduction pathway which is initiated when antigen interacts with the TCR on T cells (21). A polyphosphoinositide-specific phosphodiesterase (PPI-PDE) is activated to hydrolyze phosphatidylinositol-4,5-bisphosphate (PIP_2) producing diacylglycerol (DAG) and inositol-1,4,5-trisphosphate (IP_3). DAG causes the activation of protein kinase C (PKC) and its translocation from cytosol to the plasma membrane. IP_3 causes an intrease in intracellular Ca^{++} from the mobilization of intracellular stores followed by the influx of extracellular Ca^{++}. GTP binding proteins have been implicated in the activation of PPI-PDE (22). The phosphoinositide signal transduction pathway can be bypassed by treating cells with a calcium ionophore and PMA, a potent activator of PKC (21). Since PMA is required for the proliferative response to T cells to anti-5'-NT, soluble anti-5'-NT antibodies alone must not be capable of initiating the phosphoinositide signal transduction pathway. Experiments are in progress to determine whether they can cause an increase in intracellular [Ca^{++}].

Although 5'-NT has been regarded for many years as a classic plasma membrane marker, this enzyme appears to be covalently anchored in the membrane to glycosyl phosphatidylinositol (glycosyl-PI), or a closely related structure, rather than by conventional hydrophobic interactions (8,23). That is, ecto-5'-NT can be released from lymphocytes by treatment with PI-PLC. Although the function of this unusual membrane anchoring structure is not known, its proposed purpose is to allow the rapid release of the proteins it anchors via activation of an endogenous phospholipase (24). It is curious that five of the nine antigens listed above (24,25) that are thought to play a role in alternative pathways of T cell signaling, as well as LFA-3 (26), are also linked to the plasma membrane by covalent attachment to glycosyl-PI. A sixth glycosyl-PI-linked T cell antigen, Qa-2 (27) is secreted by Con A-activated murine T cells (28). T cell activation by anti-5'-NT antibodies is mediated by a glycosyl-PI anchored form of 5'-NT since pre-treatment of T cells with PI-PLC inhibits their ability to be activated by anti-5'-NT antibodies. The proliferation of PI-PLC-treated T cells in response to ionomycin and PMA was not inhibited (data not shown), however, demonstrating that this treatment, per se, does not inhibit the ability of T cells to be triggered by agents causing Ca^{++} mobilization and PKC translocation and activation. It will be important to determine why glycosyl-PI anchored structures are so frequently involved in lymphocyte activation and to determine the biochemical events which are initiated when antibodies interact with these structures. If ecto-5'-NT is a physiologically important membrane signaling structure, it will ultimately be necessary to identify the natural ligand whose action is mimicked by our anti-5'-NT antibodies. The results of these studies will give insight into the significance of changes in ecto-5'-NT activity which occur during T cell ontogeny, the differential expression of ecto-5'-NT in T cell subpopulations, and the reduced enzyme expression in patients with immunodeficiency diseases.

ACKNOWLEDGMENTS

This work was supported by grant AI18220 and the Biomedical Research Support Program grant RR05514 from the National Institutes of Health. This is publication no. 5414-IMM from the Research Institute of Scripps Clinic. The authors acknowledge the excellent secretarial support of Margaret Stone, and stimulating discussions with Dr. Alex Lucas.

REFERENCES

1. Trams, E.G., Lauter, C.J. On the sidedness of plasma membrane enzymes. Biochim.Biophys.Acta 345:180-197, 1974.
2. Thompson, L.F., Ruedi, J.M., Low, M.G., Clement, L.T. Distribution of ecto-5'-nucleotidase on subsets of human T and B lymphocytes as detected by indirect immunofluorescence using goat antibodies. J. Immunol. 139:4042-4048, 1987.
3. Edwards, N.L., Gelfand, E.W., Burk, L., Dosch, H.-M., Fox, I.H. Distribution of 5'-nucleotidase in human lymphoid tissues. Proc. Natl. Acad. Sci. 76:3474-3476, 1979.
4. Carson, D.A., Kaye, J., Matsumoto, S., Seegmiller, J.E., Thompson, L. Biochemical basis for the enhanced toxicity of deoxyribonucleosides toward malignant human T cell lines. Proc. Natl. Acad. Sci.USA 76:2430-2433, 1979.
5. Thompson, L.F., Ruedi, J.M. Functional characterization of ecto-5'-nucleotidase (ecto-5'-NT) positive and negative human lymphocytes. Adv. Exp. Med. Biol., this volume.
6. Thompson, L.F., Boss, G.R., Spiegelberg, H.L., Jansen, I.V., O'Connor, R.D., Waldmann, T.A., Hamburger, R.N., Seegmiller, J.E. Ecto-5'-nucleotidase activity in T and B lymphocytes from normal subjects and patients with congenital X-linked agammaglobulinemia. J. Immunol. 123:2475-2478, 1979.
7. Barankiewicz, J., Dosch, H.-M., Cohen, A. Extracellular nucleotide catabolism in human B and T lymphocytes. J. Biol. Chem. 263:7094-7098, 1988.

8. Thompson, L.F., Ruedi, J.M., Low, M.G. Purification of 5'-nucleotidase from human placenta after release from plasma membranes by phosphatidylinositol-specific phospholipase C. Biochem. Biophys. Res. Comm. 145:118-125, 1987.
9. Van Wauwe, J.P., De Mey, J.R., Goossens, J.G. OKT3: A monoclonal anti-human T lymphocyte antibody with potent mitogenic properties. J. Immunol. 124:2708-2713, 1980.
10. Rock, K.L., Yeh, E.T.H., Gramm, C.F., Haber, S.I., Reiser, H., Benacerraf, B. TAP, a novel T cell-activating protein involved in the stimulation of MHC-restricted T lymphocytes. J. Exp. Med. 163:315-333, 1986.
11. Gunter, K.C., Malek, T.R., Shevach, E.M. T cell-activating properties of an anti-thy-1 monoclonal antibody. J. Exp. Med. 159:716-730, 1984.
12. Malek, T.R., Ortega, G., Chan, C., Kroczek, R.A., Shevach, E.M. Role of Ly-6 in lymphocyte activation. Induction of T cell activation by monoclonal anti-Ly-6 antibodies. J. Exp. Med. 164:709-722, 1986.
13. Ritter, A.R., Davis, L.S., Patel, S.S., Atkinson, J.P., Lipsky, P.E. An antiserum to decay-accelerating factor (DAF) activates human T cells. FASEB J. 2:A871, 1988.
14. Hara, T., Fu, S.M., Hansen, J.A. Human T cell activation. A new activation pathway used by a major T cell population via a disulfide-bonded dimer of a 44 kilodalton polypeptide (9.3 antigen). J. Exp. Med. 161:1513-1524, 1985.
15. Spits, H., Borst, J., Tax, W., Capel, P.J.A., Terhorst, C., deVries, J.E. Characteristics of a monoclonal antibody (WT-31) that recognizes a common epitope on the human T cell receptor for antigen. J. Immunol. 135:1922-1928, 1985.
16. Meuer, S.C., Hussey, R.E., Fabbi, M., Fox, D., Acuto, O., Fitzgerald, K.A., Hodgdon, J.C., Protentis, J.P., Schlossman, S.F., Reinherz, E.L. An alternative pathway of T-cell activation: A functional role for the 50 kd T11 sheep erythrocyte receptor protein. Cell 36:897-906, 1984.
17. Wonigeit, K., Schwinzer, R. Polyclonal activation of rat T lymphocytes by RT6 alloantisera. Transplantation Proc. XIX:296-299, 1987.
18. Le, P., Denning, S., Springer, T., Haynes, B., Singer, K. Anti-LFA-3 monoclonal antibody induces interleukin 1 (IL 1) release by thymic epithelial (TE) cells and monocytes. Fed. Proc. 46:447, 1987.
19. Reinherz, E.L., Meuer, S., Fitzgerald, K.A., Hussey, R.E., Levine, H., Schlossman, S.F. Antigen recognition by human T lymphocytes is linked to surface expression of the T3 molecular complex. Cell 30:735-743, 1982.
20. Yeh, E.T.H., Reiser, H., Bamezai, A., Rock, K.L. TAP transcription and phosphatidylinositol linkage mutants are defective in activation through the T cell receptor. Cell 52:665-674, 1988.
21. Isakov, N., Scholz, W., Altman, A. Signal transduction and intracellular events in T-lymphocyte activation. Immunology Today 7:271-277, 1986.
22. Imboden, J.B., Shoback, D.M., Pattison, G., Stobo, J.D. Cholera toxin inhibits the T-cell antigen receptor-mediated increases in inositol trisphosphate and cytoplasmic free calcium. Proc. Natl. Acad. Sci. USA 83:5673-5677, 1986.
23. Low, M.G., Finean, J.B. Specific release of plasma membrane enzymes by a phosphatidylinositol-specific phospholipase C. Biochim. Biophys. Acta 508:565-570, 1978.
24. Low, M.G. Biochemistry of the glycosyl-phosphatidylinositol membrane protein anchors. Biochem. J. 244:1-13, 1987.
25. Davitz, M.A., Low, M.G., Nussenzweig, V. Release of decay-accelerating factor (DAF) from the cell membrane by phosphatidylinositol-specific phospholipase C (PIPLC). J. Exp. Med. 163:1150-1161, 1986.
26. Wallner, B.P., Frey, A.Z., Tizard, R., Mattaliano, R.J., Hession, C., Sanders, M.E., Dustin, M.L., Springer, T.A. Primary structure of lymphocyte function-associated antigen 3 (LFA-3). J. Exp. Med. 166:923-932, 1987.
27. Stiernberg, J., Low, M.G., Flaherty, L., Kincade, P.W. Removal of lymphocyte surface molecules with phosphatidylinositol-specific phospholipase C: Effects on mitogen responses and evidence that ThB

and certain Qa antigens are membrane-anchored via phosphatidylinositol. J. Immunol. 138:3877-3884, 1987.
28. Soloski, M.J., Vernachio, J., Einhorn, G., Lattimore, A. Qa gene expression: Biosynthesis and secretion of Qa-2 molecules in activated T cells. Proc. Natl. Acad. Sci. USA 83:2949-2953, 1986.

ERYTHROCYTE PYRIMIDINE 5'-NUCLEOTIDASE: HPLC ASSAY, OBSERVATION ON THE ENZYME BEHAVIOR

Giulio Magni, Adolfo Amici, *Silverio Ruggieri, *Alberto Vita, and *Paolo Natalini

Istituto di Biochimica, Università di Ancona, 60100 Ancona, and *Dipartimento di Biologia Cellulare, 62032 Camerino, Italy

INTRODUCTION

The presence of pyrimidine specific 5'-nucleotidase, which catalyzes the hydrolysis of pyrimidine nucleoside monophosphates, has been reported in 1974 by Valentine et al[1]. The enzyme is located in the soluble fraction of normal human erythrocytes. The same authors have also described that hereditary deficiency of the enzyme results in a chronic hemolytic anemia. The red cell shows basophilic stippling and contains a marked increased amount of nucleotides, several time greater than normal, 65 to 80 % of which are pyrimidine in nature. However the mechanism of accumulation of total nucleotides is still unclear[2]. In order to get insight into such a mechanism a number of studies on the mayor properties of pyrimidine 5'-nucleotidase have been conducted in erythrocytes from both normal and affected individuals. One of the problem encountered in the study of the enzyme is the development of an appropriate assay procedure suitable for measuring the activity in crude lysate. Three methodological approaches have been described for the determination of pyrimidine 5'-nucleotidase (P5N; 5'-ribonucleotide phosphohydrolase, EC 3.1.3.5). The first involves colorimetric determinations of the inorganic phosphate which is liberated from the substrates, CMP or UMP, upon 1 to 2 hours incubations[3]. The second implies setting up of radiometric procedures[4-6]. The third uses a continuous spectrophotometric assay, based upon measurement of inorganic phosphate liberation from pyrimidine nucleoside monophosphates through an ancillary system constituted of inosine, purine nucleoside phosphorylase and xanthine oxidase[7]. Although this is simpler than other methods, the fact that it is based upon the liberation of inorganic phosphate creates problems in cases of severe pyrimidine 5'-nucleotidase deficiency because of phosphate production from accumulated endogenous nucleotides. In our laboratory we have developed a continuous spectrophotometric method to determine pyrimidine 5'-nucleotidase activity , based upon measurements of uracil formation through the consecutive action of an ancillary enzyme system composed of cytidine deaminase and uridine nucleosidase (unpublished results, sub-

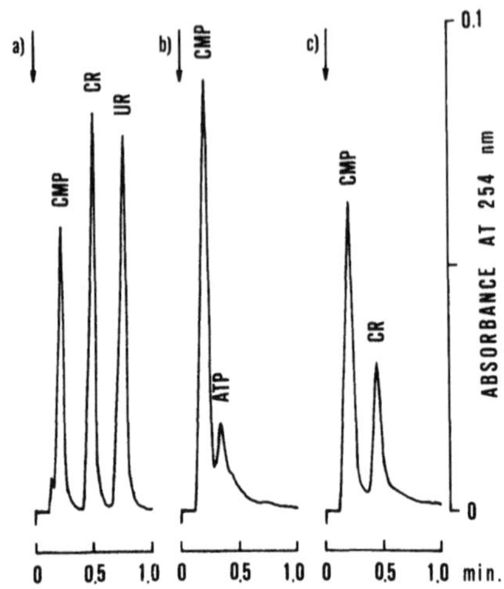

Fig. 1. HPLC assay of pyrimidine 5'-nucleotidase. Hemolysate preparation: freshly drawn blood was filtered through a column of microcrystalline cellulose and α-cellulose (Sigma) to remove leukocytes and platelets[13]. The erythrocytes were washed three times with saline, and the lysis was performed by the addition to the packed cells of four volumes of 10 mM Bis-Tris-HCl buffer solution pH 6.5 containing 1 mM DTT and 0.5 mM EDTA. The reaction mixture contained, in the final volume of 0.5 ml, 0.242 mM CMP, 50 mM Tris-HCl pH 7.5, 0.5 mM DTT, 1mM $MgCl_2$ and the appropriate amount of enzyme. After 30' incubation at 37°C 100 µl of the mixture were added to 50 µl of 1.2 M $HClO_4$, vigorously shaken, centrifuged and neutralized with 40 µl of 1 M K_2CO_3. After centrifugation 50 µl were injected into a LKB chromatographic apparatus equipped with a dual piston pump, a fixed wavelength (254 nm) detector and an Anacomp 220 integrator (Kontron Instruments). The column was a 20 mm x 4.6 mm LC-18 gard column (Supelco, Belefonte, PA). The mobile phase consisted of 0.1 M KH_2PO_4 pH 6.0 solution not containing methanol. The flow rate was 2.0 ml/min. at a pressure of 30 bars and the elution was completed within 60 seconds. (a) separation of standards (4 nanomoles); (b) separation of the reaction mixture at zero time and (c) after 30' incubation. One unit of enzyme activity is defined as the amount of the enzyme capable to hydrolyze 1 micromole of CMP per minute under standard assay conditions.

mitted for publication). In order to avoid the disadvantage due to the commercial unavailability of the ancillary enzyme system, we describe in this report an assay method based on the detection of the nucleosides formed in the enzymatic reaction through a convenient and fast HPLC separation. In the present paper the major properties of a partially purified enzyme preparation are also reported.

HPLC ASSAY

Fig.1 shows the separation obtained by injecting in the chromatograph (1a) a mixture of external standard concentrations of CMP, cytidine and uridine. The same figure depicts a typical elution of the products of the reaction (1b) at 0 time and (1c) after 30' incubation at 37°C. It should be pointed out that the low ATP amount observed at 0' incubation time (fig.1b) is completely absent in the elution chromatogram corresponding to 30 minutes of incubation (fig.1c).

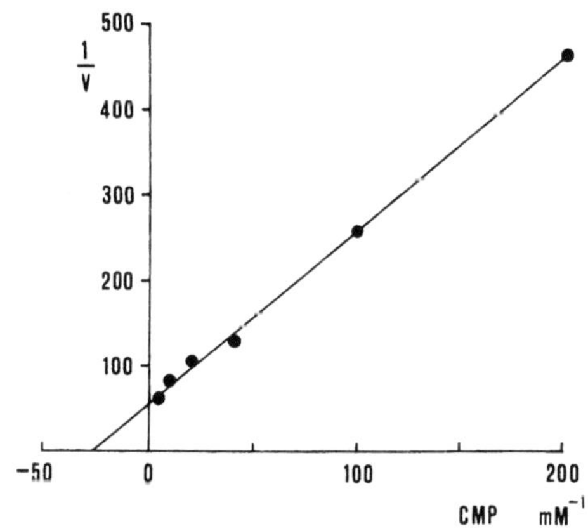

Fig. 2. Double reciprocal plot of velocity measurement versus substrate (CMP) concentration. The enzyme activity has been measured as described in the legend of fig.1 by using 20 µl of lysate.

Adopting the above assay procedure it has been studied the effect of CMP concentration on the reaction velocity in the lysate. By plotting the data according to Lineweaver and Burk a straight line has been obtained as shown in fig.2. From double reciprocal plot of the data obtained from the same experiment a Km value of 30 µM has been calculated by using CMP as the substrate. Fig.3 describes the behavior of the reaction velocity by increasing the enzyme concentration, i.e. the lysate amount. The results indicate that up to 50 µl of lysate the velocity increase is linear by using an incubation time of 30 minutes. The time course of the reaction by mixing 20 µl of the lysate whit 0.242 mM CMP shows a linearity up to 60 minutes. It should be remarked that the large amounts of both the substrate and the enzyme which can

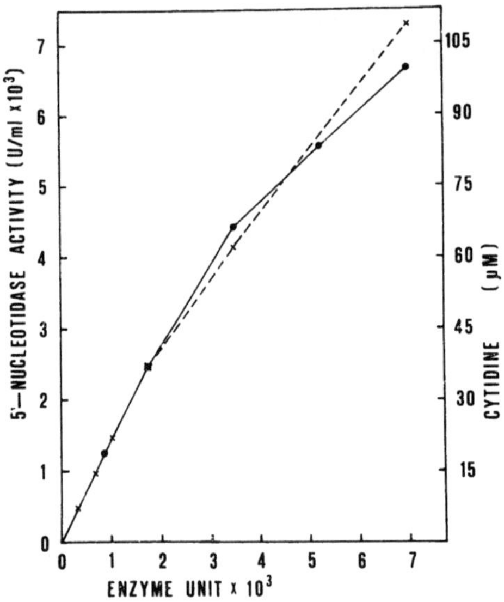

Fig. 3. Pyrimidine 5'-nucleotidase activity at different enzyme concentration. (x----x) lysate, (o----o) purified enzyme. Assay conditons were as described in the legend of fig. 1

be employed only by using an HPLC assay and the rapidity of the reaction product quantitation, less than 60' seconds, renders the present procedure superior to all other assays adopted for pyrimidine 5'-nucleotidase activity measurements. Preliminary data from our group indicate that the present methodology can be extended both to other pyrimidine nucleoside monophosphate substrates and for assaying 5'-nucleotidases capable to dephosphorylate purine and pyrimidine nucleotides.

Red cell lysate containing 14.64 enzyme units was subjected to subsequent purification steps essentially according to the procedure developed by Torrance et al[8]. The last step of the purification procedure submitted to gel electrophoresis according to Ornstein[9] and Davis[10] resulted in a series of Coomassie stainable bands. For reasons which we have not been able to understand the enzyme preparation was purified 4,420 fold instead of 252,272 as claimed by the above authors[8], being the initial specific activity of the lysate and the final recovery comparable to those already described[8]. The enzyme from the last step of the purification procedure is stable for several months if stored at -80°C. After one month at -20°C it retains 80 % of its original activity. The molec-

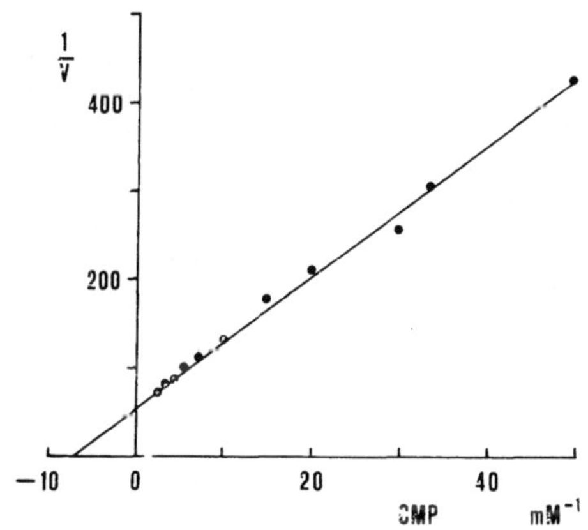

Fig. 4. Double reciprocal plot of velocity measurement versus substrate (CMP) concentration. The enzyme activity has been measured as described in the legend of fig. 1 by using 0.001 unit of partially purified enzyme.

ular weight of the enzyme estimated by gel filtration, according to Andrews[11] (1.6 x 55.5 cm S-300 superfine column equilibrated with 10 mM Bis-Tris containing 0.5 mM DTT and 1mM EDTA), is 32,000. This value partially agrees with that of 28,000 described for the same enzyme[8]. The electrophocusing experiment, corresponding to the last step of the purification procedure indicates that the isoelectric point is at pH 5.1. Fig.4 shows the behavior of the enzyme activity versus CMP concentration plotted according to Lineweaver and Burk. The calculated Km value is 140 µM instead of 30 µM as it has been found for the enzyme in the lysate (fig.2). Different forms of the enzyme activity have been described, and characterized respect to their specificity toward pyrimidine nucleotides[12]. Thus it is possible that the present purified preparation represents one of the multiple forms of the enzyme present in the lysate. However a Km value of 10 μM has been reported by others [8] by using purified enzyme preparations. Table 1 shows the effect of various compounds on the activity of the enzyme. In addition to the substances listed in the table, the following metabolites including guanosine, inosine, ribose, ribose-1-phosphate, ribose-5-phosphate, glucose-1-phosphate, glucose-6-phosphate, glucose-1,6-diphosphate and deoxyribose-5-phosphate resulted ineffective in modulating pyrimidine 5'-nucleotidase. From the results shown in the table it is remarkable that the ability to inhibit the enzyme activity in the lysate by AMP and inorganic phosphate is lost when the same compounds are tested on purified preparation. It should be pointed out that the partially purified enzyme described in [8] is inhibited by inorganic phosphate even at lower concentration. Paglia and Valentine [3] reported an inhibitory effect exerted on the lysate activity by adenosine and adenine. Both of them resulted ineffective, as shown in table 1, on the lysate activity and on the purified preparation as well. This might be due to the difference in the assay method adopted. In fact, these authors determine the inorganic phosphate liberated upon 1 to 2 hours incubation time of the lysate with the substrate and the various effectors. It would be not surprising if the phosphate generated by the enzymatic reaction is readily incorporated into various metabolites through the action of purine nucleoside phosphorylase activity which is abundantly present in human red cell.

Table 1. Effect of various compounds on pyrimidine 5'-nucleotidase activity

Addition (1 mM)	% Inhibition	
	Lysate	Purified
None	0	0
EDTA	-	94
1,10-Phenanthroline	-	91
PCMB	-	97[a]
DTNB	-	98[a]
ATP	-	60
AMP	65	0
IMP	-	36
Adenosine	15	0
Thymine riboside	-	55
Adenine	10	3
Pi (10 mM)	33	0

The enzyme activity has been assayed by HPLC as described in the legend of fig.1, by using 0.001 unit of enzyme.
[a]DTT omitted

REFERENCES

1. W. N. Valentine, K. Fink, D. E. Paglia, S. R. Harris, and W. S. Adams, Hereditary hemolytic anemia with human erythrocyte pyrimidine 5'-nucleotidase deficiency, J.Clin. Invest., 54:866 (1974).
2. E. H. Harley, and P. Berman, Diagnostic and therapeutic approaches in pyrimidine 5'-nucleotidase deficiency, Adv. Exp. Med. Biol.,165A:103 (1982).
3. D. E. Paglia, and W. N. Valentine, Characteristics of a pyrimidine-specific 5'-nucleotidase in human erythrocytes, J. Biol. Chem., 250:7973 (1975).
4. H. Buc, and J. C. Kaplan, Red cell pyrimidine 5'-nucleotidase and lead poisoning. Clin. Chim. Acta, 85:193 (1978).
5. P. H. Ellims, L. Bailey, and M. B. Van Der Weyden, An improved method for the determination of human erythrocyte pyrimidine 5'-nucleotidase activity, Clin. Chim. Acta, 88:99 (1978).
6. J. D. Torrance, C. Westand, and E. Bentler, A simple rapid radiometric assay for pyrimidine 5'-nucleotidase, J. Lab. Clin. Med., 90:563 (1977).
7. C. R. Zerez, and K. R. Tanaka, A continuous spectrophotometric assay for pyrimidine 5'-nucleotidase, Anal. Biochem., 151:282 (1985).
8. J. D. Torrance, D. Whittaker, and E. Beutler, Purification and properties of human erythrocyte pyrimidine 5'-nucleotidase, Proc. Natl. Acad. Sci. USA, 74:3701 (1977).
9. L. Ornstein, Disc-electrophoresis I. Background and theory, Ann. N. Y. Acad. Sci., 121:321 (1964).
10. B. J. Davis, Disc electrophoresis II. Method and application to human serum proteins, Ann. N. Y. Acad. Sci., 121:404 (1964).
11. P. Andrews, The gel-filtration behaviour of proteins related to their molecular weight over a wide range, Biochem. J., 96:595 (1965).
12. A. Hirono, H. Fujii, H. Natori, I. Kurokawa, and S. Miwa, Chromatographic analysis of human erythrocyte pyrimidine 5'-nucleotidase from five patients with pyrimidine 5'-nucleotidase deficiency, Brith. J. Haem., 65: 35 (1987).
13. E. Beutler, C. West, and K. G. Blume, The removal of leukocytes and platelets from whole blood, J. Lab. Clin. Med., 88:328 (1976).

STRUCTURE OF XANTHINE DEHYDROGENASE FROM CHICKEN AND RAT LIVER: CHEMICAL MODIFICATION OF NAD BINDING SITE WITH 5'-FSBA

Tomoko Nishino, Takeshi Nishino, Kumi Noda and Keizo Tsushima

Department of Biochemistry, Yokohama City University School of Medicine, Fukuura 3-9, Kanazawa-ku, Yokohama 236, Japan

INTRODUCTION

Avian and mammalian xanthine dehydrogenases have similarities in the molecular weight, the composition of cofactors (one FAD, two Fe_2S_2, one Mo-pterine per subunit) and the substrate specificity[1]. Both enzymes are known to exist mostly as NAD-dependent dehydrogenase type (D-type) in freshly prepared samples, but mammalian enzymes are converted to the O_2-dependent oxidase type (O-type) reversibly by thiol oxidation and irreversibly by proteolysis[2,3,4]. On the other hand, avian enzymes have never been converted to O-type. Structural comparison between chicken and rat liver enzymes, particularly around NAD binding site, seems to be important for understanding of the mechanism of interconversion between two forms and of a physiological significance of these two enzymes. Previously we reported the chemical modification of NAD binding site of chicken liver enzyme with an active site directed reagent, 5'-FSBA[5]. In order to know the structural difference or similarity between these enzymes limited digestion of both enzymes with protease and chemical modification of NAD-binding site of rat liver D-type enzyme have been made.

MATERIALS AND METHODS

Chicken liver xanthine dehydrogenase was purified from livers of chickens fed a high protein diet with slight modification[6]. Rat liver xanthine dehydrogenase was purified by the method of Ikegami and Nishino[7]. The ratio of D-type to O-type activity (D/O ratio) of the freshly prepared enzyme was around 2. Before experiment the enzyme was incubated at 25°C for 1.5 hrs with 5 mM DTT in the presence of 1 M salicylate followed by passing through Sephadex G-25 to remove DTT and salicylate. The D/O ratio of the obtained enzyme was 9.

Xanthine, 5'-FSBA, N-acetyltyrosine and N-acetyllysine were purchased from Sigma, NAD and NADH from Oriental Yeast Co., and methylene blue and DCI from Wako Chemicals. p-Fluorosulfonyl benzoic acid was purchased from Aldrich and DTT was from Boehringer Mannheim. CBS-tyrosine and CBS-lysine were synthesized by the procedure of Likos et al.[8]. All other reagents were of analytical grade. ^{14}C-5'-FSBA was obtained from New England Nuclear and was diluted with cold 5'-FSBA to a specific activity of 29100 dpm/nmol. Specific activity was checked by measuring radioactivity and concentration of 5'-FSBA with a molar extinction coefficient of $1.58 \times 10^4\ M^{-1}cm^{-1}$ at 259 nm.

Chicken and rat liver xanthine dehydrogenase activities were determined by the method described previously[6,7]. D/O ratio defined by Waud and Rajagopalan[4] was determined in the assay mixture described previously[7]. The concentrations of chicken and rat liver enzymes were determined photometrically using molar extinction coefficients at 450 nm of 35800 and 36500, respectively[4,9].

Modification of rat liver enzyme was performed in 0.05M potassium phosphate buffer (pH 7.8) at 25°C with various concentrations of ^{14}C-5'-FSBA. After various incubation periods, aliquots were withdrawn and residual catalytic activities were determined. For determination of incorporation of radioactivity, aliquots were chilled in an ice cold water bath followed by passing through Sephadex G-25 to remove excess ^{14}C-FSBA and the obtained samples were subjected to determination of radioactivities, absorption spectrum and catalytic activities. Radioactivity was counted in Aquasol 2 (New England Nuclear) with Beckman liquid scintillation counter.

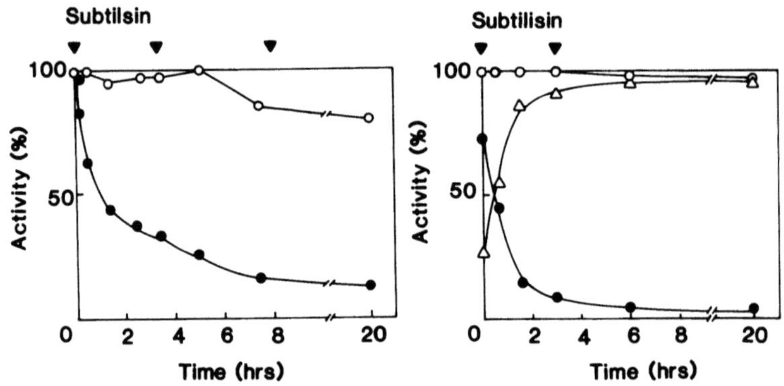

Fig. 1. Time course of subtilisin treatment of chicken (left) and rat (right) liver enzymes. Chicken and rat (73% D-type activity) liver enzyme were incubated with subtilisin as described in Materials and Methods. The points of addition of subtilisin were indicated as arrows. After removing aliquots at various periods, xanthine-NAD (●), xanthine-O_2 (△) and xanthine-DCI (○) activities were determined.

CBS-amino acids were identified essentially by the procedure of Saradambal et al.[10]. The enzyme modified with 5'-FSBA was hydrolyzed for 20 hrs in 6N HCl at 110°C. Hydrolysate or standard CBS-amino acids were spotted on cellulose thin layer chromatography plate (Eastman Kodak) and were developed two dimensionally as described previously. Spots were detected by ninhydrin reaction.

The limited digestion with subtilisin was carried out in 50 mM potassium phosphate buffer (pH 7.8) containing 0.4 mM EDTA at 30°C for 20 hrs. Subtilisin (1/10 w/w) was added to the enzyme solution for a few times (initial and after a few hrs digestion as indicated). The digested enzyme was passed through Ultro-gel filtration (type AcA 34, 1 x 60 cm) to separate subtilisin. After carboxymethylation of the digested enzyme in 6 M guanidine-HCl, the sample was subjected to HPLC using two TSK 3000 sw columns (each 0.75 x 60 cm) which were connected in series and equilibrated with 0.1 M sodium phosphate buffer (pH 7.2) containing 0.1 % SDS. The molecular weight calibration kit (Bio-Rad, low molecular type) was used for determination of molecular weights.

RESULTS AND DISCUSSION

Partial digestion of rat and chicken liver xanthine dehydrogenase with subtilisin

Fig. 1 shows the time courses of limited proteolysis of chicken and rat xanthine dehydrogenase with subtilisin at pH 7.8. Xanthine-NAD activities of both enzyme were reduced to several % of initial activities by this treatment, but xanthine-DCI activities were retained about 90% during incubation. With decrease of xanthine-NAD activity concomitant increase of xanthine $-O_2$ activity was observed with rat liver enzyme but not with chicken liver enzyme. After treatment with subtilisin, the digested enzymes were passed through Ultrogel as described in Materials and methods. The spectra of the eluted enzymes were almost same as those of native enzymes, but spectra showed that the samples digested at pH 10 contained a small amount of deflavoenzymes.

HPLC gel filtration patterns of limited digested enzymes after carboxymethylation were slightly different each other (such as molecular weight of each domain peptide), but similar in main composition of domain peptides. Molecular weights of main bands from rat liver enzyme were 80 kd, 39 kd and 20 kd and those from chicken liver enzyme were 76 kd, 33 kd and 16 kd. These values were coincident well with the results determined by SDS-PAGE.

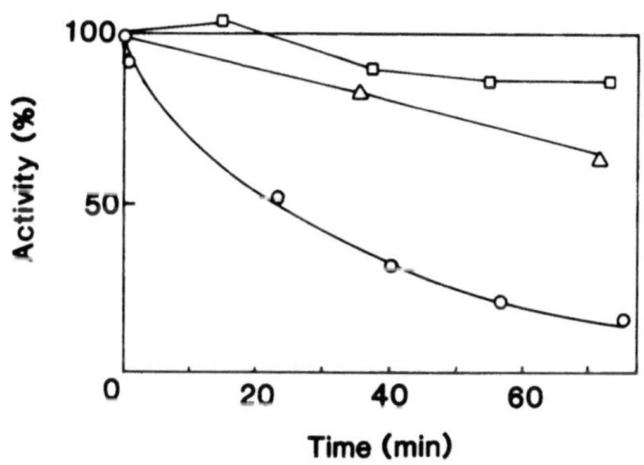

Fig. 2. Time course of inactivation of rat liver xanthine dehydrogenase. 2.7 μ mol of enzyme was incubated with 100 μ mol of 5'-FSBA under the conditions described in Materials and Methods. Xanthine-NAD (○), xanthine-O_2 (□) and xanthine-DCI (△).

Modification of rat liver D-type enzyme with 5'-FSBA

As reported previously NAD dependent activities of chicken liver enzyme were inactivated with 5'-FSBA, an active site directed reagent for nucleotide binding site. 5'-FSBA was also found to inactivate NAD-dependent activities of rat liver xanthine dehydrogenase. As shown in Fig. 2, xanthine-NAD activity was inactivated time dependently by incubation with 5'-FSBA, whereas xanthine-DCI and xanthine-O_2 activities were maintained at the initial level. These results suggest that 5'-FSBA attacked not the xanthine binding site but the NAD binding site of rat liver enzyme. During the incubation, xanthine-O_2 activity was not increased but maintained at the initial level. NAD protected the enzyme almost completely from inactivation, supporting the conclusion that the modification site was near NAD binding site.

Incorporation of radioactivity from ^{14}C-FSBA into the enzyme

Fig. 3 shows a linear correlation between the percentage of remaining activity and the incorporated radioactivity from ^{14}C-5'-FSBA into the enzyme. A stoichiometry of 1.5 mols of 5'-SBA residue per mol of enzyme-FAD was obtained by extrapolation for full inactivation.

Reactivation of 5'-FSBA inactivated enzyme by dithiothreitol

As described previously, the 5'-FSBA inactivated chicken enzyme could be reactivated very slowly but almost completely by incubation with 20 mM DTT at 37°C. However, as shown in Fig. 4, the 5'-FSBA inactivated rat liver enzyme was not reactivated at all with 20 mM DTT at 37°C. This result suggests that the residue responsible for inactivation was different from that of chicken liver enzyme. During this incubation, small amount of remaining O-type activity was converted to D-type activity.

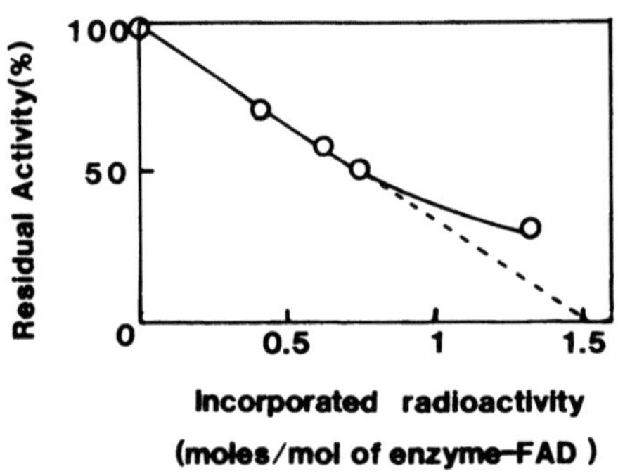

Fig. 3. Incorporation of radioactivity into rat liver xanthine dehydrogenase from ^{14}C-5'-FSBA. After incubation of 17 uM enzyme (D/O ratio 9.7) with 83 uM ^{14}C-5'-FSBA, incorporated radioactivity and enzyme activity were determined as described in Materials and Methods.

Identification of modified amino acid residues

As described in Materials and Methods, hydrolysate of the 5'-FSBA modified rat liver enzyme was applied on thin layer plate for two dimensional chromatography. Two spots of CBS amino acids were detected on the plate , i.e. CBS-lysine and CBS-tyrosine. The spot of CBS-lysine was more intense than that of CBS-tyrosine.

From the results of stoichiometry, failure of reactivation with DTT and TLC analysis, lysine is most likely to be involved in the NAD binding site of rat liver enzyme and was modified with 5'-FSBA, in contrast to the result of chicken liver enzyme[5].

Acknowledgment

This work was supported in part by a Grant-in-Aid for scientific research from the Japanese Ministry of Education, Science and Culture (62480135) and in part by a research grant for the Intractable Diseases from the Japanese Ministry of Health and Welfare.

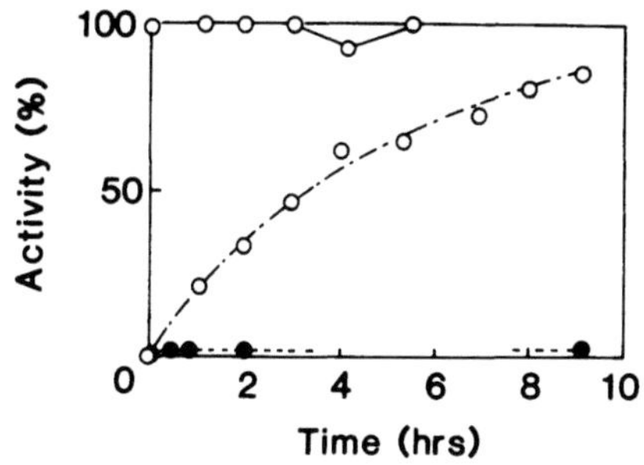

Fig. 4. Reactivation of the 5'-FSBA modified rat and chicken liver xanthine dehydrogenase with DTT. After modified enzymes were incubated in 50 mM potassium phosphate buffer (pH 7.8) containing 0.4 mM EDTA and 20 mM DTT at 37°C, aliquots were withdrawn to determine the catalytic activities. Xanthine-NAD activity (●) of modified rat liver enzyme , NADH-methylene blue activity of modified chicken liver enzyme (O—O) and of native chicken liver enzyme (O—O).

REFERENCES

1. R.C.Bray, Molybdeneum iron-sulfur flavin hydroxylases and relate enzymes, in "The Enzyme", 3rd Ed., vol.12, part B, 299-419, Academic Press, New York and London (1975)
2. E. Della Corte and F. Stirpe, The regulation of rat liver xanthine oxidase, Biochem.J. 126:739 (1972)
3. F. Stirpe and E. Della Corte, The regulation of rat liver xanthine oxidase, J.Biol.Chem. 244:3855 (1969)
4. W.R. Waud and K.V. Rajagopalan, The mechanism of conversion of rat liver xanthine dehydrogenase from an NAD^+-dependent form (type-D) to an O_2-dependent form (type-O), Arch.Biochem.Biophys. 172:365 (1976)
5. T. Nishino and T. Nishino, Evidence for the existence of a tyrosine residue in the nicotinamide adenine dinucleotide binding site of chicken liver xanthine dehydrogenase, Biochemistry 26:3068 (1987)
6. T. Nishino, Purification of hepatic xanthine dehydrogenase from chicken fed a high protein diet, Biochim. Biophys. Acta 341:93 (1974)

7. T. Ikegami and T. Nishino, The presence of desulfo xanthine dehydrogenase in purified and crude enzyme preparations from rat liver, Arch.Biochem.Biophys. 247:254 (1986)
8. J.J. Likos, B. Hess, and R. F. Colman, Affinity labeling of the active site of yeast pyruvate kinase by 5'-p-fluorosulfonylbenzoyl adenosine, J. Biol. Chem. 255:9388 (1980)
9. K. V. Rajagopalan and P. Handler, Purification and properties of chicken liver xanthine dehydrogenase, J. Biol. Chem. 242:4097 (1967)
10. K. V. Saradambal, R. A. Bednar, and R. F. Colman, Lysine and tyrosine in the NADH inhibitory site of bovine liver glutamate dehydrogenase, J. Biol. Chem. 256:11866 (1981)

INTERCONVERSION BETWEEN NAD-DEPENDENT AND O_2-DEPENDENT TYPES OF RAT LIVER XANTHINE DEHYDROGENASE AND DIFFERENCE IN KINETIC AND REDOX PROPERTIES BETWEEN THEM

Tatsuya Saito, Takeshi Nishino and Keizo Tsushima

Department of Biochemistry
Yokohama City University School of Medicine
3-9 Fukuura, Kanazawa-ku, Yokohama, Japan

INTRODUCTION

Rat liver xanthine dehydrogenase consists of two identical subunits, and each subunit contains one FAD, two iron-sulfur centers and one molybdo-pterine as cofactors[1]. This enzyme exists as NAD-dependent type (type-D) in the freshly prepared sample, but through the procedures of extraction and purification most of type-D enzyme is converted to O_2-dependent type (type-O)[2,3]. This conversion occurs either reversibly through oxidation of sulfhydryl groups[2] or irreversibly through proteolysis of enzyme molecule[3]. The number of sulfhydryl groups responsible for this interconversion was reported by Waud and Rajagopalan[4] to be 14 moles per mole of FAD when type-D enzyme was converted to type-O one using 5,5'-dithiobis-(2-nitrobenzoic acid) as SH group modifier. They also reported[4] that these residues existed close to each other. Furthermore they pointed out[4] that the EPR signal of flavin neutral semiquinone was observed in type-D enzyme-antibody complex reduced with xanthine or NADH, and not in type-O enzyme. In present work the sulfhydryl groups responsible for the D-O conversion have been investigated using purified enzyme, and kinetic and redox properties are compared between both types of enzyme.

MATERIALS AND METHODS

Purification of rat liver xanthine dehydrogenase and assays of enzyme activity were carried out by the previously reported method[5]. Type-D enzyme was prepared by incubating freshly purified enzyme with 5 mM dithiothreitol(DTT) at 20°C for 2 hours followed by passing through a small column of Sephadex G-25. In most cases the content of remaining type-O enzyme in the sample after DTT treatment was estimated to be less than 10 %(the D/O ratio = 9)[5]. Type-O enzyme was prepared by incubating purified enzyme with 1 mM 4,4'-dithiodipyridine(DTDP) at 25°C for 1 hour followed by dialysis to remove excess DTDP.

RESULTS AND DISCUSSION

Type-D enzyme was modified with DTDP by adding DTDP stepwise to enzyme at pH 7.8, at 25°C. After each addition absorbance at 324 nm was followed to quantify the formation of 4-thiopyridone using E = 19800 as molar extinction coefficient[6]. After it reached a constant level, NAD-dependent and O_2-dependent activities of the enzyme were determined. As shown in Fig.1-(a), when all type-D enzyme initially present was converted to type-O enzyme, 6.9 moles of 4-thiopyridone were formed per mole of enzyme-bound FAD. Fig.1-(b) shows that 2 moles of 4-thiopyridone were formed upon addition of one mole of DTDP per

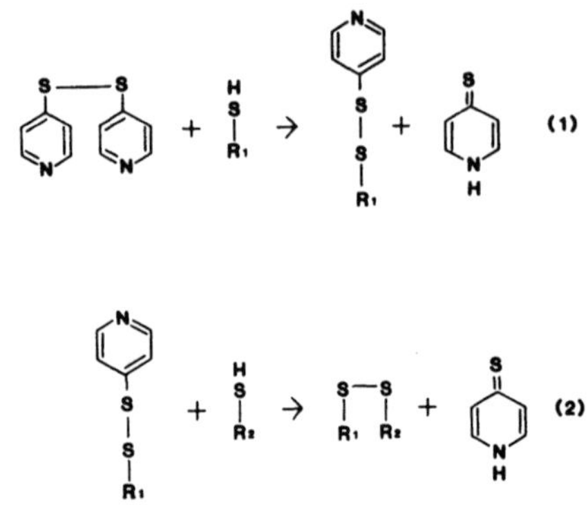

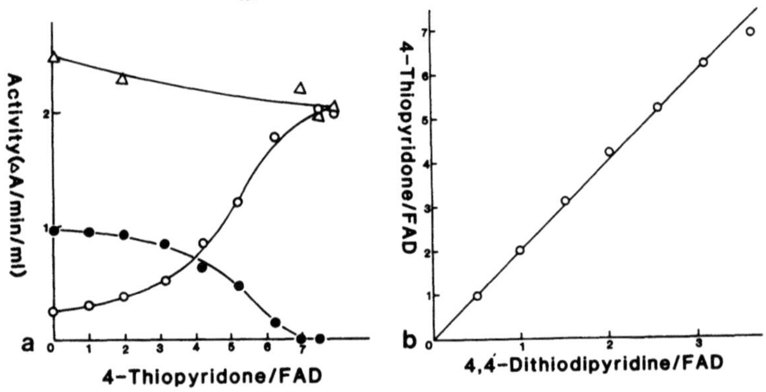

Fig.1 Modification of type-D enzyme with 4,4'-dithiodipyridine(DTDP). (a)The relationship between the activities and 4-thiopyridone formed after modification of type-D enzyme with DTDP. Enzyme assays were carried out under aerobic condition. (●), NAD-dependent activity determined by following formation of NADH at 340 nm in the presence of NAD. (○), O_2-dependent activity determined by following formation of urate at 295 nm in the absence of NAD. (△), activity determined by following formation of urate at 295 nm in the presence of NAD. (b) Relationship between the DTDP added and the 4-thiopyridone formed during the modification of type-D enzyme with DTDP, shown in (a).

enzyme-bound FAD during these modification. This suggests that the modified sulfhydryl groups are located close to each other to form disulfide bond between them.(Scheme 1)

Modification of type-D enzyme(1～3 μM) were carried out at 25°C with 2 mM of monoiodoacetic acid(pH 8.5), 2 mM of iodoacetamide(pH 8.5) and 1 mM of NEM(pH 7.0). In all three cases the conversion to type-O enzyme was not found to proceed completely by these modifiers.

Percentages of remaining type-D activity which was finally obtained after reaction with excess reagents were 70 %, 50 % and 30 % of type-D activity initially present with monoiodoacetic acid, iodoacetamide and NEM, respectively. The order of degree of conversion was coincided with that of hydrophobicity of modifier. This appears to suggest that sulfhydryl groups responsible for D-O conversion are located in hydrophobic environment of protein molecule.

Effects of oxidation-reduction of these sulfhydryl groups on function of enzyme were studied by reductive titration of both types of enzyme under anaerobic condition using sodium dithionite or xanthine as reducing reagents. The reduction of each chromophore was monitored spectrophotometrically, on the basis that the absorbance at 450 nm of the oxidized xanthine dehydrogenase was due to both FAD and iron-sulfur centers, while that at 520 nm was due only to iron-sulfur centers[7]. The condition of titration was pH 7.8 and 25°C. In type-D enzyme development of flavin neutral semiquinone was clearly shown by increase in absorbance at 620 nm during titration. But during titration of type-O enzyme formation of FAD semiquinone was not detected spectrophotometrically. The order of redox potential of chromophores of type-D enzyme is as the following(FADH* and $FADH_2$ represent neutral semiquinone form of FAD and fully reduced FAD, respectively).

$$Fe/S(II) > FAD/FADH^* = Fe/S(I) > FADH^*/FADH_2$$

On the other hand that of type-O enzyme is as the following.

$$Fe/S(II) > FADH^*/FADH_2 > FAD/FADH^* > Fe/S(I)$$

The process of reductive titration of type-D enzyme was markedly changed when it was performed in the presence of NAD. In this titration xanthine was used as a reducing reagent instead of sodium dithionite, because dithionite can directly reduce NAD. When type-D enzyme was anaerobically titrated with xanthine in the presence of 1.8 mM of ß-NAD at pH 7.8 at 25°C, the formation of FAD semiquinone, concomitant with reduction of one iron-sulfur center, was observed from the beginning, in good contrast with titration in the absence of NAD, in which it appeared only after reduction of one of two iron-sulfur centers was completed. It was also noticeable that the maximum formation of FAD semiquinone was about half of that in the absence of NAD. These results indicate that binding of NAD to type-D enzyme shifts the redox potentials of FAD/FADH* and $FADH^*/FADH_2$ in a positive direction so that difference between these two potentials becomes smaller, resulting in less stabilized FAD semiquinone.

Absorbance spectrum of type-D enzyme during turnover with xanthine and oxygen as substrates was recorded at pH 7.8 at 3°C under air saturated condition. The absorbance spectrum revealed formation of pronounced amount of FAD semiquinone. On the other hand type-O enzyme during turnover with xanthine and oxygen as substrates showed no detectable amount of FAD semiquinone. Because turnover of xanthine-oxygen reaction catalyzed by type-O enzyme was very rapid, absorbance change during turnover was studied at pH 7.8 at 4°C under air saturated condition by stopped-flow apparatus.

The effect of NAD on redox property of enzyme was also verified in study of absorbance spectrum of type-D enzyme during turnover with xanthine and NAD as substrates under anaerobic condition at pH 7.8 and 4°C. During turnover the spectrum of enzyme showed only negligible sign of FAD semiquinone. After cease of the formation of NADH the spectrum of the enzyme remained at the same level of reduction as that in turnover. Upon aeration the enzyme began to be reoxidized slowly. The order of reoxidation of chromophores was clearly distinct from that in turnover of reaction of type-D enzyme with xanthine and oxygen as substrates. In the absence of NAD, FAD semiquinone and one iron-sulfur were reoxidized followed by reoxidation of the other iron-sulfur center, while reoxidation of iron-sulfur preceded that of FAD in the presence of NAD. These facts support the order of redox potentials of chromophores of type-D enzyme in the presence or absence of NAD, estimated from the result of reductive titration.

The above mentioned redox properties of type-D enzyme, including formation of stable FAD semiquinone and its decrease by binding of NAD to the enzyme, has also been clearly demonstrated with chicken liver xanthine dehydrogenase[8], which is always NAD-dependent and can never become O_2-dependent. On the other hand the redox properties of type-O enzyme are very similar to those of milk xanthine oxidase[7], which is originally NAD-dependent in fresh milk, but is irreversibly converted to O_2-dependent form through proteolysis in purification procedures.

Steady state kinetics of type-D enzyme with xanthine and NAD or oxygen as substrates and that of type-O enzyme with xanthine and oxygen as substrates were studied at pH 7.8 at 25°C. As was previously reported[1,9,10], the Lineweaver-Burk plots for varying concentrations of one substrate at a series of fixed concentrations of the other substrate showed parallel lines. The Km for xanthine was below 5 μM in all the studied cases. The V_{max} for xanthine-NAD reaction by type-D enzyme was about 800/min/enzyme-bound FAD while that of xanthine-oxygen reaction by type-O enzyme was about 1000/min/enzyme-bound FAD, which are similar to each other. The K_m for O_2 of type-D enzyme was 260 μM, which was about five times as high as that of type-O enzyme. The V_{max} of xanthine-oxygen reaction by type-D enzyme was about one fourth of that by type-O enzyme.

These data appear to suggest that reduced sulfhydryl groups in type-D enzyme are responsible for low accessibility of oxygen molecule to reduced FAD , which is suggested by higher K_m for oxygen of type-D enzyme than that of type-O enzyme. Reduction of sulfhydryl groups possibly changes environment around FAD so that its hydrophobicity may be decreased or reaction site on isoalloxazine ring may become covered with protein loop. Difference in environment around FAD between both types of enzyme has been clearly demonstrated by experiment using 8-SH FAD and 6-OH FAD as active site probe(unpublished data).

On the other hand low V_{max} of xanthine-oxygen reaction by type-D enzyme is possibly due to higher fraction of FAD semiquinone during turnover of type-D enzyme than that of type-O enzyme. During turnover of chicken liver xanthine dehydrogenase with xanthine and oxygen as substrates, it is suggested that a considerable fraction of enzyme having FAD semiquinone transfers electrons to oxygen through sequential steps of one-electron transfer reaction to form O_2^- molecules instead of formation of H_2O_2, while enzyme having fully reduced FAD reacts with oxygen through two-electrons transfer reaction to form H_2O_2. The rate constant of the one-electron transfer reaction is estimated to be smaller than that of the two-electrons transfer reaction(unpublished data). But when NAD binds to type-D enzyme, the fraction of FAD semiquinone is decreased, resulting in an increased fraction of fully reduced FAD, which possibly leads to V_{max} of xanthine-NAD reaction by type-D enzyme as high as that of xanthine-oxygen reaction by type-O enzyme. It may be possible that NAD-dependent activity of type-D enzyme is regulated by NAD in that reactivity of reduced FAD toward NAD is induced by binding of the latter to the enzyme.

Acknowledgement

We thank to Dr. Vincent Massey, Professor of University of Michigan, for helpful discussion.

This work was supported in part by a grant-in-aid for scientific research from the Japanese Ministry of Education, Science and Culture (62480135), in part by a research grant for the intractable diseases from Japanese Ministry of Health and welfare and in part by a research grant from Gout Research Foundation of Japan.

REFERENCES

1. R.C.Bray, Molybdeneum iron-sulfur flavin hydroxylases and relate enzymes, in "The Enzyme", 3rd Ed., vol.12, part B, 299-419, Academic Press, New York and London (1975)
2. E. Della Corte and F. Stirpe, The regulation of rat liver xanthine oxidase, Biochem.J. 126:739 (1972)
3. F. Stirpe and E. Della Corte, The regulation of rat liver xanthine oxidase, J.Biol.Chem. 244:3855 (1969)
4. W.R. Waud and K.V. Rajagopalan, The mechanism of conversion of rat liver xanthine dehydrogenase from an NAD^+-dependent form (type-D) to an O_2-dependent form (type-O), Arch.Biochem.Biophys. 172:365 (1976)
5. T. Ikegami and T. Nishino, The presence of desulfo xanthine dehydrogenase in purified and crude enzyme preparations from rat liver, Arch.Biochem.Biophys. 247:254 (1986)
6. D.R. Grassetti and J.F Murray, Determination of sulfhydryl groups with 2,2'- or 4,4'-dithiodipyridine, Arch.Biochem.Biophys. 119:41 (1967)
7. J.S. Olson, D.P. Ballou, G. Palmer and V. Massey, The mechanism of action of xanthine oxidase, J.Biol.Chem. 249:4363 (1974)
8. L.M.Schopher, V. Massey and T. Nishino, Rapid reaction studies on the reduction and oxidation of chicken liver xanthine dehydrogenase by the xanthine/urate and NADH/NAD couples, J.Biol.Chem. (in press)
9. V. Massey, P.E. Brumby and H. Komai, Studies on milk xanthine oxidase, J.Biol.Chem. 244:1682 (1969)
10. M.P. Coughlan and K.V. Rajagopalan, The kinetic mechanism of xanthine dehydrogenase and related enzymes, Eur.J.Biochem. 105:81(1980)

MAMMALIAN XANTHINE OXIDOREDUCTASE — A UNIQUE ENZYME AMONG HYPOXANTHINE HYDROXYLATING ENZYMES IN VERTEBRATES

Maria M. Jeżewska, Zbigniew W. Kamiński
and Barbara Zakrzewska

Institute of Biochemistry and Biophysics
of Polish Academy of Sciences
Department of Comparative Biochemistry
Warsaw, Poland

INTRODUCTION

Xanthine:NAD^+ oxidoreductase catalyzes irreversible hydroxylation: hypoxanthine → xanthine → uric acid, which eliminates hypoxanthine from the purine compound pool, and thus may prevent its reutilization in the salvage pathway.
In vertebrates hydroxylation is associated with the reduction of NAD^+ to NADH which may be used in the respiratory chain for recovery of ATP lost during purine de novo biosynthesis.

The present report concernes the catalytic properties of xanthine oxidoreductases from ammono—, ureo— and uricotelic vertebrates which considerably differ in the intensity of purine de novo synthesis. In animals with various types of nitrogen excretion, hypoxanthine reutilization in the IMP resynthesis as well as hypoxanthine elimination differ in importance, and this may be reflected in the xanthine oxido-reductase properties.

We examined xanthine:NAD^+ oxidoreductase from fish *Cyprinus carpio* (ammonotelic), snake *Natrix natrix*, hen *Gallus gallus*, turkey *Meleagris domestica* (all three uricotelic) toad *Bufo viridis* (mainly ureotelic) and Wistar rat (ureotelic). The data for the rat, snake and fish enzyme have mostly been published[1,2,3,4,5,6], whereas those for the toad, hen and turkey were not.

MATERIALS AND METHODS

Enzyme preparations from liver (and heart in the case of rat) were obtained as described [1,2,4,5,6]. Xanthine oxidoreductase was 50 - 70 times purified as compared with the 120 000 x g supernatant, and was free from the NAD^+ reductase, aldehyde oxidase and uric acid oxidase activities.

Spectrophotometric methods for measurements of various enzymic activities in the enzyme preparations and for following the two-step hypoxanthine hydroxylation were as previously described [1,2,4,5,6].

RESULTS AND DISCUSSION

Occurrence of xanthine:O_2 oxidoreductase forms

Xanthine oxidoreductases freshly obtained from liver of all vertebrates examined by us and other authors [7,8,9] hydroxylate oxypurines, preferentially with NAD^+ as natural cosubstrate.

Mammalian xanthine:NAD^+ oxidoreductase is known to be converted by several factors to the oxidase form[10]. We found tha in liver and heart of rat this enzyme occurs in three forms: one acting with NAD^+, another one using NAD^+ and, in its absence O_2, and a third exclusively acting with O_2[1,4]; this last form is known to produce H_2O_2 and oxygen radicals. Under normal physiological conditions two forms acting with O_2 occur in minor amount Interconversion of above three forms is associated with the reduction/oxidation of one sulfhydryl group[3], probably in the vicinity of the FAD site in the active centre, because NAD^+ protects this group against modification by sulfhydryl reagents[3]

Non-mammalian vertebrates xanthine:NAD^+ oxidoreductases are nonconvertible by sulfhydryl reagents to the xanthine:O_2 forms. However, their NAD^+-dependent xanthine hydroxylating activity is destroyed by some such reagents. For example, the enzyme from snake and toad do not hydroxylate xanthine after 5-min preincubation with 50 and 200µM $CuSO_4$, respectively; the hen enzyme is resistant toward 1 mM $CuSO_4$ and 100 mM 5,5'-dithiobis-(2-nitrobenzoic acid),whereas its activity is destroyed by this reagent at a 500 µM concentration.

Some kinetic properties of xanthine:NAD^+ oxidoreductases

The differences in the kinetic properties of the enzyme studied (Table 1) seem to be consistent with the physiological role of this enzyme in animals differing in the intensity of purine metabolism, on account of various types of nitrogen excretion.

The apparent K_m's for NAD^+ are substantially lower in the case of uricotelic species. This may reflect the preferential use of NAD^+ for elimination of hypoxanthine synthesized in abudance during the protein catabolism. NADH oxidation in the respiratory chain may be of a special importance in these animals, on account of vigorous ATP-consuming synthesis of purines.

On the other hand, the differences in K_m's for xanthine may suggest that the enzyme <u>in vivo</u> operates at higher xanthine concenrations in urico- than in ureo- and ammonotelic animals. The fact that hypoxanthine is faster hydroxylated than xanthine indicates that hypoxanthine is preferentially eliminated from the purine compound pool, probably as the main intermediate of IMP

Table 1. Some kinetic parameters of xanthine:NAD^+ oxidoreductases from vertebrates

Animal	Saturating concentration (μM)		Hydroxylation rate[x] of		Apparent K_m for (μM)	
	Hyp	Xan	Hyp	Xan	NAD^+	Xan
Uricotelic						
snake[5]	100	25	>		14.4	12.3
turkey	> 100	100	>		10.2	18.0
hen	200	100	>		3.1	12.0
Ammonotelic						
fish[6]	40	40	<		21.5	4.5
Ureotelic						
toad	> 60	50	=		27.0	7.0-9.0
rat[2]	20	20	=		22.4[xx]	2.4

[x]measured at saturating concenrations

[xx]according to[11]

catabolism. Moreover, the enzyme from uricotelic animals is saturated by hypoxanthine at particularly high concentrations. The possibility of changes in the hydroxylation rate within so wide a range of hypoxanthine concentrations offers a simple mechanism of a fast response to fluctuations in the protein catabolism resulting from alternation of periods of food intake and digestion.

The properties of fish and toad enzyme seem to be intermediate between those of uricotelic species and ureotelic rat.

Xanthine oxidoreductase from rat hydroxylates both oxypurines at an equal rate and becomes saturated at their identic concentrations; thus, probably the K_m's for both substrates are also equal. Therefore xanthine arising from guanine compound po could *in vivo* effectively compete with hypoxanthine for the enzyme, and in this way spare hypoxanthine for the IMP resynthesis catalyzed by hypoxanthine-guanine phosphoribosyltransfera

Hypoxanthine → xanthine → uric acid hydroxylation

Two patterns of the time-course of the two-step hypoxanthi hydroxylation were found: the one is characteristic of enzyme from toad, snake, turkey and hen (Fig. 1a), and the other one – of enzyme from fish and rat (Fig. 1b). The most evident difference between these patterns involves the dependence of uric acid production on the initial hypoxanthine concentration.

As concerns the enzyme from uricotelic animals and toad, the uric acid production is initially independent of the initia hypoxanthine concentration and also of xanthine accumulation. At the later stages of the hydroxylation process, the rate of uric acid production is positively related to the oxypurine concentrations. Such time-course may suggest a lack of competition between hypoxanthine and xanthine for the enzyme.

In case of the fish and rat enzyme, the higher the initial hypoxanthine concentration, the greater the accumulation of xanthine, but the lower the uric acid production. Such a time-course is in accord with the known competition between both substrates for the mammalian enzyme. Only in the case of the ra enzyme, the exhaustion of both hypoxanthine and xanthine is simultaneous, and the curve representing the changes in xanthin

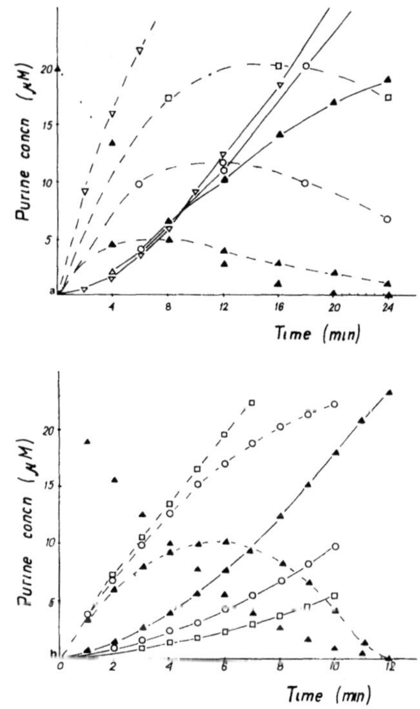

Fig.1. Time-course of hypoxanthine → xanthine → uric acid hydroxylation catalyzed by xanthine:NAD^+ oxidoreductase a. of rat (similar found for the fish enzyme), b. of hen (similar for the snake, turkey and toad enzyme). Details of spectrophotometric measurements as described[2]. The incubation mixture (3 ml) contained: 50 mM Tris-HCl buffer pH 8.0, 350 µM NAD^+, 88 (a) or 51 (b) pkat/ml of enzyme activity, lactate dehydrogenase + pyruvate, and hypoxanthine at: 20-23 (▲); 46 (○); 92 (□); 200 (◁) µM concentration. .... hypoxanthine utilization; -·-·-·- xanthine accumulation; —— uric acid production.

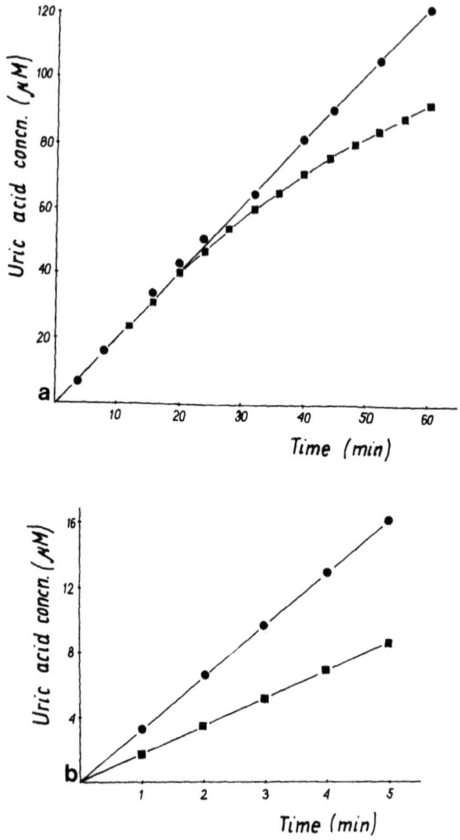

Fig. 2. Effect of NADH accumulation on xanthine → uric acid hydroxylation catalyzed by xanthine:NAD^+ oxidoreductase a. from hen (similar effect was observed in the case of snake, turkey, fish and toad enzyme); b. from rat. Details of spectrophotometric measurements of uric acid formation as described[1]. The incubation mixture contained 50 mM Tris-HCl buffer pH 8.0, 350 μM NAD^+, 33 (a) or 53 (b) pkat of enzymic activity/ ml of incubation mixture, xanthine at 146 (a) or 20 (b) μM concentration, with (●) or without (■) lactate dehydrogenase + pyruvate (NADH reoxidized or not, respectively).

concentration has a symmetrical shape (compare data in Table 1).

Inhibition of oxypurine hydroxylation by NADH

Xanthine:NAD^+ oxidoreductase from rat and chicken have been found to be inhibited by NADH[11]. In our studies on the effect of NADH accumulating during hydroxylation of oxypurines on the rate of their hydroxylation, we found that this enzyme from all non-mammalian vertebrates examined is inhibited in a different way (Fig.2a) than the enzyme from rat (Fig. 2b).

Inhibition of xanthine:NAD^+ oxidoreductase from all vertebrates, except of rat, appears when NADH is accumulated at micromolar concentrations in the reaction medium, and increases over time with further rise of the NADH concentration. The enzyme inhibition attains no more than 20 - 30 %. The low sensitivity of the enzyme from those animals to the inhibition by NADH even at high, rather nonphysiological concentrations excluded the regulatory effect of changes in the NAD^+/NADH on *in vivo* hydroxylation of oxypurines to uric acid.

Instead the rat enzyme is inhibited in 45 -50 % within some seconds by NADH accumulated at concentrations lower than 1 µM [2,4], i.e. at approximately physiological concentration[1:]. In the case of hypoxanthine hydroxylation, the accumulation of xanthine is inhibited in about 35 %, and the production of uric acid in over 60 %. We have postulated[2] that such inhibition by NADH could *in vivo* spare hypoxanthine for the salvage pathway and thus limit the IMP *de novo* biosynthesis under conditions of hypoxia.

It seems that the progressive inhibition of the enzyme by NADH in the case of nonmammalian vertebrates results from the competition between NAD^+ and NADH for the FAD site in the active centre. In contrast, some other unknown mechanism seem to be involved in the instantaneous inhibition of the mammalian xanthine:NAD^+ oxidoreductase.

REFERENCES

1. Z. W. Kamiński and M. M. Jeżewska, Intermediate dehydrogenase-oxidase form of xanthine oxidoreductase in rat liver, Biochem. J 181:177 (1979.

2. Z. W. Kamiński and M. M. Jeżewska, Effect of NADH on hypoxanthine hydroxylation by native NAD^+-dependent xanthine oxidoreductase of rat liver, and the possible biological role of this effect, Biochem. J. 200:597 (1981).
3. Z. W. Kamiński and M. M. Jeżewska, Involvement of a single thiol group in the conversion of the NAD^+-dependent activi of rat liver xanthine oxidoreductase to the O_2-dependent one, Biochem. J. 207:341 (1982).
4. Z. W. Kamiński, R. Pohorecki, Ch. L. Ballast and E. F. Domin Three Forms of xanthine:acceptor oxidoreductase in the rat heart, Circul. Res. 59:628 (1986).
5. Z. W. Kamiński and M. M. Jeżewska, Xanthine:NAD^+ oxidoreductase in the liver of grass snake *Natrix natrix*, Comp. Biochem. Physiol. 78B:447 (1984).
6. Z. W. Kamiński and M. M. Jeżewska, Xanthine:NAD^+ oxidoreductase in the liver of the teleostean fish *Cyprinus carpio*, Comp. Biochem. Physiol. 80B:371 (1985),
7. F. Stirpe and E. Della Corte, Regulation of rat liver xanthi oxidase. Conversion *in vivo* of the enzyme activity from dehydrogenase (type D) to oxidase (type O), J. Biol. Chem. 244:3855 (1969).
8. C. J. Francois, Activité oxydasique de l' oxydoréductase de la xanthine, specifique de la classe des Mammifères, Biochem. System. Ecol. 1:231 (1973).
9. W. F. Cleere and M. P. Coughlan, Avian xanthine dehydrogenases — I. Isolation and characterization of the turkey enzyme, Comp. Biochem. Physiol. 50B:311 (1975).
10. E. Della Corte and F. Stirpe, Regulation of rat liver xanthine oxidase. Involvement of thiol groups in the conversion of the enzyme activity from dehydrogenase (type D) into oxidase (type O) and purification of the enzyme, Biochem. J. 126:739 (1972).
11. E. Della Corte and F. Stirpe, The regulation of xanthine oxidase. Inhibition by reduced nicotinamide-adenine dinucleotide of rat liver xanthine oxidase type D and of chick liver xanthine dehydrogenase, Biochem. J.117:97 (1970).
12. M. E. Tischler, D. Friedericks, K. Coll and J. R. Williamson Pyridine nucleotide distribution and enzyme mass action ratios in hepatocytes from fed and starved rats. Arch. Biochem. Biophys. 184:222 (1977).

HETEROGENEITY OF THE TOXIC MECHANISMS OF METHYLTHIOADENOSIN IN METHYLTHIOADENOSINE PHOSPHORYLASE DEFICIENT MURINE LYMPHOMA CELLS

Hisashi Yamanaka, Kajander, E. Olavi, Masaru Kubota,
D. Bruce Wasson, Erik H. Willis
and Dennis A. Carson

Research Institute of Scripps Clinic
La Jolla, CA, U.S.A.

INTRODUCTION

Methylthioadenosine(MeSAdo) is an end product of the polyamine synthesis and is metabolized by MeSAdo phosphorylase(1). Because this enzyme is deficient in many malignant cell lines (2) and some leukemic patients(3), the toxic mechanisms of MeSAdo is quite important not only in the basic biochemistry but also in the clinical oncology. Exogenous addition of MeSAdo in MeSAdo phosphorylase deficient cells has been shown to exert toxic effects(4). MeSAdo directly inhibits the catalytic activity of spermine synthetase and spermidine synthetase *in vitro* (5), however, a series of evidence has shown that the depletion of polyamine synthesis cannot explain the toxic mechanism of MeSAdo(6). In this report, we have demonstrated two alternative toxic mechanisms of MeSAdo in mammalian cells, one is the inhibition of diphthamide biosynthesis and the other is the intracellular accumulation of the acetylated derivative of decarboxylated S-adenosylmethionine (AdoMet). These data show the heterogenous effects of MeSAdo in MeSAdo phosphorylase deficient cells.

MATERIALS AND METHODS

Cell lines: Murine T lymphoma cell line R1.1, a mutant(H) which is deficient in MeSAdo phosphorylase, and a secondary mutant (H3) which is deficient in MeSAdo and resistant to the exogenous addition of MeSAdo have been described previously(6,7).

Effect of diphtheria toxin and MeSAdo on protein synthesis. Since mouse cells lack the transport system of diphtheria toxin into the intracellular space, low pH treatment (8) was used to facilitate the entry of the functional subunit of diphtheria toxin into these R1.1 cells. To test the effect of MeSAdo on the toxic effect of diphtheria toxin, cells were preincubated with or without MeSAdo for overnight. Then, cells were washed and incubated for 60min at 37° C in the low pH medium(pH5.5) supplemented with various concentrations of nicked, partially purified diphtheria toxin(Calbiochem, La Jolla, CA). After washing the cells, cells were cultured for another 24hrs. For the measurement of the protein synthesis, [^{3}H]leucine incorporation into the acid insoluble fraction was used.

Effect of MeSAdo on ADP-ribosylation of Elongation factor-2 (EF-2) After the preincubation with MeSAdo for 16hrs, EF-2 was partially purified by the method described by Moehring, et al(9,10). The amount of diphthamide present in cells were measured by a incubation of crude EF-2 with 1.67mM [adenine-2,8-^{3}H] NAD and 200mg/ml diphtheria toxin in the presence of the inhibitor of poly(ADP-ribose) polymerase. The radioactivity incorporated into EF-2 was measured either by the radioactive count in the acid precipitable fraction or by SDS-polyacrylamide gel electrophoresis.

Measurement of polyamines, AdoMet and its derivatives: Polyamine, AdoMet and its derivatives in 0.4N perchloric acid soluble fractions were fractionated by reversed-phase ion-paired HPLC. After the separation, polyamines were measured fluorometrically using o-phthalaldehyde and AdoMet and its derivatives were measured spectrophotometrically(11).

Labelling study for identifying the peak: To label the acetylated derivative of AdoMet, H3 cells were incubated with either [2(n)-^{3}H] methionine or [^{3}H] acetate. After the washing, acid soluble fractions were separated by HPLC as described above, and measured the radioactivity in each fraction.

RESULTS AND DISCUSSION

Diphthamide, an unique post-translationally modified histidine, which appears in eukaryotic elongation factor-2(EF-2)(12,13). EF-2 is a ribosomal enzyme which translocate transfer RNA, and one of the key enzymes in the protein synthesis. Interestingly, this molecule is the substrate of the mono(ADP-ribose) transferase of diphtheria toxin or *Pseudomonous* toxin and this molecule is the target of lethal effect of these bacterial exotoxins(13). The structure of diphthamide, 2-[3-carboxyamido-3-(trimethylammonio)propyl]histidine(12), is complicated and the post-translational modification of histidine is also complicated. In the first step of this reaction, MeSAdo may be involved. The aminocarboxypropyl moiety of AdoMet is transferred to histidine to produce MeSAdo(14, Figure 1). In this case, intracellular accumulation of MeSAdo may inhibit the synthesis of diphthamide in MeSAdo phosphorylase deficient cells. We have investigated the effect of MeSAdo on the biosynthesis of diphthamide.

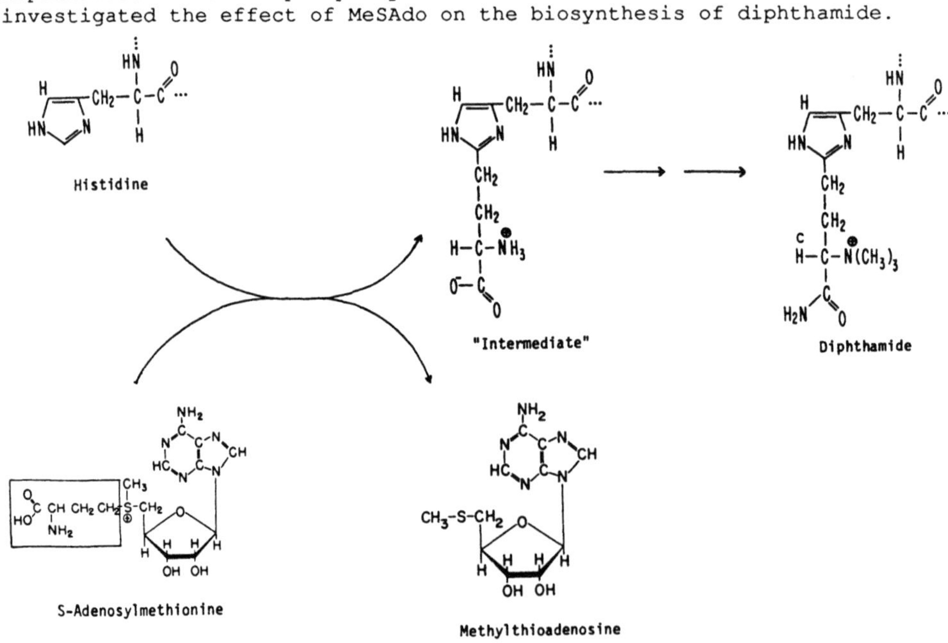

Figure 1 *Pathway of the diphthamide biosynthesis*

We have selected MeSAdo phosphorylase deficient cells from murine T lymphoblast cell line R1.1. Because diphthamide is the acceptor of mono(ADP-ribose) moiety of NAD, we can measure the cellular content of diphthamide by the incorporation of [^{3}H]NAD into EF-2. R1.1 H3 cells were cultured with or without MeSAdo for overnight. These cells were deficient in MeSAdo phosphorylase, but resistant to the exogenous addition of MeSAdo, and this treatment inhibited neither the cell proliferation nor the protein synthesis. Then, cells were extracted, EF-2 was partially purified, and the amount of diphthamide is measured semiquantitatively . The incorporation of [^{3}H]NAD into EF-2 ,whose molecular weight is 94kDa (9), was ascertained by SDS-polyacrylamide gel electrophoresis. The preincubation of R1.1 H3 cells with MeSAdo apparently reduced the cellular content of diphthamide(Figure 2). Although diphthamide is the established target of the diphtheria toxin, the physiological role of diphthamide has not been elucidated well. However, from our result, it is possible to consider that biosynthesis of diphthamide is involved in the toxic mechanism of MeSAdo.

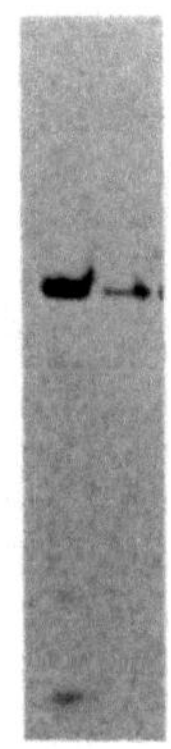

Figure 2. *Intracellular content of diphthamide as shown by the incorporation of [^{3}H]-NAD into elongation factor 2 by diphtheria toxin.*

Cellular content of diphthamide was demonstrated in R1.1 H3 cells preincubated without(lane1) or with 500 mM MeSAdo (lane2) for overnight.

As an alternative mechanism for the toxicity of MeSAdo, we investigated the formation of acetylated derivatives of the decarboxylated AdoMet in mammalian cells. MeSAdo inhibit spermine and spermidine synthetase but activate ornithine decarboxylase and AdoMet decarboxylase(6). As a result, decarboxylated AdoMet accumulates in the cells which are treated by MeSAdo(4). In the cell extract of MeSAdo treated cells, we had observed an unidentified peak in the high performance liquid chromatogram (Figure 3). Because the characteristics of this peak was very similar to that was reported by Pegg and colleagues in difluoromethylornithine (DFMO) treated cells, this peak was characterized by the labelling experiment. MeSAdo phosphorylase deficient H3 cells were cultured with MeSAdo and DFMO in the presence of either [2(n)-^{3}H]-labelled AdoMet or [^{3}H]acetate. In both system, the radioactivities were specifically incorporated into this unidentified peak on the chromatography. This data strongly suggests that the peak reprsents an acetylated derivative of decarboxylated S-adenosylmethionine, which is consistent with the previous report(15).

We have determined the amount of acetylated decarboxylated AdoMet in MeSAdo treated R1.1 H3 cells. Decarboxylated AdoMet accumulated first intracellularly, and subsequently the amount of acetylated decarboxylated AdoMet increased. Finally, the amount of these compounds reached to a plateau, at the amount of approximately equivalent to that of cellular AdoMet(Figure 4). We have to emphasize that AdoMet is quite abundant in cells because it serves not only as a source of the polyamine but also as a donor of cellular methylation reactions(1). The accumulation of acetylated decarboxylated AdoMet was

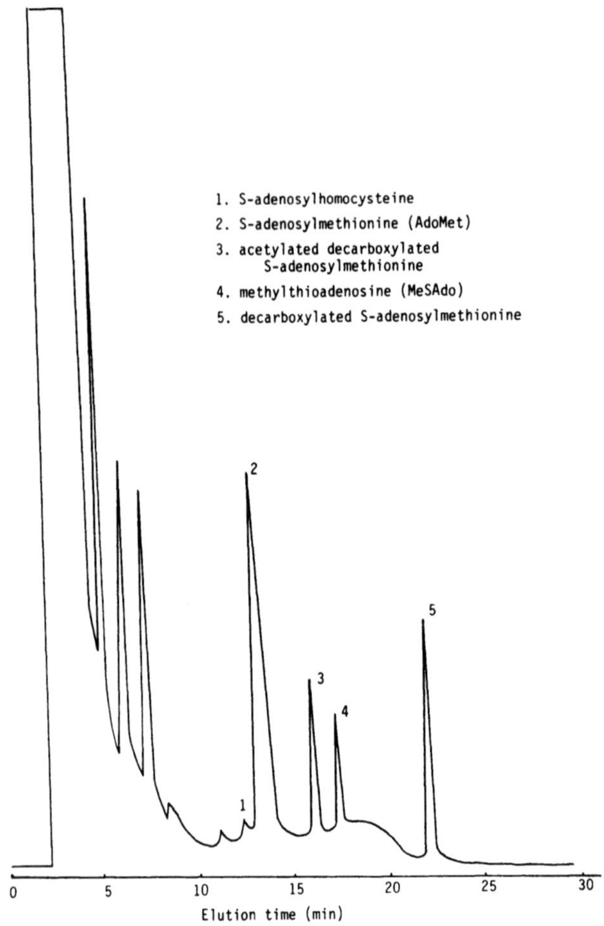

Figure 3 *HPLC pattern of the nucleosides in the cells treated with MeSAdo.*

enhanced by the addition of DFMO in the culture medium together with MeSAdo. In this case, the cellular amount of acetylated decarboxylated AdoMet was almost 3-fold larger than that of AdoMet (data not shown). This accumulation should have significances in the cellular metabolism in MeSAdo deficient cells. Pegg and colleagues have reported that the acetylation of decarboxylated AdoMet is catalyzed by histone acetylase, and they suggested that accumulation of acetylated decarboxylated AdoMet in the cells treated with difluoromethylornithine (DFMO) may inhibit the acetylation of histone(15). The acetylation of histone is an essential post-translational reaction in the rapid activation of genes(16). Thus, the accumulation of the acetylated form of decarboxylated AdoMet can be harmful for rapidly dividing cells.

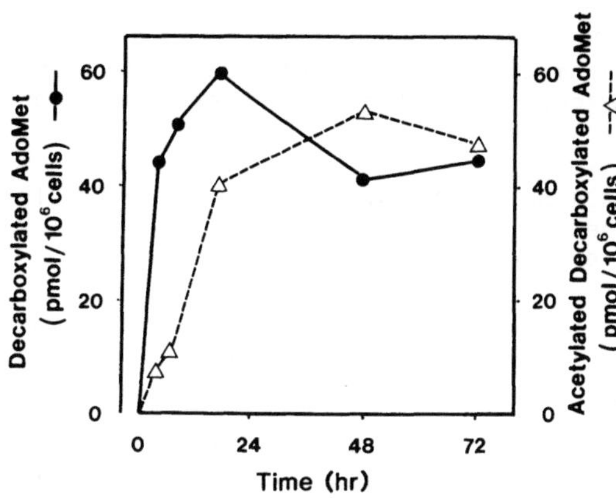

Figure 4 *Accumulation of decarboxylated S-adenosylmethionine and acetylated S-adenosylmethionine by the preincubation with methylthioadenosine.*

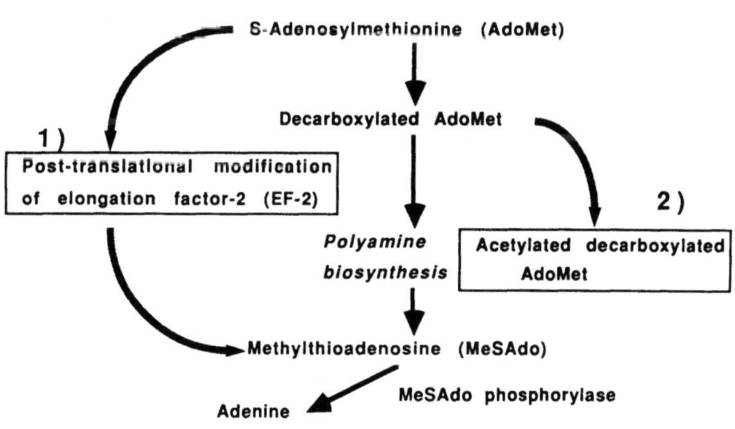

Figure 5 *Metabolic pathways that make methylthioadenosine.*

In summary, two possible toxic mechanisms of MeSAdo have been proposed. One is the modulation of the diphthamide synthesis and other is the perturbation of the histone acetylation due to the accumulation of the acetylated decarboxylated AdoMet. We cannot demonstrate the physiological significance of these metabolic pathways, however, the wide variety of the biological activities of MeSAdo should be emphasized (Figure 5).

REFERENCES

1. H.G. William-Ashman, J. Seidenfeld, and P. Galletti: Trends in the biochemical pharmacology of 5'-deoxy-5'-methylthioadenosine. *Biochem. Pharmacol.* 31:277 (1982).

2. N. Kamatani, W.A.Nelson-Rees, and D.A. Carson: Selective killing of human malignant cells deficient in methylthioadenosine phosphorylase, a purine metabolic enzyme. *Proc. Natl. Acad. Sci.U.S.A.* 78:1219 (1981).

3. N. Kamatani, A.L. Yu and d.A. Carson: Deficiency of methylthioadenosine phosphorylase in human leukemic cells *in vivo*. *Blood,* 60:1387 (1982)

4. N. Kamatani and D.A. Carson: Abnormal regulation of methylthioadenosine and polyamine metabolism in methylthioadenosine deficient human leukemic cells lines. *Cancer Res.* 40:4178 (1980).

5. R-L, Pajula and A. Raina: Methylthioadenosine, a potent inhibitor of spermine synthase from bovine brain. *FEBS lett.* 99:343 (1979).

6. M. Kubota, E.O. Kajander and D.A. Carson: Independent regulation of ornithine decarboxylase and S-adenosylmethionine decarboxylase in methylthioadenosine phosphorylase-deficient malignant murine lymphoblasts. *Cancer Res.* 45:3567 (1985).

7. M. Kubota, N. Kamatani and D.A. Carson: Biological genetic analysis of the role of methylthioadenosine phosphorylase in a murine lymphoid cell line. *J. Biol. Chem.* 258:7288 (1983).

8. K. Sadvig and S. Olsnes:Rapid entry of nicked diphtheria toxin into cells at low pH. *J. Biol. Chem.* 256:9068 (1981).

9. J.M. Moehring and T.J. Moehring: Diphthamide in elongation factor 2: ADP-ribosylation, purification and properties. *Methods Enzymol.* 106:388 (1984).

10. J.M.Moehring and T.J. Moehring: Characterization of the diphtheria toxin-resistance system in Chinese Hamster Ovary cells. *Somatic Cell Mol. Genet.* 77:1010 (1980).

11. J. Wagner, C. Danzin and O. Mamont: Reversed-phase ion-pair liquid chromatographic procedure for the simultaneous of S-adenosylmethionine, its metabolites and natural polyamines. *J. Chromatogr.* 227:349 (1982).

12. B.G. Van Ness, J.B. Howard and J.W. Bodley: ADP-ribosylation of elongation factor 2 by diphtheria toxin. *J. Biol. Chem.* 255:10710 (1980).

13. N.J. Oppenheimer and J.W. Bodley: Diphtheria toxin. Site and configuration of ADP-ribosylation of diphthamide in elongation factor 2. *J. Biol. Chem.* 256:8579 (1981).

14. Nishimura, s. Taya, Y., Kuchino, Y. and Ohashi, Z: *Biochem. Biophys. Res. Commun.* 57:702 (1974).

15. A.E. Pegg, R.S. Wechter, R.S. Clark, L. Wiest and B.G. Erwin: Acetylation of decarboxylated S-adenosylmethionine by mammalian cells. *Biochemistry* 25:379 (1986).

16. M. Perry and R. Chalkey: Histone acetylation increases the solubility of chromatin and occurs sequentially over most of the chromatin. *J. Biol. Chem.* 256:3313 (1981).

IMPROVED ASSAY METHOD OF dTMP SYNTHASE IN RAT LIVER AND ITS APPLICATION TO HUMAN LUNG CANCER CELLS

Yasuko Hashimoto, Taiichi Shiotani, Jiro Fujita, Masami Bungo, Norikc Yamauchi and Shozo Irino

1st Department of Internal Medicine, Kagawa Medical School
Kita, Kagawa 761-07, Japan

INTRODUCTION

dTMP synthase (EC 2.1.1.45) is the enzyme responsible for the final reaction in the de novo synthesis of dTMP by the methylation of dUMP with concomitant conversion of the methylene group of CH_2-THF to the 5-carbon of dUMP. In the dTMP synthase reaction, the tritium atom is released and exchanges with water according to the following equation:

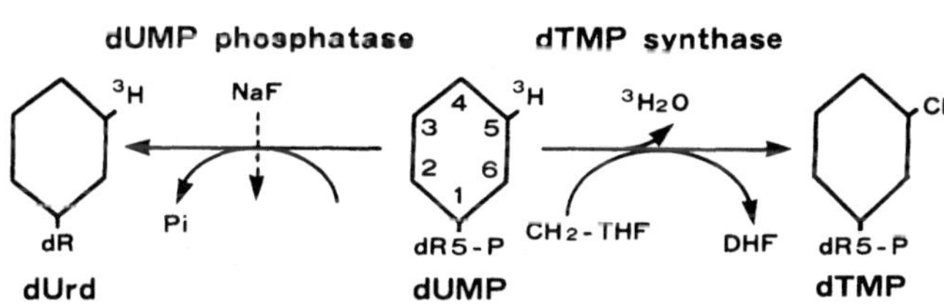

The enzyme activity has been measured by the amount of tritium atom released to H_2O from (5-^{3}H)dUMP after absorption of (5-^{3}H)dUMP with charcoal treatment[1-3] or Dowex 1 chloride column.[4] However, (5-^{3}H)dUrd which is a side-product by dUMP phosphatase during the dTMP synthase reaction cannot be separated from tritiated water by methods reported previously.This fact might lead to overestimation of dTMP synthase activity, causing the discrepancy in the reported values of rat liver dTMP synthase activity.[5-8] Recently we reported an improved method for separation of tritiated water from (5-^{3}H)dUMP.[9] The present report describes further improved method for separation of tritiated water from (5-^{3}H)dUrd by 100% KOH absorption for dTMP synthase assay and provides a comparison between dTMP synthase and dUMP phosphatase activities in the crude extracts of rat liver, hepatoma 3924A and human lung cancer cells.

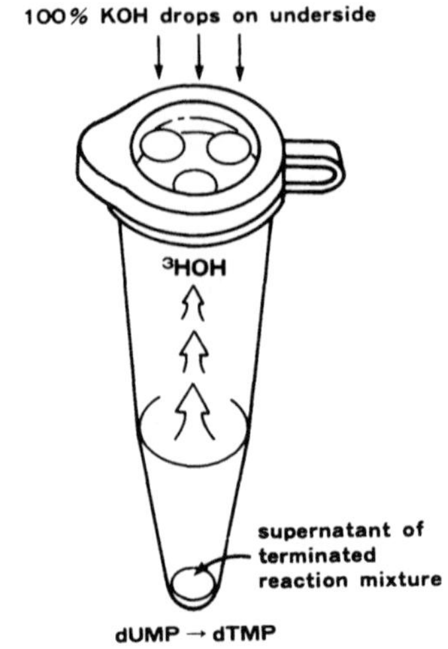

Fig.1. Vessel used to trap tritiated water generated in the dTMP synthase reaction.

EXPERIMENTAL PROCEDURES

Enzyme Extracts

Preparation of enzyme extracts from rat liver and hepatoma 3924A were reported elsewhere.[9] Both the original human lung cancer cell line; PC-9 and a CDDP-resistant subline; PC-9/CDDP[10] and other cell lines were provided by Dr. N.Saijo, National Cancer Center, Tokyo, Japan. At 72 hr after seeding with cell concentration of 1 x 10^6 in 10 ml of RPMI 1640 medium supplemented with 10 % fetal calf serum, cells were harvested by centrifugation and washed with PBS for 2 times. The cells suspended in 70 mM Tris-HCl pH7.4 were disrupted by rapid freeze and thaw in liquid nitrogen for 5 times. The extracts were centrifuged at 100,000 x g at 4 °C for 60 min and the resulting supernatants were used for the assay of enzyme activity. Protein concentration was measured by a routine method[11] with BSA as standard.

Standard Assay Mixture

CH_2-THF was prepared as described previously[9]. The assay mixture contained 62.5 mM Tris-HCl pH7.4, 1 mM CH_2-THF,50 mM NaF, 100 µM (5-^{3}H)dUMP (125.2 mCi/mmol,0.1 mCi/ml) and enzyme solution in a final volume of 80 µl. In the control, the enzyme solution was omitted from the complete reaction mixture. The reaction was initiated by the addition of (5-^{3}H)dUMP.

dTMP Synthase Assay

Method A; After incubation at 37 °C, the reaction was terminated by the addition of 400 µl of charcoal suspension (100 mg/ml) in 2 % TCA. After standing for 1 hr at room temperature, the mixture was centrifuged at 10,000 rpm for 5 min. Fifteen microliters of the resulting supernatant was transferred to the microcentrifuge tubes (size,1.5 ml) and 7.5 µl of 100 % KOH solution was spotted in three droplets on the underside of the lids. The tubes were capped and stood for 2 hr at 80 °C in order to absorb ^3H_2O (Fig.1). The lids were removed into counting vials containing 100 µl of glacial actetic acid. Finally 5 ml of ACS-II (NEN) was added and counted in a liquid scintillation counter. These radioactivities represent the dTMP synthase activity. The tube from which the lid was removed was placed into a counting vial containing 5 ml of ACS-II and counted for (5-^{3}H)dUrd production by dUMP phosphatase.

Method B; For routine assay of dTMP synthase, the enzyme reaction was terminated by the addition of 20 µl of 10 % TCA After centrifugation 15 µl of the resulting supernatant was treated with the same method of A. Only the lids were counted for dTMP synthase dactivity.

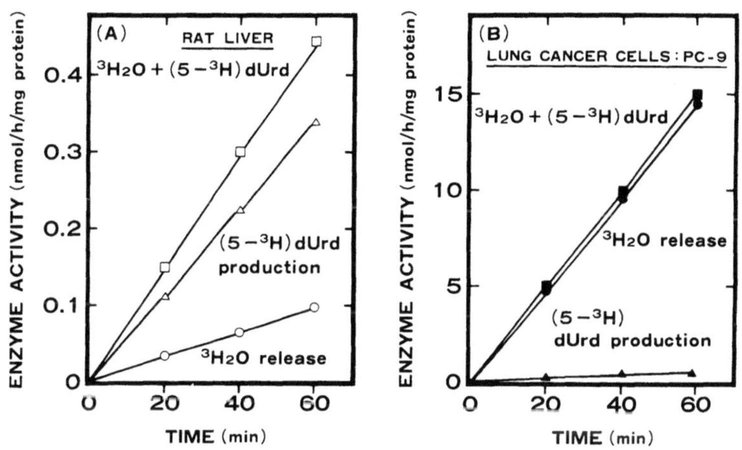

Fig.2. dUrd production during dTMP synthase reaction. Proportionality with time was examined in rat liver (A) and human lung cancer cells PC-9 (B): ^3H_2O release (○,●) (5-^{3}H)dUrd production (△,▲) and the sum of them (□,■).

RESULTS AND DISCUSSION

(5-^{3}H)dUrd Production in Rat Liver and PC-9 Human Lung Cancer Cells during dTMP Synthase Reaction

Fig.2 shows the rate of production of (5-^{3}H)dUrd and the release of ^3H_2O measured simultaneously by method A. Proportionality of (5-^{3}H)dUrd (dUMP phosphatase) and ^3H_2O (dTMP synthase) productions with time was observed. The radioactivity in the supernatant of charcoal treatment was the sum of the (5-^{3}H)dUrd and ^3H_2O production in rat liver(Fig.2-A). By contrast, negligible activity of phosphatase was detected and the activity of charcoal supernatant was almost the same to that of ^3H_2O release in PC-9 cells (Fig.2-B). These results indicate that there was significant (5-^{3}H)dUMP phosphatase activity present during dTMP synthase assay in rat liver, and the formed

Table 1. Requirement of dTMP synthase and dUMP phosphatase.

	% activity of complete system							
	dTMP synthase				dUMP phoshatase			
Incubation conditions	Rat liver	Hepatoma 3924A	Lung cancer PC-9	Lung cancer PC-9/CDDP	Rat liver	Hepatoma 3924A	Lung cancer PC-9	Lung cancer PC-9/CDDP
Complete[a]	100 (0.10)[b]	100 (2.19)	100 (14.86)	100 (37.14)	100 (0.34)	ND[c]	100 (0.13)	100 (0.23)
$-CH_2$-THF	45	1	1	2	112	ND	101	98
-BSA	104	90	102	99	85	ND	97	102
-NaF	99	98	98	97	310	ND	98	102

[a] The "complete" reaction mixture, in a final volume of 80 µl, contained 1 mM CH_2-THF, 62.5 mM Tris-HCl pH 7.4, 0.6 % BSA, 50 mM NaF, 100 µM (5-^{3}H)dUMP and enzyme solution.
[b] Numbers in parentheses are the specific activity(nmol/h/mg protein) of complete system of each tissue.
[c] ND: not detectable.

(5-^{3}H)dUrd and ^3H_2O were easily separated by the method of 100 % KOH absorption (Fig.2-A). Thus in tissues which contain a significant dUMP phosphatase activity, the activity of dTMP synthase measured by charcoal treatment[5-8] would be overestimated.

Requirement of dTMP Synthase Activity

Table 1 shows the role of several factors for optimal assay conditions on dTMP synthase. The presence of CH_2-THF was required for maximal dTMP synthase activity in all tissues examined. Only 1 to 2 % of activity of complete system was observed without CH_2-THF in hepatoma 3924A, human lung cancer cell lines PC-9 and PC-9/CDDP. However, the rat liver enzyme showed significant activity (45 % of complete system). without CH_2-THF, suggesting the presence of endogeneous CH_2-THF in extracts. Neither BSA nor NaF had effect on dTMP synthase. Several factors examined for dTMP synthase did not affect on the activity of dUMP phosphatase except that 3.1-fold increase in activity was detected without NaF in rat liver, indicating the significant activity of phosphatase in rat liver crude extracts.No activity of phosphatase was found in hepatoma 3924A.

dTMP Synthase Activity Measured by Different Methods of Termination of the Reaction

The enzyme reaction was terminated by the addition of 10 % TCA (method B) instead of charcoal suspension (method A) in order to simplify the assay procedure. Only dTMP synthase was measurable by method B, whereas dUMP phosphatase, dTMP synthase and the sum of their activities can be determined by method A. The sum of the activities of phosphatase and synthase was equal to that of charcoal supernatant in rat liver and lung cancer cell lines (Table 2 and Fig.2). Moreover the activity of dTMP synthase was the same with method A and B. Therefore, method B, i.e., termination of the reaction with 10 % TCA followed by 100 % KOH absorption of the supernatant, was used as a routine assay.

Table 2. dTMP synthase activity by different methods of termination of the reaction.

	Termination of reaction (nmol/h/mg protein)			
	Charcoal suspension : method A			10% TCA : method B
Tissues	Charcoal supernatant (control)[a]	dUMP phosphatase	dTMP synthase	dTMP synthase
Rat liver	0.44 ± 0.02 (100)[a]	0.34 ± 0.01 (77)[b]	0.10 ± 0.01 (23)	0.10 ± 0.01 (23)
Hepatoma 3924 A	2.21 ± 0.01 (100)[a]	ND[c]	2.19 ± 0.02 (99)	2.19 ± 0.02 (99)
Lung cancer PC-9	15.01 ± 0.76 (100)[a]	0.13 ± 0.04 (1)[b]	14.59 ± 0.86 (97)	14.86 ± 0.41 (99)
PC-9/CDDP	37.53 ± 0.20 (100)[a]	0.23 ± 0.08 (1)[b]	37.15 ± 0.40 (99)	37.14 ± 0.10 (99)

[a] Numbers in parentheses are percentages of the activity of charcoal supernatant.
[b] Significantly different from control value. ($P < 0.05$)
[c] ND: not detectable.

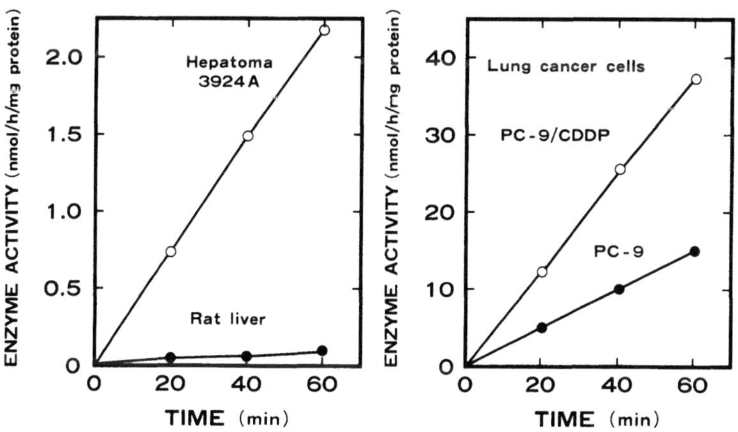

Fig. 3. Proportionality of dTMP synthase activity with time.

Table 3. dTMP synthase activity in human lung cancer cell lines

Cells	PC-9	PC-9/CDDP	PC-7	PC-14	H-69	A-549
Enzyme activity (nmol/h/mg protein)	14.86	37.14	10.80	9.13	20.55	5.92

dTMP Synthase Activity in Rat Liver, Hepatoma 3924A and Human Lung Cancer Cells

Good proportionaltiy was achieved with length of incubation time over 60 min (Fig.3) and with amount of enzyme added (data not shown). The minimal amount of the product detected was 10 pmol. The enzyme activity in liver and hepatoma 3924A were 0.10 $\pm$ 0.01 and 2.19 $\pm$ 0.02 nmol/h/mg protein,respectively. The activity in CDDP-resistant lung cancer cells, PC-9/CDDP (37.14 $\pm$ 0.10 nmol/h/mg protein) was elevated by 2.5-fold of its parent cell line PC-9 (14.86 $\pm$ 0.41 nmol/h/mg protein). Table 3 shows dTMP synthase activity in various human lung cancer cells. The activities of H-69, PC-7, PC-14 and A-549 cells were 20.55, 10.80, 9.13 and 5.92 nmol/h/mg protein,respectively.

Apparent Km Values for dTMP synthase

The apparent Km values for dUMP were almost the same in all tissues and cell lines examined (Table 4). Km values for CH_2-THF were different in rat liver (5 μM) and hepatoma 3924A (333 μM) crude extracts. However, this difference might not be significant because rat liver extracts contained significant endogeneous level of CH_2-THF (Fig.2 and Table 2). A slight better affinity for CH_2-THF in PC-9/CDDP was observed than that in PC-9, suggesting one of a possible cause for increased activity of dTMP synthase in CDDP-resistant cells.

Our results repoted in this paper indicate that it is important to eliminate the dUMP phosphatase activity in crude extracts of rat liver and human lung cancer cells for the accurate determination of

Table 4. Apparent K_m for dTMP synthase.

Tissues	K_m (μM) for	
	dUMP	CH_2-THF
Rat liver	2.9	5
Hepatoma 3924A	8.3	333
Lung cancer cells		
PC-9	9.1	49
PC-9/CDDP	9.3	13

the dTMP synthase activity. Charcoal treatment repoted previously [1-3] can not eliminate the (5-^{3}H)dUrd which is the side-product during dTMP synthase reaction. Our method requires only a low enzyme volume and can process over 100 samples.
Moreover whole assay can be done in 3 hr if ^3H_2O absorption by 100 % KOH is performed in the room of 80 °C.

In summary, the major advantages of our procedure are (1) the elimination of the phosphatase activity which interferes with the estimation of dTMP synthase in crude extracts, (2) one-step separation of ^3H_2O, (3) high sensitivity (with a limit of detection of 10 pmol of ^3H_2O production), (4) capability to measure the activity in small amount of samples (35-45 µg), (5) whole assay can be done in 3 hr and (6) adaptation of this method for cultured lung cancer cells.

REFERENCES

1. H.O. Kammen, Anal. Biochem. 17:553 (1966).
2. D. Roberts, Biochemistry 5:3546 (1966).
3. M.I.S. Lomax and G.R. Greenberg, J. Biol. Chem. 242:109 (1969).
4. A.H. Calvert, T.R. Jones, P.J. Dady, B. Grzelakowska-Sztabert, R.M. Paine, G.A. Taylor and K.R. Harrap, Eur. J. Cancer 16:713 (1980).
5. A. Herzfeld and S.M. Raper, Cancer Res. 40:744 (1980).
6. H.L. Elford, M. Freeze, E. Passamani and H.P. Morris, J. Biol. Chem. 245:5228 (1970).
7. R.C. Jackson and D. Niethammer, in: "Chemistry and Biology of Pteridines", R.L. Kisliuk and G.M. Brown, Eds., Elsevier/North Holland, Amsterdam, 665 (1979).
8. R. Labow, G.F. Maley and F. Maley, Cancer Res. 29:366 (1968).
9. Y. Hashimoto, T. Shiotani and G. Weber, Anal. Biochem. 167:340 (1987).
10. W.S. Hong, N. Saijo, Y. Sasaki, H. Takahashi, H. Nakano, K. Nakagawa, Y. Fujiwara, K. Nomura and P.R. Twentyman, International J. Cancer. 41:462 (1988).
11. O.H. Lowry, N.J. Rosebrough, A.L. Farr and R.J. Randall, J. Biol.Chem. 193:265 (1951).

ADENOSINE DEAMINASE ACTIVITY IN RECIPIENTS OF BONE MARROW FROM IMMUNODEFICIENT MICE HOMOZYGOUS FOR THE WASTED MUTATION

Erik H. Willis[1], Dennis A. Carson[1], and Leonard D. Shultz[2]

[1] Scripps Clinic and Research Foundation, La Jolla, California
[2] The Jackson Laboratory, Bar Harbor, Maine

INTRODUCTION

Given the many distinct syndromes of immune deficiency in humans and the limits imposed on experimentation, an animal model would aid considerably the study of the relationship between immunity and disease. Many investigations of the mammalian immune system in normal and pathologic states have been facilitated by the genetically-determined immunologic disorders in experimental animals. Heritable immunologic defects in mice have been studied extensively as models for immunodeficiency and autoimmunity diseases of humans. The spontaneous murine mutation called *wst* is just such an example. Homozygotes for the autosomal recessive mutation *wasted* (*wst*) are distinguished by a number of neurological and immunological abnormalities that in some respects resemble the multifaceted, genetically determined human syndrome ataxia-telangiectasia (1,2). The pathological changes found in the lymphoid systems of *wst/wst* mice include significantly decreased spleen, thymus, and lymph node to body weight ratios, marked lymphocytopenia affecting both T- and B-cell populations, and increased levels of spontaneous and gamma-ray induced chromosomal damage. Affected mice develop progressive paralysis and die by 30 days of age with degeneration of neurons in the brain and spinal cord.

Immunologic abnormalities found in *wst/wst* mice are also similar to those of the genetically determined human disorder, severe combined immunodeficiency disease (SCID) (3). The most debilitating of human lymphoid deficiency diseases, SCID impairs the differentiation of both T and B lymphocytes. Infants with the disorder are very susceptible to recurring viral, fungal, and bacterial infections, and invariably die within two years of birth. This syndrome can be inherited in an X-linked recessive or autosomal recessive fashion (3). Consequently, 75 percent of the patients with the disorder are male. Approximately one-half of the autosomal recessive cases are due to a deficiency in the purine metabolic enzyme adenosine deaminase (ADA) (4), important in regulating cellular levels of purines. The murine gene coding for ADA has been mapped to chromosome 2, and likewise, the gene for *wst* has been shown to reside in the distal region of the same chromosome (5).

Recently, Abbott *et al.* have reported low levels and altered kinetic parameters of erythrocyte ADA in *wst/wst* mice, and postulated that *wst* is a mutation within the ADA structural gene locus (6). In the present work we transplanted marrow from *wst/wst* and littermate control mice into normal, gamma-irradiated recipients in order to investigate the possible association between the *wst* mutation and ADA. The results indicate that the *wst* gene is not a defective gene coding for ADA.

MATERIALS AND METHODS

Mice. The original autosomal recessive mutation to *wst* occurred on the inbred HRS/J strain at the Jackson Laboratory. To increase viability the autosomal recessive mutation was crossed onto a segregating background (C57BL/6JxC3HeB/FeJ). Subsequently, *wst* was backcrossed 6 generations onto the C57BL/6J strain background by ovary transplantation from *wst/wst* females to histocompatible recipients and mating to C57BL/6J males.

Cell Suspensions. Bone marrow cells from 26 day-old C57BL/6J *wst/wst* and +/- sex-matched littermate control mice were obtained by flushing the medullary cavities of femurs and tibias with RPMI 1640 medium (M.A. Bioproducts, Walkersville, MD). Marrow clumps were passed gently through a 25 gauge needle to prepare single- cell suspensions. The marrow cell suspensions were then strained through 100-mesh nylon cloth, washed, and resuspended in RPMI. Nucleated cells were counted in a Coulter ZB1 electronic cell counter, and percent viability determined by trypan blue exclusion.

Irradiation and Bone Marrow Injection. C57BL/6J mice at 16 weeks of age were exposed to 1000R whole body gamma irradiation at a rate of 230R per minute from a Shepard Mark 1 irradiator loaded with 10,000 Ci of cesium-137. Groups of 4-5 irradiated mice were injected in the lateral tail veins with either 5×10^6 *wst/wst* or +/- littermate control cells. At 89 days after injection, the recipients were sacrificed. Thymuses, spleens, livers, brains, and packed erythrocytes were collected and frozen immediately at -20°C.

Adenosine Deaminase Assay. The enzyme was assayed spectrophotometrically as described previously (7), with modifications. Briefly, tissue extracts were prepared by homogenizing individual organs using a Serval Omni-Mixer with a Mini-Max attachment. Homogenization buffer consisted of 50 mM Tris/HCl, pH 7.4 with 0.5 mM dithiothreitol (DTT) at 4°C. Homogenates were centrifuged at 25,000 x g for 60 min, and supernatants dialyzed overnight against 25 mM Tris/HCl with 1 mM DTT. Ten microliters of tissue extract was added to a 1 ml cuvette containing 0.99 ml adenosine solution buffered with 50 mM potassium phosphate, pH 7.4, and the change in absorbance at 265 nm, reflecting the conversion of adenosine to inosine, was recorded. Enzyme activity was expressed per milligram protein, as determined by the method of Bradford (8), using bovine serum albumin as standard.

For the determination of kinetic constants, assays were conducted at five adenosine concentrations between 0.003 mM and 0.243 mM. Kinetic constants were determined graphically from Eadie-Hofstee plots. Differences in mean values were tested for significance with a two-tailed t-test.

RESULTS

In order to examine the possible relationship between the mutation *wst* and alterations in the ADA gene, bone marrow transplants were performed after lethal irradiation of normal (+/+) recipients. The normal irradiated mice were transplanted either with marrow from two *wst/wst* or from two littermate control mice. If there was an inherent ADA deficiency in the hematopoietic stem cells of the *wst/wst* mutant, one would expect this defect to be passed on to the irradiated marrow

Table 1. Kinetics of ADA in tissue extracts from irradiated recipients of *wst/wst* or control bone marrow.

Donor Marrow	Recipient Tissue Homogenate	Km (mM)	Vmax (nmol/min.mg)
+/-	Thymus	27.7 ± 2.6	362 ± 112
wst/wst		29.2 ± 6.9	387 ± 83
+/-	Liver	27.7 ± 5.1	12.9 ± 5.7
wst/wst		32.0 ± 3.9	5.8 ± 2.6
+/-	Spleen	31.9 ± 3.9	97.0 ± 18.0
wst/wst		27.8 ± 6.6	112 ± 17
+/-	Brain	51.3 ± 14.7	9.8 ± 6.1
wst/wst		32.1 ± 7.9	5.8 ± 1.6
+/-	RBCs	26.6 ± 8.5	1.9 ± 0.6
wst/wst		16.1 ± 6.3	1.3 ± 0.2

All values are expressed as mean ± SEM.

chimeras, resulting in decreased Vmax values for ADA and/or a change in the Km for adenosine. The results in Table 1 show no significant change in the Vmax, nor a change in the apparent Km for adenosine, between recipients of littermate +/- marrow and recipients of *wst/wst* marrow.

Confirming these results, we also found no significant differences in ADA activities between lymphoid and liver tissue from untreated 24-28 day old *wst/wst* and normal littermate control mice. Tissue ADA activities of both *wst/wst* and control mice were inhibited to non-detectable levels with the addition of 5uM deoxycoformycin, a potent inhibitor of ADA (9).

DISCUSSION

The murine *wasted* mutant recently described by Shultz *et al* (1) has immune abnormalities which are similar to the genetically determined human disease SCID (6). The present study was conducted to re-examine the reported deficiency of ADA in tissues of *wst/wst* mice, and to investigate more carefully any association between the *wst* mutation and ADA.

We have shown on the basis of specific activity measurements of tissue homogenates, no significant changes in ADA activity in the thymus, spleen, liver, brain, and red blood cells of *wst/wst* mice. This is in agreement with findings of Geiger *et al.* (10) except that they observed slightly elevated levels of ADA, expressed per mg protein, in *wst/wst* spleens. Abbott *et al.* (6) found significantly decreased levels of erythrocyte ADA in *wst/wst* homozygotes. However, the latter authors used six week old mice as controls, and adult levels of ADA in erythrocytes may not be reached in mice prior to the age of weaning. In addition, quantitation and characterization of ADA activity in red cells is imprecise due to the normally low enzyme activity. It is also likely that the cross breeding of their *wasted* animals with ones carrying the visible-marker gene *ragged*, of which the intercrossed mice died before weaning, may have introduced a genetic variance which accounts for the disparate results (10).

To assess more conclusively the association between the *wst* mutation and alterations in ADA, bone marrow was removed from *wst/wst* and littermate control mice and transplanted into lethally irradiated recipients. If hematopoietic stem cells from the mutant have reduced ADA activities and altered kinetic parameters, one would expect these defects to be passed on to the irradiated marrow chimeras. Our results show no significant difference in the Vmax nor Km between recipients of *wst/wst* and normal marrow in spleen, thymus, and red blood cells. An analysis of the donor mice similarly showed no significant differences in the Vmax or Km values for ADA, using adenosine as a substrate.

Our findings of normal ADA activity and unaltered kinetic parameters in *wst/wst* mice suggest that the animal is not a model for ADA deficiency and SCID. Previous observations of cerebral nerve fiber degeneration in *wst/wst* mice (1), prior to progressive hypoplasia of lymphoid organs, also imply that the immune deficit is not the sole or even primary abnormality. It is possible that the immune dysfunction may occur as a consequence of the neural degeneration, with accompanying stress, and corticosteroid release (11).

ACKNOWLEDGEMENTS

This research was supported in part by NIH grants GM23200 and CA35048. The Jackson Laboratory is fully accredited by the American Association for Accreditation of Laboratory Animal Care.

REFERENCES

1. L. D. Shultz, H. O. Sweet, M. T. Davisson, and D. R. Coman, Nature. 297:402-404 (1982).
2. D. Kaiserlian, W. Savino, J. Uriel, J. Hassid, M. Dardenne, and J. F. Bach, Clin. Exp. Immunol. 63:562-569 (1986).
3. I. W. Gelfand, W. D. Biggar, and R. P. Orange, Pediatr. Clin. North Am. 21:745-776 (1974).
4. R. Hirschorn, G. E. Vawter, J. A. Kirkpatrick Jr., and F. S. Rosen, Clin. Immunol. Immunopathol. 14:107-120 (1979).
5. H. O. Sweet, Mouse Newsletters 71:31 (1984).
6. C. M. Abbott, C. J. Skidmore, A. G. Searle, and J. Peters, Proc. Natl. Acad. Sci. USA 83:693-695 (1986).
7. H. M. Kalckar, J. Biol. Chem. 167:416-475 (1947).
8. M. M. Bradford, Anal. Biochem. 72:248-254 (1976).
9. R. P. Agarwal, T. Spector, and R. E. Parks, Jr., Biochem. Pharmacol. 26:359-367 (1977).
10. J. D. Geiger, and I. Nagy, FEBS Lett. 208:431-434 (1986).
11. D. G. Goldowitz, P. M. Shipman, J. F. Porter, and R. R. Schmidt, J. Immunol. 135:1806-1812 (1985).

UTILIZATION OF 2'-DEOXYNAD FOR ADP-RIBOSE TRANSFER REACTIONS

D. Bruce Wasson, Hisashi Yamanaka, and Dennis A. Carson

Department of Basic and Clinical Research
Scripps Clinic & Research Foundation, La Jolla, California

SUMMARY

2'-deoxyNAD was examined as a substrate for both mono(ADP-ribosyl)ation and poly(ADP-ribosyl)ation reactions. 2'-deoxyNAD is a substrate for the diphtheria toxin-catalyzed mono(ADP-ribosyl)ation of elongation factor-2, inactivating its function to enhance protein synthesis. On the other hand, 2'-deoxyNAD is a poor substrate for poly(ADP-ribose) polymerase. 2'-deoxyNAD was not synthesized intracellularly from deoxyATP, even when deoxyATP content was markedly increased by incubation of cells with deoxyadenosine and an adenosine deaminase inhibitor. 2'-deoxyNAD, because of its specificity, could be a quite useful reagent for the investigation of cellular mono(ADP-ribosyl)ation reactions.

INTRODUCTION

Mono(ADP-ribosyl)ation is a post-translational modification of proteins, initially described as the mechanism of action of certain bacterial enzymes (Reviewed in 1). Recently, similar enzymes have been identified in eukaryotic cells, and mono(ADP-ribosyl)ation has been suggested to have an important role in the regulation of metabolism (2,4). A major problem in research concerning mono(ADP-ribosyl)ation is the necessity to eliminate the effects of poly(ADP-ribosyl)ation reactions. Poly(ADP-ribose) polymerase is a very abundant enzyme in eukaryotic cells. Both mono(ADP-ribosyl)ation and poly(ADP-ribosyl)ation use NAD as the common substrate. Moreover, no specific inhibitor of poly(ADP-ribosyl)ation has been described. 3-Aminobenzamide has been used widely as an inhibitor of poly(ADP-ribosyl)ation. However, it also inhibits mono(ADP-ribosyl)ation reactions at high concentrations (5). For these reasons, a specific and functional substrate for mono(ADP-ribosyl)ation but not for poly(ADP-ribosyl)ation would be quite useful for studies of mono(ADP-ribosyl)ation reactions in crude cell extracts. In this report, we have examined 2'-deoxyNAD (2'-dNAD) as a functional substrate of mono(ADP-ribosyl)ation in experiments using the eukaryotic elongation factor-2 (EF-2)-diphtheria toxin system.

MATERIALS AND METHODS

Preparation 2'-dNAD and [^{3}H]2'-dNAD

2'-dNAD and [^{3}H]2'-dNAD were synthesized enzymatically from dATP and [8-^{3}H]-deoxyATP(ICN, Costa Mesa,Ca), respectively, and nicotinamide mononucleotide (NMN) by NAD pyrophosphorylase (E.C.2.7.7.1, Boehringer Mannheim Biochemicals, Indianapolis, IN) as described by Kornberg (6). The reaction mixture consisted of 100 mM glycylglycine buffer pH 7.4,10 mM $MgCl_2$, 7 mM NMN, 0.65 mM dATP, and 0.005 U/ml NAD pyrophosphorylase. After 6 hr at 37° C, the reaction was quenched at 100° C for 2 minutes. The reaction mixture was purified first on a P10-SAX column (Altex,San Ramon,Ca) using a linear gradient from buffer A (3.5mM KH_2PO_4, pH 3.5 and 2% acetonitrile) to buffer B (250mM KH_2PO_4, pH 3.5, 500mM KCl and 2% acetonitrile). The peak corresponding to 2'-dNAD was lyophilized and further purified on a uBondapak C_{18} column (Waters Milford,MA) using a gradient from 0 to 10% acetonitrile. Yields of these compounds were from 19% to 23%.

Mono(ADP-ribosyl)ation reactions

We examined the effects of 2'-dNAD on mono(ADP-ribosyl)ation of elongation factor 2 (EF-2) catalyzed by diphtheria toxin, as described previously (7). As a source of EF-2, a 100,000 g supernatant of a cytosol extract was prepared from the murine lymphoma cell line R1.1 H3. 50mM thymidine was included in the reaction mixture as an inhibitor of poly(ADP-ribosyl)ation.

Cell-free assay for the function of elongation factor-2 (EF-2)

To examine the function of mono(deoxyADP-ribosyl)ated EF-2, a cell-free protein synthesis system was employed by the modified method of Iglewski and Lee (8). In the first step, rabbit reticulocyte lysate (Promega Biotec, Madison, HI) was reacted with 200 μg/ml diphtheria toxin (Calbiochem. La Jolla, CA) and either NAD and 2'-dNAD to mono(ADP-ribosyl)ate or mono(deoxyADP-ribosyl)ate the EF-2 in the reticulocyte lysate. The reaction mixture was incubated with hemin and 1mM $CaCl_2$ for 15 min at 20°C. 10 U/ml creatine kinase was then added (pretreated lysate). In the second step, the pretreated lysate was incubated at 37°C for 1 hr with [^{3}H]leucine in the presence of translational cocktail(9), 50mM KCl, 0.4M $Mg(C_2H_3O_2)_2$, 1,000U/ml RNAsin (Promega Biotec). The reaction was terminated by the addition of 1N NaOH and 1.5% H_2O_2. After a further incubation at 37°C for 15 min, the protein was precipitated with cold 10% trichloroacetic acid. The radioactivity incorporated into newly synthesized protein was collected onto glass fiber filters (Whatman GF/C) and after extensive washing, the radioactivity was assayed by liquid scintillation counting.

Poly(ADP-ribosyl)ation reactions

Poly(ADP-ribose) polymerase was partially purified from calf thymus by ammonium sulfate precipitation and DNA cellulose column chromatography, followed by hydroxyapatite column chromatography (10,11). Poly(ADP-ribosyl)ation was assayed radiochemically with histone as an acceptor(11).

Intracellular production of NAD and 2'-dNAD

Peripheral blood lymphocytes (PBL) were isolated from blood of normal volunteers on a Ficoll-Hypaque density gradient, followed by the depletion of monocytes by the surface adherance method. The human T-lymphoblastoid cell line CCRF-CEM (CEM) was described previously (12). Peripheral blood lymphocytes or CEM cells were cultured in nicotinamide free RPMI 1640 medium supplemented with 10% fetal bovine serum, 1.8 μM (0.1 $\mu Ci/ml$) [^{14}C]-nicotinamide with or without 5μM 2'-deoxycoformycin and 100 μM deoxyadenosine. After 4 or 20 hrs, the cells were washed with cold phosphate buffered saline and the cellular protein was precipitated with 0.4N perchloric acid. After centrifugation, the supernatant was neutralized with KOH, and analyzed by HPLC with a P10 SAX column (Altex,San Ramon,Ca). A linear gradient from buffer A (3.5mM KH_2PO_4, pH 3.5 and 2% acetonitrile) to buffer B (250mM KH_2PO_4, pH 3.5, 500mM KCl and 2% acetonitrile) was used as a mobile phase. The cellular levels of ATP, deoxyATP, NAD and 2'-dNAD were quantitated by comparison to standard curves of authentic compounds.

RESULTS AND DISCUSSION

Diphtheria toxin has been shown to specifically mono(ADP-ribosyl)ate "diphthamide", a post-translationally modified histidine in EF-2 (13). Crude EF-2 was prepared from murine lymphoma cells, and [^3H]2'-dNAD was apparently utilized in the mono(deoxyADP-ribosyl)ation of EF-2 by diphtheria toxin (Table 1). Mono(ADP-ribosyl)ated EF-2 is known to lose its ability to synthesize protein. Therefore it was important to determine if the mono(deoxyADP-ribosyl)ated EF-2 also lost this function. For this reason, a simple assay to examine the function of EF-2 was employed, using an *in vitro* protein synthesis system. As shown in Table 2, the mono(deoxyADP-ribosyl)ation of EF-2 by 2'-dNAD inhibited the subsequent incorporation of [^3H]leucine into protein, just as did the

TABLE 1. 2'-dNAD as a substrate for mono(ADP-ribosyl)ation

1. [3H]2'-dNAD + Thymidine	390 *
2. [3H]2'-dNAD + Thymidine + Diphtheria toxin	2,944
3. [3H]2'-dNAD	386
4. [3H]NAD + Thymidine + Diphtheria toxin	11,748

* mean cpm of duplicate samples

100,000g supernatant of cytosol extract of R1.1 cells (source of elongation factor-2) was incubated either with [^3H]NAD (sp.act. 1.75Ci/mmol) or with [^3H]2'-dNAD (sp.act. 2.0 Ci/mmol), in the presence or absence of 50mM thymidine and 200$\mu g/ml$ diphtheria toxin. Then the radioactivity in total cellular protein was determined. These combinations indicate 1)background, 2)mono(deoxyADP-ribosyl)ation, 3)poly(deoxyADP-ribosyl)ation, 4)mono(ADP-ribosyl)ation.

mono(ADP-ribosyl)ation by NAD. Together, these data show that 2'-dNAD is a substrate for diphtheria toxin catalyzed mono(deoxyADP-ribosyl)ation of EF-2, and furthermore, is a functional inactivator of EF-2. The effect of 2'-dNAD on poly(ADP-ribosyl)ation was also examined *in vitro*. The incorporation of [^3H]NAD into poly(ADP-ribose) polymer was actually inhibited in the presence of high concentrations of 2'-dNAD (Table 3). On the other hand, the incorporation of [^3H]2'-dNAD into the polymer was quite small, and was only detectable when the specific activity of [^3H]2'-dNAD was high (Table 4). These data suggest that 2'-dNAD is a poor substrate for poly(ADP-ribosyl)ation reactions. Presumably, 2'-dNAD cannot be utilized in the elongation of the ADP-ribose polymer (15).

TABLE 2. Mono(ADP-ribosyl)ation of EF-2 and in vitro protein synthesis

1. Background	792*
2. Untreated lysate	41,084
3. Pretreated with Diphtheria toxin	27,696
4. Pretreated with Diphtheria toxin and NAD	4,453
5. Pretreated with Diphtheria toxin and 2'-dNAD	4,193

* mean cpm of duplicate samples
Reticulocyte lysates were initially pretreated with diphtheria toxin and either NAD or 2'-dNAD (pretreated lysate). Next, each of the pretreated lysates were reacted with [^{3}H]leucine in the translational mixture, and the [^{3}H]leucine incorporation into protein was measured.

We have also examined the possibility that 2'-dNAD might be produced intracellularly. In patients with adenosine deaminase deficiency, 2'-deoxyadenosine and 2'-dATP are known to accumulate intracellularly and cause lymphocyte dysfunction(reviewed in 16). Because 2'-dATP is an established inhibitor of mammalian ribonucleotide reductase, and hence of DNA synthesis, the accumulation of 2'-dATP has been implicated in the pathogenesis of the lymphocyte dysfunction(16). When 2'-dATP concentration increases greatly in the cell, 2'-dATP might be utilized by NAD pyrophosphorylase to produce 2'-dNAD. Earlier, the production of 2'-dNAD *in vitro* was described following incubation of HeLa cell nuclei with 2'-dATP and NMN (16). To examine the possible production of 2'-dNAD in viable cells, both non-dividing T cells (PBL) and dividing T lymphoblasts (CEM cells) were cultured with [^{14}C]nicotinamide. Under these conditions, NAD was efficiently labelled from [^{14}C]nicotinamide in both cell types. However no detectable incorporation of [^{14}C]nicotinamide into 2'-dNAD was observed, even when cells were cultured with 2'-deoxyadenosine and 2'-

TABLE 3. Inhibition of Poly(ADP-ribosyl)ation by 2'-dNAD

2'-dNAD (μM)		0 (control*)	1	10	100	334
				% control		
NAD (μM)	0.2	0.50	76	100	48	32
	1.2	1.66	189	125	112	53
	10	16.2	93	114	84	48
	100	209	96	83	77	46
	334	475	107	97	85	67

* pmol[^{3}H]NAD incorporated
Partially purified Poly(ADP-ribose) polymerase was incubated with histone and [^{3}H]NAD in the presence of various concentrations of NAD and 2'-dNAD. Incorporation of [^{3}H]ADP into acid precipitable material was measured.

TABLE 4. 2'-dNAD as a substrate for Poly(ADP-ribose)polymerase

2'-dNAD (μM)		0.2	10	334
		pmol [^{3}H]2'-dNAD incorporated		
NAD (μM)	0	0.006	< 0.05	< 0.15
	10	0.012	0.14	< 0.15
	334	0.003	< 0.05	< 0.15

Partially purified Poly(ADP-ribose) polymerase was incubated with histone and [^{3}H]deoxyNAD in the presence of various concentrations of NAD and 2'-dNAD. Incorporation of [^{3}H]deoxyADP into acid precipitable material was measured.

TABLE 5. Nucleotide and Pyridine Nucleotide Content

		dAdo+dCF	ATP	deoxyATP	[^{14}C]NAD	[^{14}C]2'-dNAD
			pmol/10^6 cells			
PBL	4 hr	-	453	< 6	5	< 0.06
		+	527	33	13	< 0.06
	20 hr	-	659	< 6	17	< 0.06
		+	354	23	9.2	< 0.06
CEM	4 hr	-	1181	< 6	28.6	< 0.06
		+	625	661	14.4	< 0.06
	20 hr	-	1242	< 6	53.4	< 0.06
		+	97	18	11.2	< 0.06

PBL or CEM cells were cultured in nicotinamide free RPMI1640 with 10% fetal calf serum containing 1.8 μM [^{14}C]nicotinamide (0.1μCi/ml) for 4 or 20 hours with (+) or without (-) 100μM deoxyadenosine plus 5μM deoxycoformycin. After harvesting, acid-soluble fractions of cell extracts were analyzed by HPLC.

deoxycoformycin, a potent inhibitor of adenosine deaminase that causes the intracellular dATP pool to increase to as much as 661 pmol/10^6 cells. (Table 5). Possibly this is due to the much higher reported Km value of NAD pyrophosphorylase for deoxyATP (180 uM) compared to that for ATP (25 uM) (16). These results indicate that 2'-NAD is not involved in the pathogenesis of adenosine deaminase deficiency, but do show that 2'-dNAD is a useful experimental tool to specifically assess mono(ADP-ribosyl)ation reactions.

ACKNOWLEDGMENTS

This research was supported in part by grants GM23200 and AI24466 from the National Institutes of Health, by the State of California, as allocated by the recommendation of the Universitywide Task Force on AIDS, and by the American Foundation for AIDS Research. This is publication 5462BCR from the Research Institute of Scripps Clinic.

REFERENCES

1. K. Ueda, and O. Hayaishi, Ann. Rev. Biochem. 54:73-100 (1985).
2. D.A. Yost, and J. Moss. J. Biol. Chem. 258:4926-4929 (1983).
3. H. Lee, and W.J. Iglewski, Proc. Natl. Acad. Sci. USA 81:2703- 2707 (1984).
4. C. Jacquemin, H. Thibout, B. Lambert, and C. Correze, Nature 323:182-184 (1986).
5. M.R. Duncan, P.R. Rankin, R.L. King, M.K. Jacobson, and R.T. Dell'orco, J. Cell. Phys. 134:161-165 (1988).
6. A. Kornberg, J. Biol. Chem. 182:779-793 (1950).
7. H. Yamanaka, E.O. Kajander, and D.A. Carson, Biochim. Biophys. Acta 888:157-162 (1986).
8. W.J. Iglewski, and H. Lee, Eur. J. Biochem. 134:237-240 (1983).
9. T. Maniatis, E.F. Fritsch, and Sambrook in: "The Molecular Cloning," Cold Spring Harbor Laboratory, New York, pp. 344-348 (1982).
10. H. Yamanaka, E.H. Willis, C.P. Penning, C.P. Peebles, E.M. Tan, and D.A. Carson, J. Clin. Invest. 80:900-904 (1987).
11. S.G. Carter, and N.A. Berger, Biochemistry 21:5475-5481 (1983).
12. G.E. Foley, H. Lazarus, S. Farber, B.g. Uzman, B.A. Boone, and R.E. McCarthy, Cancer 18:522-529 (1965).
13. D.M. Gill, and L. Dinius, J. Biol. Chem. 248:654-658 (1973).
14. R.J. Suhadolnik, in: "ADP-Ribosylation Reaction," O. Hayaishi, and K. Ueda, ed., pp.65- 75, Academic Press, New York (1982).
15. R.J. Suhadolnik, R. Baur, D.M. Lichtenwalner, T. Uematsu, J.H. Roberts, S. Sudhakar, and M. Smulson, J. Biol. Chem. 252:4134-4144 (1977).
16. C.J. Carrera, and D.A. Carson, in: "The Molecular Basis of Blood Diseases," G. Stamatoyannopoulos, A.W. Nienhuis, P. Leder, and P.W. Majerus, eds., pp.407-449, Saunders, Philadephia. (1987).

PROFOUND TOXICITY OF DEOXYADENOSINE AND 2-CHLORODEOXYADENOSINE TOWARD HUMAN MONOCYTES IN VITRO AND IN VIVO

Carlos J. Carrera, Hisashi Yamanaka, Lawrence D. Piro
Martin Lotz, and Dennis A. Carson

Department of Basic and Clinical Research
Scripps Clinic & Research Foundation, La Jolla, California

SUMMARY

Deoxyadenosine is known to be toxic to both proliferating and resting lymphocytes that lack adenosine deaminase (ADA) activity. We now show that human monocytes are also highly sensitive in vitro to nanomolar concentrations of deoxyadenosine plus the ADA inhibitor deoxycoformycin, and to the the ADA-resistant analogue 2-chlorodeoxyadenosine (CdA). Monocytes exposed to deoxyadenosine or to CdA in vitro accumulate massive DNA damage detectable within 1 hour. The DNA damage in monocytes exposed to CdA is associated with a decrease in protein synthesis and with inhibitions of phagocytosis and IL-6 secretion. However, unlike lymphocytes with similar DNA damage, the monocytes show no significant NAD or ATP depletion until cell viability declines. The selective toxicity of CdA to monocytes was confirmed by in vivo studies. In almost all patients receiving CdA infusion chemotherapy for cutaneous lymphoma, the blood monocyte counts fell to near 0 during one week of therapy. Our results suggest that CdA and related compounds may have potential clinical use in the therapy of immune disorders associated with monocyte/macrophage activation.

INTRODUCTION

Selective lymphocyte depletion underlies the severe combined immunodeficiency associated with inherited adenosine deaminase (ADA) deficiency (1). Accordingly, deoxyadenosine (dAdo) plus ADA inhibitors, and the congener drug 2-chlorodeoxyadenosine (CdA) have been shown to be selectively toxic towards proliferating T-cells and towards normal or malignant non-dividing lymphocytes (2,3). The substitution of chlorine at the 2 position of the purine ring renders CdA highly resistant to deamination.

Both dAdo and CdA cause DNA strand break accumulation in lymphocytes, followed by the rapid consumption of NAD for poly(ADP-ribose) synthesis (4,5). Based on these findings, CdA has been developed for clinical use in the treatment of chronic lymphocytic leukemia, lymphoma, and immune hemolytic anemia (6,7).

We now find that normal human monocytes are also sensitive to dAdo and to CdA at the same nanomolar concentrations that lyse lymphoid cells. In addition, we observed a marked decrease in the circulating monocyte count among cutaneous T-cell lymphoma patients receiving CdA chemotherapy. Monocytes and tissue macrophages are important effector cells in immune cytopenias, and may also play a pathogenetic role in rheumatoid arthritis. Thus, the following in vitro experiments were performed to characterize the biochemical basis for the effects of dAdo and CdA on human monocyte function and viability.

METHODS

Normal human monocytes were purified (>85%) from heparinized blood by Hypaque-Ficoll sedimentation, followed by adherence to gelatin-coated plastic flasks (8). Neutrophils were isolated following hypotonic lysis of erythrocytes in the high density fraction. Cells were cultured in medium supplemented with 20% autologous plasma. For biochemical studies, monocytes were detached from plastic surfaces by brief incubation in medium containing xylocaine 0.4% and EDTA 5mM. GM 1380 human fetal fibroblasts were cultured to confluence in DMEM plus 20% fetal bovine serum.

The viability of cells exposed to various compounds was determined by erythrosin B dye exclusion or by a spectrometric assay for the reduction of methylthiazolyl tetrazolium (MTT). DNA strand breaks were measured by the fluorometric assay of Birnboim and Jevcak (9). In this assay, the residual double-stranded DNA is quantitated by ethidium bromide fluorescence following a controlled period of DNA unwinding in alkaline solution. The rate of DNA unwinding is proportional to the number of strand breaks. Cellular NAD was measured by an enzyme cycling assay, and ATP content was measured by HPLC. RNA and protein synthesis by monocytes in vitro was measured by the incorporations of 3H-uridine 50 uCi/ml and 3H-leucine 20 uCi/ml, respectively, added during the final hour of culture.

Phagocytosis of rabbit anti-human red cell antibody-coated red cells (EA) was determined by a spectrometric assay for the hemoglobin catalyzed peroxidation of diaminobenzidine, occurring in sodium dodecylsulfate lysates of washed phagocytes (10). Dialyzed monocyte culture supernatants were assayed for IL-6 present as a growth factor activity for the IL-6-sensitive B9.9 hybridoma cell line (11).

RESULTS AND DISCUSSION

In Vitro Toxicity

The combination of dAdo plus the ADA inhibitor deoxycoformycin was highly toxic to normal human monocytes during short-term culture (ID50 at 3 days = 150 nM). Table 1 shows that this toxicity required the presence of the ADA inhibitor, and could be prevented by simultaneous addition of deoxycytidine. Thus, the toxicity of dAdo appears to be mediated by dAdo nucleotides produced via the activity of monocyte deoxycytidine kinase. In contrast, inhibitors of poly(ADP-ribose) synthetase showed no protective effect, and perhaps caused a slight enhancement of dAdo toxicity.

The profound toxicity of CdA towards human monocytes in vitro is shown in Fig. 1. In contrast, confluent human fibroblasts were resistant. This experiment confirms the monocyte selectivity of dAdo and CdA toxicity, and rules out a non-specific effect of the deoxynucleosides on cell attachment or on mitochondrial reduction of MTT.

DNA Strand Breaks and Biochemical Sequelae

Both dAdo plus deoxycoformycin, and CdA caused the rapid appearance of DNA breaks in monocytes in vitro. In contrast, the DNA of neutrophils

Table 1. Effect of Various Compounds on the Toxicity of dAdo plus Deoxycoformycin Towards Human Monocytes

Addition	dAdo IC50 at 3 Days (μM)
None	>100
Deoxycoformycin 5 μM (dCF)	0.15
dCF + Deoxycytidine 100 μM	>100
dCF + Nicotinamide 5 mM	0.08
dCF + 3-Aminobenzamide 5 mM	0.06

Compounds and dCF were added 1 hour prior to the addition of dAdo. Viability was determined by the MTT reduction assay after a 3 day exposure to the indicated drugs.

was more than 500-fold resistant (data not shown). The DNA damage to monocytes was dose- and time-dependent, and could easily be detected as early as 1 hour after drug exposure. Cell viability was not affected until 16 hours had elapsed. Table 2 shows that one of the earliest events following CdA-induced DNA damage was a reduction in RNA synthesis, which declined further as cell viability decreased. Protein synthesis remained at control levels for 8 hours, and then declined in concert with changes in RNA synthesis (data not shown). Cellular NAD and ATP pools were also well maintained until cell lysis occurred. This latter result is in contrast to the prompt fall in lymphocyte NAD and ATP following dAdo or CdA exposure in vitro (5). In lymphocytes, DNA damage is known

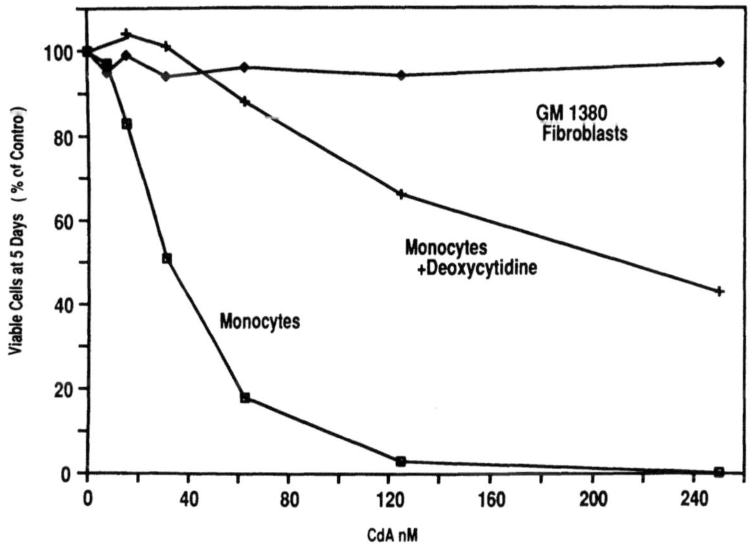

Fig. 1. Selective toxicity of CdA towards human monocytes, and protective effect of deoxycytidine.

Table 2. Biochemical Changes in Monocytes Exposed to 2-Chlorodeoxyadenosine

CdA 1 μM Exposure	Viable Cells	Double-Stranded DNA	RNA Synthesis
(Hours)	(Percent of control cultures without CdA)		
1	100	79	78
2	95	65	87
4	98	45	83
8	98	33	74
16	91	13	26

Double-stranded DNA was measured by the alkaline unwinding assay. 3H-Uridine incorporation into TCA precipitates was assaued during the final hour of culture. Viability was assessed by dye exclusion.

to stimulate nuclear poly(ADP-ribose) synthetase, which causes a lethal consumption of substrate NAD. However, human monocytes and neutrophils are virtually devoid of poly(ADP-ribose) synthetase by western blotting techniques (data not shown) (12). Thus, the mechanism of dAdo or CdA toxicity to human monocytes does not appear to involve excessive ADP-ribosylation and NAD depletion.

Impairment of Monocyte Function

Monocytes were cultured for 3 days in the presence of low concentrations of CdA. Dialyzed culture supernatants were then assayed for IL-6 activity, and the phagocytic capacity of the adherent cells was tested using antibody-coated red cell targets. Fig. 2 shows that both IL-6 release and phagocytic function were severely impaired by 10-20 nM CdA, at which doses cell viability was unaffected. The increase in medium IL-6 found at higher levels of CdA correlated with cell lysis, and presumably reflected release of intracellular monokines.

CdA-Induced Monocyte Depletion In Vivo

Eight patients with refractory cutaneous T-cell lymphoma received CdA chemotherapy (0.1 mg/kg per day) by continuous infusion for 7 days. Therapy did not cause nausea, vomiting, hair loss, or other metabolic side effects. As shown in Fig. 3, circulating absolute monocyte and lymphocyte counts decreased in each patient during the infusion period. In contrast, granulocytes were unaffected. Hemoglobin and platelet counts also did not change. The effect of CdA on monocytes was striking: 4 patients had complete disappearance of circulating monocytes by 7 days. Monocyte counts retured to pre-treatment levels within 2 weeks after the end of the CdA infusion.

CONCLUSIONS

Human monocytes are exquisitely sensitive to dAdo plus deoxycoformycin, and to CdA in vitro. In contrast, non-dividing fibroblasts and neutrophils are resistant. The toxicity of dAdo and CdA towards

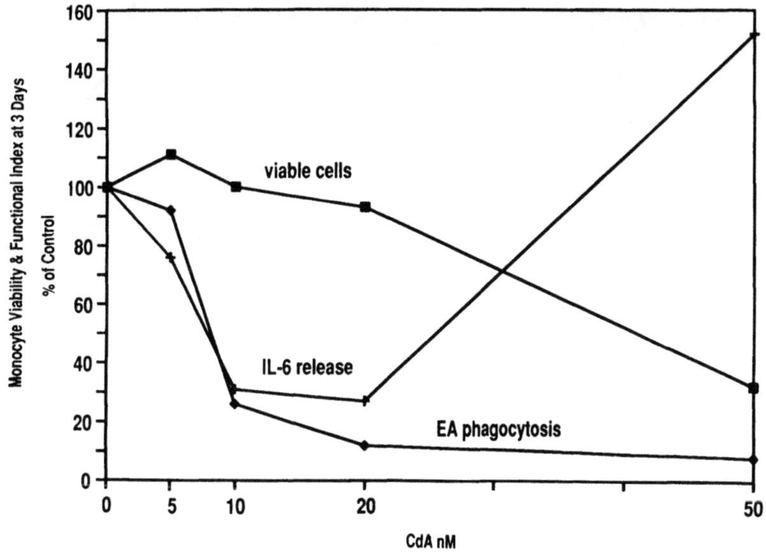

Fig.2 Inhibition of in vitro monocyte function by low-dose CdA.

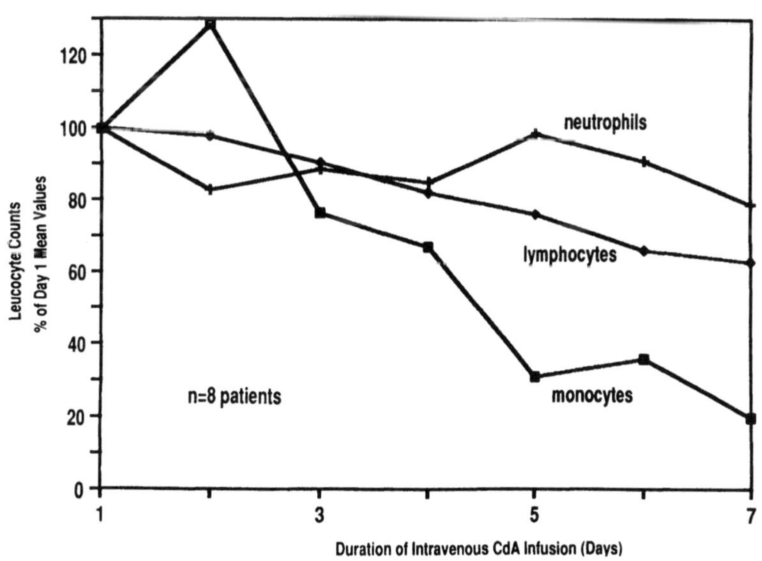

Fig. 3. Changes in blood leukocyte counts during CdA chemotherapy.

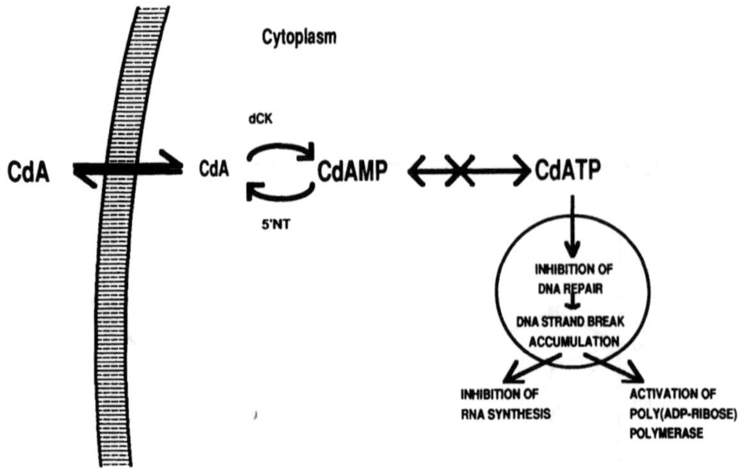

Fig. 4. Intracellular metabolism of CdA and mechanisms of toxicity in non-dividing cells.

monocytes requires the formation of deoxynucleotides via the action of deoxycytidine kinase (Fig. 4). Exposure of monocytes to dAdo or to CdA in vitro initiates a rapid, progressive accumulation of DNA strand breaks. Decreased RNA synthesis is an early result of CdA-induced DNA damage. However, unlike lymphocytes, the monocytes maintain adequate NAD and ATP pools until cell lysis ensues. Certain monocyte functions, such as phagocytosis and the release of monokines, may be inhibited at sub-toxic concentrations of CdA. In vivo, plasma levels of CdA achieved during chemotherapy (5-10 nM) are sufficient to produce a profound, reversible fall in the circulating monocyte count. The selectivity of CdA toxicity towards monocytes and lympyhocytes, together with the demonstrated efficacy of a relatively non-toxic CdA regimen to affect monocyte production in vivo, provide a rationale for the use of this agent in the management of rheumatoid arthritis and other immune disorders associated with the activation of monocytes.

ACKNOWLEDGMENTS

Funding for this research was provided in part by grants CA01100, GM23200, and RR00833 from the National Institutes of Health, and by a Federal Food and Drug Administration grant FD-4-000280. Dr. Carrera was a Special Fellow of the Leukemia Society of America. This is publication 5447BCR from the Research Institute of Scripps Clinic.

REFERENCES

1. E. R. Giblett, J. E. Anderson, F. Cohen, B. Pollara, and H. J. Meuwissen, Adenosine-deaminase deficiency in two patients with severely impaired cellular immunity, Lancet 2:1067 (1972).
2. D. A. Carson, D. B. Wasson, J. Kaye, B. Ullman, D. W. Martin, Jr., R. K. Robins, and J. A. Montgomery, Deoxycytidine kinase-mediated toxicity of deoxyadenosine analogs toward malignant human lymphoblasts in vitro and toward murine L1210 leukemia in vivo, Proc. Natl. Acad. Sci. USA 77:6865 (1980).
3. D. A. Carson, D. B. Wasson, R. Taetle, and A. Yu, Specific toxicity of 2-chlorodeoxyadenosine toward resting and proliferating human lymphocytes, Blood 62:737 (1983).
4. L. Brox, A. Ng, E. Pollock, and A. Belch, DNA strand breaks induced in human T-lymphocytes by the combination of deoxyadenosine and deoxycoformycin, Cancer Res. 44:934 (1984).

5. S. Seto, C. J. Carrera, M. Kubota, D. B. Wasson, and D. A. Carson, Mechanism of deoxyadenosine and 2-chlorodeoxyadenosine toxicity to nondividing human lymphocytes, J. Clin. Invest. 75:377 (1985).
6. D. A. Carson, D. B Wasson, and E. Beutler, Antileukemic and immunosuppressive activity of 2-chloro-2'-deoxyadenosine, Proc. Natl. Acad. Sci. USA 81:2232 (1984).
7. L. D. Piro, C. J. Carrera, E. Beutler, and D. A. Carson, 2-Chlorodeoxyadenosine: an effective new agent for the treatment of chronic lymphocytic leuekemia, Blood 72:(in press, 1988).
8. B. Freundlich and N. Avdalovic, Use of gelatin/plasma coated flasks for isolating human peripheral blood monocytes, J. Immunol. Meth. 62:31 (1983).
9. H. C. Birnboim and J. J. Jevcak, Fluorometric method for rapid detection of DNA strand breaks in human white blood cells produced by low levels of radiation, Cancer Res. 41:1889 (1981).
10. T. W. Jungi, A rapid and sensitive method allowing photometric determination of erythrophagocytosis by mononuclear phagocytes, J. Immunol. Meth. 82:141 (1985).
11. L.A. Aarden, E.R. DeGroot, O.L. Schaap, and P.M. Lansdorp, Production of hybridoma growth factor by human monocytes,l Eur. J. Immunol. 17:1411 (1987).
12. H. Yamanaka, E. H. Willis, C. A. Penning, C. L. Peebles, E. M. Tan, and D. A. Carson, Human autoantibodies to poly(adenosine diphosphate-ribose) polymerase, J. Clin. Invest. 80:900 (1987).

THE MECHANISM OF 2-CHLORODEOXYADENOSINE-INDUCED CELL DEATH

Y. Wataya[1], Y. Hirota[1], A. Hiramoto-Yoshioka[1], S. Tanaka[1], T. Otani[2], J. Minowada[2], A. Matsuda[3], and T. Ueda[3].

[1]Faculty of Pharmaceutical Sciences, Okayama University, Tsushima, Okayama 700,; [2]Fujisaki Cell Center, Hayashibara Biochemical Laboratories Inc., 675 Fujisaki, Okayama 702,; and [3]Faculty of Pharmaceutical Sciences, Hokkaido University, Sapporo, Japan

INTRODUCTION

2-Chlorodeoxyadenosine (2-CldAdo), an adenosine-deaminase resistant 2'-deoxypurine nucleoside analog, is strongly toxic to cells in culture (1-7). Carson et al. reported that 2-CldAdo had significant therapeutic activity for mice with L1210 leukemia (1) and in a phase 1 clinical trial in patients with intractable hematologic malignancies (8). It is reported that 2-CldAdo is similarly cytotoxic to resting T cells as compared with that of deoxyadenosine (dAdo) (3, 5). The considerable cytotoxicity of 2-CldAdo is accounted for by resistance toward adenosine deaminase and, consequently, by more prolonged effect on cell metabolism compared with that of deoxyadenosine (4). However, the mechanisms of action of the compound is still not clear.

In this paper, we report that the *in vitro* cytotoxicities of 2-CldAdo were tested using 34 human-leukemia-lymphoma cell lines and we have now measured the levels of intracellular dNTP pool, DNA strand breaks and cell viability in 2-CldAdo treated CCRF-HSB-2 cells. We found that the dNTP imbalance caused by 2-CldAdo is followed by DNA double strand breaks and subsequent cell death.

MATERIALS AND METHODS

Human leukemia-lymphoma cells were grown in RPMI-1640 containing 10% heat-inactivated fetal bovine serum at 37 °C under 5% CO_2. The cell cultures were maintained by twice weekly passage into subcultures. For growth inhibition studies, cells were seeded at about 5 x 10^4 cells/ml, and treated with various concentrations of drugs. Cell numbers were counted 48 and 72 hr after the start of incubation. EC_{50}

value refers to the concentration of drug necessary to reduce the growth rate of cells by 50% of the control. Cell viability during treatment with 2-CldAdo was determined in a hemocytometer by the trypan blue dye exclusion method.

2-CldAdo was added to an exponentially growing cells in suspension culture at a concentration of 95% cell-growth inhibition. The preparation of cell extracts was described previously (9,10,11). The quantitative determination of deoxyribonucleoside triphosphate (dNTP) and ribonucleoside triphosphate (rNTP) pools were performed by high performance liquid chromatography (HPLC) as previously described (9,10,11). The alkaline and neutral elution methods were done by the previously described methods (10,12). Orthogonal-field-alternation gel electrophoresis (OFAGE) was as described by Carle and Olson (13): electrophoresis was performed with 0.6 x TBE (53 mM Tris, 53 mM boric acid, 1.2 mM EDTA, pH 8.3) at 13°C for 18 hr including 50 to 100 second linear gradient pulses. The buffer was recirculated and changed at approximately the midpoint of the run for optimal resolution of the DNA samples. The OFAGE units that were used have been described (13). Agarose blocks containing intact chromosomal DNA from *S. cerevisiae* (BO133-3B) were prepared as described previously (14). Agarose blocks containing DNA from 2-CldAdo treated CCRF-HSB2 cells were prepared by slightly modifying the procedure of Schwartz and Canter (14). The gel was stained with ethidium bromide. We used DNA from λ phage (48.5 Kbp), T4 phage (166 Kbp), HSV-1 (150 Kbp), and yeast (260 Kbp - 2000 Kbp) as size markers

RESULTS

The *in vitro* cytotoxicities of 2-chlorodeoxyadenosine (2-CldAdo) were tested using 34 human leukemia-lymphoma cell lines. 2-CldAdo was found to be broadly cytotoxic to human T- (EC_{50} value: 0.005-5.3 μM), B- (0.015-4.8 μM), myeloblastic (0.009-12 μM), and monocytic (2.6-5.4 μM) cell lines. We showed the distribution of the 34 human cell lines in our experiments based on the maker profiles of the cells as described by Minowada (15) (Fig. 1). The cell lines show striking differences in the sensitivity to the cytostatic activity as shown in Table 1. This toxic effect of 2-CldAdo was partially reversed by addition of deoxycytidine, and the apparent EC_{50} of 2-CldAdo in the presence of 1.0 μM and 10 μM deoxycytidine was 0.17 μM and 7.1 μM in CCRF-HSB-2 cells, respectively. Deoxyadenosine, deoxyguanosine, and thymidine (at the concentration of 10 μM) did not affect the toxic effect of 2-CldAdo (data not shown). When cells were exposed to 1 μM 2-CldAdo for 24 hr, washed twice and incubated 2 days in fresh medium, no growth was observed.

Since it is known that cycloheximide can prevent thymineless death (16) and dNTP imbalance death (9, 10) in FM3A cells, we examined the effect of cycloheximide on cell death

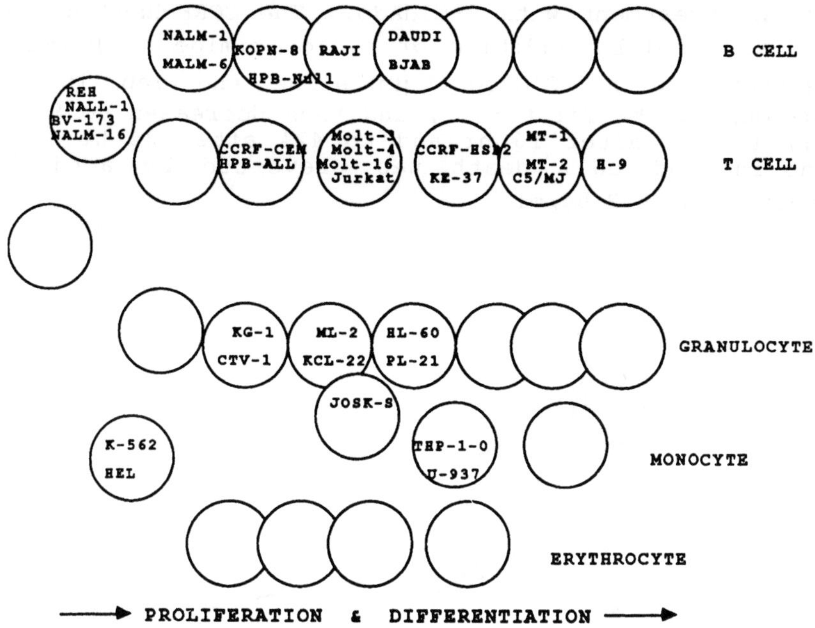

Figure 1. Distribution of 34 human leukemia-lymphoma cell lines examined

Table 1. Toxicity of 2-CldAdo in human leukemia-lymphoma cells

Cell	EC_{50} (μM)	Cell	EC_{50} (μM)
Non-T, Non-B		**B-cell**	
REH	0.45	NALM-1	0.13
NALL-1	0.0052	NALM-6	0.064
BV-173	0.001	KOPN-8	0.15
NALM-16	0.04	HPB-Null	0.043
T-cell		RAJI	0.015
CCRF-CEM	0.0074	DAUDI	4.8
Molt-3	5.3	BJAB	0.015
Molt-4	0.96	**Myelomonocyte**	
Molt-16	0.005	CTV-1	2.3
Jurkat	0.16	ML-2	0.0091
CCRF-HSB2	0.011	KCL-22	12
KE-37	1.6	HL-60	1.1
MT-1	0.082	PL-21	2.5
MT-2	0.13	JOSK-S	2.6
C5/MJ	3.4	THP-1-0	5.4
H-9	3.2	U-937	4.1
HPB-ALL	0.72	KG-1	0.011
Erythrocyte			
K-562	0.028		
HEL	0.0022		

caused by treatment with 2-CldAdo. The CCRF-HSB2 cells were growth-arrested by addition of cycloheximide (2 μg/ml). In cells treated with 2-CldAdo (1 μM) alone, the viability did not change during the first 8 hr, and then decreased to 86% after 12 hr, to 67% after 16 hr and to 42% after 24 hr. The 2-CldAdo-induced cell death was prevented by addition of cycloheximide at 2 μg/ml.

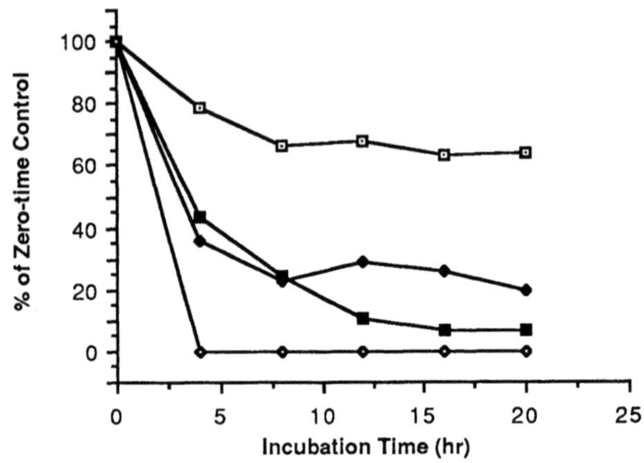

Figure 2. dNTP pool changes in CCRF-HSB2 cells treated with 1 μM 2-CldAdo

CCRF-HSB2 cells at a cell density of 2.0×10^5 cells/ml were treated with 1 μM 2-CldAdo. At the indicated times, aliquots of 50 ml were removed and dNTP pools were measured by HPLC as described in Materials and Methods. ■, dATP; ◆, dGTP; ◇, dCTP; and □ , dTTP. Retention times of dNTP were as follows (min): dCTP, 8.0; dTTP, 9.4; dATP, 12.4; and dGTP, 20.7. dNTP pool size at zero time was as follows (pmol/10^6 cells): dATP, 34.7; dCTP, 11.0; dGTP, 18.8; and dTTP,37.6.

CCRF-HSB2 cells were seeded into the culture medium at 5×10^4 cells/ml, and when the cell density became 2×10^5 cells/ml, 2-CldAdo (1 μM) was added. At desired times, aliquots were removed and cellular dNTP- and rNTP-pools were measured . Fig. 2 shows the dNTP pool changes in the presence of 2-CldAdo. Within 4 hr of treatment dCTP pool size became lower than the limit of measurement. At 20 hr, dATP, dCTP, and dTTP decreased below the amounts of zero-time control. At 12 hr treatment with 2-CldAdo, intracellular 2-CldAdoTP pool size became 4 pmol/10^6 cells.

DNA single- and double-strand breaks, as measured by alkaline and neutral filter elution methods, were detected at

23 hr after treatment with 1 μM 2-CldAdo. The level of the single-strand breaks [SSB] was approximately equal to that found with γ-irradiation at 95 rad. In contrast, the DNA double-strand breakage [DSB] was approximately equivalent to that of the γ-irradiation at 1,475 rad. The measured ratio of [SSB]/[DSB] was 0.0644. Therefore, as described previously (10), the double strand breaks were not a result of an accumulation of single strand breaks, but were formed by direct double strand cuts. The DNA strand breaks, inducible by 2-CldAdo, were abolished on addition of cycloheximide (2 μg/ml) which is an inhibitor for protein biosynthesis.

2-CldAdo-induced DNA double-strand breaks have been directly examined by means of OFAGE. Asynchronously growing CCRF-HSB2 cells were treated with 1 μM 2-CldAdo for the indicated periods. In untreated cells, DNA was retained at the origin of the gel. After the administration of 2-CldAdo for 24 hr, fragmented DNA bands were observed by the analysis. The band appeared to be approximately 50 Kbp - 100 Kbp in size, which is close to the postulated size of replication units in mammalian cells (17).

DISCUSSION

Our hypothesis for the mechanism of 2-CldAdo-induced cell death is shown in Fig. 3. 2-CldAdo, after conversion to 2-CldAdoTP, may interfere with the regulation of ribonucleotide reductase; the resultant formation of 2-CldAdoTP may cause dNTP pool imbalance by subsequent effects on ribonucleotide reductase. The dNTP pool imbalance induced by 2-CldAdo would be a signal for the production of an endonuclease which would cause double strand breaks in DNA. Accumulation of DNA double strand breaks is known to be lethal for cells. We investigated the induction of DNA double strand breaks observed on addition of 2-CldAdo to CCRF-HSB2 cells. The results suggest that the double strand breaks were not due to the accumulation of single strand breaks, but rather were formed by direct double strand cuts. The DNA double strand breaks induced by dNTP imbalance were not repaired, and both the DNA breakage and the cell death induced by 2-CldAdo were prevented by addition of cycloheximide, an inhibitor of protein biosynthesis. Previously, we reported that a double strand DNA endonuclease activity can be detected in cells treated with 5-fluorodeoxyuridine (10) and dAdo (9). On treatment of cells with 2-CldAdo, fragmented DNA band was observed by the OFAGE analysis. The band appeared to be approximately 50 Kbp - 100 Kbp in size, which is close to the postulated size of replication units in mammalian cells (17). We assumed that the DNA breaks must have occurred in a specific fashion and were localized in specific regions of the genome. We assume that the double strand breaks play an important role on the process of 2-CldAdo-mediated cell death and the intracellular dNTP imbalance is the trigger of these events.

The ctyostatic activity and the 2-CldAdo induced dNTP pool imbalance have been measured in several cell lines. Striking differences in drug potency (Table 1) and intracellular dCTP pool depletion were observed among different cell lines examined. 2-CldAdo was more cytotoxic to dCTP pool depletion-induced cell lines (data not shown). The differences in the cytostatic action of 2-CldAdo on human leukemia-lymphoma cells could be attributed to differential affinities of 2-CldAdo for nucleoside-nucleotide metabolism systems.

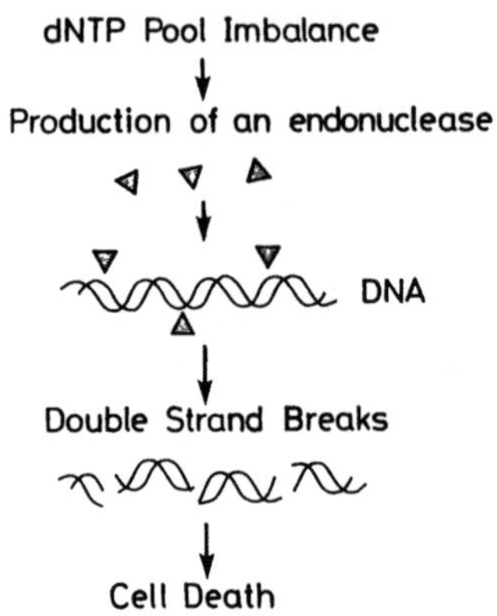

Figure 3. Mechanism of dNTP imbalance death

ACKNOWLEDGMENTS

Professor H. Hayatsu of the Okayama University is gratefully acknowledged for his encouragement throughout this research. This work was supported by a grant (Y. W.) for Scientific Research from the Ministry of Education and Culture.

REFERENCES

1. Carson, D. A., Wasson, D. B., Kaye, J., Ullman, B., Martin, D. W.Jr., Robins, R. K., and Montgomery, J. A. Deoxycytidine kinase-mediated toxicity of deoxyadenosine analogs toward malignant human lymphoblasts in vitro and toward murine L1210 leukemia in vivo. Proc. Natl. Acad. Sci. USA, 77 : 6865 (1980).
2. Hung, M-C., Hatfield, K., Roetker, A. W., Montgomery, J. A., and Blakley, R. L. Analogs of 2'-deoxyadenosine: Facile enzymatic preparation and growth inhibitory effects on human cell lines. Biochem. Pharmac., 30: 2663 (1981).
3. Carson, D. A., Wasson, D. B., Taetle, R., and Yu, A. Specific toxicity of 2-chlorodeoxyadenosine toward resting and proliferating human lymphocytes. Blood, 62 : 737 (1983).

4. Montgomery, J. A. The chemistry and biology of nucleosides of purines and ring analogues. In:"Nucleosides, Nucleotides, and Their Biological Applications," J. Ridout. D. W. Henry, and L. M. Beacher III ed., pp19-46. Academic Press, Inc., New York (1983).
5. Seto, S., Carrera, C. J., Kubota, M., Wasson, D. B., and Carson, D. A. Mechanism of deoxyadenosine and 2-chlorodeoxyadenosine toxicity to nondividing human lymphocytes. J. Clin. Invest., 75 : 377 (1985).
6. Huang, M-C., Ashmun, R. A., Avery, T. L., Kuehl, M., and Blakley, R. L. Effects of cytotoxicity of 2-chloro-2'-deoxyadenosine and 2-bromo-2'-deoxyadenosine on cell growth, clonogenicity, DNA synthesis, and cell cycle kinetics. Cancer Res., 46: 2362 (1986).
7. Parsons, P. G., Bowman, E. P. W., and Blakley, R. L. Selective toxicity of deoxyadenosine analogues in human melanoma cell lines. Biochem. Pharmacol. 35: 4025.(1986).
8. Carson, D. A., Wasson, D. B., and Beutler, E. Antileukemic and immunosuppressive activity of 2-chloro-2'-deoxyadenosine. Proc. Natl. Acad. Sci. USA, 81 :2232 (1984).
9. Yoshioka, A., Tanaka, S., Hiraoka, O., Koyama, Y., Hirota, Y., and Wataya, Y. Deoxyribonucleoside-triphosphate imbalance death : Deoxy- adenosine-induced dNTP imbalance and DNA double strand breaks in mouse FM3A cells and the mechanism of cell death. Biochem. Biophys. Res. Commun., 146 : 258 (1987).
10. Yoshioka, A., Tanaka, S., Hiraoka, O., Koyama, Y., Hirota, Y., Ayusawa, D., Seno, T., Garrett, C., and Wataya, Y. Deoxyribonucleoside triphosphate imbalance. 5-Fluorodeoxyuridine-induced DNA double strand breaks in mouse FM3A cells and the mechanism of cell death. J. Biol. Chem., 262 : 8235 (1987).
11. Tanaka, K., Yoshioka, A., Tanaka, S., and Wataya, Y. An improved method for the quantitative determination of deoxyribonucleoside triphosphates in cell extracts. Anal. Biochem., 139 : 35 (1984).
12. Kohn,K. W., Erickson, L. C., Ewig, R. A. G., and Friedman, C. A. Fractionation of DNA from mammalian cells by alkaline elution. Biochemistry, 15 : 4629 (1976).
13. Carle, G. F., and Olson, M. V. Separation of chromosomal DNA molecules from yeast by orthogonal-field-alternation gel electrophoresis. Nucleic Acids Res., 12 : 5647 (1984).
14. Schwartz, D. C., and Cantor, C. R. Separation of yeast chromosome-sized DNAs by pulsed field gradient gel electrophoresis. Cell, 37 : 67 (1984).
15. Minowada, J., Drexler, H. G., and Menon, M. Significance of the multiple marker analysis of leukemia cells, in:"Cellular, Molecular and Genetic Approaches to Immunodiagnosis and Immunotherapy," Kano, K., Mori, S., Sugisaki, T., and Torisu, M., ed.,pp. 355-364, University of Tokyo Press, Tokyo (1987).
16. Ayusawa, D., Shimizu, K., Koyama, H., Takeishi, K., and Seno, T. Accumulation of DNA strand breaks during

thymineless death in thymidylate synthase-negative mutants of mouse FM3A cells. J. Biol. Chem., 258 : 12448 (1983).
17. Hand, R., Eucaryotic DNA: Organization of the genome for replication. Cell, 15: 317 (1978).

ARTEFACTS DUE TO RADIATION-INDUCED CELL DAMAGE IN PULSE-LABELING EXPERIMENTS USING TRITIATED NUCLEIC ACID PRECURSORS

Juha Salonen, Kari O. Raivio, and Lauri Saxén

Department of Pathology and Children's Hospital
University of Helsinki
Finland

INTRODUCTION

Precursors labeled with low energy beta-emitting 3H, 14C, and 35S radioisotopes are widely used to follow metabolic pathways or to label intra- or extracellular macromolecules. Doses for pulse-labeling are often in the order of 10 uCi/ml or even more for several hours. However, radioactivity in amino acid at concentrations as low as 1 uCi/ml has been shown to cause an histologically evident injury already by 6 hours in organ cultures (Minor 1982).

We have studied the embryonic development of mouse kidney, specifically the incorporation of 14C- and 3H-adenine and 3H-thymidine into nucleotides, DNA, and RNA during the differentiation and growth of the tissue. Of the main tissue components of the developing kidney, the epithelial ureteric bud branches to form the collecting tubules, and the mesenchyme is induced by the ureteric bud to differentiate into kidney tubules and glomeruli. We show here the differential sensitivity of these two tissue components to damage by radiation and explore the mechanisms of this damage.

MATERIALS AND METHODS

Organ culture conditions. Metanephrogenic kidneys were dissected from 11- to 12-day old hybrid mouse embryos (CBA x NMRI) and grown on Nuclepore polycarbonate filters with 1.0 um pore size, placed on stainless-steel grids to keep them on the medium/gas interphase (Saxén and Saksela, 1971; Lehtonen, 1976). Tissues were cultured in Eagle's minimum essential medium (MEM) supplemented with 10% fetal calf serum, in 95% air-5%CO2 atmosphere, for 6 or 24 hr to allow adaptation to culture conditions (Saxén et al., 1983). The cultures were then labeled for 2-3 hr using different concentrations of 3H- or 14C-adenine or 3H-thymidine. After three washes with medium, culture was continued for various lengths of time, after which the tissue samples were collected and stored frozen until analysis. In rescue experiments, some of the cultures were labeled and grown in the presence of 5 mM 3-aminobenzamide (ABA).
Measurement of DNA content and 3H-adenine incorporation. Tissue samples, each consisting of an individual kidney, were extracted with 50ul of ice cold 0.5 M perchloric acid (PCA) for 1 hr, washed once with PCA, and radioactivity in the supernatant was analyzed. The proteins of the precipitate were digested overnight at +37C in 200 ul of a solution of 100ug/ml proteinase-K, 0.1M NaCl, 0.1M EDTA, and 0.1% N-lauroylsarcosine detergent in 0.1M Tris-buffer, pH 8.2. Specimens were then shaken vigorously before taking paired 20 ul aliquots for the analysis of DNA content spectrophotometrically in 2 ml of buffer containing 0.1M NaCl, 0.1M sodium phosphate,

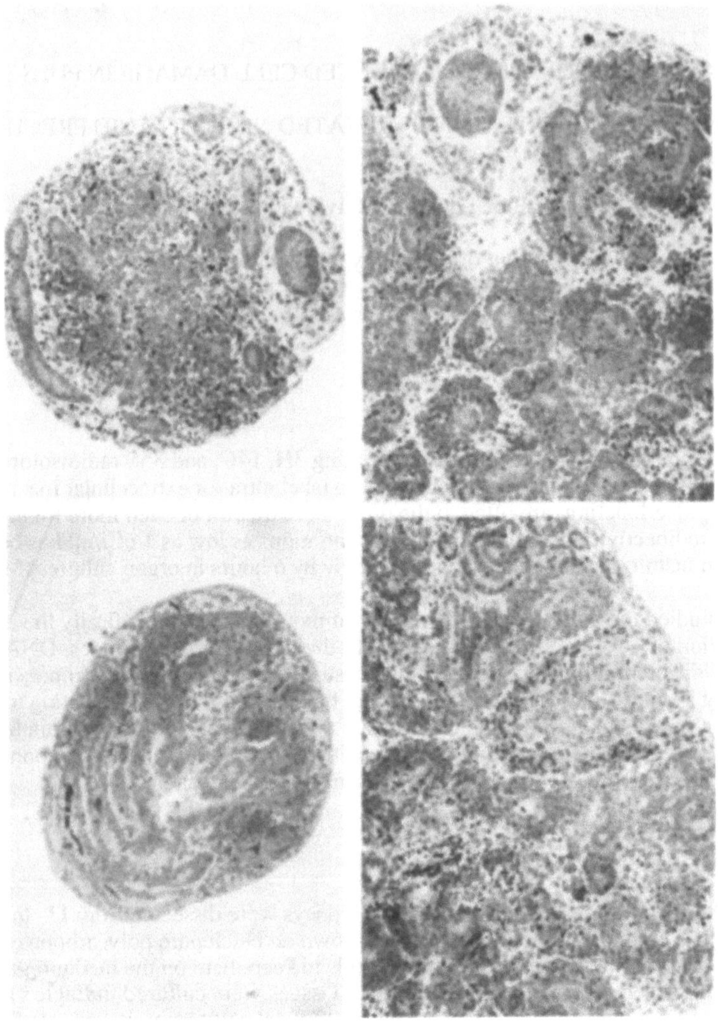

Fig. 1. Effect of a 3 hr pulse of 10 uCi/ml 14C-adenine on the morphology of the developing kidney. After 1 day of culture (top left), numerous dark pyknotic nuclei are seen in the metanephrogenic mesenchyme which has almost dissappeared leaving mainly ureteral tissue after 4 days (bottom left). Corresponding unlabeled kidneys show slight stromal necrosis after 1 day (top right), but good growth and tubule formation after 4 days (bottom right). Toluidine blue, x200.

pH 7.3, and 100 ng/ml of a fluorochrome binding specifically to DNA (Hoechst 33258). Salmon testis DNA was used as a standard.
To the rest of the digest 35 ul of 2M KOH were added and incubated in room temperature for 3 hr to hydrolyze RNA, after which DNA was precipitated by adding 50 ug of unlabeled carrier DNA and 200 ul of ice cold 3M trichloroacetic acid. After standing on ice for 15 min, specimens were sentrifuged at 13000 RPM for 5 min at +5 C and the radioactivities in the supernatant (RNA) and the precipitate (DNA solubilized in 0.3M NH4OH) were analyzed separately.
Radioactivity in adenine nucleotides. Soluble nucleotides in the initial PCA extract were separated with thin-layer chromatography (Crabtree and Henderson, 1971), and their radioactivities determined in a liquid scintillation counter.
Histology. The kidneys were fixed with 2.5% glutaraldehyde in 0.1 M phosphate buffer, pH 7.3, for 45 minutes at room temperature, mounted in Epon, and 1 um sections were cut for staining with toluidine blue.

RESULTS

Morphological changes during radiation-induced cell damage. 12-day old kidneys were precultivated for 24 hr and then labeled for 3 hr with 10uCi/ml 14C-adenine. Immediately after labeling some dark hematopoietic cell nuclei in the stromal areas and some diffusely distributed dark pyknotic dead cells were seen. When the kidneys were cultured further for 1 day, innumerable pyknotic degenerating cells can be seen in the metanephrogenic mesenchyme; in the ureteral epithelium (in cross section seen as tubules) and in the loose stromal mesenchyme very few damaged and necrotic cells are seen (Fig. 1, top left). In contrast, in unlabeled kidneys cultured for 1 day, healthy and abundant metanephrogenic mesenchyme can be seen around the branching ureteral epithelium, with some central necrosis in the loose stromal mesenchyme. The kidney is markedly bigger than the labeled one (Fig 1, top right). After 2 days of culture almost all of the metanephrogenic mesenchyme exposed to radioactivity was either necrotic or lost, whereas ureteral epithelium and loose stromal mesenchyme seemed quite intact. After 4 days of culture the necrotic cells have vanished, as has almost all of the metanephrogenic mesenchyme, the amount of ureteral epithelium and loose stromal mesechyme is approximately the same as at the time of labeling and the size of the whole explant has diminished somewhat during the cultivation (Fig. 1, bottom left). In the unlabeled kidney of the same age, there are some central necrotic areas, but otherwise the tissue seems healthy and many induced condensates and differentiating tubules derived from the metanephrogenic mesenchyme can be seen among ureteral branches. Loose mesechyme on the margins of the kidney also seems healthy. The size of the explant is considerably larger than that of the labeled one (Fig 1, bottom right).

Effect of radiation dose on DNA content. The DNA content of the control kidney decreases slightly during the first day of culture but then increases linearly. When pulse-labeled with 20 uCi/ml of 3H-thymidine or adenine after 6 hr of preculture, the explants show a similar increase in DNA content but starting 1-2 days later than the growth of the control kidney (Fig. 2, top). If the experiments are done after 1 day of precultivation, the DNA content of the untreated kidney increases linearly, whereas that of the labeled explants indicates a progressive inhibition of growth in a dose-dependent way (Fig. 2 bottom).

Effect of ABA on DNA content. In order to test, whether radiation-induced inhibition of kidney growth could be prevented by blocking poly-ADP ribosylation, experiments were done in the presence of ABA. This compound itself, in concentrations of 1 or 5 mM, had no effect on the increase of kidney DNA during culture (data not shown). In kidneys labeled with 3H-adenine and then cultured for 1 to 4 days, ABA had no consistent effect on the DNA content (Fig. 3, top). If anything, there was a tendency to lower DNA levels with 5 mM ABA than without it.

Metabolism of 3H-adenine. Kidneys precultivated for 6 hours and then incubated with 3H-adenine showed an increase of label in the acid-insoluble pool with doses of 0.3, 1, and 3 uCi/ml, but no further increase with 10 or 30 uCi/ml, when analyzed immediately after the

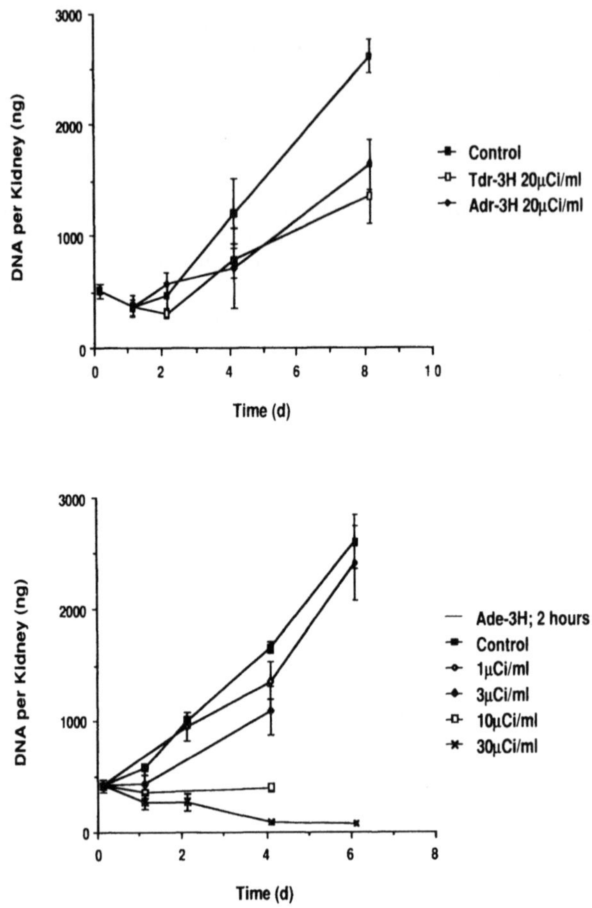

Fig. 2. Change in DNA content of developing kidneys exposed to 3H-adenine or -thymidine (2 hr) and then cultured. Top: labeling after 3 hr of precultivation. Bottom: 24 hr of precultivation.

pulse or after 1 more day of culture. However the acid-soluble nucleotides showed a linear dose-dependent increase of radioactivity at all these doses after 1 day of culture. The effect of ABA, studied only with the 30 uCi/ml dose, was to increase significantly the incorporation only into the acid-soluble fraction but not into nucleic acids (Fig 3, middle)
.

Kidneys precultivated for 1 day showed a linear increase of the label in the acid-soluble fraction and in RNA and DNA for the whole range of 1 to 30 uCi/ml 3H-adenine 1 day after labeling. However, 3 days later the labeling remained at the same level only with the dose of 1 uCi/ml, whereas with higher doses the kidneys had lost an increasing proportion of their label (Fig. 3, bottom). ABA made no difference to these labeling patterns (data not shown).

In order to evaluate, whether radiation damage resulted in nucleotide depletion or a disturbance in the adenine nucleotide ratios, the label in ATP, ADP, and AMP was analyzed separately immediately after pulse-labeling of kidneys precultivated for 1 day. Energy charge was 0.74 after 1 uCi/ml and 0.75 after 30 uCi/ml. ABA had no effect on the latter value. Incorporation of label into the adenine nucleotide pool was 30 times higher after 30 uCi/ml than after 1 uCi/ml.

DISCUSSION

In the central stromal areas of unlabeled kidneys, some necrosis appears during cultivation, when the explants grow bigger. This may be due to tissue culture conditions, possibly insufficient nutrition or oxygenation. In the explants labeled with 10 uCi/ml of 14C-adenine for 3 hr, massive necrosis is seen in the metanephrogenic mesenchyme. The loose stromal mesechyme and the ureteral epithelium survive with no marked damage and comprise almost all of the kidney at day 4 (Fig. 1). It seems that the metanephrogenic mesenchyme is clearly more sensitive to damage by radiation than the other tissue components. This may be due to larger proportion of rapidly dividing cells. Selective damage induced by radiation may thus result in artefacts during culture.

The DNA content of unlabeled kidneys declines during the first day of cultivation and increases linearly thereafter. The first day seems to be required for adaptation to culture conditions, involving some cell death and a partial standstill of proliferation (Saxen et al., 1983). In explants labeled with 3H-adenine or -thymidine, the beginning of growth is postponed or, with higher doses, the DNA content remains stationary or gradually decreases. This can be explained by the necrosis during the first and second day after labeling seen in the histology. The growth of explants precultivated for 6 hr prior to labeling with 20 uCi/ml is inhibited to a similar degree than of those precultivated for 24 hr and labeled with 3 uCi/ml. This difference may be due to a smaller number of dividing cells or to decreased intake of label during the first day of adaptation (Fig. 2).

In the kidneys precultivated for 6 hr, an upper limit for incorporation of 3H-adenine seems to be reached at a dose of 3 to 10 uCi/ml, after which there is no more increase of the label in the acid-insoluble fraction. No such upper limit can be observed in any of the fractions, if the kidneys are labeled after 1 day of precultivation. The incorporation is many times higher even at the lowest doses, when labeling is done after 24 hr compared to only 6 hr of precultivation (Fig. 3). This suggests that there are marked differences in adenine metabolism depending on the adaptation time and on the induction and differentiation stage. Artefacts can again result from such differences.

One of the major mechanisms of radiation-induced cell damage is DNA strand breakage (Clever et al, 1974). The repair of these breaks is associated with increase of poly(ADP)-ribosylation of nuclear proteins to facilitate DNA ligation. The stimulation of poly(ADP-ribose) polymerase has been shown by various DNA-damaging agents, including X irradiation and gamma irradiation (Benjamin &Gill, 1980; Goodwin et al.,1978; Campagnari et al., 1966). The substrate for poly(ADP-ribose) polymerase is NAD+ (Hayaishi & Ueda, 1977), which is rapidly consumed after massive DNA damage, accompanied by ATP depletion. The depletion of NAD+ and ATP pools can be prevented

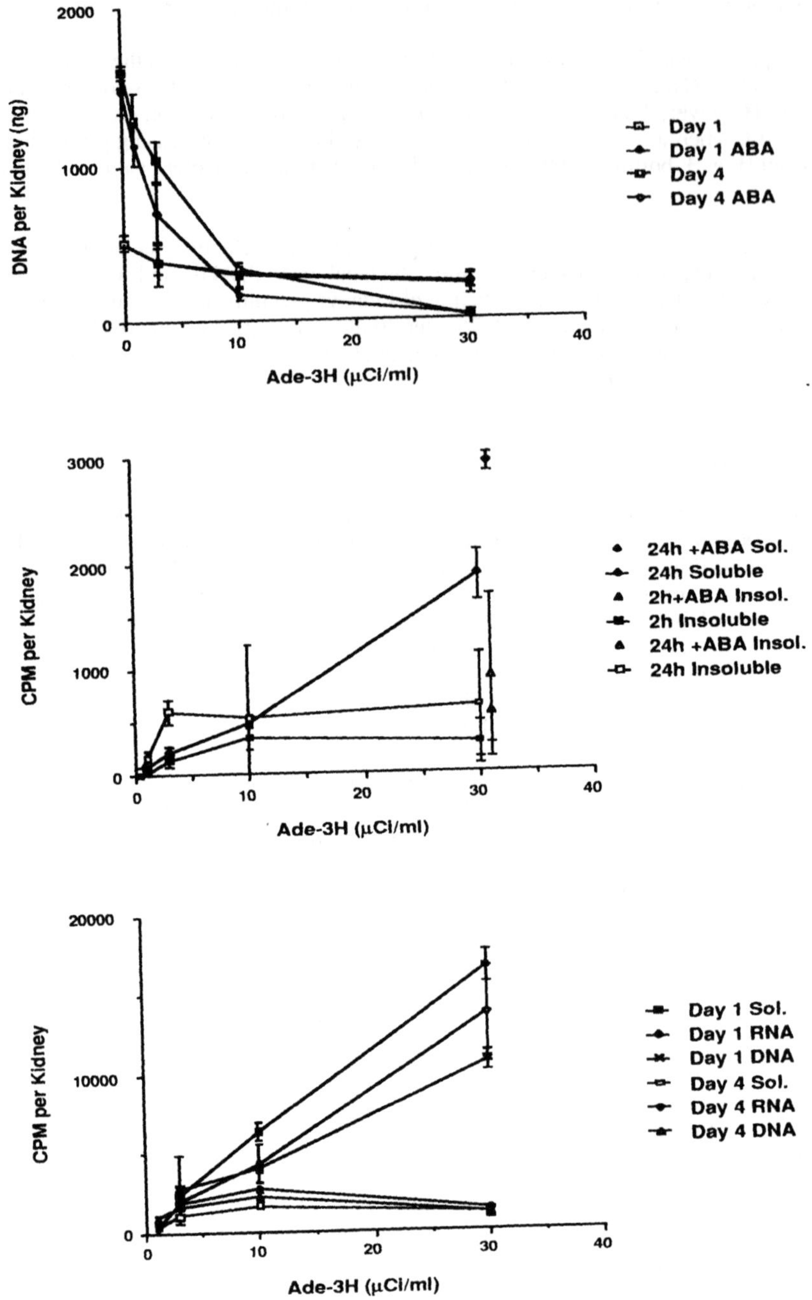

Fig. 3. Effect of varying doses of 3H-adenine, with or without 3-aminobenzamide (ABA, 5 mM), on DNA content (top) and labeling pattern after 3 hr (middle) or 24 hr (bottom) of precultivation.

by 3-aminobenzamide (ABA), an inhibitor of poly(ADP-ribose) polymerase (Schraufstatter et al., 1986).

In our experiments ABA had no effect on the growth of the kidneys. Neither did it prevent the decrease in DNA content nor alter the labeling pattern of nucleotides or nucleic acids after exposure to radiation (Fig. 3). This suggests that ATP depletion may not be the most significant mechanism of radiation-induced cell damage under our conditions. This conclusion is supported by the finding that immediately after a pulse of 30 uCi/ml 3H-adenine the adenine nucleotide pool contained 30 times more radioactivity than after 1 uCi/ml, with a similar energy charge.

ACKNOWLEDGMENTS

We are grateful to Anja Tuomi and Birgitta Tjäder for technical assistance. This work was supported by grants from the Sigrid Jusélius Foundation.

REFERENCES

Benjamin, R. C. and Gill D. M., 1980, ADP-ribosylation in mammalian cell ghosts, J. Biol. Chem., 255:10493.
Campagnari, F., Whitfield, J. F., and Bertazzoni, U., 1966, The effect of X-irradiation on the nicotinamide adenine dinucleotides (NAD-NADH) content of rat thymocytes, Exp. Cell Res., 42:646.
Cleaver, J. E. and Burki, H. J., 1974, Biological damage from intranuclear carbon-14 decays: DNA single-strand breaks and repair in mammalian cells, Int. J. Radiat. Biol., 26:399.
Crabtree, G. W. and Henderson, J. F., 1971, Rate-limiting steps in the interconversion of purine ribonucleotides in Ehrlich ascites tumor cells, Cancer Res., 31:985.
Goodwin, P. M., Lewis, P. J., Davies, M. I. Skidmore, C. J., and Shall, S., 1978, The effect of gamma radiation and Neocarzinostatin on NAD and ATP levels in mouse leukaemia cells, Biochim. biophys. Acta, 543:576.
Hayaishi, O. and Ueda, K., 1977, Poly(ADP-ribose) and ADP-ribosylation of proteins, Ann. Rev. Biochem., 46:95.
Lehtonen, E., 1976, Transmission of signals in embryonic induction, Med. Biol., 54:108.
Minor, R. R., 1982, Cytotoxic effects of low levels of 3H-, 14C-, and 35S-labeled amino acids, J. Biol. Chem., 257:10400.
Saxén, L., Salonen, J., Ekblom, P., and Nordling, S., 1983, DNA synthesis and cell generation cycle during determination and differentiation of the metanephric mesechyme, Dev. Biol., 98:130.
Saxén, L. and Saksela, E., 1971, Transmission and spread of embryonic induction. II. Exclusion of an assimilatory transmission mechanism in kidney tubule induction, Exp. Cell Res., 66:369.
Schraufstatter, I., U., Hyslop, P., A., Hinshaw, D., B., Spragg, R., G., Sklar, L., A., and Cochrane, C., G., 1986, Hydrogen peroxide-induced injury of cells and its prevention by inhibitors of poly(ADP-ribose) polymerase, Proc. Natl. Acad. Sci. USA, 83:4908.

REDUCED PURINE NUCLEOTIDE CONTENT AND POLY(ADP-RIBOSE) SYNTHETASE ACTIVITY IN HPRT-DEFICIENT HUMAN LYMPHOBLASTS

Gwyn M. McCreanor

Division of Inherited Metabolic Diseases
MRC Clinical Research Centre, Watford Road
Harrow HA1 3UJ, UK

INTRODUCTION

HPRT-deficiency or Lesch-Nyhan syndrome is a deficiency of HPRT enzyme activity which salvages the purine base hypoxanthine to IMP and guanine to GMP.

HPRT-deficient EB-virus transformed human lymphoblasts have shown reduced concentrations of purine nucleotides compared with control lymphoblasts (McCreanor et al., 1987) but this decrease was not consistently found in all cells.

So investigation of the nucleotide content of HPRT-deficient lymphoblasts under various conditions was undertaken to establish whether ATP concentration decreases with changes in the nutritional state of the cells and whether this decrease in ATP concentration affects poly(ADP-ribose) synthetase activity in these cells.

Poly(ADP-ribosyl)ation is one of several mechanisms for the post-translational modification of chromosomal proteins in eucaryotic cells (Ueda and Hayaishi, 1985).

Cell nuclei contain the enzyme poly(ADP-ribose) synthetase which catalyses the transfer of ADP-ribose from NAD to form poly(ADP-ribosyl)ated protein conjugate, mainly chromatin proteins. This enzyme is associated with DNA repair, and the appearance of DNA damage is accompanied by an increase in its activity.

The covalent modification of chromatin proteins by this enzyme is thought to make the damaged area more accessible to other DNA repair enzymes by promoting a loosening of the protein matrix near the site of the break.

It has been suggested that histone H1 and poly(ADP-ribose) synthetase itself are the major acceptor proteins for poly(ADP-ribosyl) groups in a variety of cell types although other chromatin associated proteins such as DNA ligase II can also be ADP-ribosylated (Creisson and Shall, 1982).

METHOD

EB-virus transformed lymphoblasts from HPRT-deficient patients and controls were grow under increasingly unfavourable nutritional conditions by varying the length of time post feeding. The purine nucleotide levels and poly(ADP-ribose) synthetase activity of the cells were measured at intervals of 1,3,5 and 7 days post feeding.

Purine nucleotides were extracted from lymphoblast cells using ice-cold fresh 14% (w/v) trichloroacetic acid (TCA) solution. The precipitate was removed by centrifugation (800g for 10 min) and the supernatant was extracted with water-saturated diethyl ether to remove the TCA.

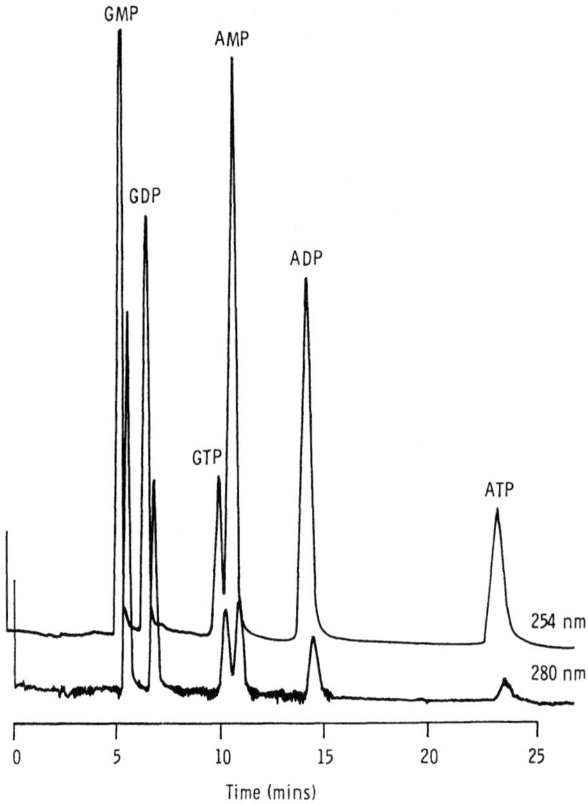

Figure 1. Separation of purine nucleotides by reversed phase HPLC

Nucleotides were assayed by reverse phase HPLC using an isocratic variation of the method of Willis et al. (1986), using 5% (v/v) methanol in 83.3 mM triethylammonium-phosphate, pH 6.0, at a flow rate of 1 ml/min. Peaks were identified by co-chromatography of pure standards and by characteristic absorbance ratios at 254 nm and 280 nm. (Figure 1).

Poly(ADP-ribose) synthetase activity in intact lymphoblasts was assayed by permeabilising the cells to NAD. The method of permeabilisation was essentially that of Berger et al. (1978). Cells were harvested by centrifugation at 800g for 2 min, they were then resuspended at 2 x 10^6 cells/ml in permeabilising buffer consisting of 10mM Tris/HCl (pH 7.8 at 4°C), 0.25M sucrose, 1mM EDTA, 30mM 2-mercaptoethanol and 4mM MgCl. They were incubated at 4°C for 15 min.

Poly(ADP-ribose) synthetase activity was assayed by the method of Furneaux and Pearson (1980). 200μl of permeabilised cells (5 x 10^6) were added to 100μl of assay buffer giving a final concentration of 33mM Tris/HCl, pH 7.8, 40mM MgCl, 3mM [^{14}C] NAD^+ (600mCi/mmol). After incubation at 30°C for 30 min the reaction was terminated by adding ice-cold TCA and filtering onto glass fibre discs (Whatman GF/C). The discs were washed and counted (Gaal and Pearson, 1986). The amount of radiolabelled NAD converted to acid-insoluble product, poly(ADP-ribose), per mg protein was determined.

RESULTS

Feeding cells on day 0 and assaying them on days 1,3,5 and 7 showed that there is a significant decrease in ATP content and poly(ADP-ribose) synthetase activity in the HPRT-deficient cells compared with the controls. (Figures 2 and 3).

There is a significant positive corelation between ATP content of lymphoblasts and poly(ADP-ribose) synthetase activity and an inverse correlation between NAD content and poly(ADP-ribose) synthetase activity of the lymphoblasts. (Figures 4 and 5).

DISCUSSION

The Lesch-Nyhan syndrome is a rare sex-linked recessive neurological disorder associated with the complete absence of the purine salvage enzyme, hypoxanthine phosphoribosyltransferase (HPRT). The main clinical and biochemical features include delayed motor development, generalised muscle hypotonia, choreoathetosis, progressive mental retardation, hyperuricaemia, urinary stones and gout. The brain does not show any characteristic structural abnormalities in the Lesch-Nyhan syndrome (Kelley and Wyngaarden, 1978).

It has been suggested that HPRT catalysed purine salvage becomes more important relative to purine *de novo* synthesis during the later stages of brain maturation, this alteration being related to functional rather than structural changes (Allsop and Watts, 1980, 1984; Watts et al., 1982).

The present study shows decreased ATP levels in HPRT-deficient cells. The brain has one of the lowest levels of purine *de novo* synthesis and therefore depends to a greater extent on the purine salvage pathway and HPRT enzyme activity than other tissues, so the brain may be highly affected by HPRT deficiency and ATP depletion.

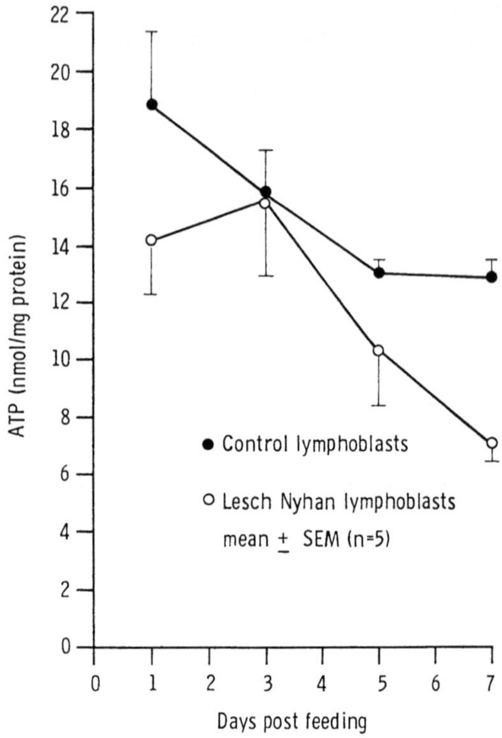

Figure 2. ATP content of control and Lesch Nyhan lymphoblasts against days post feeding

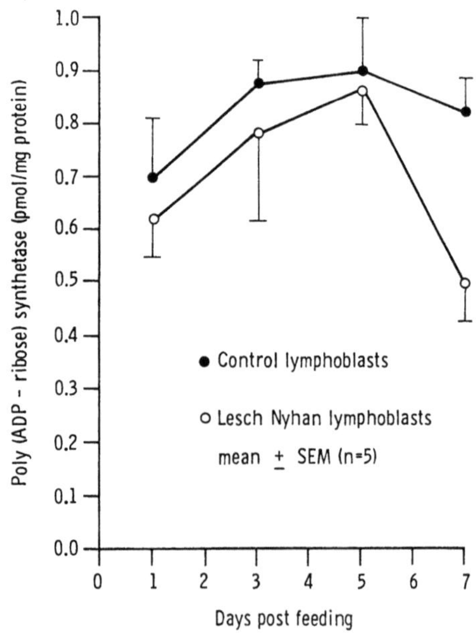

Figure 3. Poly(ADP-ribose) synthetase activity of control and Lesch Nyhan cells against days post feeding.

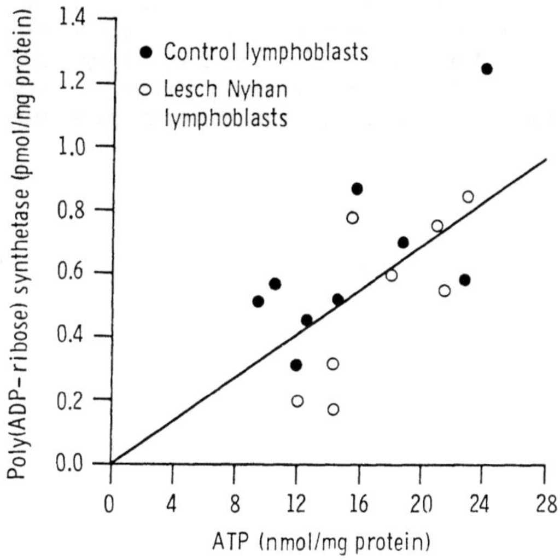

Figure 4. Correlation between Poly(ADP-ribose)synthetase activity and ATP content in control and Lesch-Nyhan lymphoblasts

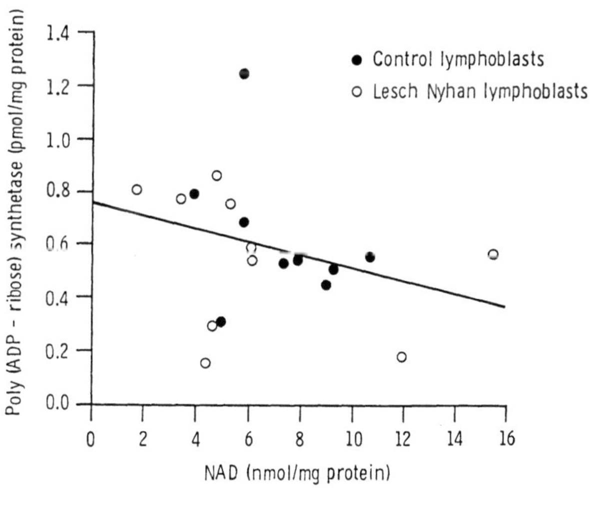

Figure 5. Correlation between poly(ADP-ribose)synthetase activity and NAD content in control and Lesch-Nyhan lymphoblasts

The mechanism by which the progressive neurological dysfunction occurs in HPRT-deficiency is not clear but the fact that it is progressive suggests that it may be due to intermittent episodes of energy depletion. The decreased ATP concentration in HPRT-deficient lymphoblasts, which have adequate purine *de novo* synthesis, occurs when conditions are unfavourable and the ATP depletion may be intermittent, whereas the brain of Lesch Nyhan patients have lower levels of purine *de novo* synthesis and therefore may be even more likely to suffer ATP depletion.

ATP depletion correlates with a decrease in poly(ADP-ribose) synthetase activity, which is involved in DNA repair. The respiratory centres of the brain cannot reduce their use of ATP since they must continue to function, so other metabolic processes are more likely to be affected, and an accumulation of defects in the DNA of the cells involved may occur as a result of episodes of ATP depletion.

Other disorders which involve defects of DNA repair include Fanconi's anaemia and Cockayne syndrome. The symptoms of these disorders include progressive mental retardation. Fanconi's anaemia is caused by the impaired metabolism of ADP-ribose and Cockayne syndrome involves a defect in DNA ligase (Schweiger et al., 1987).

CONCLUSION

There is a decrease in ATP concentration in HPRT-deficient lymphoblasts compared with control lymphoblasts. This decrease is seen when the cells are subjected to unfavourable conditions, in this case 7 days post feeding.

There is also a decrease in poly(ADP-ribose) synthetase activity in these cells compared with the control lymphoblasts.

There is a positive correlation between ATP content of lymphoblasts and poly(ADP-ribose) synthetase activity of control and HPRT-deficient lymphoblasts.

Decreased poly(ADP-ribosyl)ation of nuclear protein may lead to decreased DNA repair in these cels. This may be a factor in explaining the long term deterioration of brain and testicular function in the Lesch-Nyhan syndrome.

REFERENCES

Allsop, J., and Watts, R.W.E., 1980, Activities of amidophosphoribosyltransferase (EC 2.4.2.14) and the purine phosphoribosyltransferases (EC 2.4.2.7 and 2.4.2.8) and the phosphoribosylpyrophosphate content of rat central nervous system at different stages of development, J.Neurol.Sci.,46:221-232.

Allsop, J., and Watts, R. W. E., 1984, Purine synthesis and salvage in brain and liver, in: "Purine Metabolism in Man IV, Part B," C. De Bruyn, H. A. Simmonds and M. Muller, eds., Plenum Publishing Corporation.

Berger, N. A., Weber, G., and Kaichi, A. S., 1978, Characterisation and comparison of poly(ADP-ribose) synthesis and DNA synthesis in nucleotide-permeable cells, Biochim.Biophys.Acta., 519:87-104.

Creisson, D., and Shall, S., 1982, Regulation of DNA ligase activity by poly(ADP-ribose), Nature, 196:271-272.

Furneaux, H. M., and Pearson, C. K., 1980, Adenosine diphosphate ribose transferase from baby-hamster kidney cells (BHK-21/C13). Characterisation of the reaction and product. Biochem.J., 187:91-103.

Gaal, J. C., and Pearson, C. K., 1986, A gel-electrophetic analysis of protein ADP-ribosylation in polyoma virus-transformed and non-transformed BHK-21/C13 fibroblasts, Biochim.Biophys.Acta, 881:196-209.

Kelley, W. N., and Wyngaarden, J. B., 1978, The Lesch-Nyhan Syndrome, in: "The Metabolic Basis of Inherited Disease," 4th edition, McGraw Hill Inc., New York, Ch.40, pp.1011-1036.

McCreanor, G. M., Harkness, R. A., and Watts, R. W. E., 1987, Reduced concentrations of purine nucleotides in hypoxanthine phosphoribosyltransferase-deficient human lymphoblasts, Biochem.Soc.Trans., 15:555-556.

Schweiger, M., Auer, B., Burtscher, H. J., Hirsch-Kauffmann, M., Klocker, H., and Schneider, R., 1987, DNA repair in human cells. Biochemistry of the hereditary diseases Fanconi's anaemia and Cockayne syndrome, Eur.J.Biochem., 165:235-242.

Ueda, K., and Hayaishi, O., 1985, ADP-Ribosylation, Ann.Rev.Biochem., 54:73-100.

Watts, R. W. E., Spellacy, E., Gibbs, D. A., Allsop, J., McKeran, R. O., and Slavin, G. E., 1982, Clinical post-mortem, biochemical and therapeutical observations of the Lesch-Nyhan syndrome with particular reference to the neurological manifestations, Quart.J.Med., N.S., 51:43-48.

Willis, C. L., Lim, C. K., and Peters, T. J., 1986, Separation of ribonucleotides, deoxynucleotides, cyclic nucleotides and deoxycyclic nucleotides by reverse-phase high performance liquid chromatography, J.Pharm.Biomed.Anal., 4: 247-252.

INABILITY OF POLY-ADP-RIBOSYLATION INHIBITORS TO PROTECT PERIPHERAL BLOOD LYMPHOCYTES FROM THE TOXIC EFFECTS OF ADA INHIBITION

K. Ganeshaguru[1], A. Piga[2], L. Latini[2] and A. V. Hoffbrand[1]

1. Haematology Department, Royal Free Hospital School of Medicine, Rowland Hill, London, UK
2. Oncologia Clinica, Ospedale Regionale Torrette, Ancona Italy

SUMMARY

We have evaluated the effectiveness of two inhibitors of poly-ADP-ribosylation, nicotinamide and 3-aminobenzamide as rescue agents in resting and PHA-stimulated lymphocytes damaged by the combination of deoxycoformycin (dCF), plus deoxyadenosine (dAdo). Incubation with dCF (10^{-5}M) and dAdo (10^{-4}M) for 18 hours, inhibited protein and RNA synthesis in unstimulated lymphocytes and impaired the ability of the cells to respond to PHA stimulation or to give rise to T-cell colonies in methyl-cellulose. Predominantly dead cells using trypan blue exclusion were observed at day 4, in both unstimulated and PHA-stimulated lymphocytes, whether or not the drugs were removed at 18 hours. The number of viable cells at day 4 increased from 13.7% to 41.1% with the addition of 5 mM nicotinamide, and to 28.8% with 5 mM 3-aminobenzamide added with dCF and dAdo. Although nicotinamide was able to prevent a fall in NAD concentration for 24h (but not for 48h) and to reduce the fall of cell ATP concentration, the inhibition by dCF and dAdo of protein synthesis, RNA synthesis, ability of cells to form colonies or to respond to PHA was not reversed. We conclude that inhibition of NAD utilisation by inhibiting ADP-ribosylation with nicotinamide or 3-aminobenzamide does not protect cells *in vitro* from deoxyadenosine toxicity with ADA inhibition and is not likely to give significant clinical benefit in ADA deficiency.

INTRODUCTION

Congenital absence of the enzyme adenosine deaminase (ADA) or its inhibition by the drug deoxycoformycin (dCF) leads to death of lymphoid cells, especially of thymic origin. This is thought to be largely due to the inability of T-lymphocytes to degrade deoxyadenosine triphosphate (dATP).(1) This is the biochemical basis of the marked depression of immune system in severe combined immunodeficiency (SCID) syndromes(2) as well as the response of lymphoid leukaemias of T-cell origin to treatment with dCF(3-5). Inhibition of ribonucleotide reductase by excess dATP with consequent arrest of DNA synthesis due to lack of the other deoxyribonucleoside triphosphates may kill proliferating cells but cannot explain the toxicity of ADA absence or inhibition to resting lymphocytes. This has been suggested to occur by suicide inactivation of S-adenosylhomocysteine hydrolase(6), ATP depletion(7), inhibition of RNA synthesis(8) and induction of DNA strand breakage(9). More recently, Seto et al(10) have suggested that death of resting lymphocytes incubated with dCF and deoxyadenosine (dAdo) is due to depletion of NAD consequent on activation of DNA repair mechanisms. In their study, cell death (assessed by inability to exclude erythrocin B) was reversed by co-incubating with nicotinamide and, to a lesser extent, 3-aminobenzamide (3-ABA), which inhibit poly-ADP-ribose synthetase, an enzyme involved in DNA repair and whose substrate is NAD. They also suggested that this rescue mechanism may be exploited therapeutically. In a previous study, we found no difference in NAD, ATP, dATP and DNA strand breaks between different forms of lymphoid leukaemias (including those sensitive or resistant to killing by dCF) incubated with dCF and dAdo, in vitro(11).

We present here evidence that whilst nicotinamide, and to a lesser extent 3-ABA, prolong viability of lymphocytes (measured by dye exclusion) incubated with dCF and dAdo, this rescue effect is only apparent as the cells are metabolically inert, protein and RNA synthesis remain inhibited and the cells are unable to proliferate, presumably due to lesions (e.g. DNA damage) other than NAD depletion.

MATERIALS AND METHODS

Drugs and Reagents

Deoxycoformycin (dCF) was obtained from NIH (Dr M Suffness); deoxyadenosine (dAdo), nicotinamide, 3-aminobenzamide (3-ABA), NAD and alcohol dehydrogenase were purchased from Sigma, L-[4.5-^{3}H]-leucine, (45Ci/mmol, TRK170), [5-^{3}H]-uridine (28.5Ci/mmol, TRK 178) and [methyl-^{3}H]-thymidine (5Ci/mmol, TRA 120) were obtained from Amersham International. Nucleoside-free alpha medium was purchased from Gibco. RPMI 1640 medium (RPMI) and fetal calf serum (FCS) from Flow Labs.; phytohaemagglutinin (PHA) from Wellcome; PHA-lymphocyte conditioned medium (PHA-LCM) was prepared by incubating peripheral blood lymphocytes in alpha medium + 10% FCS + 1% PHA, harvesting the supernatant after seven days in culture and filter-sterilising prior to storage at -20°C(12).

Cell Culture

Peripheral blood mononuclear cells were isolated by Lymphoprep (Nyegaard, Norway) sedimentation of heparinised blood from normal donors, resuspended at a concentration of 1 x 10^6 cells/ml in RPMI containing 10% dialysed FCS, 100U/ml penicillin, 100ug/ml streptomycin, 2mM L-glutamine and dispensed in 180ul triplicate aliquots into 96-well flat-bottomed microtiter plates containing 20ul/well of drug solution in RPMI. The final drug concentrations were as follows: dCF 10^{-5}M, dAdo 10^{-4}M, nicotinamide 5 x 10^{-3}M, 3-ABA 5 x 10^{-3}M. Some experiments were performed with dCF at 10^{-6}M and dAdo at 10^{-6}M in order to reproduce conditions used in other studies(11) and to achieve concentrations comparable to those in vivo during dCF therapy. The plates were incubated for 18h at 37°C in humidified atmosphere containing 7.5% CO_2 washed twice and the incubation prolonged for the appropriate times, either in the presence or in the absence of 1% PHA. In some experiments, the washing procedure was omitted and the drugs left in culture. In all experiments time zero followed the 18h pre-incubation period. Protein, RNA and DNA synthesis were studied daily by the addition of either 1uCi/well of ^{3}H-leucine for 24h, or 1uCi/well of ^{3}H-uridine for 4h, or 0.4uCi/well of ^{3}H-thymidine for 1h; and the cells harvested onto fibre-glass filters. The standard deviations of triplicate values was between 5 and 10% except for values below 2000 cpm. The number of viable cells was assessed daily on unlabelled cultures by the trypan blue exclusion method. For DNA strand breaks, NAD and ATP measurements, cells were cultured in similar conditions in centrifuge tubes in 1ml of medium containing 5-15 x 10^6 cells.

Colony Forming Assay

Peripheral blood mononuclear cells were preincubated for 18h with drugs as above and washed. T-lymphocyte colony forming units (CFU-TL) were cultured in duplicate in 35mm Petri dishes (Falcon) at 2 x 10^5 cells/ml of alpha medium supplemented with 0.8% methyl-cellulose, 20% dialysed FCS, 20% PHA-LCM, 1% PHA, antibiotics and L-glutamine, and drugs when indicated. Cultures were grown at 37°C in 7.5 % CO_2, and colonies of greater than 50 cells were counted at day 6(13).

DNA Strand Breaks

At the end of the incubation, the cells were washed twice and processed for fluorescence analysis of DNA unwinding(14). Cells were resuspended in an isotonic buffer and lysed in an urea/detergent solution at 0°C. Alkali (final pH 11.8) was added and cell extracts were incubated for 1h at 15°C to allow partial unwinding of the DNA. Ethidium bromide was added and fluorescence measurements made in an electrofluorometer. Each point was done in triplicate.

NAD and ATP Extraction and Assay

Cells were washed and the cell pellet incubated with cold 0.5 M perchloric acid to extract NAD and ATP. The supernatant was neutralised with 0.5 M KOH, and the resultant cell extract stored at -20°C until assayed. NAD was measured using the enzymatic cycling assay of Jacobson and Jacobson(15). To a 1ml reaction mixture, 0.2ml of either NAD standards or cell extract and 0.1ml of alcohol dehydrogenase (16 units) was added and incubated in the dark at 37°C for 20 min. The reaction was stopped by adding 0.5ml of 12 mM iodoacetate and the absorbance of aliquots measured in an automatic scanner. Standards and extracts were assayed in duplicates. ATP measurements were done by Dr R Hutton, Royal Free Hospital, London, using an LKB Bioluminescence Assay kit on neutralised perchloric acid cell extracts.

RESULTS

Unstimulated Lymphocytes

Continuous exposure of unstimulted lymphocytes to 10^{-5}M dCF and 10^{-4}M dAdo, caused a progressive decrease in the number of viable cells to 13.7% of the control at day 4 (Fig 1); 5 x 10^{-3}M nicotinamide and to a lesser extent 5 x 10^{-3}M 3-ABA caused a reduction in the toxic effect of dCF and dAdo with respectively 41.1% and 28.6% of cells surviving at day 4 (Fig.1). Neither nicotinamide nor 3-ABA on its own had any effect on cell viability (not shown). Protein and RNA synthesis were dramatically inhibited (almost completely by day 2) by dCF and dAdo, with no rescue from either nicotinamide or 3-ABA (Fig. 1). When all the drugs were removed after the first 18h of incubation comparable results were obtained (Fig. 1). At dCF 10^{-6}M and dAdo 10^{-6}M the effects on cell viability, protein and RNA synthesis were less pronounced. However, even at these low concentrations there was no rescue with nicotinamide.

DNA Strand Breaks, NAD and ATP Measurement

The incubation of peripheral blood lymphocytes for 24h and 48h with the combination of dCF and dAdo, markedly reduced the percentage of double-stranded DNA (Fig. 2), and the addition of nicotinamide on its own caused moderate DNA damage and did not rescue from DNA strand breaks (Fig. 2). NAD levels, following treatment with dCF and dAdo were dramatically reduced to 15% of control values after 24h and to 9% after 48h; the addition of nicotinamide supported sustained levels at 24h (110% of the control) but subsequently NAD concentration dropped to 27% of the control at 48h (Fig. 2). ATP concentration also fell progressively as a

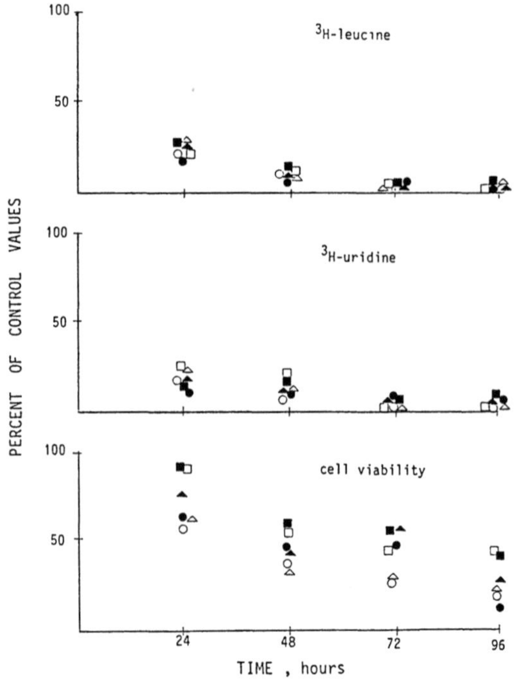

Fig.1 Effect of the combination of dCF + dAdo on its own (●) or in association with nicotinamide (■) or 3-ABA (▲) on ^{3}H-leucine, ^{3}H-uridine and cell viability in unstimulated lymphocytes. Cells were preincubated with drugs for 18 h before time 0, and the drugs left in culture thereafter (each point is the mean of 2 experiments, each run in triplicate). Open symbols represent experiments in which all the drugs were removed at time 0.

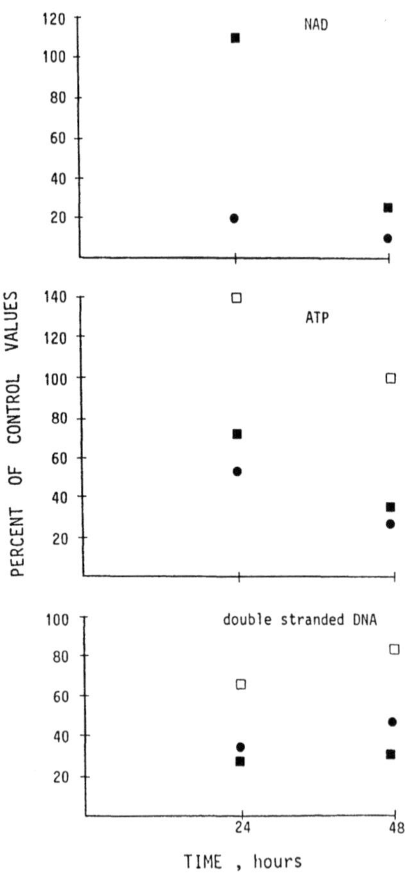

Fig.2 Effect of the combination dCF + dAdo on its own (●) or in association with nicotinamde (■) on DNA strand breaks, NAD and ATP concentrations in unstimulated lymphocytes. Cells were exposed to the drugs for the indicated times. Open squares represent the effect of nicotinamide on its own on ATP concentration and DNA strand breaks; NAD concentrations following incubation with nicotinamide alone were 342% at 24 h and 48 h of the control.

consequence of dCF/dAdo treatment, to 51% of control at 24h and 28% at 48h; the addition of nicotinamide reduced the fall in the ATP concentration at 24h (73%) and at 48h (37%) (Fig. 2). Results on NAD levels and DNA strand breaks at low concentrations of dCF and AdR (10^{-6}M) were no different from those shown in Fig. 2 with higher concentrations. Neither was there any rescue effect with nicotinamide.

PHA-Stimulated Lymphocytes

Continuous exposure of the peripheral blood lymphocytes to 10^{-5}M dCF and 10^{-4}M dAdo in the presence of 1% PHA caused a gradual decrease in the number of viable cells (Fig. 3), comparable to that obtained in unstimulated lymphocytes. Similarly, a rescue effect was seen with nicotinamide and to a lesser extent with 3-ABA (Fig. 3) which on their own did not have any effect on viability (not shown). Protein, RNA and DNA synthesis were rapidly inhibited, almost completely by day 2, by dCF and dAdo; no rescue effect was obtained with the addition of either nicotinamide or 3-ABA. When dCF and dAdo were removed after 18h, immediately before the addition of PHA, a gradual, partial recovery in the capacity of the cells to incorporate ^{3}H-leucine, ^{3}H-uridine and ^{3}H-thymidine was observed following the first 24h; in no case was there any advantage from the use of either nicotinamide or 3-ABA. The number of viable cells gradually decreased throughout the experiment and only marginal advantage was obtained with the use of nicotinamide (Fig. 3).

CFU-TL Assay

The combination of 10^{-5}M dCF and 10^{-4}M dAdo produced almost complete inhibition of T-colony formation at day 6, whether the drugs were removed following the first 18h of incubation (before plating) or left permanently in culture (Table I). No advantage was obtained with the addition of either nicotinamide or 3-ABA. These latter drugs when used alone did not have any effect on the growth of CFU-TL, whether the drugs were removed or not following the first 18h incubation (Table I).

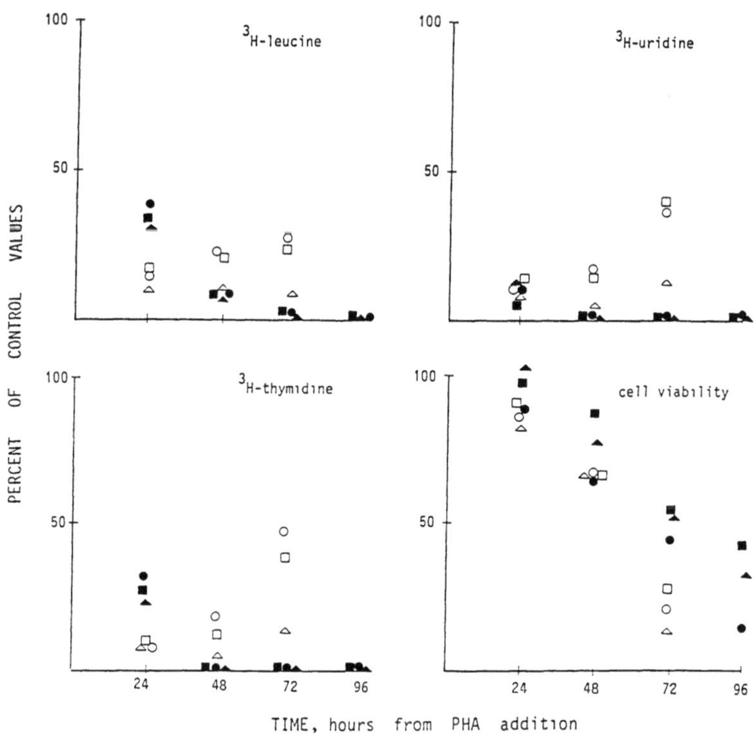

Fig.3 Effect of the combination dCF + dAdo on its own (●) or in association with nicotinamide (■) or 3-ABA (▲) on ^{3}H-leucine, ^{3}H-uridine, ^{3}H-thymidine uptake and cell viability in PHA-stimulated lymphocytes. PHA was added to the cultures at time 0, after 18 h exposure of the cells to the drugs. Closed symbols = drugs left in culture. Open symbols = drugs removed at 0h.

TABLE I Lack of rescue effect by nicotinamide and 3-ABA on the inhibition of CFU-TL induced by the combination of dCF and dAdo.

	EXPERIMENT 1		EXPERIMENT 2		EXPERIMENT 3	
	No. colonies	%	No. colonies	%	No. colonies	%
Control	124.5 (120-129)	100	122.5 (120-125)	100	500.5 (476-525)	100
dCF + dAdo	1.5 (1 - 2)	1.2	6.5 (6- 7)	5.3	42.0 (39- 45)	8.4
dCF + dAdo + nicotinamide	0.5 (0 - 1)	0.4	6.0 (6- 6)	4.9	47.5 (49- 50)	9.5
dCF + dAdo + 3-ABA	0.5 (0 - 1)	0.4	5.5 (5- 6)	5.5	29.5 (26- 33)	5.9
Nicotinamide	123.5 (120-127)	99.2	124.0 (118-130)	101.2	- -	-
3-ABA	120.5 (111-130)	96.8	121.5 (115-128)	99.2	- -	-

Lack of rescue effect by nicotinamide and 3-ABA on the inhibition of CFU-TL induced by the combination of dCF and dAdo. In Exp. 1, all drugs were left permanently in culture (7 days). In Exp. 2, dCF and dAdo were present in the incubation medium for 18h only, whilst nicotinamide and 3-ABA were left permanently in culture. In Exp. 3, all drugs were removed after 18h.

DISCUSSION

The results here show that incubation of human peripheral blood lymphocytes with the combination of dCF and dAdo induces: (a) dramatic reduction of protein and RNA sysnthesis, (b) inability of the cells to proliferate in response to PHA, or (c) to give rise to T-cell colonies (CFU-TL) in methyl-cellulose medium; after four days in liquid culture the vast majority of cells are dead.

Addition of nicotinamide or 3-ABA as well as dCF and dAdo only prolonged the ability of the cells to exclude trypan blue. No protective effect could be seen as a result of nicotinamide or 3-ABA on protein or RNA synthesis in unstimulated lymphocytes, or on the ability of the cells to synthesise protein, RNA or DNA in response to PHA, or on the ability of the cells to give rise to CFU-TL. From a biochemical point of view, nicotinamide was able, at least for 24h to preserve NAD levels and to partially reduce the severity of ATP depletion due to dCF and dAdo but DNA strand breakage was, if anything, increased.

Seto et al (10) have recently shown that exposure of resting lymphocytes to the combination of dCF and dAdo rapidly causes accumulation of DNA strand breaks and inhibition of RNA synthesis, and that a significant fall in NAD levels appears at 8h with severe NAD depletion at 24h. At this time ATP depletion becomes evident, with cell death following 24h later. They showed that this cell death (assessed by erythrocin B) could be delayed by the incubation of the lymphocytes with nicotinamide, and to lesser extent with 3-ABA. Both these compounds are known inhibitors of poly-ADP-ribose synthetase, an enzyme involved in the process of DNA repair, in which large amounts of NAD are converted on poly-ADP-ribose. In addition, nicotinamide is a precursor of NAD. All these data support the hypothesis that depletion of NAD during DNA repair triggered by dATP-induced DNA strand breakage is the ultimate cause for cell death due to ADA deficiency or inhibition in resting cells.

However, our data show that the rescue effect exerted by nicotinamide and 3-ABA against toxicity from the dCF and dAdo combination is only apparent, since although cells retain their viability, assessed by trypan blue exclusion, they are otherwise metabolically dead, being unable to synthesise proteins or RNA, or to respond to PHA stimulation or to give rise to T cell colonies in the clonogenic assay. Co-incubation with nicotinamide or 3-ABA seems only to prolong the period for which the cell membrane retains its integrity and so is able to exclude the dye. The cells are, however, doomed and their inability to proliferate after dCF and dAdo incubation at all concentrations tested is not reversed by nicotinamide.

The present results are not unexpected since both nicotinamide and 3-ABA, by inhibiting poly-ADP-ribose synthetase, presumably slow DNA repair. As reported recently(16) deoxyadenosine itself inhibits DNA repair in resting lymphocytes; the addition of nicotinamide or 3-ABA would therefore be expected to exaggerate the damage. NAD depletion may be prevented by these compounds but the cells are presumably irreparably damaged, consequent on DNA strand breaks induced by dCF and dAdo. Indeed, nicotinamide was found here to increase DNA damage due to dCF and dAdo.

The possibility of preventing toxicity of dCF and dAdo in resting cells is relevant clinically in SCID due to ADA deficiency and to the treatment of T-cell malignancies with ADA inhibitors. However, our in vitro data suggest no clinical benefit from nicotinamide treatment in SCID. Nevertheless, in a disease like SCID, which is usually fatal, dietary supplementation with safe concentrations of the vitamin may still be justified. Long-term intake of the vitamin may succeed in maintaining normal intracellular levels of NAD and help maintain DNA repair without raising intracellular nicotinamide concentration to levels which inhibit DNA repair.

ACKNOWLEDGEMENTS

We are indebted to Mrs S Stanborough for word processing and the Leukaemia Research Fund for financial support.

REFERENCES

1. Carson DA, Kaye J, Seegmiller JE: Differential sensitivity of human leukaemic T cell lines and B cell lines to growth inhibition by deoxyadenosine. J Immunol 121:1726 (1978).
2. Giblett EW, Anderson JE, Cohen F, Pollara B, Meuwissen H: Adenosine deaminase deficiency in two patients with severely impaired immune deficiency. Lancet ii:1067, (1972).
3. Smyth JF, Paine RM, Jackman AL, Harrap KR, Chassin MM, Adamson RH, Johns DG: The clinical pharmacology of the adenosine deaminase inhibitor 2'deoxycoformycin. Cancer Chemother Pharmacol 5:93, (1980).
4. Prentice HG, Smyth JF, Ganeshaguru K, Wonke B, Bradstock KF, Janossy G, Goldstone AH, Hoffbrand AV: Remission induction with the adenosine deaminase inhibitor 2'deoxycoformycin in Thy-acute lymphoblastic leukaemia, Lancet ii:170, (1980).
5. Prentice HG, Russell NH, Lee N, Piga A, Ganeshaguru K, Smyth JF, Hoffbrand AV: Therapeutic selectivity of and prediction of response to 2'deoxycoformycin in acute leukaemia. Lancet ii:1250, (1981).
6. Hershfield MS: Apparent suicide inactivation of human lymphoblasts 5-adenosylhomocysteine hydrolase by 2'deoxyadenosine and adenosine arabinoside. J Biol Chem 250:22, (1979).
7. Siaw MFE, Mitchell BS, Koller CA, Coleman MS, Hutton JJ: ATP depletion as a consequence of adenosine deaminase inhibition in man. Proc Natl Acad Sci USA 77:6157, (1981).
8. Matsumoto SS, Yu J, Yu AL: Inhibition of RNA synthesis by deoxyadenosine plus deoxycoformycin in resting lymphocytes. J Immunol 131:2762, (1983).
9. Brox L, Ng A, Pollock E, Belch A: DNA strand breaks induced in human T lymphocytes by the combination of deoxyadenosine and deoxycoformycin. Cancer Res 55:934, (1984).
10. Seto S, Carrera CJ, Kubota M, Wasson W, Carson DA: Mechanism of deoxyadenosine and 2-chlorodeoxyadenosine toxicity to non-dividing human lymphocytes. J Clin Invest 75;377, (1985).
11. Ganeshaguru K, Ho AD, Piga A, Catovsky D, Hoffbrand AV: Biochemical mechanism of deoxycoformycin toxicity in chronic leukaemias. Leukemia Res. 11: 941 (1987).
12. Aye MT, Till JE, McCulloch EA: Interacting populations affecting proliferation of leukaemic cells in culture. Blood 45: 485, (1975).
13. Russell NH, Carron J, Hoffbrand AV, Bellingham AJ; The relative sensitivity of periphral blood T-lymphocyte colony forming cells and bone marrow CFU-GM to deoxyadenosine and 2'deoxycoformycin. Leuk Res 9:315, (1985).
14. Birnboim HC, Jevcak JJ: Fluorometric method for rapid detection of DNA strand breaks in human white blood cells produced by low doses of radiation. Cancer Res 41:1889, (1981).
15. Jacobson EL, Jacobson MK: Pyridine nucleotide levels as a function of growth in normal and transformed 3T3 cells. Arch Biochem Biophys 175:627, (1976).
16. Seto S, Carrera CJ, Wasson DB, Carson DA: Inhibition of DNA repair by deoxyadenosine in resting human lymphocytes. J Immunol 136:2839, (1986).

CHARACTERIZATION OF POLY(ADP-RIBOSE) POLYMERASE WITH HUMAN AUTOANTIBODIES

Hisashi Yamanaka, Carol A. Penning, Erik H. Willis
D. Bruce Wasson and Dennis A. Carson

Research Institute of Scripps Clinic
La Jolla, CA, U.S.A.

INTRODUCTION

On the toxic mechanism of the deoxyadenosine in patients of adenosine deaminase deficiency, the inhibition of DNA synthesis due to the inactivation of ribonucleotide reductase is one of the possible mechanisms in rapidly dividing cells(1). However, this explanation is not likely to be responsible for the toxicity of deoxyadenosine in resting lymphocytes. Recently, DNA strand break formation has been proposed as a possible mechanism of deoxyadenosine toxicity in resting lymphocytes(2,3). DNA strand breaks cause the rapid activation of a nuclear enzyme, poly(ADP-ribose) polymerase. Poly(ADP-ribose) polymerase catalyzes the covalent attachment of (ADP-ribose) units on nuclear proteins(4). Extensive activation of poly(ADP-ribose) polymerase potentially consumes the cellular NAD, the substrate of this reaction. However, it has not been well demonstrated whether human lymphocytes have enough amount of poly(ADP-ribose) polymerase to deplete the cellular NAD so rapidly. In this regard, accurate cellular content and the turn over rate of poly(ADP-ribose) polymerase have not been investigated. Recently, we have identified human autoantibodies to poly(ADP-ribose) polymerase in the sera of rheumatic patients(5). These autoantibodies specifically recognize poly(ADP-ribose) polymerase, and bind the substrate binding domain of this nuclear enzyme. These autoantibodies enabled us to investigate the characteristics of poly(ADP-ribose) polymerase as a protein. In this report, we show the quantitative characteristics of the poly(ADP-ribose) polymerase.

MATERIALS AND METHODS

<u>Purification of poly(ADP-ribose) polymerase, and determination of the intracellular content of poly(ADP-ribose) polymerase:</u> Poly(ADP-ribose) polymerase was purified from calf or human thymus by the modified methods as described before(6). After purification, this enzyme was almost 100% pure by both protein staining and immunoblotting. Using this protein as standard, the amount of cellular poly(ADP-ribose) polymerase was determined in the extract of various cells. Serially diluted purified poly(ADP-ribose) polymerase and crude cell extract were fractionated by SDS-polyacrylamide gel electrophoresis, and immunoblotted by human autoantibodies to poly(ADP-ribose) polymerase. A standard curve was established based on the measurement of the density of immunoreactive 116kDa band by

densitometer. The amount of cellular poly(ADP-ribose) polymerase was determined by this standard curve.

Cellular amount of poly(ADP-ribose) polymerase in cell growth cycle: Cell cycle study was investigated either with phytohemagglutinin (PHA) stimulated human peripheral blood lymphocytes (PBL) or with synchronized cell line cells. Human T-lymphoblastoid cell line CEM cells were synchronized at G_0 phase by incubation in isoleucine-free RPMI1640 medium with low serum concentration (7). Synchronized cells were released into G_1 phase by the addition of isoleucine and serum. Cells were harvested at the indicated time point, and cell extracts were prepared for either the immunoblotting analysis or the Northern blot analysis. For Northern blot, total RNA (15 mg per each time point) was obtained by lithium chloride-urea method (8) and fractionated on 1% agarose gel containing 2% formaldehyde. Northern transfer was done by the method as described(9). DNA-RNA hybridization was carried out for 30 min at 42 ^{o}C, and the filters were washed with 0.5X SSC buffer containing 0.1% SDS at 48 ^{o}C for 1 hour.

Cell cycle distribution of poly(ADP-ribose) polymerase was also demonstrated by fluorescent activated cell sorting(FACS). Cells were fixed by 1% paraformaldehyde/ 0.1% Triton-X100, and were stained by autoantibodies to poly(ADP-ribose) polymerase followed by fluoresceine labelled goat anti-human IgG. Thereafter, cellular DNA were stained by propedium iodide, and analyzed by FACS IV (Becton Dickinson, Mountain View, CA, U.S.A.) (10).

Cloning of cDNA which encodes poly(ADP-ribose) polymerase : cDNA clones (11) were isolated from a human placental lgt11 cDNA expression library(Clonetech, Palo Alto, CA) according to the method of Young and Davis(12). Immunopositive colonies were detected after processing the filters with serum of a patient which contains autoantibodies to poly(ADP-ribose) polymerase, exactly as immunoblotting. The serum used for the detection was preabsorbed extensively with *E-coli* lysates. The antibody-reactive clones were subcloned into Bluescript(Stratagene, San Diego, CA). For Northern blotting analysis, the DNA probe (1.8kb) was labelled with [a-^{32}P]dATP (Amersham Co, Arlington Heigts, IL) by Klenow fragment of DNA polymerase 1.

Turn over rate of poly(ADP-ribose) polymerase: To examine the turn over rate of poly(ADP-ribose) polymerase, cultured CEM cells were labelled with [^{35}S]-methionine for 16 hours. Then cells were washed, cultured for the indicated time in the regular medium without radioactive compounds and thereafter subjected to make the cell extract. Poly(ADP-ribose) polymerase in the cell extract was immunoprecipitated by the autoantibodies to this nuclear enzyme, and a radioactive band (116kDa) was fractionated in SDS-polyacrylamide gel electrophoresis followed by autoradiography. The turn over rate of poly(ADP-ribose) polymerase was determined by the disappearance of radioactivity from 116kDa band in comparison to that of total cellular protein.

RESULTS AND DISCUSSION

The intracellular amount of poly(ADP-ribose) polymerase has not been well characterized mainly because of the absence of the useful ligand for the quantitation. As shown in Figure-1, an immunoblotting technique was used for the quantitative analysis of poly(ADP-ribose) polymerase in the crude human cell extract with autoantibodies (anti-human). The amount of the poly(ADP-ribose) polymerase was measured by the density of the 116kDa band. A standard curve was established by the purified poly(ADP-ribose) polymerase. The amount of poly(ADP-ribose) polymerase in human PBL, CEM lymphoblastoid cells and U937 histiocytic cells were 9.8, 38.5 and 40.7 ng/mgDNA respectively, showing larger content in rapidly dividing cell line cells than resting lymphocytes. In addition, peripheral blood monocytes contained little, if any, poly(ADP-ribose) polymerase. Based on these data, the approximate amount of poly(ADP-ribose) polymerase was estimated as 1×10^6 molecules per rapidly dividing cells and 5×10^5 molecules in resting lymphocytes. These amounts are in the same order of the magnitude as other abundant nuclear proteins such as topoisomerase II or nuclear lamins (13), although each amount is only 1% of that of total histones.

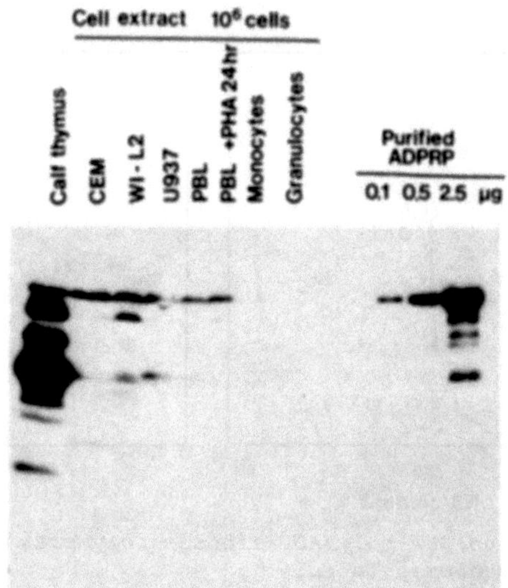

Figure 1. *Quantitative analysis of cellular poly(ADP-ribose) polymerase by immunoblotting*

To evaluate the possible role of poly(ADP-ribose) polymerase in the toxic mechanism of deoxyadenosine, the amount of NAD which can be consumed in poly(ADP-ribosyl)ation was estimated. Poly(ADP-ribose) polymerase can potentially consume 560 pmol NAD per min in one million cells, and this is approximately 100 times larger than the amount of NAD consumed in steady state growing human cells. This is the case not only in dividing cells but also in resting lymphocytes. In other words, human cells have large excess of poly(ADP-ribose) polymerase in comparison to that is required for their steady cellular metabolism. Furthermore, cellular amount of poly(ADP-ribose) polymerase in the cell growth cycle was investigated to define the function of this nuclear enzyme in the cellular metabolism more clearly. In this study, PHA-stimulated PBL was examined by both immunoblotting (data not shown) and FACS technique (Figure-2). The amount of poly(ADP-ribose) polymerase in PBL increased during the course of G_0 to G_1 phase without changing the amount of cellular DNA, suggesting the requirement of this protein for the re-entry into the cell growth cycle. Then, cellular amount of poly(ADP-ribose) polymerase reached to the maximum in G_2-M phase. Similarly, synthesis of poly(ADP-ribose) polymerase was examined in synchronized cell line cells by immunoblotting, showing the maximum content in G_2-M phase. (data not shown).

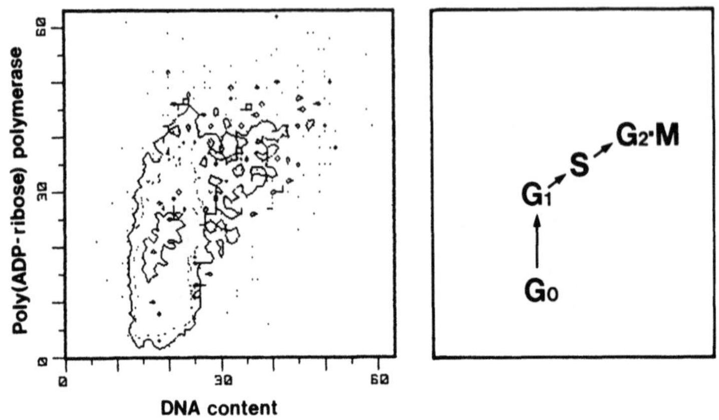

Figure 2. *The amount of poly(ADP-ribose) polymerase in PBL activated by phytohemagglutinin.*

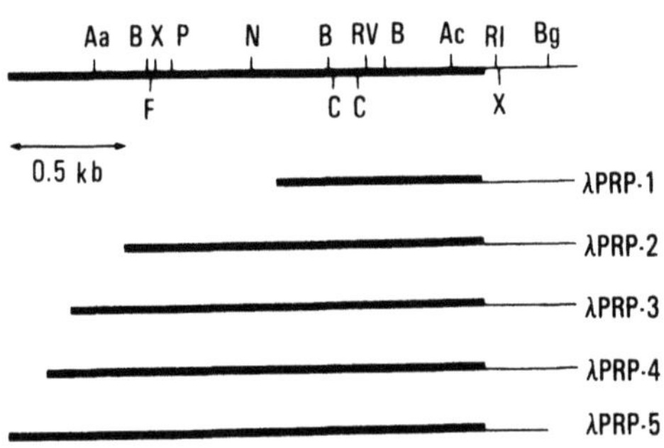

Figure 3. *Schematic drawing of five overlapping cDNA clones encoding poly(ADP-ribose) polymerase*

For further characterization of poly(ADP-ribose) polymerase, specific cDNA clones which encode the gene of poly(ADP-ribose) polymerase were isolated from lgt11 expression library. The cDNA insert from each recombinant was analyzed by electrophoresis after restriction endonuclease digestion. Five distinct size classes of cDNA inserts (0.9 - 2.1 kb) were identified (figure 3). The specificity of these cDNA clones were confirmed by the autoantibody responsiveness to the lysogen made from each cDNA clone.

Northern blotting analysis was carried out with RNA from synchronized human lymphoblastoid cells and cDNA of poly(ADP-ribose) polymerase. Single transcript of approximately 4.0kb, mRNA of poly(ADP-ribose) polymerase, was detectable throughout the cell cycle, showing the constant production of poly(ADP-ribose) polymerase.

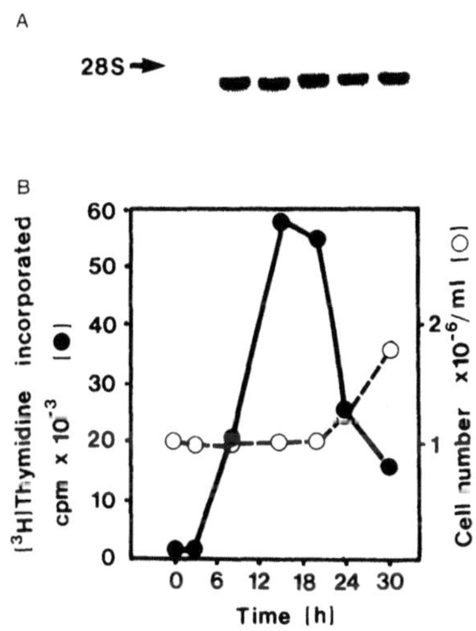

Figure 4. *Northern blot analysis of mRNA of poly(ADP-ribose) polymerase in synchronized human lymphoblastoid cells.*

The turn over rate of poly(ADP-ribose) polymerase was estimated by the disappearance of radioactivity from 116kDa protein and the rate of the disappearance of label from poly(ADP-ribose) polymerase was remarkably similar as that from total cellular protein(data not shown). These results suggest poly(ADP-ribose) polymerase does not turn over rapidly but is synthesized constantly all through cell cycle(14). Considering the large amount of poly(ADP-ribose) polymerase throughout the cell cycle, this DNA-binding protein may have a role as a structural protein in the chromatin. The enzymatic reaction may serve to cause the separation of this structural protein from the chromatin. In addition, abundant poly(ADP-ribose) polymerase in human cells can easily kill the cells upon being activated. Conceivably, poly(ADP-ribose) polymerase is involved in the toxic mechanism of DNA-damaging agents including deoxyadenosine not only in resting lymphocytes but also in dividing cells.

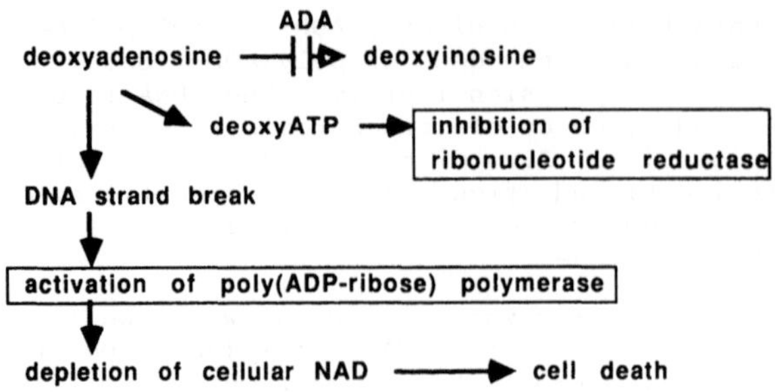

Figure 5. *Schematic diagram of the toxic mechanism of the deoxyadenosine*

References

1. B. Ullman, L. J. Gudas, A. Cohen, and D. W. Jr. Martin. Deoxyadenosine metabolism and cytotoxicity in cultured mouse T lymphoma cells: A model for the immunodeficiency disease. *Cell* 14:365, 1978.

2. S. Seto, C. J. Carrera, M. Kubota, D.B. Wasson, and D. A. Carson. Mechanism of deoxyadenosine and 2-chloroadenosine toxicity to non-dividing human lymphocytes. *J. Clin. Invest.* 75:377-381, 1985.

3. S. Seto, C. J. Carrera, D. B. Wasson and D. A. Carson. Inhibition of DNA repair by deoxyadenosine in resting human lymphocytes. *J. Immunol.* 136:2839-2843, 1986.

4. K. Ueda, and O. Hayaishi. ADP-ribosylation. *Annu. Rev. Biochem.* 54:73-100, 1985.

5. H. Yamanaka, E. H. Willis, C. A. Penning, C. L. Peebles, E. M. Tan and D. A. Carson. Human autoantibodies to poly(adenosine diphosphate-ribose) polymerase. *J. Clin. Invest.* 80:900-904, 1988

6. S.G. Carter and N. A. Berger. Purification and characterization of human poly(adenosine diphosphate ribose) polymerase. *Biochemistry* 21:5475-5481, 1982.

7. W. T. Melvin, J. F. Burke, A. A. Slater, and H. M. Keir. Effect of amino acid deprivation on DNA synthesis in BHK-21/C13 cells. *J. Cell Physiol.* 98:73-80, 1979.

8. C. Auffray and F. Rougeon. Purification of mouse immunoglobuline heavy-chain messenger RNAs from tptal myeloma tumor RNA. *Eur. J. Biochem.* 107:303-314, 1980

9. L. G. Davis, M.D. Dibner and J.F. Battey, Formaldehyde gel for electrophoretic separation of RNA and Northern blot. in *Basic Methods in Molecular Biology*, pp143-151, Elsevier/North Holland, New York, 1986.

10. P. Kurki, M. Vanderlaan, F. Dolbeare, J Gray, and E. M. Tan. Expression of proliferating cell nuclear antigen(PCNA)/cyclin during the cell cycle. *Exp. Cell Res.* 166:209-219, 1986.

11. C. A. Penning, H. Yamanaka, D. A. Carson, C. L. Peebles, E. M. Tan and K. F. Sullivan. Molecular cloning of poly(ADP-ribose) polymerase using human autoantibodies. *Clin. Res.* 35:566A, 1987.

12. R. A. Young and R. W. Davis. Yeast RNA polymerase II genes: Isolation with antibody probes. *Science* 222:778-782, 1983.

13. M. M. S.Heck, and W. C. Earnshaw. Topoisomerase II: A specific marker for cell proliferation. *J. Cell Biol.* 103:2569-2581, 1986.

14. H. Yamanaka, C. A. Penning, E. H. Willis, D. B. Wasson, and D. A. Carson. Characterization of human poly(ADP-ribose) polymerase with autoantibodies. *J. Biol. Chem.* 263:3879-3883, 1988.

MECHANISM OF ATP CATABOLISM INDUCED BY DEOXYADENOSINE AND OTHER NUCLEOSIDES IN ADENOSINE DEAMINASE - INHIBITED HUMAN ERYTHROCYTES

Françoise Bontemps and Georges Van den Berghe

Laboratory of Physiological Chemistry, International Institute of Cellular and Molecular Pathology, B-1200 Brussels, Belgium

INTRODUCTION

Accumulation of dATP, accompanied by depletion of ATP and total adenine ribonucleotides, has been found in the erythrocytes [1,2], lymphocytes and bone marrow [3] of children with adenosine deaminase (ADA) deficiency, as well as in the erythrocytes and lymphoblasts of leukaemia patients treated with the adenosine deaminase inhibitor, deoxycoformycin (dCF) [4-6]. Bagnara and Hershfield [7] have explained these changes as follows : *(1)* deoxyadenosine (dAdo), which accumulates in both clinical conditions, is phosphorylated to dAMP and subsequently to dATP, a process which utilizes ATP and generates ADP and AMP (Fig. 1); *(2)* the accumulation of dATP, which is greatly facilitated by the fact that dAMP is a poor substrate for AMP deaminase, stimulates both the deamination of AMP to IMP and the dephosphorylation of IMP to inosine, thereby provoking adenine ribonucleotide catabolism. Two observations, when combined, are, nevertheless, difficult to reconcile with this mechanism : *(1)* dATP and ATP are equipotent as stimulators of AMP deaminase [7,8]; *(2)* the sum of dATP and ATP in ADA-deficient or in ADA-inhibited cells is not or only slightly higher than that of ATP in control cells [2], which contain only traces of dATP [1]. These data, and uncertainties with respect to the concentration of AMP in ADA-deficient or -inhibited cells, prompted a reinvestigation of the mechanism of the depletion of ATP induced by dAdo. Normal human erythrocytes, in which ADA was inhibited by dCF, were used as experimental model.

METHODS

Fresh blood from normal human volunteers, collected on heparin, was centrifuged to remove plasma and buffy coat. The red cells were washed three times with Krebs-Ringer-bicarbonate (KRB) buffer, pH 7.4, containing

5 mM glucose and gassed with 95 % O_2- 5 % CO_2. They were resuspended in the same medium at an haematocrit of 20-25 %, and their adenine nucleotides were labelled by a 60 min preincubation at 37 °C with 1 µM [U-^{14}C]adenine [9]. This was followed by two washes and resuspension in KRB buffer containing 5 mM glucose. Incubations were performed in carefully regassed and stoppered vials. In all the experiments, suspensions were preincubated during 20 min with 1 µM dCF, before addition of nucleosides. This concentration of dCF was found to inhibit by 95 % the deamination of up to 0.5 mM dAdo. In some experiments, 10 µM iodotubercidin (ITu) was used to inhibit adenosine kinase [10]. This concentration was found to prevent any detectable synthesis of nucleotides from 0.5 mM dAdo, tubercidin and 6-methylmercaptopurine riboside (6-MMPR).

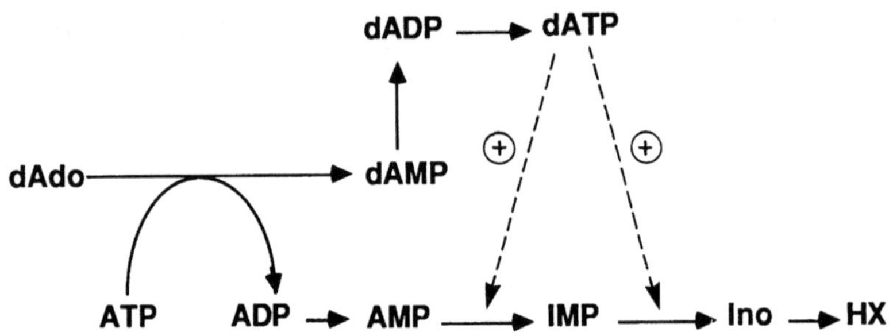

Figure 1. Mechanism of dAdo-induced catabolism of ATP, as proposed by Bagnara and Hershfield [7]. Explanations are given in the text.

For analytical determinations on the red cell suspensions as a whole, 0.5 ml aliquots were transfered into 0.25 ml of ice-cold 10 % $HClO_4$, and neutralized as described in [9]. For the measurement of intracellular Pi, 0.5 ml of cell suspension were centrifuged through a 0.5 ml layer of silicone as given in [9]. Nucleotide concentrations were measured by HPLC, and their radioactivity as well as that in nucleosides and bases by TLC, as described previously [9]. The concentration of the latter compounds was calculated from the specific radioactivity of the adenine nucleotides, which remained constant over the duration of the experiments.

AMP deaminase from human erythrocytes was purified approx. 100,000-fold by the following, two-step procedure : *(1)* binding of the enzyme to a DEAE-Trisacryl column, followed by elution with a 0-300 mM NaCl gradient; *(2)* binding of the enzyme eluted from DEAE-Trisacryl to a GTP-agarose affinity column, followed by elution with 10 mM 2,3-bisphosphoglycerate (2,3-BPG). The activity of the enzyme was measured by the formation of [8-^{14}C]IMP from [8-^{14}C]AMP. Incubations were performed at 37 °C in 50 mM HEPES buffer pH 7.2, 100 mM KCl, 10 mM $MgCl_2$, with various concentrations of substrate and additions as indicated. At appropriate time intervals, 10 µl of the incubation medium were spotted on

PEI cellulose TLC plates, on which carrier solutions (50 nmol) of AMP and IMP had been applied. After development in 1.4 M LiCl, activity was calculated from the radioactivity appearing in IMP.

RESULTS AND DISCUSSION

Effects of dAdo on nucleotide concentrations and on the production of purine ribonucleotide catabolites. As shown in Fig. 2, incubation of ADA-inhibited erythrocytes with 0.5 mM dAdo decreased their ATP content by about 50 % after 2 h. Concomitantly there was an accumulation of dATP, which reached approx. 90 nmol/ml of cell suspension over the same time interval. The concentration of ADP increased about 2-fold and that of AMP nearly 10-fold, from a basal value of 2 to around 18 nmol/ml of cell

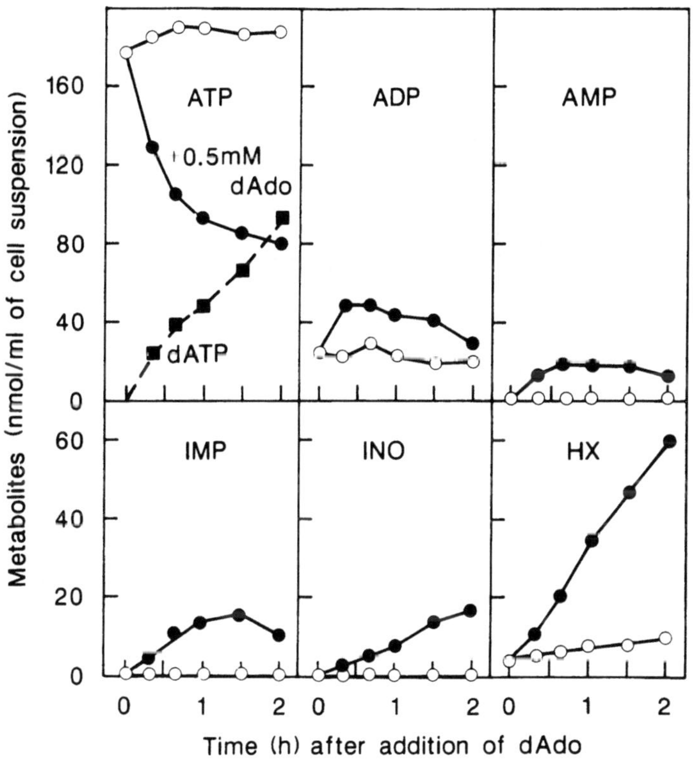

Figure 2. Effects of the addition of dAdo in ADA-inhibited erythrocytes. Cells were incubated in the absence (open symbols) and in the presence (closed symbols) of dAdo.

suspension. These data accord with the proposal of Bagnara and Hershfield [7], that the phosphorylation of dAdo brings about a generation of ADP and of AMP. In addition, they show that this generation results in an increase of the concentration of both nucleotides. There was also a progressive increase in IMP, from a basal value of 1, to 15 nmol/ml of cell suspension after 1 h of incubation. This accumulation most likely reflects the increased rate of degradation of AMP by way of AMP deaminase. Both radioactive inosine and hypoxanthine, derived from the catabolism of the prelabelled adenine ribonucleotides, accumulated in the medium, at 12-fold the rate recorded under control conditions. That an accumulation of inosine took place may be explained by the decrease inside the cells, recorded concomitantly with the phosphorylation of dAdo, of Pi, the other substrate of purine nucleoside phosphorylase, (see below). With lower concentrations of dAdo, down to the range of those measured in the plasma of ADA-deficient or -inhibited patients (around 20 µM), similar, although less pronounced effects were observed.

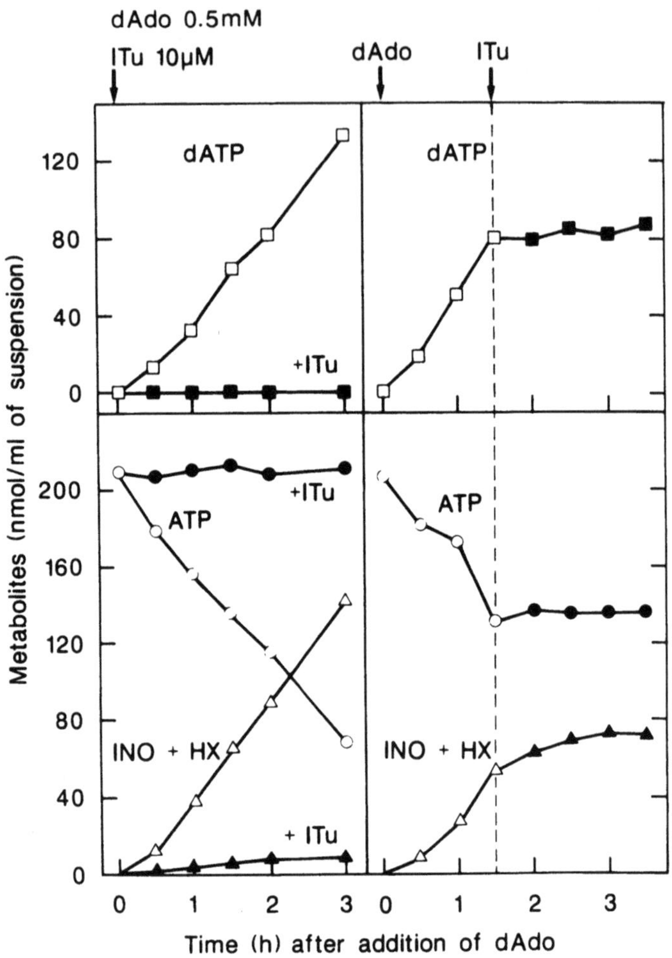

Figure 3. Influence of ITu on the effects of dAdo in ADA-inhibited erythrocytes. The cells were incubated in the absence (open symbols) and in the presence of ITu (closed symbols)

Influence of adenosine kinase inhibition on the accumulation of dATP, the depletion of ATP and the production of purine catabolites. Similarly to that of adenosine, the phosphorylation of dAdo can be inhibited by the addition to the cell suspension of ITu, a potent inhibitor of adenosine kinase. As depicted in Fig. 3 (left pannels), when 10 µM ITu was added together with dAdo, at the beginning of the incubation, the accumulation of dATP, as well as the depletion of ATP and the increase in the production of purine catabolites were completely abolished. These results accord with the observation that dAdo did not stimulate adenine nucleotide catabolism in an adenosine kinase- and deoxycytidine kinase-deficient cell line [7]. As expected, addition of ITu 90 min after dAdo, halfway the incubations,

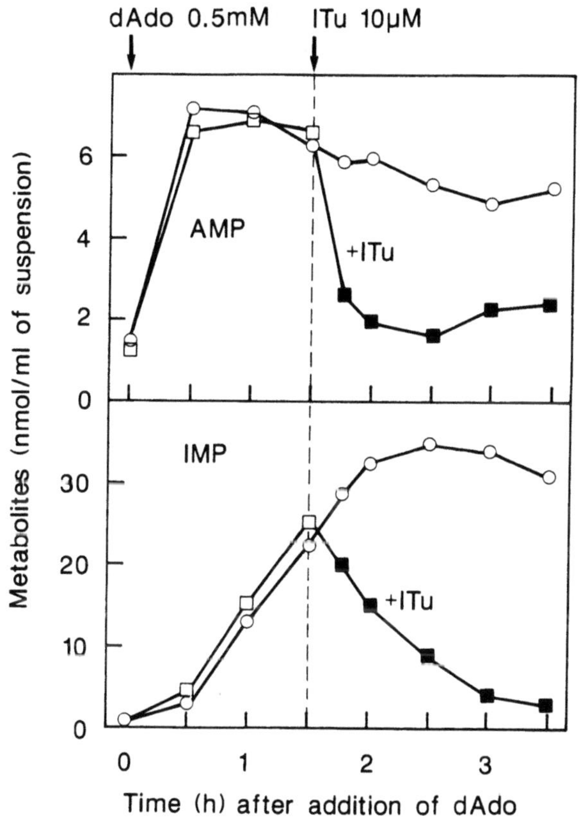

Figure 4. Influence of ITu on the concentrations of AMP and IMP in ADA-inhibited erythrocytes incubated with dAdo. Cells were incubated in the absence (open symbols) and in the presence (closed symbols) of ITu.

completely arrested the build-up of dATP (right pannels). Strikingly, however, the addition of ITu also brought about a complete arrest of the degradation of ATP, and nearly suppressed the increased rate of production of inosine and hypoxanthine. These observations indicate that the accumulation of dATP is not sufficient by itself to increase the catabolism of the adenine ribonucleotides in ADA-inhibited erythrocytes.

Influence of adenosine kinase inhibition on the accumulation of AMP and IMP. Fig. 4 (upper pannel) shows that the concentration of AMP, which had increased about 6-fold following the addition of 0.5 mM dAdo, returned to nearly basal values 15 min after the addition of 10 µM ITu. That the increase in the rate of production of purine catabolites, provoked by the addition of dAdo, was associated with an increase in the concentration of AMP, and that the decrease in AMP was accompanied by a return to normal production rates, strongly suggests that the concentration of AMP plays a determining role in the dAdo-induced catabolism of the adenine ribonucleotides. Following the addition of ITu, the concentration of IMP (Fig.4, lower pannel), which had increased progressively to about 25 nmol/ml of suspension over 90 min, and continued to increase in the absence of the adenosine kinase inhibitor, also decreased, but more progressively than AMP. This corroborates the conclusion that the catabolism of the adenine nucleotides induced by dAdo results from an increased activity of AMP deaminase.

Effects of dAdo and ITu on the concentrations of intracellular Pi and 2,3-BPG . Human red cell AMP deaminase is known to be stimulated by ATP and alkali metal cations, and inhibited by Pi and 2,3-BPG [8,11], the most abundant phosphate ester in these cells. Both inhibitors were therefore measured in ADA-inhibited erythrocytes, incubated in the presence of dAdo. Intracellular Pi decreased from a basal value of around 1 mM, to about 0.4 mM, 1 h after the addition of 0.5 mM dAdo, and remained at that level over the 3 h duration of the experiment. Upon addition of 10 µM ITu 90 min after dAdo, the concentration of Pi reincreased slightly, to about 0.5 mM at 3 h (results not illustrated). Nevertheless, as shown before (Fig. 3, right pannels), under the latter condition the catabolism of the adenine ribonucleotides was nearly completely arrested. This indicates that the increased catabolism is not caused by the decrease in intracellular Pi. The addition of dAdo did not modify the concentration of 2,3-BPG in the erythrocytes.

Influence of physiological concentrations of substrate and effectors on the activity of erythrocytic AMP deaminase. To our knowledge, the kinetics of erythrocytic AMP deaminase had hitherto not been investigated at the micromolar concentrations of AMP that prevail in intact red cells. Experiments were therefore performed with the aim to assess the influence on the enzyme of the increase in AMP, as well as of the decrease in intracellular Pi, recorded during the metabolism of dAdo. In the presence of 100 mM KCl but in the absence of other effectors, erythrocytic AMP deaminase displayed hyperbolic kinetics with K_m AMP = 0.4 mM. Addition of 1 mM ATP, the concentration of the stimulator in control cells, decreased K_m AMP to 0.15 mM. Further addition of the inhibitors at their physiological concentrations, namely 1 mM for Pi and 3 mM for unbound 2,3-BPG, rendered the kinetics sigmoid and the enzyme nearly inactive up to 10 µM AMP. This led to the conclusion that the limited production of purine catabolites by control erythrocytes (approx. 4 nmol/h per ml of packed cells, ref. 9) is most likely due to their very low concentration of AMP. At 50 µM AMP, and in the presence of 1 mM ATP and 3 mM 2,3-BPG, Pi inhibited the activity of the enzyme by 40 % at 1 mM and by 95 % at 10 mM concentration. A decrease in Pi, from 1 to 0.5 mM, as recorded in ADA-inhibited erythrocytes incubated with 0.5 mM dAdo, increased the activity only by 20 %. This finding corroborates the conclusion that the approx. 50 % decrease in intracellular Pi, recorded concomitantly with the phosphorylation of dAdo, but which was maintained upon its arrest, plays nearly no role in the markedly increased catabolism of the adenine ribonucleotides. It can thus be concluded that the main factor in the dAdo-induced catabolism of the adenine ribonucleotides, is the increase in AMP brought about by the utilization of ATP in the phosphorylation of the nucleoside.

Effects of other nucleosides on the catabolism of ATP. To further validate the proposed mechanism, it was investigated if the phosphorylation of other nucleosides was also accompanied by an increase in AMP and by an acceleration of the catabolism of the adenine ribonucleotides. Experiments were performed with the nucleoside analogues, tubercidin, which similarly to dAdo is phosphorylated to a triphosphoderivative, and 6-MMPR, which is not further metabolized than the monophosphate level. Tubercidin, at 0.5 mM, nearly totally depleted the adenine ribonucleotides within 3 h, increased the concentration of AMP 16-fold, and the production of inosine and hypoxanthine about 50-fold. The same concentration of 6-MMPR decreased total adenine ribonucleotides by 30 % over 3 h, increased AMP 8-fold, and the production of the purine catabolites 20-fold. With both analogues, the decrease in adenine ribonucleotides and the increase in the production of their catabolites were nearly completely abolished by the addition of 10 μM ITu, 90 min after the nucleoside.

CONCLUDING REMARKS

Taken together, our results indicate that the main factor in the depletion of the adenine ribonucleotides, accompanying the phosphorylation of dAdo and of nucleoside analogues in human erythrocytes, is the elevation of the concentration of AMP above its physiological level (5-10 μM). The elevation of AMP is the consequence of the utilization of ATP in the phosphorylation process. This elevation increases the activity of AMP deaminase, the limiting enzyme of purine catabolism in erythrocytes and in other tissues. Stimulation of AMP deaminase by dATP and by the triphosphates generated from the nucleoside analogues may contribute to the degradation process, but only in the presence of elevated concentrations of AMP. Owing to the kinetic characteristics of erythrocytic AMP deaminase, the decrease of its inhibitor, Pi, accompanying the phosphorylation of the nucleosides, decreases only minimally the physiologic inhibition of the enzyme, and therefore contributes only marginally to the degradation process.

ACKNOWLEDGEMENTS

We thank Mrs A. Delacauw for expert technical assistance. This work was supported by grant 3.4539.87 from the Fund for Medical Scientific Research (Belgium). G.V.D.B. is Director of Research of the Belgian National Fund for Scientific Research.

REFERENCES

1. A. Cohen, R. Hirschhorn, S.D. Horowitz, A. Rubinstein, S.H. Polmar, R. Hong and D.W. Martin jr., Deoxyadenosine triphosphate as a potentially toxic metabolite in adenosine deaminase deficiency, *Proc. Natl. Acad. Sci. USA* 75 : 472 (1978).

2. H.A. Simmonds, A. Sahota, C.F. Potter, D. Perrett, K. Hugh-Jones and J.G. Watson, Purine metabolism in adenosine deaminase deficiency,*in:* Enzyme defects and immune dysfunction, Ciba Symposium 68 (new series), Excerpta Medica, 255 (1979).
3. J. Donofrio, M.S. Coleman, J.J. Hutton, A. Daoud, B. Lampkin, and J. Dyminski, Overproduction of adenine deoxynucleosides and deoxynucleotides in adenosine deaminase deficiency with severe combined immunodeficiency disease, *J. Clin. Invest.* 62 : 884 (1978).
4. C.A. Koller, B.S. Mitchell, M.R. Grever, E. Mejias, L. Malspeis and E.N. Metz, Treatment of acute lymphoblastic leukemia with 2'-deoxycoformycin : clinical and biochemical consequences of adenosine deaminase inhibition, *Cancer Treat. Rep.* 63 : 1949 (1979).
5. B.S. Mitchell, C.A. Koller and R. Heyn, Inhibition of adenosine deaminase activity results in cytotoxicity to T lymphoblasts in vivo, *Blood* 56 : 556 (1980).
6. C.A. Koller and B.S. Mitchell, Alterations in erythrocyte adenine nucleotide pools resulting from 2'-deoxycoformycin therapy, *Cancer Res.* 43 : 1409 (1983).
7. A.S. Bagnara and M.S. Hershfield, Mechanism of deoxyadenosine-induced catabolism of adenine ribonucleotides in adenosine deaminase-inhibited human T lymphoblastoid cells, *Proc. Natl. Acad. Sci USA* 79 : 2673 (1982).
8. C.Y. Lian and D.R. Harkness, The kinetic properties of adenylate deaminase from human erythrocytes, *Biochim. Biophys. Acta* 341 : 27 (1974).
9. F. Bontemps, G. Van den Berghe and H.G. Hers, Pathways of adenine nucleotide catabolism in erythrocytes, *J. Clin. Invest.* 77 : 824 (1986).
10. J.F. Henderson, A.R.P. Paterson, I.C. Caldwell, B. Paul, M.C. Chan and K.F. Lau, Inhibitors of nucleoside and nucleotide metabolism, *Cancer Chemother. Rep. Part II* 3 : 71 (1972).
11. A. Askari and S.N. Rao, Regulation of AMP deaminase by 2,3-diphosphoglyceric acid : a possible mechanism for the control of adenine nucleotide metabolism in human erythrocytes, *Biochim. Biophys. Acta* 151 : 198 (1968).

Deoxy GTP ACCUMULATES IN THYMOCYTES, BUT NOT IN T OR B LYMPHOCYTES IN SIMULATED PNP DEFICIENCY

Anna Taddeo*, Lynette D. Fairbanks**, H. Anne Simmonds**, John A. Duley**, and George S. Morris**

* Institute of Rheumatology, University of Siena, Italy, ** Purine Research Laboratory UMDS Guy's Hospital, London UK

INTRODUCTION

A variety of in vitro models have been employed to investigate the toxicity to the immune system of deoxyguanosine (dGR), the lymphotoxic metabolite accumulating in patients with purine nucleoside phosphorylase (PNP) deficiency[1-8]. The defect affects predominantly T cells. Although the toxicity of dGR has received less attention than that of deoxyadenosine (dAR) - the metabolite accumulating in the companion disorder adenosine deaminase (ADA) deficiency, associated with severe combined immunodeficiency (SCID) - existing data are equally conflicting[9].

Evaluation of the different studies relating to the in vitro toxicity of dGR has been complicated by the fact that some did not include any PNP inhibitor, or employed high levels of substrate, sometimes over long time scales[1-3]. Others used only cultured lymphocytes, or focussed exclusively on toxicity[4-7]. Very few studies have addressed the situation in unstimulated cells over a short time scale using low substrate levels and also followed the complete metabolism of substrate in cell and medium. In this report the metabolism of dGR by unstimulated lymphoid cells has been compared under normal conditions, and under conditions simulating the in vivo situation in PNP deficiency[10] by preincubating with guanosine to inhibit PNP.

MATERIAL AND METHODS

Materials : [8-^{14}C]-deoxyguanosine (dGR) was prepared from [U-^{14}C] dGR (504 mCi/mmole) purchased from Amersham (UK). The other unlabelled purine nucleosides and bases were the highest analytical grade available and were purchased from Sigma (UK). The incubation medium consisted of RPMI (Gibco, Europe) supplemented with glutamine plus heat inactivated foetal calf serum (Sera Labs), or horse serum (Gibco, Europe).

METHODS

Cell source: Control peripheral blood cells were obtained from healthy laboratory personnel. Tonsils and thymuses were obtained at operation from children undergoing tonsillectomy or cardiac surgery and processed as rapidly as possible. The methods used for the separation of peripheral blood mononuclear cells (PBMs) and thymocytes, as well as the preparation of enriched B lymphocyte populations, have been published[9].

Incubation studies with [8-^{14}C] dGR: In all instances 10-20 x 10^6 cells), resuspended in Iscove's medium plus heat inactivated 10% horse serum, were incubated for 2 hours with [8-^{14}C] dGR at a final concentration of 10µM (0.625 µCi/ml). PNP deficiency was simulated by preincubation with guanosine (500µM). Following incubation, cells and medium were separated and the reactions terminated as described[5,9], with the exception that 200 µl of acetonitrile (ACN) was used to terminate the reaction in the medium and thus avoid any potential degradation of dGR. ACN was extracted using water-saturated diethyl-ether. Extracts of cells and medium were stored at -20°C if not processed immediately.

Measurement of [8-^{14}C] accumulation in nucleotides, nucleosides, and bases: Incorporation of radiolabel into cellular nucleotides, as compared with distribution in nucleosides and bases in the medium, was determined using two different HPLC (high performance liquid chromatography) systems, anion exchange for separating the nucleotides in the cell extracts and reversed-phase for the nucleosides and bases in the mediums, coupled with in-line radiodetection as described[5,9,10].

RESULTS

Viability remained constant during the incubations and was greater than 95% for all cell types. This was confirmed by the high ATP/ADP ratios found in nucleotide extracts made at the beginning and end of the incubation period and demonstrated in the UV traces at 254nm (Figure 1).

A. Metabolism of [8-^{14}C] dGR by different cell types under physiological conditions without any PNP inhibitor: In all experiments the highest % of counts were found in the medium (~70%) as guanine and unmetabolised dGR.

The principal route of dGR metabolism for all cell types was degradation by PNP to guanine, with subsequent conversion to GTP by HGPRT, confirming active conversion of dGR into the guanine nucleotide pool in the absence of any PNP inhibitor. However, the metabolism of dGR by thymocytes, tonsillar T & B-cells and PBMs, was never complete, even at the low substrate concentration used, with up to 30% of the counts still being found as unmetabolised dGR in the medium.

Thymocytes accumulated detectable amounts of dGTP even under physiological conditions, but were the only uncultured lymphoid cells to do so. The fact that neither tonsillar lymphocytes nor PBMs formed any dGTP (despite the presence of unmetabolised dGR), while high levels were formed by the thymocytes, is consistent with a much higher activity of deoxycytidine kinase in thymocytes than in any other cell type or tissue in man[1,2].

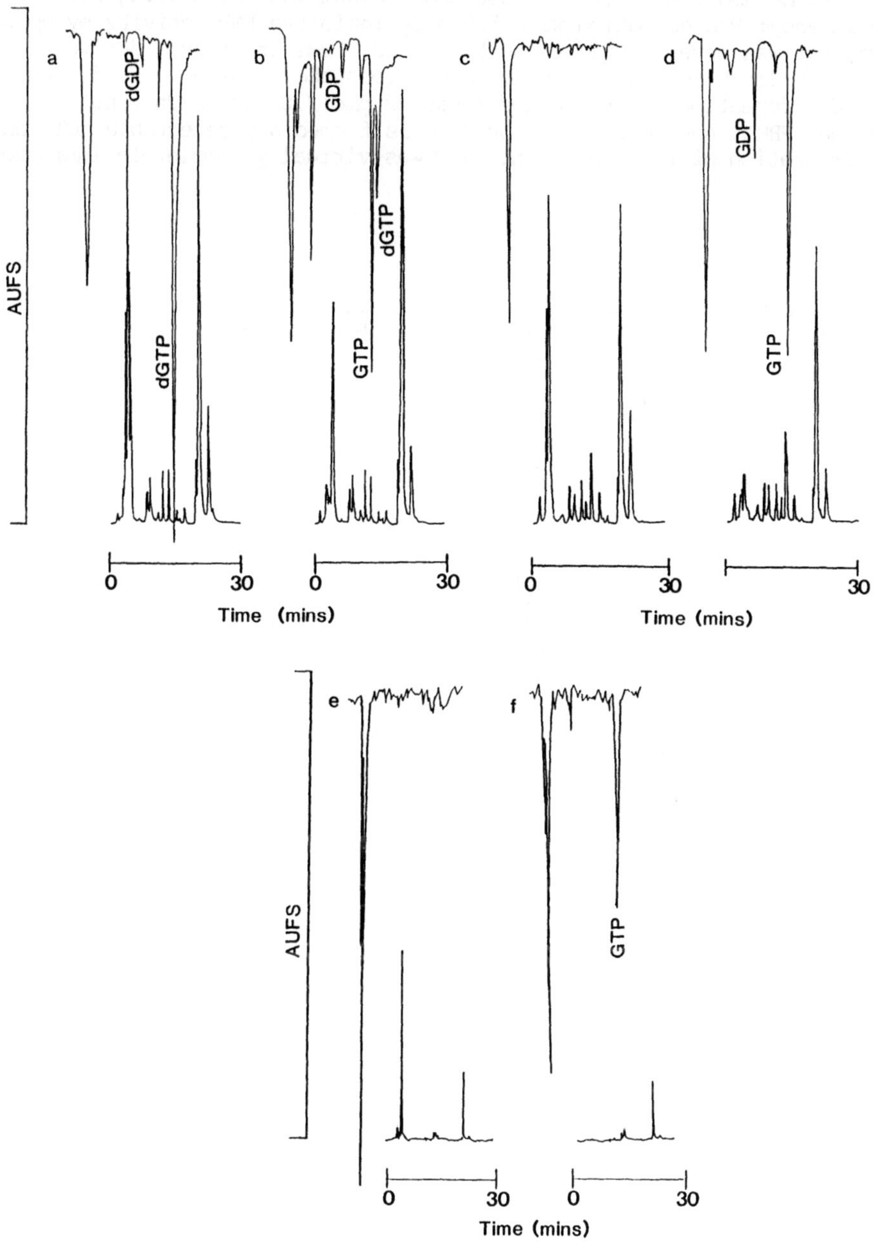

Figure 1. Charts b, d and f, demonstrate the radiotrace (upper) and the UV trace at 254 nm (lower) obtained by HPLC following injection of extract from thymocytes, tonsillar B-cells and PBMs respectively incubated with dGR alone as described in methods. Charts a, c and e, show the radiotraces for the same cells incubated with dGR plus GR. Identical results to those shown for the B cells were found using Tonsillar T cells (not shown).

B. Metabolism of [8-^{14}C] dGR under conditions simulating PNP deficiency: Preincubation with 500μM GR inhibited PNP activity by up to 90%, but was never complete in thymocytes, PBMs and tonsil derived T and B cells. Under these PNP inhibited conditions there was a 10-fold increase in dGTP formation by thymocytes (mean 33 pmol/10^6 cells/2h), but neither PBMs, nor tonsillar T or B cells formed any detectable dGTP and incorporation of radiolabel into GTP was virtually completely inhibited.

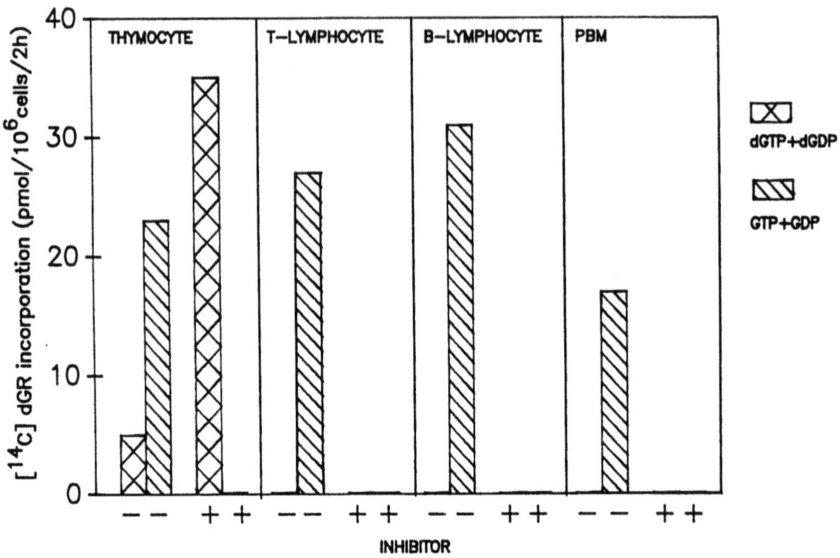

Figure 2. Histograms comparing the incorporation of [8-^{14}C] dGR (pmols/10^6cells /2h) into guanosine or deoxyguanosine nucleotides in thymocytes, T and B lymphocyes, and unfractionated PBM's.

DISCUSSION

These studies of dGR metabolism in different lymphoid cell-types indicate that in the absence of any PNP inhibitor the major cellular metabolite was GTP, even over the short period and low substrate concentrations employed. The fact that the toxicity of dGR to stimulated peripheral blood lymphocytes related not to dGTP formation, but rather rapid metabolism to GTP, has already been demonstrated conclusively by Spaapen et al[8]. These workers pointed out, moreover, that dGR was apparently equally inhibitory of both T and B lymphocyte differentiation and proliferation in vitro, which contrasted with the inherited disorder, where only T cells were affected. Identical results were found by ourselves and others in cultured T and B lymphocytes[5,6]: toxicity also apparently resulted from GTP rather than dGTP accumulation in these cell types.

There are several reasons for the apparent discrepancies in earlier studies. The first relates to the fact that early investigators based their hypotheses on the differing activity of enzymes capable of phosphorylating and degrading deoxynucleotide triphosphates in lysed T as compared with B cells [1,2], supported by toxicity studies in intact cultured lymphoid cells, which have also proved a poor model for the *in vivo* situation [5,6]. The second is that some toxicity studies did not attempt to measure dGTP levels, or employ any PNP inhibitor [7], and the possibility of alternative routes of dGR metabolism was overlooked; the absence of an effective PNP inhibitor has compounded the problem. The third complication has related to the variable amount of substrate and time scales used. A careful assessment of results demonstrating enhancement of lymphocyte transformation and differentiation indicate that this probably related to the use of dGR at concentrations below 50 uM which would not have increased GTP levels greatly; those demonstrating inhibition used from 100 μM up to levels in excess of 1mM, which was obviously highly toxic, presumably due to the large increment in GTP concentrations produced, although the effect of high levels of guanine or unmetabolised substrate in the culture medium may have been contributory.

Our previous study of the companion disorder ADA deficiency - which presents as severe combined immunodeficiency (SCID) - using uncultured lymphoid cells showed that, in contrast to earlier reports, tonsillar B cells as well as thymic cells accumulated high dATP levels from deoxyadenosine (dAR) [9] under conditions simulating ADA deficiency. Moreover, both thymic cells and tonsillar B cells accumulated significant amounts of dATP without ADA inhibition. We now report that thymocytes also accumulate dGTP from dGR even without PNP inhibition and, more importantly, very high dGTP levels *in vitro* under conditions simulating PNP deficiency. However, in contrast to our similar study using dAR, they were the only lymphoid cells to do so. The present results confirm that, as with dAR in ADA deficiency, inability to metabolise dGR in PNP deficiency may be extremely damaging to T lineage stem cells at an early intrathymic stage of differentiation.

The simplest conclusion to be drawn from our combined studies [9] must be that the major mechanism of lymphotoxicity in both inherited ADA and PNP deficiency relates primarily to deoxynucleotide triphosphate accumulation, with subsequent inhibition of ribonucleotide reductase. The fact that significant deoxynucleotide accumulation can occur in both thymocytes and circulating B cells in ADA deficiency, but only thymocytes in PNP deficiency, would explain the differential toxicity in these disorders. This hypothesis is supported by the variable tissue distribution of the different kinases responsible for the formation of dATP and dGTP. The deoxycytidine kinase capable of phosphorylating both dGR and dAR is restricted almost exclusively to the human thymus, whilst adenosine kinase, which readily phosphorylates dAR when ADA is inhibited or defective, has a widespread distribution. This contrasts with the very low activity of the dGR specific kinase. Such an hypothesis does not exclude the possible contribution of additional mechanisms of toxicity in either disorder.

Acknowledgements

These studies were supported by The Wellcome Trust, The Dunhill Trust, The Special Trustees of Guy's Hospital and the Italian Society of Rheumatology.

REFERENCES

1. D.A. Carson, J. Kaye, D.B Wasson. Differences in deoxyadenosine metabolism in human and mouse lymphocytes. J Immunol 124:8 (1980).

2. A. Cohen, J.W.W. Lee, H-M. Dosch, E.W. Gelfand. The expression of deoxyguanosine toxicity in T lymphocytes at different stages of maturation. J Immunol 124:1578 (1980).

3. L.J.M. Spaapen, G.T. Rijkers, G.E.J. Staal, G. Rijksen, S.K. Wadman, J.W. Stoop, B.J.M. Zegers. The effect of deoxyguanosine on human lymphocyte function 1. Analysis of inteference with lymphocyte proliferation in vitro. J Immunol 132:2311 (1984).

4. D. Albert, H.G. Bluestein, R.C. Willis, K. Nette, J.E. Seegmiller. The mechanism of inhibition and 'reversal' of mitogen-induced lymphocyte activation in a model of purine-nucleoside phosphorylase deficiency. Cell Immunol 86:501 (1984).

5. A. Goday, H.A. Simmonds, G.S. Morris, L.D. Fairbanks. B cells as well as T cells form deoxynucleotides from either deoxyadenosine or deoxyguanosine. Clin exp Immunol 56:39 (1984).

6. Y. Sidi, B. Mitchell. 2'dGR toxicity for B and mature T lymphoid cell lines is mediated by guanine ribonucleotide accumulation. J Clin Invest 74:1640 (1984).

7. A. G. Hanglow, P.M. Lydyard. The effect of 2'-deoxyguanosine on human lymphocyte responses.I. 2'-deoxyguanosine enhances T lymphocyte responses. Clin Exp Immunol 59:653 (1985).

8. J.G.M. Scharenberg, L.J.M. Spaapen, G.T. Rijkers, M. Duran, G.E.J. Staal, B.J.M. Zegers. Functional and mechanistic studies on the toxicity of deoxyguanosine for the in vitro proliferation and differentiation of human peripheral blood B lymphocytes. Eur J Immunol 16:381 (1986).

9. A. Goday, H.A. Simmonds, G.S. Morris, L.D. Fairbanks. Human B lymphocytes and thymocytes but not peripheral blood mononuclear cells accumulate high dATP levels in conditions simulating ADA deficiency. Biochem Pharmacol 34:3561 (1985).

10. H.A. Simmonds, L.D. Fairbanks, G.S. Morris, G. Morgan, A.R. Watson, P. Timms , B. Singh. CNS dysfunction and erythrocyte GTP depletion in purine nucleoside phosphorylase deficiency. Arch Dis Child 62:385 (1987).

THE MECHANISM OF DEOXYGUANOSINE MEDIATED TOXICITY IN PROLIFERATING HUMAN PERIPHERAL BLOOD T LYMPHOCYTES

John G.M. Scharenberg, Ger T. Rijkers, Gerard E.J. Staal* and Ben J.M. Zegers

Dept. of Immunology, University Hospital for Children and Youth "Het Wilhelmina Kinderziekenhuis, P.O. Box 18009, 3501 CA Utrecht, The Netherlands and *Dept. of Medical Enzymology, University Hospital, Utrecht

INTRODUCTION

Deoxyguanosine (dGuo) is the toxic metabolite in purine nucleoside phosphorylase (PNP) deficiency which is associated with a (selective) cellular immunodeficiency[1,2,3]. DeoxyGuo mediated toxicity in PNP deficiency is caused by interference of accumulated deoxyguanosine triphosphate (dGTP) with DNA synthesis[4,5]. We previously showed that dGuo is toxic not only to PNP deficient T cells but also to normal peripheral blood T and B lymphocytes activated in vitro with mitogens or antigens[6,7,8]. The dGuo mediated toxicity of proliferating normal T and B lymphocytes is accompanied by intracellular accumulation of dGTP but also by increases of GDP and GTP[6,7,8]. It was shown that in normal cells the toxicity was mainly caused by the accumulated guanine ribonucleotides and that dGTP only plays a minor role[6,7,8].
In order to get more insight in the mechanism of dGuo (GTP) toxicity in normal lymphoid cells we previously analyzed the kinetics of dGuo mediated inhibition of proliferation of peripheral blood T cells.
DeoxyGuo does not interfere with early events of T cell activation, since addition of dGuo as late as 24-48 hours after initiation of the culture still abrogated the mitogen induced activation process[9]. In order to delineate more precisely the dGuo sensitive stages of T cell activation we have studied its effect on interleukin-2 (Il-2) receptor expression.

MATERIALS AND METHODS

Cell cultures

Peripheral blood T cells were cultured in RPMI 1640 (Gibco, Grand Island, NY) with 10% ($^{V}/v$) heat inactivated (2 hours, 64°C) fetal calf serum (FCS; Gibco). Deoxyguanosine (dGuo; Sigma, St. Louis, MO) and recombinant Il-2 (rIl-2; Sandoz, Geneva, Switzerland), when present, are added at the initiation of the cultures.
T cell proliferation was assessed by ^{14}C-TdR (Amersham International, Amersham, UK) incorporation during the final 18 hours of 4 days cultures of phytohaemagglutinin (PHA (40 µg/ml); Burroughs Wellcome, Beckenham, UK) activated T cells[9].

Il-2R expression

Expression of Il-2R was assessed by staining of T cells, cultured with PHA for 48 hours, with a monoclonal antibody against the Tac antigen (αTac; Becton Dickinson, Mountain View, CA) and analysis on a FACS analyzer (Becton Dickinson).

Il-2 binding assay

T cells ($1x10^6$ cells/ml) are activated with PHA (40 µg/ml) in 14 ml tubes (Greiner) in total volume of 2.5 ml/tube culture medium (RPMI 1640 with 5% FCS) during 72 hours at 37°C in 5% CO_2 in air. Cells are harvested, washed with RPMI 1640, 1% BSA, and incubated ($5x10^6$ cells/ml) twice during 1 hour at 37°C in 5% CO_2 in air to promote dissociation and/or degradation of endogenously bound IL-2.
Serial dilutions of radiolabeled Il-2 (Amersham; spec. act. > 600

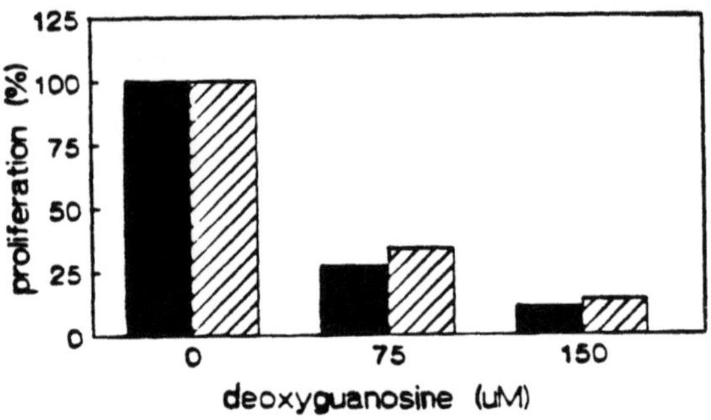

Figure 1.
Effect of extracellular Il-2 on dGuo mediated inhibition of T cell proliferation. Data expressed as percentage of control are results of one experiment out of three. DeoxyGuo and 2 U recombinant Il-2 are added at initiation of the culture. Control cultures ■ ; rIl2 containing cultures ▨

Ci/mmol) are added to $1\text{-}2x10^6$ cells in a total volume of 100 µl RPMI 1640, 1% BSA, and the mixture was incubated for 30 min at 37°C.
Next, cells were washed twice with 1 ml icecold RPMI 1640, 1% BSA. Unbound Il-2 was quantitated in the supernatant from the first wash. The cells are resuspended in 200 µl culture medium and centrifugated through a 150 µl layer of a mixture of 20% olive oil and 80% di-n-butyl phtalate. The tips of the tubes containing the cell pellet were cut off and bound Il-2 was determined by counting radioactivity in a gamma counter. The number of Il-2 binding sites and the affinity of the Il-2R was estimated by Scatchard analysis.

RESULTS AND DISCUSSION

Table I shows that dGuo in doses which completely block T cell proliferation has no effect whatsoever on expression of the Tac antigen, the α-chain of the Il-2 receptor. However, the dGuo mediated inhibition of T cell proliferation cannot be rescued by addition of recombinant Il-2 (Fig. 1). The α-chain (50-55 KD) of Il-2 receptors participates in the formation of both high- and low affinity Il-2 receptors. The β-chain (70 KD) lacks anti-Tac affinity and it is thought that the β-chain is associated with the Tac antigen to form the high-affinity Il-2 receptor

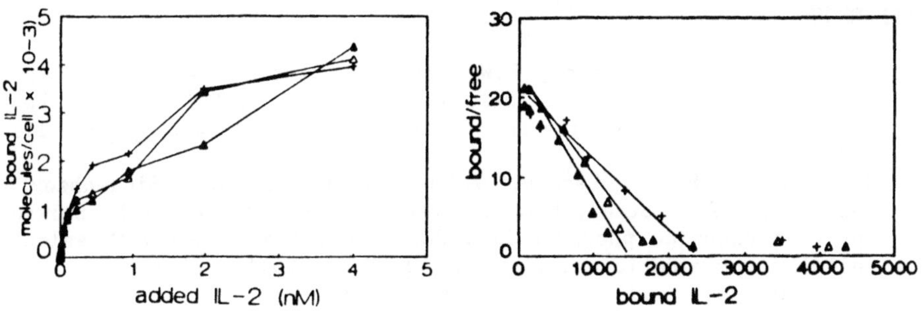

Figure 2.
A: Effect of dGuo on binding of ^{125}I-labeled Il-2 to PHA activated T cells. Culture conditions: control culture (+); 100 μM dGuo (△); 200 μM dGuo (▲).
B: Scatchard transformation of data of Figure A. The calculated numbers of binding sites are: control 2548 sites per cell; 100 μM dGuo 1769 sites per cell; 200 μM dGuo 1440 sites per cell.

complex. Only high-affinity Il-2 receptors mediate Il-2 induced T cell growth and internalization of Il-2[11]. Since no monoclonal antibodies against the β-chain are available we have analyzed the structural integrity of Il-2 receptors by a ligand-binding assay.
Figure 2 shows that the binding of Il-2 to its receptor decreases in the presence of dGuo (Fig. 2A). Transformation of these data into a Scatchard plot shows that the decrease of Il-2 binding is caused by a decrease in the number of high-affinity Il-2 receptors (Fig. 2B and Table I). DeoxyGuo has no effect on the expression of low-affinity Il-2 receptors (4,000-5,000/cell) which confirms the results of the expression of Tac antigen (Table I).
Although treatment of T cells with dGuo does lead to a reduction in the number of high-affinity Il-2 receptors, it is not clear whether this completely explains the negative effect of dGuo on T cell proliferation. In view of the known role of G-proteins in signal transduction, and the regulatory role of GTP in this system we also should envisage that the signal generated by Il-2 is not properly transduced when GTP accumulates in the cell. Therefore next to a (GTP mediated?) decrease in high affinity receptors for Il-2 an additional defect in Il-2 induced signal transduction may be the basis of dGuo mediated toxicity in normal T and B lymphocytes.

Table 1. Effect of dGuo on T cell proliferation, on Tac antigen expression and on the number of high affinity Il-2 receptors per cell

Culture conditions	proliferation (% TdR of control)	% Tac$^+$ cells	number of h.a. Il-2R/cell
Control (PHA)	100%	45%	2548
PHA + 100 μM dGuo	30%	42%	1769
PHA + 100 μM dGuo	16%	42%	1440
PHA + 300 μM dGuo	9%	40%	1410

REFERENCES

1. E.R. Giblet, A.J. Ammann, D.W. Wara, R. Sandman, and L.K. Diamond, Nucleoside phosphorylase deficiency in a child with severely defective T-cell immunity and normal B-cell immunity, Lancet, 1:1010 (1975).
2. J.W. Stoop, B.J.M. Zegers, G.F.M. Hendrickx, L.H. Siegenbeek van Heulekom, G.E.J. Staal, P.K. de Bree, S.K. Wadman, and R.E. Ballieux, Purine nucleoside phosphorylase deficiency associated with selective cellular immunodeficiency, N. Eng. J. Med, 296:651 (1977).
3. D.W. Martin, and E.W. Gelfand, Biochemistry of diseases of immunodevelopment, Ann. Rev. Biochem., 50:845 (1981).
4. A. Cohen, L.J. Gudas, A.J. Ammann, G.E.J. Staal, and D.W. Martin, Deoxyguanosine triphosphate as a possible toxic metabolite in the immunodeficiency associated with purine nucleoside phosphorylase deficiency, J. Clin. Invest, 61:1405 (1978).
5. B. Ullman, L.J. Gudas, S.M. Clift, and D.W. Martin, Isolation and characterization of purine phosphorylase deficient T lymphoma cells and secondary mutants with altered ribonucleotide-reductase: genetic model for immunodeficiency disease, Proc. Nat. Acad. Sci. USA, 76:1074 (1979).
6. L.J.M. Spaapen, G.T. Rijkers, G.E.J. Staal, G. Rijksen, S.K. Wadman, J.W. Stoop, and B.J.M. Zegers, The effect of deoxyguanosine on human lymphocyte function. I. Analysis of the interference with lymphocyte proliferation in vitro, J. Immunol., 132:2311 (1984).
7. L.J.M. Spaapen, G.T. Rijkers, G.E.J. Staal, G. Rijksen, S.K. Wadman, J.W. Stoop, and B.J.M. Zegers, The effect of deoxyguanosine on human lymphocyte function. II. Analysis of the interference with B lymphocyte differentiation in vitro, J. Immunol., 132:2318 (1984).
8. J.G.M. Scharenberg, L.J.M. Spaapen, G.T. Rijkers, M. Duran, G.E.J. Staal, and B.J.M. Zegers, Functional and mechanistic studies of the toxicity of deoxyguanosine for the in vitro proliferation and differentiation of human peripheral blood B lymphocytes, Eur. J. Immunol., 16:381 (1986).
9. J.G.M. Scharenberg, G.T. Rijkers, E.A.H. Toebes, G.E.J. Staal, and B.J.M. Zegers, Expression of deoxyadenosine and deoxyguanosine at different stages of lymphocyte activation, Scand. J. Immunol., (1988, in press).
10. J.G.M. Scharenberg, G.T. Rijkers, L.J.M. Spaapen, E.A.H. Toebes, G. Rijksen, M. Duran, G.E.J. Staal, and B.J.M. Zegers, Different pathways for deoxyguanosine toxicity in T lymphocytes of various developmental stages, Int. J. Imm. Pharmacol., (1988, in press).
11. M. Dukovich, Y. Wano, L. thi Bich Thuy, P. Katz, B.R. Cullen, J.H. Kehrl, and W.C. Greene, A second human interleukin-2 binding protein that may be a component of high-affinity interleukin-2 receptors, Nature, 327:518 (1987).

METABOLISM OF GUANOSINE AND ITS C-8 SUBSTITUTED ANALOGUES IN HUMAN LYMPHOCYTE CYTOPLASM

N. Lawrence Edwards* and James J. Mond[+]

*Department of Medicine, University of Florida
and VAMC, Gainesville, Florida, USA
[+]Uniformed Services University Health Center
Bethesda, Maryland, USA

INTRODUCTION

The 8-substituted guanosine analogues--8-bromoguanosine, 8-mercaptoguanosine, 8-aminoguanosine, 8-hydroxyguanosine and 8-methoxyguanosine have different effects on B cells. 8-Bromoguanosine and 8-mercaptoguanosine have polyclonal activating properties in B lymphocytes of mice. 8-Aminoguanosine, an inhibitor of PNP, has no immunomodulatory effect by itself on cells in culture (1). 8-Hydroxyguanosine causes both proliferative and differentiation of cells, while 8-methoxyguanosine causes differentiation without proliferation (2). These substances have been found to act primarily on B cells rather than T cells.

8-Bromoguanosine has been widely studied by Goodman and Weigle, and has been shown to act intracellularly in a series of elegant experiments (3). It is transported by carrier-mediated transport to act at a yet to be determined sites intracellularly. Immobilization of this 8-substituted nucleoside by sepharose beads caused a total loss of adjuvancy. Additionally, decreased plasma membrane fluidity or cross-linking of surface membrane proteins by lgM caused a decrease in activation by 8-substituted nucleosides. These studies support the hypothesis that these substances act from within the cell.

Previous studies compared the activation of B cells by 8-bromoguanosine vs. 8-bromo-GMP vs. 8-bromo cGMP (4). It was found that 8-bromoguanosine was the most efficient activator of B cells compared to 8-Br-GMP and 8-Br-cGMP, which had the least activity. 8-Bromoguanosine is not metabolized appreciably to 8-Br-cGMP or even to cGMP, and there is no increase of cGMP seen intracellularly in these activated cells. In fact, cGMP does not cause appreciable proliferation unless there is a bromine in the 8th position. At one time, it was thought that cGMP regulated the inductive pathway of the B cell and served as a "second messenger". Some early studies showed that mitogenic agents caused an increase in cGMP content in B cells and no increase in cAMP. However, this finding has been difficult to corroborate recently, and cGMP might actually inhibit mitogenicity in some cases. Thus, the role of cGMP as a second messenger has fallen out of favor. In looking at the 8-brominated guanosines, Goodman and Weigle postulated that 8-Br-cGMP may be stimulatory because of a structural analogy to 8-Br-guanosine or its ability to be metabolized to 8-Br-Guo. They thought the descending order of activity (8-Br-guo > 8-Br-GMP >

8-Br-cGMP) was perhaps due to an ability to fit with complimentary sites on DNA, culminating in DNA replication.

8-Mercaptoguanosine, like 8-bromoguanosine, caused stimulation as well as differentiation of B lymphocytes of mice (5). In rats, however, this substance does not cause cells to proliferate, but can induce IgM and IgG production if added to mitogen-stimulated cells. This suggests that there are two distinct effects of this substance-one that causes mitogenicity, and another effect that causes differentiation. 8-Mercaptoguanosine seems to act late in B cell differentiation. It can be added late in cell culture--on day 3 of a 4 day period--and still cause significant immune globulin synthesis. This seems to suggest that 8-mercaptoguanosine may act at the posttranscriptional level. The interaction of 8-mercaptoguanosine with B cells was studied using various concentrations of the nucleoside. It was found to augment the primary immune response to sheep RBC in a dose-dependent fashion, with maximal response occurring at 0.3mM. In other experiments it was shown that 8-SH-guo could enhance the secondary immune response, causing increased IgM/IgG by antigen-experienced B cells in response to SRBC (again at 0.3 mM concentration). There was an enhanced response to T cell independent antigens. In addition, 8-mercaptoguanosine could cause enhancement of the immune response even in the absence of intact T cells. By incubating the cells with supernatant from MLC cultures, and exposing them to Ag, an enhanced Ab response was obtained by Goodman and Weigle. Even without the supernatant of MLC being present, 8-mercaptoguanosine could still cause an enhanced immune response. Goodman and Weigle postulated that there was synergism between the T cell factors in the MLC supernatant and the 8-mercaptoguanosine.

Another facet of immune adjuvancy by these 8-substituted nucleosides was studied--the interaction with tolerant lymphocytes. There is a tolerance model in which lymphocytes that are exposed to deaggregated human gamma globulin (DHGG) become unresponsive later on to aggregated human gamma globulin on subsequent challenge. DHGG can cause tolerance in T_H and B lymphocytes. 8-Bromoguanosine can terminate this tolerance and can transform the tolerant B cell into Ab producing cells (6). Scheuer et al. postulated that there may be 2 signals involved in the activation of B cells and that exposure to one signal in the absence of the second may result in a tolerant state. T cells that are transplanted from mice exposed to DHGG and 8-bromoguanosine can demonstrate T_H function when transplanted with primed B cells into irradiated recipients. These experiments are useful in determining how tolerance is induced, and how it is overcome by 8-bromoguanosine--but, more experiments need to be performed to determine the "intracellular specifics" of these events.

METHODS

Human splenic tissue removed at surgery for non-malignant indications was immediately placed in cold Tris-buffered saline and subsequently minced, homogenized and fractionated by differential centrifugation. The particulate-free supernatant of the final one hour ultracentrifugation at 105,000 xg is considered the splenic cytoplasm. Total preparation was completed within 5 hours of surgical removal. The cytosol was dialyzed overnight and was determined to be free of purines by high performance liquid chromatography (HPLC). Metabolism of guanosine and the C-8 substituted analogues was monitored over a 2 hour incubation at 37° with samples taken at 0,5,10,30,60,90, and 120 minutes and inactivated in perchloric acid. The nucloside substrates were added at initial concentrations of 480μM with other cofactors including 0.5mM PRPP, 0.5mM ATP and 2mM $MgCl_2$.

Purine metabolites were quantitated by HPLC. Standard HPLC elution profiles were developed for bases, nucleosides and nucleotides of guanosine, deoxyguanosine, 8-bromoguanosine and 8-mercaptoguanosine. With isocratic techniques (50 mM KH_2Po_4, 2% acetonitrile at pH 4.5 on a C-18 radial compression column (Waters Assoc.) and a flow rate of 2.0 ml/min) and with standards dissolved in buffer, the retention times (RT) are as follows: uric acid (4.8), hypoxanthine (6.0), guanine (6.25), xanthine (6.9), inosine (13.4), guanosine (15.0), 8-NH_2 guanosine (19.3) and 8-mercaptoguanosine (25.2). 8-Bromoguanosine requires a gradient elution technique (50 mM KH_2Po_4, pH 4.5 as buffer A and 60% methanol as buffer B and a linear gradient to 40% B by 35 minutes). Ribo- and deoxyribonucleotides are separated by HPLC on a radial compression anion-exchanged column using an ammonium phosphate elution gradient (0.3 - 0.45M, pH 3.3 - 3.6) over 35 minutes at a flow rate of 2 ml/min. Eluted purines are detected by their dual absorbance at 254 and 280 mm and quantitated by comparison to calibration curves of nanomole amounts of standards.

RESULTS

In this study we examined the metabolic fate of guanosine and the C-8 substituted analogues in dialyzed human splenic cytosol. Required cofactors for the involved enzyme systems were added to the cytosol. The metabolism of the ribonucleoside substrate to other purine forms was monitored by reverse-phase and anion exchange HPLC at 30 minute intervals linear for 2 hours (Table 1). Guanosine declines linearly over 90 minutes with accumulations in guanine, xanthine, GTP and GDP when incubated at 37°C in the presence of PRPP. 8-SGuo and 8-BrGuo remain essentially non-metabolized to other forms after 120 minutes. 8-NH_2Guo is slowly metabolized to 8-NH_2guanine. These studies in human spleen demonstrate that the 2 C-8 substituted purine analogues (8-SGuo and 8-BrGuo) that stimulate B-cell proliferation and differentiation are not substantially metabolized by cytoplasmic purine enzymes. Conversely, the naturally occurring purine ribonucleoside, guanosine, is extensively metabolized to both higher (GTP and GDP) and lower (guanine and xanthine) forms in this assay system.

DISCUSSION

Research on the disorders of purine metabolism has demonstrated a causal relationship between the accumulation of various purine metabolites and both immunoregulation and lymphocytotoxicity. This research has also revealed that the various classes of lymphocytes have different purine metabolic potentials. These metabolic differences have been utilized to formulate more precise pharmacologic modulators of the immune system. Recently Goodman and Weigle (3-5) demonstrated that bromination or thiolation of guanosine at the C-8 position endowed it with potent biological activities, including the capacity to induce blast transformation of murine B-cells and to enhance the secretion of specific antibody when given together with antigen. These demonstrations that derivatized purine nucleosides represent a new class of adjuvant stimulation for *in vitro* antibody responses constitutes another solid connection between purine metabolism and immune responsiveness. The kinetic studies reported here contribute new information on how human immune cells metabolize these newly developed guanosine analogues. These findings may help define the molecular basis for both immune-suppressive and immune-potentiating effects of guanosine and its C-8 substituted analogues.

This work was supported by the Florida Chapter of the Arthritis Foundation.

Table 1. METABOLISM OF GUANOSINE AND THE C-8 SUBSTITUTED GUANOSINE ANALOGUES IN HUMAN SPLEEN CYTOSOL

A. GUANOSINE

Time (min)	Xanthine	Guanine	Guanosine (μM)	GMP	GDP	GTP
0	-	-	480	-	-	-
30	4	52	407	4	4	6
60	6	93	322	14	8	14
90	15	120	312	2	8	30
120	18	128	300	20	11	40

B. 8-BROMOGUANOSINE

Time (min)	8-Br Xanthine	8-Br Guanine	8-Br Guanosine (μM)	8-Br GMP	8-Br GDP	8-Br GTP
0	-	-	480	-	-	-
30	-	-	474	6	-	-
60	-	-	470	10	-	-
90	-	-	468	10	-	-
120	-	-	464	18	-	-

C. 8-MERCAPTOGUANOSINE

Time (min)	8-S Xanthine	8-S Guanine	8-S Guanosine (μM)	8-S GMP	8-S GDP	8-S GTP
0	-	-	480	-	-	-
30	-	-	472	8	-	-
60	-	-	465	14	-	-
90	-	-	464	15	-	-
120	-	-	460	22	-	-

D. 8-AMINOGUANOSINE

Time (min)	8-NH^2 Xanthine	8-NH^2 Guanine	8-NH^2 Guanosine (μM)	8-NH^2 GMP	8-NH^2 GDP	8-NH^2 GTP
0	-	-	480	-	-	-
30	-	5	475	-	-	-
60	-	15	465	-	-	-
90	-	22	450	3	-	-
120	-	28	449	3	-	-

REFERENCES

1. Sidi Y and BS Mitchell. 1984. 2'-Deoxyguanosine Toxicity for B and Mature T Cell LInes is Mediated by Guanine Ribonucleotide Accumulation. J. Clin. Invest. 24:1640-1648.
2. Ahnad A and JJ Mond. 8-Hydroxyguanosine and 8-Methoxyguanosine Posses Immunostimulating Activity for B Lymphocytes. (in press).
3. Goodman MG and WO Weigle. 1981. Intracellular Lymphocyte Activation and Carrier-Mediated Transport of C-8 Substituted Guanine Ribonucleosides. Proc. Natl. Acad. Sci. 81:862-866.
4. Goodman MG and WO Weigle. 1981. Activation of Lymphocytes by Brominated Nucleosides and Cyclic Nucleotide Analogues: Implications for the "Second Messenger" Function of cGMP. Proc. Natl. Acad. Sci. 78:7604-7608
5. Goodman MG and WO Weigle. 1984. Mechanism of 8-Mercaptoguanosine-mediated Adjuvancy: Role of Clonal Expansion and Cellular Recruitment. J.Immunol. 133:2910-2914.
6. Scheuer WV, Goodman MG, Parks DE, and Weigle WO. 1985. Active Transformation of Tolerogenic to Immunogenic Signals in T and B cells by 8-Bromoguanosine. J. Immunol. 135:2962-2966.

SELECTIVE TOXICITY OF PURINE NUCLEOSIDES TO HUMAN LEUKAEMIC CELLS

A. Piga*, K. Ganeshaguru, E.S. Green, B. Sheridan and A.V. Hoffbrand

Academic Department of Hematology, Royal Free Hospital School of Medicine, Rowland Hill Street, London, NW3 2PF

*Oncologia Clinica, Ospedale Regionale Torrette, 60020 Ancona, Italy

SUMMARY

The <u>in vitro</u> cytotoxicity of various purine nucleosides and purine enzyme inhibitors, alone or in combination, and of the alkylating agent mafosfamide (Asta Z7557), incubated for 4 and 24 h have been studied in 17 leukaemic cell lines and normal bone marrow (BM). The purine nucleosides and their analogues included: 2'chlorodeoxyadenosine (CdA), 2'deoxyadenosine (AdR), 3'deoxyadenosine (3'AdR) (cordycepin), adenosine (AR), adenine arabinoside (Ara-A), deoxyguanosine (GdR) and guanine arabinoside (Ara-G). Purine enzyme inhibitors included 2-deoxycoformycin (dCF) and 8-aminoguanosine (8-AG). Cytotoxicity was based on inhibition of (i) incorporation of ^{3}H-leucine into cell proteins and (ii) colony forming units - granulocytic/monocytic (CFU-GM) and for mixed cell colonies (CFU-GEMM). Marked and selective inhibition of T-cell growth was shown by the combinations dCF with either AdR or Ara-A, 8-AG and GdR and by CdA or Ara-G alone; these compounds even at high concentrations produced only partial inhibition of the growth of normal bone marrow CFU-GM and CFU-GEMM except for CdA which completely inhibited the formation of CFU-GEMM colonies. The combination dCF + cordycepin and alkylating agent mafosfamide were, however, toxic to all the cell lines at the concentrations employed, as well as to CFU-GM and CFU-GEMM. The high therapeutic index of some of the purine nucleosides with a relatively short exposure time makes them candidates for selective <u>in vitro</u> removal of residual neoplastic cells in autologous bone marrow transplantation (ABMT) for T-ALL.

INTRODUCTION

ABMT is being increasingly used in acute leukaemia. Success depends on effective removal of residual leukaemia in the patient and presumably, although not proven, on effective elimination of residual leukaemic cells from the BM being reinfused. Several techniques have been suggested for BM 'purging', including physical, immunological, pharmacological, immunopharmacological and biochemical. In experimental models, leukaemic cell kill of the order of 3-4 logs has been obtained, with monoclonal antibodies and complement or monoclonal antibodies coupled with toxins (1,2). T cells appear to be particularly dependent on the enzyme adenosine deaminase (ADA) to prevent toxicity from accumulating adenine and deoxyadenine nucleotides (3). We and others have successfully used the ADA inhibitor (dCF) in vivo to treat T-ALL (4-6). In vitro, the combination of dCF with AdR is effective at achieving a 10^3 or 10^4 log cell kill in T-ALL cell lines after as little as 4-h, while leaving normal BM CFU-GM and CFU-GEMM intact (7,8).

We have now evaluated the of activity of several purine analogues on a large number of leukaemic cell lines of different lineage and compared the results with those obtained with the alkylating agent mafosfamide, a cyclophosphamide derivative with in vitro cytotoxicity and commonly used for in vitro "purging" of leukaemic cells in human ABMT.

MATERIALS AND METHODS

CdA was from Dr. D.A. Carson; dCF from NIH (Dr. M. Suffness), ara-G from the Wellcome Research Laboratories (Dr. T.A. Krenitsky), and mafosfamide (Asta Z7557) from Asta Werke (Drs. N. Brock and P. Hilgard). AdR, AR, cordycepin (3'AdR), ara-A were purchased from Sigma; 8-AG was synthesised in this laboratory (9). L-[4,5-^{3}H]-leucine (45 Ci/mmol, TRK170) was obtained from Amersham International. PHA lymphocyte conditioned medium (PHA-LCM) was the supernatant harvested at day 7 from cultures of normal peripheral blood lymphocytes in alpha medium with 1% PHA (10); PHA-LCM was filter sterilised prior to storage at -20°C. All other reagents and culture material were of the purest quality commercially available.

The effect of the drugs was studied on 8 T-ALL cell lines (JM1, MOLT 3, MOLT 4, HPB-ALL, HPB-MLT, T-ALL-1, CCRF-HSB2, CCRF-CEM), 2 B-cell lines (RIVA, NALM-6), 2 non-T, non-B lymphoid cell lines (REH, KM3) and 5 myeloid cell lines (ML2, HL60, K562, U937, KG1). The cells were grown at 37°C in an humidified atmosphere containing 7.5% CO_2, in RPMI containing 10% FCS, 2mM L-glutamine, 1000 U/ml penicillin and 100 µg/ml streptomycin. Cultures were always fed the day before each experiment to ensure exponential growth. BM samples were obtained from 4 normal volunteer donors.

Cytotoxicity was based on inhibition of incorporation of ^{3}H-leucine into cell proteins (11). The final cell concentration was 0.5 x 10^6/ml in 96-well flat-bottomed microtitre plates containing RPMI alone or the drug to be tested, and incubated at 37°C for 4 or 24 h. At the end of incubation the plates were washed twice, fresh medium and ^{3}H-leucine (1µCi/well) was added and incubated for a further 24 h. The plates were washed and the cells harvested onto fibre-glass filters. Control cultures, incubated in the absence of drugs, had ^{3}H-leucine incorporation values ranging from 6,327 cpm (HPB-MLT) to 33,000 cpm. Triplicate wells were set up for each point and the standard error of the mean was lower than 5% except in the presence of very low counts which followed incubation with some of the drugs.

BM cells were incubated with the drugs for 24 h, washed twice and cultured for colony-forming units-granulocytic/monocytic (CFU-GM), and for mixed cell colonies CFU-GM colonies were scored at 10-14 days and for GEMM colonies and burst-forming units-erythroid (BFU-E) colonies were scored at day 14.

RESULTS

dCF 10^{-5}M plus AdR 10^{-5}M to 10^{-4}M for 4 h produced substantial inhibition of ^{3}H-leucine uptake in half the T-cell lines, (Fig. 1a); by prolonging the incubation with the drugs for 24 h most of the cell lines

(Figs. 3 and 4a) were sensitive to the combination. Among the T-cell lines, however, T-ALL-1 and HPB-ALL appeared resistant even after 24 h of incubation. Incubation of BM cells with dCF (10^{-5}M) and AdR (10^{-4}M) for 24 h induced 25% inhibition of growth of CFU-GM and 55% inhibition of growth of CFU-GEMM (Fig. 5).

The dCF 10^{-5}M plus ara-A 10^{-5}M to 10^{-4}M induced substantial inhibition of cell growth in almost all the T-cell lines and also in the non-T non-B line REH and in the B-cell line NALM-6 (Fig. 1b). With 24 h incubation more of the cell lines appeared sensitive (Fig. 4b) but T-ALL-1 and HPB-ALL were insensitive. dCF with ara-A (10^{-4}M) induced a mean inhibition of 55% of normal BM CFU-GM after 24 h incubation while ara-A (10^{-4}M) alone induced a mean inhibition of 25% (Fig. 5). In methyl-cellulose cultures dCF and ara-A produced only slight decrease CFU-GM and substantial decrease CFU-GEMM; BFU-E, however, appeared particularly sensitive to this combination (Fig. 5).

dCF (10^{-5}M) + 3'AdR (10^{-5} to 10^{-4}M) induced moderate to severe inhibition of growth of most of the cell lines (Fig. 1c). The combination for 24 h induced almost complete inhibition of ^{3}H-leucine uptake even in the most insensitive of the cell lines (MOLT-3) (Fig 3b). CFU-GM, CFU-GEMM and BFU-E was also severely or completely inhibited (Fig. 5).

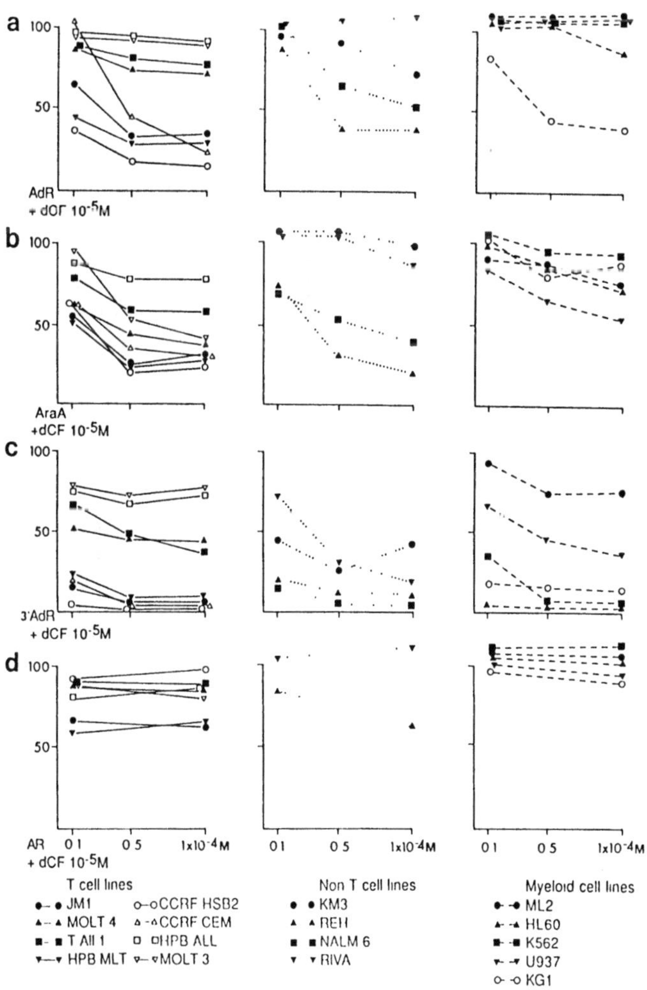

Fig. 1. The effects of 4h incubation of various cell lines with the combinations of dCF plus either AdR, ara-A, 3'AdR or AR on the incorporation of ^{3}H-leucine. Results are expressed as percent of control cultures in the absence of drugs.

The combination of dCF (10^{-5}M) + AR (10^{-4}M) for 4 h produced only moderate inhibition in a few of the leukaemic cell lines (Fig. 1d) and had no effect on normal BM cell colonies (Fig. 5).

CdA (10^{-4}M) inhibited to a variable extent the growth of all the T-cell lines after 4 h and especially after 24 h incubation (Figs. 2a & 4c) and also inhibited the growth of some of the other cell lines. Among the T-cell lines, T-ALL-1 alone was resistant to this drug. However, CdA was also toxic to normal BM progenitor cells. Incubation for 24 h with 10^{-4}M CdA resulted in no growth in the mixed colony assay (Fig. 5).

Ara-G caused marked inhibition in the proliferation of the T-cell lines, except T-ALL-1 and HPB-ALL, with moderate to no inhibition of growth of the other lines (Fig. 2b). 24 h incubation resulted in little additional effect, and particularly T-ALL-1 and HPB-ALL showed no increase in sensitivity; ara-G for 24 h caused 66.5% inhibition of CFU-GM colonies and only moderate inhibition of CFU-GEMM (Fig. 5).

The combination of GdR (10^{-4}M) with the PNP inhibitor 8AG (10^{-4}M) for 4 h, resulted in substantial cytotoxicity against all the T-cell lines with the exception of T-ALL-1 (Fig. 2c); moderate toxicity also occurred with some of the other cell lines. There was only a slight increase in inhibition with 24 h incubation. The contribution of 8AG to

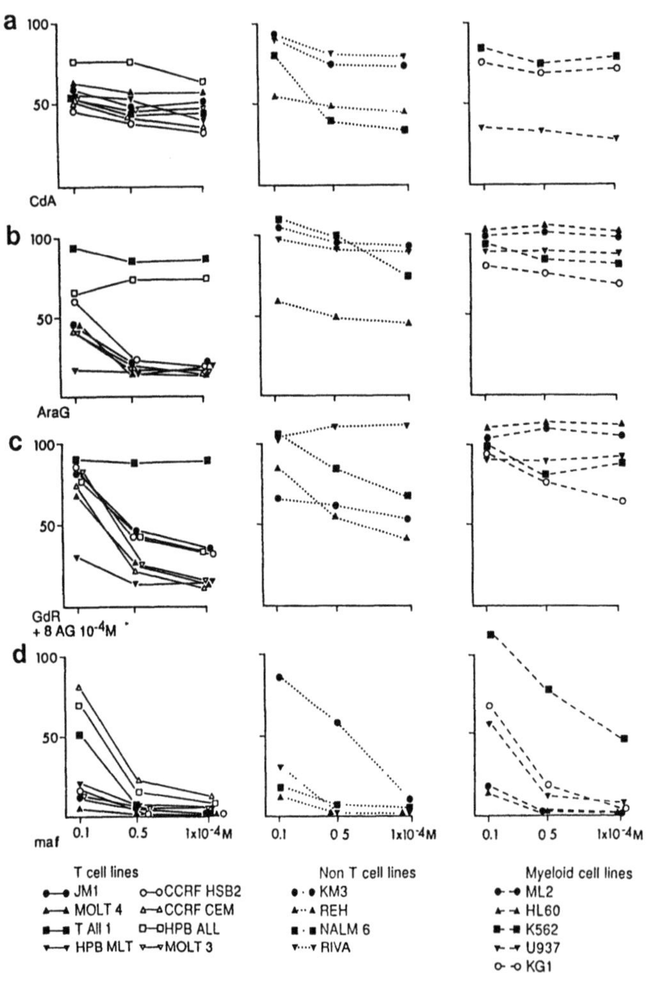

Fig. 2. The effects of 4h incubation of the cell lines with either CdA, Ara-G, mafosfamide or the combination 8AG + GdR on ^{3}H-leucine incorporation. Results are expressed as percent of control cultures in the absence of drugs.

the toxicity of GdR appeared slight since GdR alone was virtually equally toxic as the combination (data not shown). The combination only slightly inhibited growth of CFU-GM and CFU-GEMM (Fig. 5).

Following incubation for 4 h with mafosfamide (10^{-4}M) growth of all cell lines was substantially inhibited, except K562 (Fig. 2d). The effect was already apparent at 10^{-5}M in most instances. Incubation of BM cells with 10^{-4}M mafosfamide for 24 h resulted in virtually complete inhibition of colony formation; at 10^{-5}M 5-10% colony growth was observed except for BFU-E which was still completely inhibited (Fig. 5).

DISCUSSION

The results of this study have shown a wide spectrum of killing of leukaemic cell lines and normal bone marrow progenitors by a number of purine analogues alone or in combination with enzyme inhibitors. In general, these results are in agreement with many previous studies showing that T-lymphoid cells are peculiarly sensitive to these compounds. The mechanisms by which they exert their toxic effects on proliferating cells have been only partially elucidated. The toxicity of the ADA inhibitor dCF against proliferating cells is thought to be due to accumulation of toxic concentrations of dATP (3). In vitro, dCF alone is ineffective, however, and addition of AdR to the medium is required. As shown here, the other purine analogues, ara-A and cordycepin also become toxic in vitro when ADA is inhibited by dCF. On the other hand, CdA is a derivative of AdR, which is not subjected to deamination by ADA (12) and is toxic even in the absence of dCF. Toxicity of guanine nucleosides is produced by a similar mechanism, GdR is degraded by PNP and inhibition of this enzyme might enhance the toxicity of GdR, through dGTP accumulation. 8-AG, however, causes only weak and reversible inhibition of PNP and only marginally increases the in vitro toxicity of GdR for T-ALL cells (9,13). The arabinoside analogue ara-C is not subjected to catabolism by the action of PNP and alone is toxic to the cell lines (14).

Recent studies have shown that short-term (4 h) incubation with these drugs is sufficient to induce substantial inhibition of leukaemic cell growth (7). In preliminary experiments in animal models (15) we have obtained survival of mice injected with T leukaemias compatible

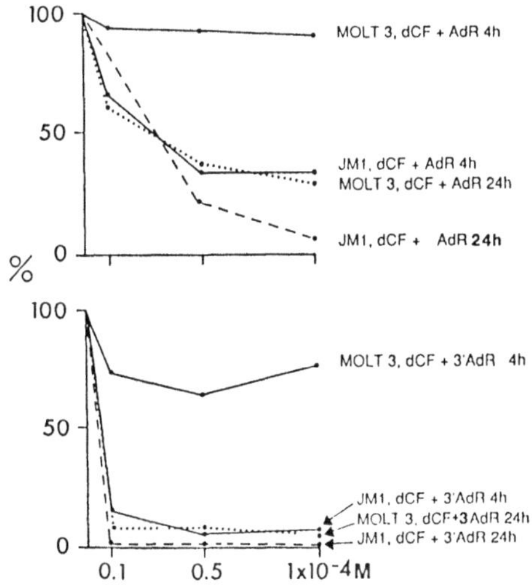

Fig. 3. Dose-response curves comparing the effect of prolonged incubation time to short incubation on the sensitivity of two T-cell lines (one sensitive, the other 'resistant') to dCF + AdR and to dCF + 3'AdR. Results are expressed as percent of untreated controls.

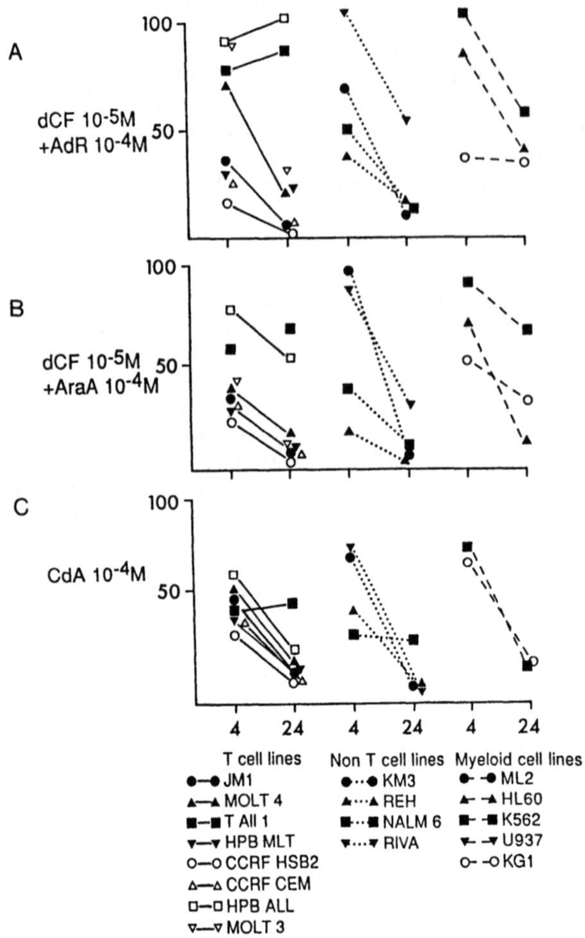

Fig. 4. The effect of prolonging the time of incubation on the sensitivity of the various cell lines to the drugs. Results are expressed as percent of untreated controls.

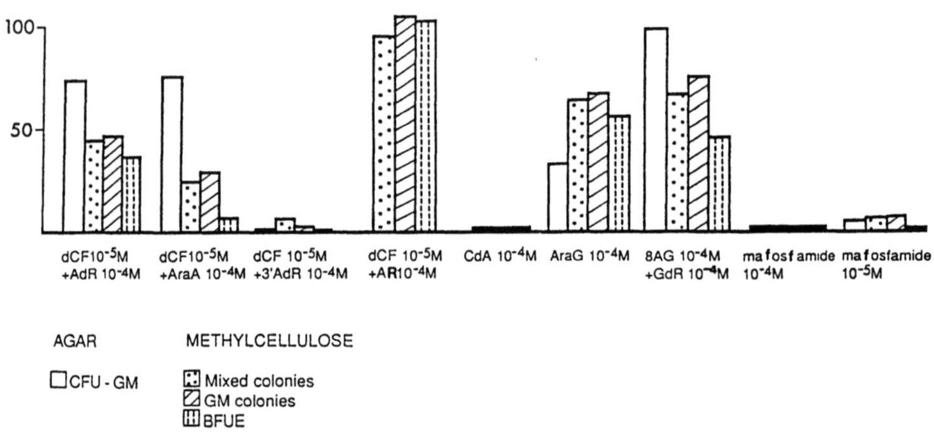

Fig. 5. Effects of the drugs on the growth of colonies from normal bone marrow cells (CFU-GM assay not performed with dCF + AR or with CdA.

with approximately four log cell kill for the most active combination (dCF + AdR), and slightly lower for other combinations. The pronounced in vitro activity of these drugs, together with the observation that they usually spare the normal haemopoietic cells supports their potential experimentation as drugs for in vitro removal of residual leukaemic cells in ABMT.

The T-cell lines are sensitive to the toxic effects of purine nucleosides, while normal BM cells appear generally only moderately damaged. The pattern of responsiveness of the various cell lines showed limited differences between the various drugs, suggesting that their mechanisms of toxicity may be similar. Although in some instances the resistance to the drugs was shown to be relative, and cells were killed by prolonging the incubation period, some cell lines, e.g. T-ALL-1 and HPB-ALL, were resistant to nearly all the drugs and combinations tested.

Most of the nucleosides employed appeared only moderately if at all toxic to normal BM cells, with two exceptions, CdA and the combination of dCF and cordycepin. CdA has previously been shown to be toxic to normal BM cells (12). On the other hand the toxic effect on the cell lines has not been particularly impressive compared to other drugs or combinations. The combination of dCF plus 3'AdR was the most effective against leukaemic cells of all varieties. Unfortunately, similarly to mafosfamide, the combination also caused marked inhibition of normal colony formation. The relevance of the colony data to the in vivo situation is still a matter of dispute, and successful in vivo restoration of the normal haemopoiesis has been reported in the presence of total in vitro inhibition of colony formation after incubation with mafosfamide (16). However, on the basis of these in vitro results the combination dCF + 3'AdR does not seem to offer particular advantages compared with the other nucleosides in terms of selectivity, or in respect to mafosfamide in terms of activity. Mafosfamide appeared equally active against all cell types studied and its use in ABMT could go beyond the traditional application as a single agent. Uckun et al. (17), have used mafosfamide in association with immunological methods in ALL.

A possible advantage of at least some of these nucleoside combinations with respect to conventional pharmacological agents, is their activity against resting as well as proliferating cells. More pre-clinical studies, in particular in vivo experiments in animals, may have to be done before further conclusions can be made on the advantages of the use of purine nucleosides and enzyme inhibitors for removal of residual leukaemic cells from transplant marrows. Our early studies in mice using some of the purine nucleosides with dCF has, however, been encouraging (15).

ACKNOWLEDGEMENTS: We thank Mrs. Megan Evans for word processing.

REFERENCES

1. Trigg ME, Polpack DG: Transplantation of leukaemic bone marrow incubated with cytotoxic antibodies. In: W.Knapp (ed) Leukaemia Markers, pp. 425-428, Academic Press, London (1981).
2. Bast RC Jr, Ritz J, Lipton, JM, Feeney M, Lipton J, Sallan SE, Nathan DG, Schlossman SF: Elimination of leukaemic cells from human bone marrow using monoclonal antibodies and complement. Cancer Res. 43:1389 (1983)
3. Carson DA, Kaye J, Seegmiller JE: Differential sensitivity of human leukaemic T cell lines and B cell lines to growth inhibition by deoxyadenosine. J Immunol 121:1726 (1978).

4. Smyth JF, Paine RM, Jackman AL, Harrap KR, Chassin MM, Adamson RH, Johns DG: The clinical pharmacology of the adenosine deaminase inhibitor 2'deoxycoformycin. Cancer Chemother Pharmacol 5:93 (1980).
5. Prentice HG, Smyth JF, Ganeshaguru K, Wonke B, Bradstock KF, Janossy G, Goldstone AH, Hoffbrand AV: Remission induction with the adenosine deaminase inhibitor 2'deoxycoformycin in Thy-acute lymphoblastic leukaemia. Lancet ii:170 (1980).
6. Prentice HG, Russell NH, Lee N, Ganeshaguru K, Blacklock H, Piga A, Smyth JF, Hoffbrand AV: Therapeutic selectivity of and prediction of response to 2'-deoxycoformycin in acute leukaemia. Lancet ii:1250 (1981).
7. Russell NH, Carron J, Hoffbrand AV, Bellingham AJ: The relative sensitivity of peripheral blood T-lymphocyte colony forming cells and bone marrow CFU-GM to deoxyadenosine and 2'deoxycoformycin. Leuk Res 9:315 (1985).
8. Russell NH, Hoffbrand AV, Bellingham AJ: Potential use of purine nucleosides and enzyme inhibitors for selective depletion of Thy-lymphoblasts from human bone marrow. Leuk Res 10:325 (1986).
9. De Fouw NJ, Ma DDF, Michalevicz R, Gray DA, Hoffbrand, AV: Differential cytotoxicity of deoxyguanosine and 8-aminoguanosine for human leukaemic cells lines and normal bone marrow progenitor cells. Hematol Oncol 2:189 (1984).
10. Aye MT, Till JE, McCulloch EA: Interacting populations affecting proliferation of leukaemic cells in culture. Blood 45:484 (1975).
11. Thorpe PE, Cumber AJ, Williams N, Edwards DC, Ross WCJ, Davies AJS: Abrogation of the non-specific toxicity of abrin conjugated to anti-lymphocyte globulin. Clin Expt Imm 43:195 (1981).
12. Carson DA, Wasson DB, Beutler E: Antileukemic and immunosuppressive activity of 2-chloro-2'-deoxyadenosine, PNAS, USA, 81;2232 (1984).
13. Kazmers IS, Mitchell BS, Daddona PE, Wotring LL, Townsend LB, Kelly WN: Inhibition of purine nucleoside phosphorylase by 8-aminoguanosine: selective toxicity for T lymphoblasts. Science 214:1137 (1981).
14. Cohen A, Lee JWW, Gelfand DW: Selective toxicity of deoxyguanosine and arabinosyl guanine for T-leukaemic cells. Blood 61:660 (1983).
15. Piga A, Ganeshaguru K, Sheridan B, Hoffbrand AV, Thorpe P. Selective killing of leukaemic T lymphoblasts with purine nucleosides and enzyme inhibition: implications for bone marrow purging prior to autotransplantation. Klinische Wochenschrift 65(Suppl.10):33 (1987).
16. Uckun FM, Ramakrishnan S, Haag D, Houston LL: Ex vivo elimination of lymphoblastic leukaemia cells from human marrow by mafosfamid: Leuk Res 9:83 (1985).
17. Uckun FM, Strong RC, Youle RJ, Vallera DA: Combined ex vivo treatment with immunotoxins and mafosfamid: a novel immunochemotherapeutic approach for elimination of neoplastic T-cells from autologous marrow grafts. J Immunol 134;3504 (1985).

RECOVERY OF THE ACTIVITIES OF IMP DEHYDROGENASE AND GMP SYNTHASE AFTER TREATMENT WITH TIAZOFURIN AND ACIVICIN IN HEPATOMA CELLS IN VITRO

Tadashi Ikegami, Yutaka Natsumeda and George Weber

Laboratory for Experimental Oncology, Indiana University School of Medicine, Indianapolis, IN 46223, U.S.A.

INTRODUCTION

Systematic studies carried out with the guidance of the molecular correlation concept pinpointed IMP dehydrogenase and de novo guanylate pathway as sensitive targets for chemotherapy.[1] This hypothesis is supported by the following observations: IMP dehydrogenase has the lowest activity in purine biosynthesis and is the rate-limiting enzyme in GTP production;[1] the activities of IMP dehydrogenase and GMP synthase are increased in transformation- and progression-linked fashion;[2,3] these enzyme activities were elevated in all examined animal and human tumors; the concentrations of GMP and dGTP were increased in cancer cells;[4] dGTP has the smallest pool among deoxynucleoside triphosphates in tissues[5] and thereby might be the rate-limiting factor of DNA synthesis; recently it was shown by measuring the metabolic fluxes that in rat hepatoma cells during the transition of plateau into exponential growth phase IMP utilization for guanylate synthesis was markedly enhanced;[6] also the concentration of GTP expanded in the cells prior to the logarithmic cell growth;[7] GTP plays significant roles in signal transduction, microtubule polymerization, biosynthesis of c-GMP, AMP, CTP, biopterins, and proteins, as well as RNA, and in regulation of several important metabolic pathways.[8,9]

Tiazofurin, a potent inhibitor of IMP dehydrogenase, and acivicin, an anti-glutamine agent which inhibits GMP synthase, are under clinical trial as anticancer drugs. Acivicin and tiazofurin are rapidly excreted from the blood stream in rat with half lives of 1.5 and 2.5 h, respectively;[10] no acivicin could be detected at 12 h after i.p. injection to rats.[11] However, drug action by tiazofurin or acivicin in vivo persisted at least 24 h after i.p. injection because of tight binding of the drug to the target enzymes.[10] The present study was undertaken to elucidate the mechanisms of recovery of the enzyme activities from inhibition by acivicin and tiazofurin. The information would assist in designing a rational treatment schedule and combination chemotherapy.

MATERIALS AND METHODS

Chemicals

[8-^{3}H]XMP and [8-^{14}C]IMP were purchased from Amersham and cycloheximide was obtained from Sigma. Tiazofurin was kindly provided by Dr. R. K. Robins, Nucleic Acid Research Institute, Costa Mesa, CA. Acivicin was from Upjohn Co., Kalamazoo, MI. Both drugs were dissolved in phosphate buffered saline (PBS) on the day of use. All other reagents were of the highest purity available.

Cell culture

Hepatoma 3924A cells were grown in monolayer culture in McCoy's 5A medium with 10% fetal calf serum and incubated in a 5% CO_2: 95% air atmosphere. To keep the cells in logarithmic growth phase, they were subcultured three times a week. For the experiments, 2.5×10^6 hepatoma 3924A cells were seeded into plastic flasks of 75 cm^2 with 15 ml of McCoy's 5A medium containing 10% dialyzed fetal calf serum and incubated for 21 h. Then either 100 μM tiazofurin or 10 uM acivicin was added and incubated for 3 h. The control group received PBS. The medium was removed and the cells were gently washed with warm, fresh medium and incubated with 15 ml of fresh medium. In parallel experiments cycloheximide (6.7 uM) was added to inhibit protein synthesis in the cells. The cells were harvested at indicated time points by scraping, centrifuged and washed with ice-cold PBS. Cell pellets were homogenized in 0.4 ml to 0.8 ml of 0.25 M sucrose containing 40 mM Tris-HCl, pH 7.4, by a teflon-glass homogenizer. The homogenate was centrifuged at 100,000 x g for 30 min and the supernatant was used for enzyme assays.

Enzyme Assays

IMP dehydrogenase activity was measured as previously described.[12] GMP synthase assay was carried out by measuring [8-^{3}H]-XMP conversion to [8-^{3}H]GMP. The reaction mixture contained 1.8 mM ATP, 1.8 mM L-glutamine, 130 mM Tris-HCl, pH 7.6, 16 mM $MgCl_2$, 3 mM creatinine phosphate, 10 μg creatinine phosphokinase (50-100 unit/mg), 0.13 mM [^{3}H]XMP (5.4.Ci/mmol) and enzyme solution in 25 μl. The reaction was started by the addition of 5 μl of enzyme and stopped by immersing the reaction tube into boiling water for 20 sec. The substrate and the products were separated by high voltage electrophoresis as described.[12]

RESULTS AND DISCUSSION

Recovery of GMP synthase activity after acivicin treatment

Acivicin has been shown to inhibit and inactivate glutamine-utilizing enzymes in purine and pyrimidine metabolism, amidophosphoribosyltransferase, phosphoribosylformylglycinamidine synthase, GMP synthase, carbamoyl-phosphate synthase II and CTP synthase.[10] Among these enzymes GMP synthase was the most sensitive to acivicin in experiments in vivo using hepatoma 3924A-bearing rats.[10] In cultured hepatoma 3924A cells GMP synthase activity was completely inactivated after treatment with 10 μM acivicin and remained at a neglibible level until removal of acivicin from the medium (Fig. 1). After the medium replacement GMP synthase activity slowly recovered and at 12, 24 and 36 h the activities were 55, 63 and 78%, respectively, of control activity (Fig. 1). However, cycloheximide (6.7 μM) completely

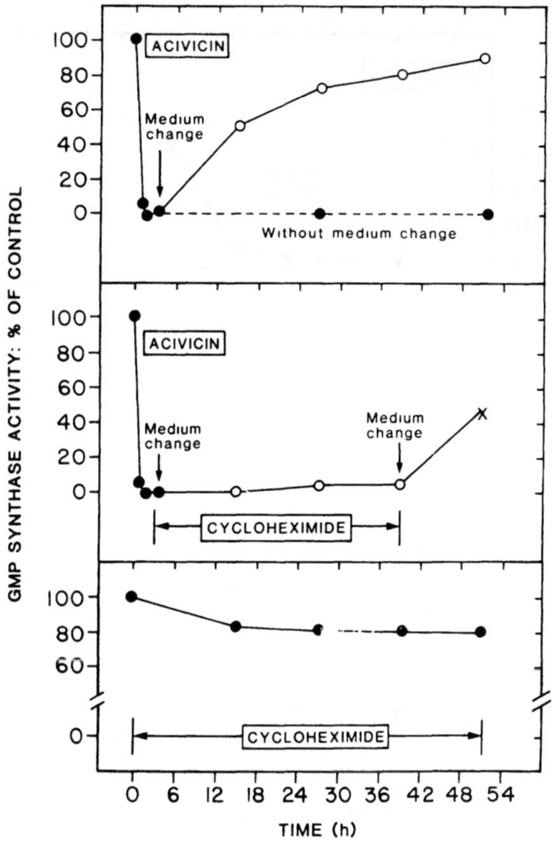

Fig. 1. Effects of acivicin and cycloheximide on GMP synthase activity in hepatoma 3924A cells. Upper panel: the cells were treated with 10 uM acivicin for 3 h and then medium was replaced by acivicin-free fresh medium. Middle panel: after treatment with acivicin medium was replaced by the fresh medium containing 6.7 μM cycloheximide. Lower panel: the cells were treated with 6.7 μM cycloheximide alone. Values are means of triplicate assays and standard errors of means did not exceed ± 5%.

blocked restoration of the enzyme activity. When cycloheximide was removed from the culture medium, the enzyme activity started to recover (Fig. 1). Cycloheximide alone did not markedly affect GMP synthase activity in the hepatoma cells (Fig. 1). These results indicate that the recovery of GMP synthase activity after acivicin treatment is due to the synthesis of the new enzyme protein. The time-course of the recovery in vitro in cultured hepatoma 3924A cells was comparable with that observed in vivo in solid hepatoma.[13]

Recovery of IMP dehydrogenase activity after tiazofurin treatment

Tiazofurin is phosphorylated in cells to tiazofurin 5'-monophosphate and then through NAD pyrophosphorylase-catalyzed reaction an NAD analogue (TAD, thiazole-4-carboxamide adenine dinucleotide) is formed. TAD is a potent inhibitor of IMP dehydrogenase.[14] Fig. 2 shows the time-course of the inhibition and the recovery of IMP dehydrogenase activity in hepatoma 3924A cells after tiazofurin

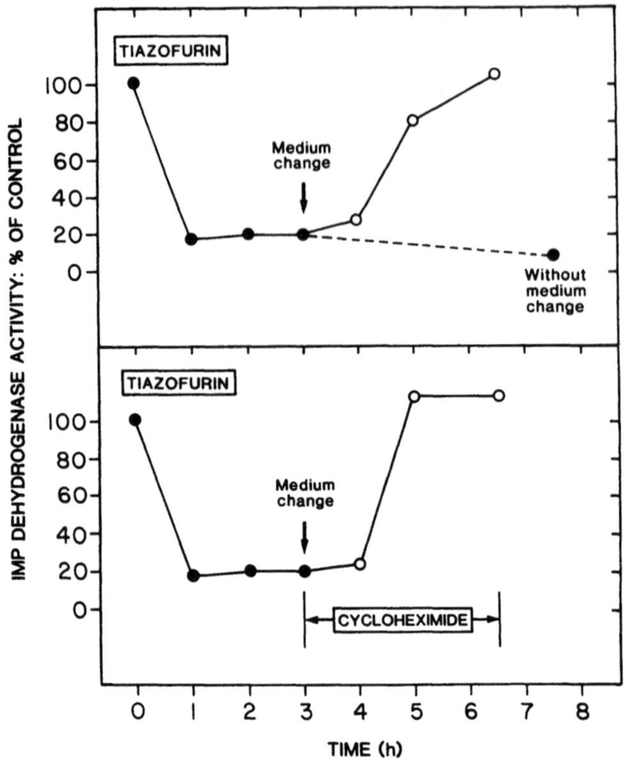

Fig. 2. Effects of tiazofurin and cycloheximide on IMP dehydrogenase activity in hepatoma 3924A cells. Upper panel: the cells were treated with 100 μM tiazofurin for 3 h and then medium was replaced by tiazofurin-free fresh medium. Lower panel: after treatment with tiazofurin medium was replaced by the fresh medium containing 6.7 μM cycloheximide. Values are means of triplicate assays and standard errors of means did not exceed ± 5%.

treatment. Tiazofurin (100 μM) inhibited IMP dehydrogenase activity markedly and rapidly. The removal of tiazofurin by replacing the medium restored the enzyme activity to the control level within 2 h. Cycloheximide (6.7 μM) did not affect this recovery (Fig. 2) or inhibit IMP dehydrogenase activity (data not shown). These results suggest that the inhibition of IMP dehydrogenase by tiazofurin treatment could be reversed after removal of the drug by a rapid process, e.g., dissociation of the drug from the enzyme or degradation of the active metabolite, other than the time-requiring process, synthesis of new enzyme protein.

The simplified experimental system in cultured hepatoma cells was used to study mechanisms of recovery in the activities of the target enzymes after treatment with acivicin and tiazofurin. Similar trials in vivo in hepatoma-bearing rats were not completed because of toxic deaths of animals by cycloheximide at the dose effective to block protein synthesis. Acivicin has been shown in bacterial anthranilate synthase to alkylate cysteine residue involved in glutamine binding.[15]

Similarly acivicin might irreversibly inactivate the other glutamine-utilizing enzymes by alkylating the glutamine-binding site. The present studies provide evidence that the recovery of GMP synthase activity after acivicin treatment requires the synthesis of the enzyme protein. This observation supports our previous results that a combination of acivicin and actinomycin D, a blocker of protein synthesis, exerted in vivo synergistic inhibition of the growth of hepatoma 3924A.[11]

The present study on tiazofurin action suggests that the recovery of IMP dehydrogenase activity after the drug treatment does not require synthesis of the enzyme protein. It is unlikely that TAD dissociates from IMP dehydrogenase, since it binds to the enzyme so tightly that the enzyme inhibition by the drug remains after homogenization of cells or tissues and even after partial purification of the tissue extract by $(NH_4)_2SO_4$ fractionation and dialysis for 4 h.[10] However, incubation of the tiazofurin-treated cells in the fresh medium quickly restored IMP dehydrogenase activity, suggesting that the intact cells catabolized TAD. Thus, the degradation of TAD should be one of the critical factors responsible for duration of tiazofurin action as well as for sensitivity to tiazofurin.[16] The applicability of the observations obtained in the in vitro cell culture system to animals and in human chemotherapy is under investigation.

ACKNOWLEDGEMENTS

This investigation was supported by UHPHS Outstanding Investigator Grant CA-42510 to G. W.

REFERENCES

1. G. Weber, Biochemical strategy of cancer cells and the design of chemotherapy: G. H. A. Clowes Memorial Lecture, Cancer Res. 43: 3466 (1983).
2. R. C. Jackson, G. Weber, and H. P. Morris, IMP dehydrogenase: an enzyme linked with proliferation and malignancy, Nature 256: 331 (1975).
3. T. J. Boritzki, R. C. Jackson, H. P. Morris, and G. Weber, Guanosine 5'-phosphate synthetase and guanosine 5'-phosphate kinase in rat hepatomas and kidney tumors, Biochim. Biophys. Acta 658: 102 (1981).
4. R. C. Jackson, M. S. Lui, T. J. Boritzki, H. P. Morris, and G. Weber, Purine and pyrimidine nucleotide patterns of normal, differentiating and regenerating liver and of hepatomas in rats, Cancer Res. 40: 1286 (1980).
5. P. Reichard, Regulation of deoxyribotide synthesis, Biochemistry 26: 3245 (1987).
6. Y. Natsumeda, T. Ikegami, K. Murayama, and G. Weber, De novo guanylate synthesis in the commitment to replication in hepatoma 3924A cells, Cancer Res. 48: 507 (1988).
7. G. Weber, E. Olah, J. E. Denton, M. S. Lui, E. Takeda, D. Y. Tzeng, and J. Ban, Dynamics of modulation of biochemical programs in cancer cells, Adv. Enzyme Regul. 19: 87 (1981).
8. G. Weber, Y. Natsumeda, and K. Pillwein, Targets and markers of selective action of tiazofurin, Adv. Enzyme Regul. 24: 45 (1985).
9. G. Weber, H. N. Jayaram, E. Lapis, Y. Natsumeda, Y. Yamada, Y. Yamaji, G. J. Tricot, and R. Hoffman, Enzyme-pattern-targeted chemotherapy with tiazofurin and allopurinol in human leukemia, Adv. Enzyme Regul. 27: 405 (1988).

10. G. Weber, Y. Natsumeda, M. S. Lui, M. A. Faderan, J. J. Liepnieks, and W. L. Elliott, Control of enzymic programs and nucleotide pattern in cancer cells by acivicin and tiazofurin, Adv. Enzyme Regul. 22: 69 (1984).
11. G. Weber, N. Prajda, M. S. Lui, J. E. Denton, T. Aoki, J. Sebolt, Y.-S. Zhen, M. E. Burt, M. A. Faderan, and M. A. Reardon, Multi-enzyme-targeted chemotherapy by acivicin and actinomycin, Adv. Enzyme Regul. 20: 75 (1982).
12. T. Ikegami, Y. Natsumeda, and G. Weber, Direct assay method for inosine 5'-monophosphate dehydrogenase activity, Anal. Biochem. 150: 155 (1985).
13. J. E. Denton, M. S. Lui, T. Aoki, J. Sebolt, and G. Weber, Rapid in vivo inactivation by acivicin of CTP synthetase, carbamoyl-phosphate synthetase II and amidophosphoribosyltransferase in hepatoma, Life Sci. 30: 1073 (1982).
14. Y. Yamada, Y. Natsumeda, and G. Weber, Action of the active metabolites of tiazofurin and ribavirin on purified IMP dehydrogenase, Biochemistry 27: 2193 (1988).
15. J. Y. Tso, S. G. Bower, and H. Zalkin, Mechanism of inactivation of glutamine amidotransferases by the antitumor drug L-(αS,5S)-α-amino-3-chloro-4,5-dihydro-5-isoxazoleacetic acid (AT-125), J. Biol. Chem. 255: 6734 (1980).
16. G. S. Ahluwaria, J. N. Jayaram, J. P. Plowman, D. A. Cooney, and D. G. Jones, Studies on the mechanism of action of 2-B-D-ribofuranosylthiazole-carboxamide-V: factors governing the response of murine tumors to tiazofurin, Biochem. Pharmacol. 33: 1195 (1984).

CHANNELING OF IMP INTO GUANYLATE SYNTHESIS IN THE GROWTH PROGRAM OF HEPATOMA 3924A CELLS

Yutaka Natsumeda, Tadashi Ikegami, Yasukazu Yamada, Yasufumi Yamaji and George Weber

Laboratory for Experimental Oncology, Indiana University School of Medicine, Indianapolis, IN, U.S.A.

INTRODUCTION

Extensive studies on purine metabolic enzymes in this Laboratory have suggested that in cancer cells the important pathway is the one for de novo production of IMP and its channeling into guanylate biosynthesis.[1] The purpose of this study was to test this hypothesis by measuring the metabolic flux in intact cancer cells during the display of the proliferative program. The results indicate that purine de novo synthesis increased in hepatoma 3924A cells as they expressed their replicative program from resting to logarithmic growth phases. With the rise in the rate of purine de novo pathway the synthesis of both adenylates and guanylates was enhanced; however, in the lag and log phases the relative utilization of IMP to guanylate synthesis increased and that to adenylates decreased. This reciprocal behavior in the fluxes of the two pathways that compete for IMP utilization emphasizes the importance of the linkage of de novo guanylate synthesis with proliferation and supports the potential significance of this pathway as a target of chemotherapy.

MATERIALS AND METHODS

The assay for purine de novo synthesis is designed to determine absolute rates in terms of nmol of purine nucleotides synthesized per h per cell to make it possible to compare with fluxes of the other metabolic pathways, especially with purine salvage synthesis for our current study. Materials and Methods were described previously in detail.[2] However, the following observations require attention. (a) The substrate [^{14}C]formate labels not only purines but serine, methionine, cysteine, thymidylates, RNA, DNA and protein which must be separated from purine compounds. After the hydrolysis of whole purine compounds in nucleosides, nucleotides and nucleic acids to purine bases the samples were applied on Dowex 50-H$^+$ column.[2,3] Among purine bases and the degradative metabolites, uric acid and allantoin passed through the column, whereas adenine, hypoxanthine, xanthine and guanine were collected with 6N HCl in one fraction. However, this fraction contained [^{14}C]methionine and its oxidized compound, [^{14}C]methionine sulfoxide, in the assays in hepatoma 3924A cells.[2] These contaminants were separated from purines by two-dimensional chromatography on Whatman 3MM paper (Fig. 1).[2] The use of butanol:methanol:water:ammonia (60:20:20:1) in one-dimension reported earlier[3] provided overlapping of the guanine spot with methionine sulfoxide and methionine sulfone (Fig. 1). (b) Assay media should be free from serine which is the major source of one-carbon unit, dilutes the specific activity of [^{14}C]formate and causes underestimation of purine synthesis. In the serine-free assay media excess [^{14}C]formate diluted out incorporation of serine into

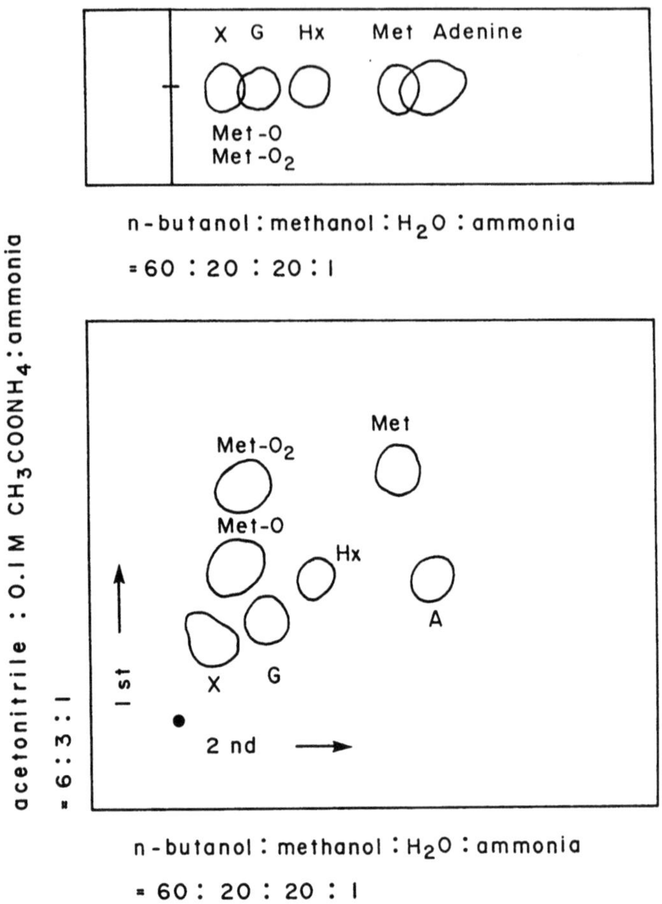

Fig. 1. Separation of purines and other labeled products by Whatman 3MM paper. Upper and lower panels show one- and two-dimensional chromatography. Met-O, methionine sulfoxide; Met-O_2, methionine sulfone; X, xanthine; G, guanine; HX, hypoxanthine; A, adenine.

purines.[2] Substrate concentration curves for purine de novo synthesis with [^{14}C]formate and L-[3-^{14}C]serine both of which provide 10-[^{14}C]formyl-H_4folate, the direct precursor for purine synthesis, followed Michaelis-Menten kinetics with the same V_{max} value.[2] (c) [^{14}C]Formate initially labels IMP among purine compounds. IMP is metabolized by adenylosuccinate synthase, IMP dehydrogenase, 5'-nucleotidase and hypoxanthine phosphoribosyltransferase (reverse reaction). However, IMP is convertible to adenine and guanine compounds only at the nucleotide level. In the time-course studies the labelings of adenine and guanine, which reflect de novo synthesis of adenylates and guanylates from IMP, were linear with incubation time up to 60 min (Fig. 2).[4] Therefore, overall synthesis of adenylates and guanylates in hepatoma cells could be calculated from the rates of total purine de novo synthesis and from the distribution of label in adenine and guanine in the standard assays.

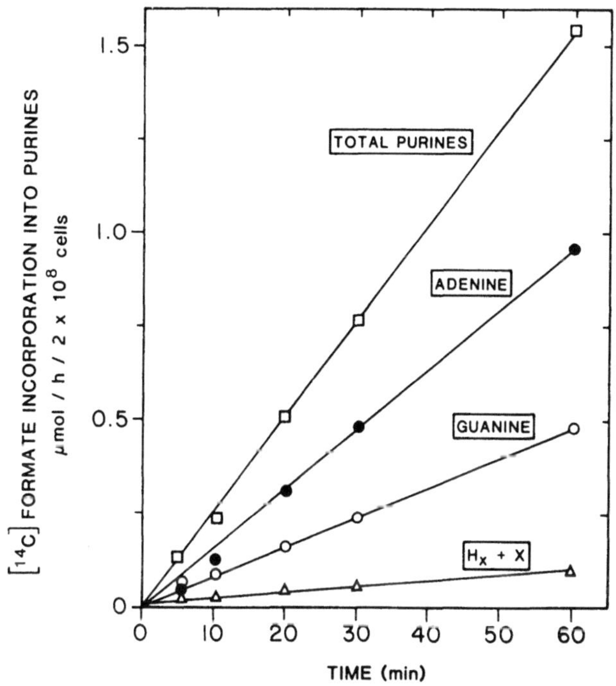

Fig. 2. Time course of labelings of total purine, adenine and guanine compounds with $[^{14}C]$formate in hepatoma 3924A cells.

RESULTS AND DISCUSSION

When hepatoma 3924A cells in plateau phase were seeded in the fresh culture medium at a density of 2 x 10^4 cells/cm^2 the cells were allowed to grow from lag (0-24 h after seeding) into log (24-72 h) and plateau (72-96 h) phases (Fig. 3A). An early increase was observed for PRPP synthase activity with a peak of 2.2-fold rise at 6 h after plating and the increase remained at the level of 1.4-fold rise over 48 h (Fig. 3B). The concentration of PRPP, the substrate for amidophosphoribosyltransferase, was 3.3 ± 0.2 nmol/g cells in the plateau phase; it markedly increased in lag phase, reached a peak of 14-fold rise at 24 to 40 h, declined with the increase of cell number in late log phase and returned to the resting phase value (Fig. 3C). The expansion of the PRPP pool was followed by a marked rise in purine de novo synthesis which increased 8-fold in log phase (Fig. 3D). The activity of amidophosphoribosyltransferase rose 1.3- to 1.4-fold in lag and log phases (Fig. 3B).

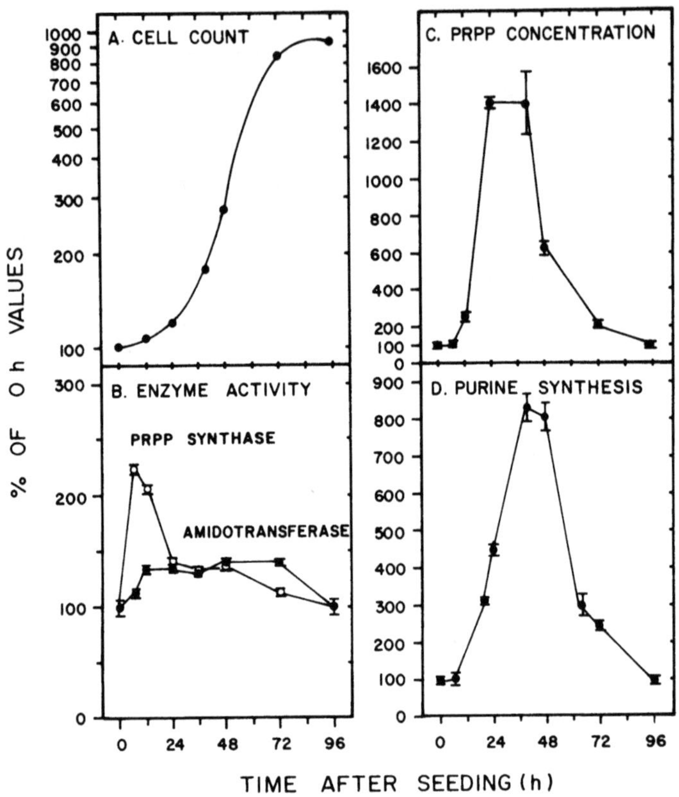

Fig. 3. Biochemical alterations in purine metabolism in hepatoma 3924A cells in various growth phases; (A) cell counts; (B) activities of PRPP synthase and amidophosphoribosyltransferase; (C) concentrations of PRPP and (D) rates of purine de novo synthesis.

With the rise in the rate of purine de novo synthesis the overall synthesis of both adenylates and guanylates increased in log phase (Fig. 4). However, there was a readily detectable change in the distribution of label during the culture growth. Fig. 5 shows the reciprocal change of the label distribution in adenine and guanine. At 96 (0) h the label was distributed 71% in adenine, 16% in guanine, and the rest in hypoxanthine and xanthine. Thus, adenylates were produced predominantly in plateau phase. After replating the cells in fresh culture medium the relative labeling of guanine sharply increased and that of adenine inversely decreased with the peak and nadir at 20 to 24 h. The subsequent decline of the relative labeling of guanine was accompanied with the restoration in that of adenine with the increase in the cell density. This mirror picture indicates a metabolic switch at the IMP branching point for the preferential synthesis of guanylates during the transition of plateau into exponential growth phase. The regulatory mechanisms which markedly change the relative utilization of IMP in growth stimulation remain to be elucidated. However, it is interesting that the concentration of GTP was elevated more markedly than that of ATP in the hepatoma cells in proliferating phase.[5] The preferential synthesis of guanylates from IMP and the expansion of GTP pool, which preceded the onset of exponential growth, might play important roles in the biochemical commitment of the hepatoma cells in replication.

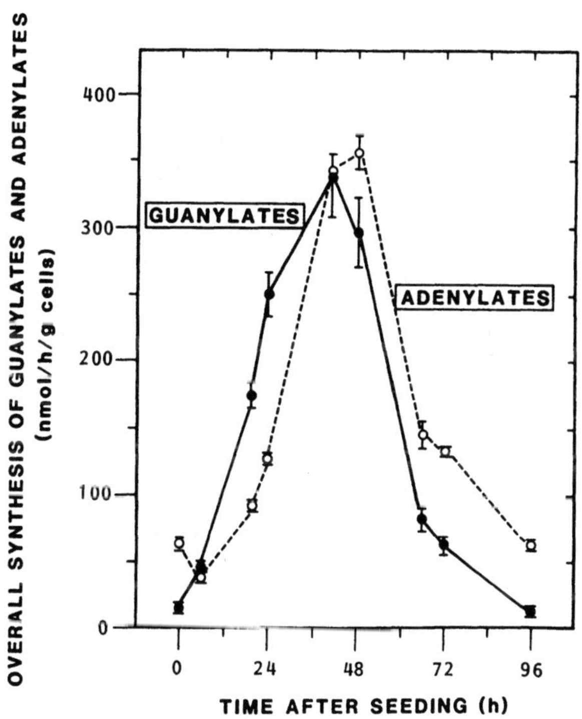

Fig. 4. Activities of overall synthesis of adenylates and guanylates in hepatoma 3924A cells in various growth phases.

Recently evidence was provided in patients with refractory acute myeloid leukemia[6] that a marked inhibition of IMP dehydrogenase by tiazofurin through its active metabolite TAD (thiazole-4-carboxamide adenine dinucleotide)[7] caused a striking depresssion in GTP concentration of the blast cells and subsequently a profound decline in the peripheral blast cell count.[8] Tiazofurin has also been shown to induce erythroid differentiation of K562 human leukemia cells.[9]

The tiazofurin-induced differentiation was associated with down-regulation of c-ras oncogene which is involved in the expression of GTP-dependent signal transducers.[9] Furthermore, the decrease of GTP pool was stringently linked with erythroid differentiation, since guanosine prevented the action of tiazofurin.[9] These observations and the results presented here suggest that the regulation of de novo guanylate pathway and GTP pool may serve significant roles in mediating signal transduction that leads to the expression of growth or differentiation program in cancer cells.

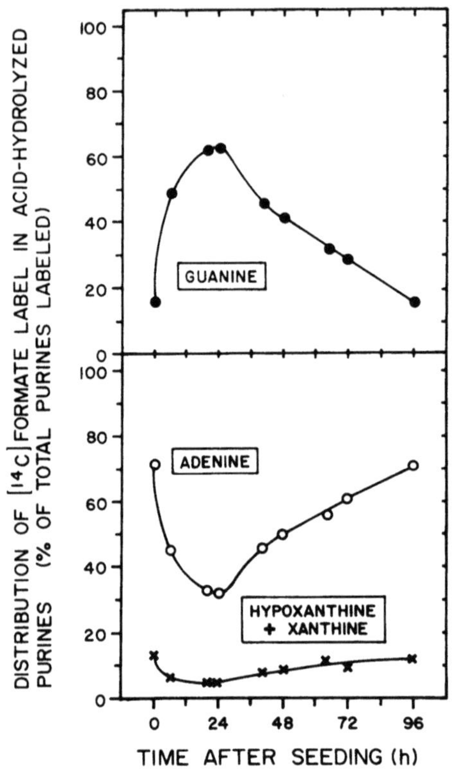

Fig. 5. Relative labeling of purines with $[^{14}C]$formate in hepatoma 3924A cells in various growth phases. Data are from the same experiments as those shown in Fig. 4.

ACKNOWLEDGEMENTS

This investigation was supported by USPHS Outstanding Investigator Grant CA-42510 awarded to G. W. The authors are grateful to Maria C. Briones-Ikegami for technical assistance.

REFERENCES

1. G. Weber, Biochemical strategy of cancer cells and the design of chemotherapy: G. H. A. Clowes Memorial Lecture, Cancer Res. 43: 3466 (1983).
2. Y. Natsumeda, T. Ikegami, K. Murayama, and G. Weber, De novo guanylate synthesis in the commitment to replication in hepatoma 3924A cells, Cancer Res. 48: 507 (1988).
3. M. S. Hershfield and J. E. Seegmiller, Regulation of de novo purine biosynthesis in human lymphoblasts: coordinate control of proximal (rate-determining) steps and the inosinic acid branch point, J. Biol. Chem. 251: 7348 (1976).
4. Y. Natsumeda, Y. Yamada, Y. Yamaji, and G. Weber, Synergistic cytotoxic effect of tiazofurin and ribavirin in hepatoma cells, Biochem. Biophys. Res. Commun. 153: 321 (1988).
5. G. Weber, E. Olah, J. E. Denton, M. S. Lui, E. Takeda, D. Y. Tzeng, and J. Ban, Dynamics of modulation of biochemical programs in cancer cells, Adv. Enzyme Regul. 19: 87 (1981).
6. G. J. Tricot, H. N. Jayaram, C. R. Nichols, K. Pennington, E. Lapis, G. Weber and R. Hoffman, Hematological and biochemical action of tiazofurin (NSC 286193) in a case of refractory acute myeloid leukemia, Cancer Res. 47: 4988 (1987).
7. Y. Yamada, Y. Natsumeda, and G. Weber, Action of the active metabolites of tiazofurin and ribavirin on purified IMP dehydrogenase, Biochemistry 27: 2193 (1988).
8. G. Weber, H. N. Jayaram, E. Lapis, Y. Natsumeda, Y. Yamada, Y. Yamaji, G. J. Tricot, and R. Hoffman, Enzyme-pattern-targeted chemotherapy with tiazofurin and allopurinol in human leukemia, Adv. Enzyme Regul. 27: 405 (1988).
9. E. Olah, Y. Natsumeda, T. Ikegami, Z. Kote, M. Horanyi, J. Szelenyi, E. Paulik, T. Kremmer, S. R. Hollan, J. Sugar, and G. Weber, Induction of erythroid differentiation and modulation of gene expression by tiazofurin in K562 leukemia cells, Proc. Natl. Acad. Sci. USA 85: in press (1988).

INCORPORATION OF 5-FLUOROURACIL INTO NUCLEOTIDE SUGARS AND THE EFFECT ON GLYCOCONJUGATES IN RAT HEPATOMA CELLS AND HEPATOCYTES

Pels Rijcken, W.R., Telleman, F.,*Peters, G.J. and Ferwerda, W.

Dept. Med.Chemistry and *Dept. Oncology, Vrije Universiteit, P.O.Box 7161, NL-1007 MC Amsterdam,The Netherlands

INTRODUCTION

The antimetabolite 5-fluorouracil[1] is widely used for the treatment of patients with various types of cancer, such as breast, colon, head and neck cancer [for review see 1]. The metabolism of 5FU is complex. It is anabolized to FUTP, leading to the incorporation of FUMP into RNA, and to FdUMP, which is a potent inhibitor of thymidylate synthase [for review see 1,2]. These events probably are responsible for the antitumor activity. The catabolism of 5FU to CO_2, NH_3 and α-fluoro-β-alanine occurs prodominantly in liver and kidney, but is reduced in tumors [3,4]; this might account at least in part for the chemotherapeutic selectivity. For colon cancer 5FU is the only active drug, although the response dose is lower than 20%. Depending on scheduling, either myeloid or gastrointestinal toxicity is dose-limiting [1]. Therefore,combinations of 5FU with other compounds have been used, either to increase the antitumor activity or to reduce cytotoxicity to normal tissues, e.g. leucovorin can potentiate the antitumor effect using the same 5FU dose [5], while uridine prevents myeloid toxicity [6]. It is still not completely clear, how toxicity of 5FU is mediated and there are indications that not only the above mentioned events are responsible for the antitumor effect [1]. For a better understanding of the suitability of 5FU and possible complications, more information is needed on its exact mode of action. Heretofore few studies have been published on the effect of 5FU on the synthesis of nucleotide sugars and subsequently the glycoconjugate synthesis. Such effects may be important, because oligosaccharides on the outside of the

---

[1]Abbreviations used: 5FU, 5-fluorouracil; FUMP,FUDP,FUTP, 5-fluoroUMP,-UDP,-UTP; FdUMP, 5-fluoro-2'deoxyUMP; FUDP-Hex, FUDP-hexoses (FUDP-glucose + FUDP-galactose); FUDP-HexNAc, FUDP-*N*-acetylhexosamines (FUDP-*N*-acetylglucosamine + FUDP-*N*-acetyl-galactosamine); FUDP-Glc*A*, FUDP-glucuronic acid; sr*a*, specific radioactivity; PCA, perchloric acid; HPLC, High Performance Liquid Chromatography.

plasmamembrane play an important role in communication of the cell with other cells and the environment [7]. Weckbecker and Keppler [8] investigated the synthesis of FUDP-sugars in AS-30D hepatoma cells and the substrate properties of these compounds for various enzymes *in vitro.* McDowell et al. [9] studied the substrate properties of FUDP-sugars for lipid-linked saccharide formation in chick embryo cells. We observed a high synthesis of FUDP-sugars in one melanoma cell line in contrast to two other melanoma cell lines [10]. Some of our studies on the metabolism of 5FU in tumor cell lines were directed to the relation with thymidylate synthase [11,12] and RNA [10,11,13]. Recently we studied the effect of uridine and cytidine on the synthesis of nucleotide sugars and glycoconjugates in rat hepatocytes [14]. Now we extended our studies to a detailed analysis of the metabolism of 5FU in monolayers of rat hepatoma cells and hepatocytes, and its effect on the incorporation of radioactive sugars into glycoconjugates in relation to the nucleotide sugar synthesis.

MATERIALS AND METHODS

Chemicals

5-Fluoro-[6-^{3}H]uracil (sra 1.1 Ci/mmol), L-[U-^{14}C]leucine (sra 344.0 mCi/mmol), D-[6-^{3}H]glucosamine hydrochloride (sra 30.0 Ci/mmol), L-[6-^{3}H]fucose (sra 86.3 Ci/mmol) and N-acetyl-D-[6-^{3}H(N)]mannosamine (sra 30.0 Ci/mmol) were products of NEN Research Products, Boston, USA. 5-FU was obtained from Sigma Chemical Co, St.Louis, MO, USA. New Born Calf serum was from Gibco BRL, USA. Trichlorotrifluoroethane, tri-*N*-octylamine and perchloric acid were obtained from Merck, Darmstadt, West Germany. Soluene-350 was from Packard, USA. All other chemicals were of the highest quality commercially available.

Cell preparations

Hepatocytes were isolated from adult male Wistar rats (specific-pathogen free, 250 - 300 g body weight, home bred) essentially according to Berry and Friend (15), using retrograde perfusion of the liver. The hepatocytes were cultured in modified Waymouth medium in monolayer as described (16). $2*10^{6}$ cells in 3 ml medium, supplemented with 5% (v/v) heat inactivated Newborn Calf Serum, were plated in collagen precoated dishes (6 cm diameter) and were allowed to settle for 20 h before incubation. Rat hepatoma cells (H35) were maintained in Dulbecco's medium. For each experiment H35 cells were plated in culture dishes (6 cm diameter) and the incubation was started when about $3*10^{6}$ exponentially growing cells were present. The cells were exposed to various concentrations of 5FU (0, 17, 50 and 100 μM), with or without 0.5 mM thymine in the medium, for 4 hours in the presence of 1 μCi of [^{14}C]leucine and either 2 μCi of [^{3}H]5FU or 2-4 μCi of a [^{3}H]sugar.

Extraction of nucleotides

To stop the incubation the medium was expired and the cells were washed twice with 1 ml of icecold phosphate-buffered saline (10 mM sodium phosphate, 0.154 M NaCl), pH 7.4. The medium and the washes were used for the determination of the radioactivity incorporated into secreted (glyco)proteins. The cells were extracted by addition of 1 ml of icecold 0.4 M PCA. Soluble and insoluble material was scraped off the dishes with a teflon "policeman". After centrifugation the pellet was washed twice with 250 µl 0.4 M PCA. The combined supernatant and washes was neutralized and desalted by extraction with two volumes of a mixture of trichlorotrifluoroethane and tri-N-octylamine (4:1, v/v) as described (13,17). After lyophilization of the aqueous phase, the residue was redissolved in bidistilled water and injected onto the HPLC. The acid-insoluble pellet, containing the cellular and matrix proteins and glycoconjugates, was solubilized in 250 µl Soluene 350 for 3 h at 65 °C. After mixing with 12 ml scintillation cocktail the solution was neutralized with 250 µl 2.0 M HCl and counted. Secreted (glyco)proteins were precipitated from the medium in the presence of bovine serum albumin. 50 µl albumin solution (1%, w/v) was added to 500 µl of the medium, mixed, acidified with 100 µl 2.0 M PCA, and left on ice for 15 min. After centrifugation and washing twice with 250 µl 0.4 M PCA, the pellet was treated as described above.

HPLC procedures

HPLC was performed on a Spectroflow 400 solvent delivery system equipped with a Spectroflow 480 injection/valve module from Kratos analytical instruments, New Jersey, USA. The absorbance was detected with a Kratos Spectroflow 757 variable wavelength detector, operating at 280 nm, combined with a Hitachi, D-2000 chromato integrator. The separation of the soluble nucleotides was performed on a Whatman Partisphere SAX cartridge (4.6 x 125 mm, particle size 5 µm). Separation was achieved with gradient elution; the low concentration buffer (A) was 0.005 M KH_2PO_4, pH 4.0, and the high concentration buffer (B) was 0.5 M KH_2PO_4, pH 4.5. A three step linear gradient was started with 100% buffer A and increased with B by 0.3%/min for the first 30 min, by 2.5%/min from 30 to 44 min and by 5%/min from 44 to 55 min until 100% B, which was maintained until the end of the run (total time 70 min). The flow rate was 1.0 ml/min. Before starting a new run the column was equilibrated for 20 min with 100% A. For rechromatography fractions containing the UDP-nucleotide sugars were collected, lyophilized and redissolved in bidistilled water. The samples were rechromatographed on a Lichrosorb RP-18 column (4 x 250 mm, particle size 4 µm, Merck). Starting with 0.1 M KH_2PO_4 pH 4.0 (buffer C) for 5 min, the separation was accomplished with a linear gradient of the same buffer, containing 30% (v/v) methanol (buffer D), with an increase of 4% D/min. The flow rate was 0.5 ml/min. Before injection of the next sample the column was equilibrated for 15 min with buffer C. For measurement of the radioactivity 0.5 min fractions were collected and counted in a Berthold liquid scintillation counter BF 8000 equipped with an automatic external standard for calculation of the dpm. The concentration of the 5FU-derivatives was calculated from the incorporated dpm and the sra of the added 5FU. The concentration of the nucleotides was determined by comparison of the UV-absorbance with standards of known concentrations.

RESULTS AND DISCUSSION

In hepatocytes 5FU is incorporated into nucleotides to a low extent because of the high degrading capacity of the liver cells (3,18). In our experiments the concentration of 5FU-nucleo

Table 1. Concentration of 5FU-nucleotides in rat hepatocytes and H35 hepatoma cells.

Cells were incubated during 4 hours in the presence of various concentrations of unlabeled 5FU containing 2 μCi [^{3}H]5FU, with or without 0.5 mM thymine. The cell extract was analyzed with anion-exchange HPLC and the concentrations were calculated from the recovered dpm per peak. After anion-exchange HPLC, the fractions containing FUDP-HexNAc (uncorrected values given in parentheses) were collected and rechromatographed on a reverse phase column to remove the unknown radioactive compound(s). All values are the mean of two incubations and expressed as pmoles/10^6 cells.

Experimental conditions: 5FU	Thymine	FUDP-HexNAc		FUDP-Hex	FUDP-GlcA + FUDP	FUTP	FUΣP*
Hepatocytes:							
17 μM	-	0.3	(13)	1.6	1.5	3.2	8
	+	3.4	(15)	18	13	25	67
50 μM	-	2.7	(32)	10	8.5	16	44
	+	9.6	(44)	55	42	92	220
100 μM	-	11	(64)	36	33	67	167
	+	18	(88)	106	84	198	441
Hepatoma cells:							
17 μM	-	11		22	11	50	98
50 μM	-	36		84	40	170	336

*. FUΣP is the sum of the soluble 5FU-nucleotides, including FUMP.

tides/10^6 hepatocytes was 8 pmoles in the presence of 17 μM 5FU after 4 hours, while 27 nmoles 5FU/10^6 cells were degraded. With higher concentrations up to 100 μM, the concentration of 5FU-nucleotides increased to 167 pmoles (Table 1, column 5) and the degradation was 106 nmoles/10^6 cells. In the presence of thymine the catabolism of 5FU was inhibited by 90%. This resulted in a substantial increase of the synthesis of 5FU-nucleotides, especially at lower 5FU-concentrations (Table 1). In H35 hepatoma cells the concentration of 5FU-nucleotides was approximately 10 times higher. Because of the much lower degrading capacity of these cells thymine was not included in the medium. With thymine in the medium the concentration of FU-nucleotides in hepatocytes was enhanced except that of FUDP-HexNAc (Table 1, column 1, values in parentheses). Therefore w decided to check the purity of the FUDP-HexNAc peak by reverse phase chromatography (see Materials and Methods). This could easily be done without desalting of the collected and concentrated fractions. Fig.1 shows the chromatograms for the hepatocytes and H35 hepatoma cells after reverse phase chromato graphy, with in the inset the profiles of the collected peaks after Partishere-Sax passage. The hepatocytes contained an unknown radioactive component (peak 3, left figure) with a retention time of 10 min in this system. In the presence of thymine, this component accounted for 75-80% of the radioactivity of the FUDP-HexNAc peak, and in the absence of thymine for 98%, 91% and 83% of the radioactivity for 17, 50 and 100 μM 5FU respectively. Without thymine a small extra radioactive peak wa present, as a shoulder of peak 3, about 1 min before the top of peak 3 (not shown). In H35 hepatoma cells the unknown radioactive component was absent (Fig.1, right panel). For compariso the FUDP-Hex peak was also collected and rechromatographed, but in these fractions additional radioactive compounds were not found, either in hepatocytes, or in H35 hepatoma cells. In addition to the unknown radioactive component, an unknown UV-absorbing component was also detected (Fig.1, UV-absorbing peak on 10.92 (left) / 10.82 min (right)). Approximately 30% and 10% of the UV-absorbing material was recovered in this peak, for hepatocytes and hepatoma cells, respectively. This material did not show the typical absorption spectrum of uridine derivatives, but the top of the spectrum was below the 200 nm (not shown).

We did not investigate further the nature of the unknown components, but the radioactive derivative might be 5FU-glucuronide, as reported by Sommadossi et al. (18), although its retention time on Partishere-Sax is unexpected. In addition, even at 17 µM 5FU, our component could already be detected, while thymine did not enhance its concentration. Sommadossi et al. (18) did not find any 5FU-glucuronide after incubation at 30 µM 5FU in the absence of thymine. However, they incubated hepatocytes in suspension, in a relatively poor medium, immediately after preparation, while our experiments were

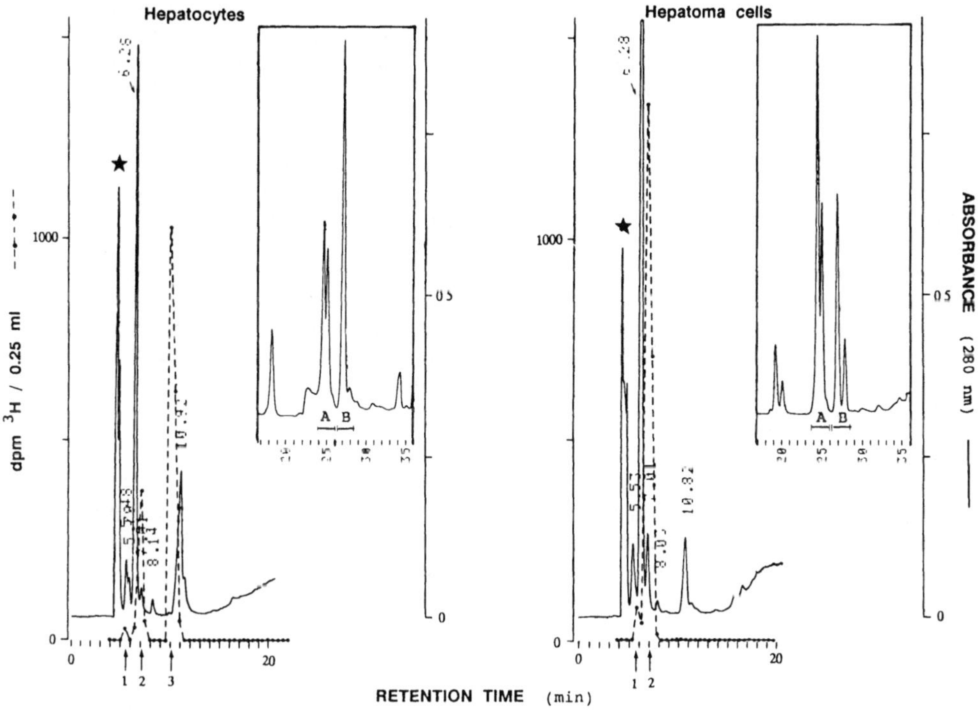

<u>Fig.1. Reversed phase HPLC of the (F)UDP-HexNAc containing peaks collected after anion-exchange HPLC of soluble nucleotides extracted from rat hepatocytes and hepatoma cells.</u>
Soluble nucleotides extracted from hepatocytes (left panel) or hepatoma cells (right panel) after incubation with [^{3}H]5FU were separated on Partisphere-SAX. Fractions containing (F)UDP-HexNAc (A, inset) or (F)UDP-Hex (B, inset) were collected and counted for radioactivity. The radioactive peaks were identified as: 1, FUDP (retention time 5.8 min); 2, FUDP-GlcNAc and FUDP-GalNAc (7.0 min); 3, unknown (10.1 min). The UV-absorbing peaks eluting at 5.5, 6.28 and 8.1 min were identified as UDP, UDP-GlcNAc/GalNAc and UMP respectively. The UV-absorbing peak at the void volume (*) originated from phosphate. The peak at 10.9 (left)/10.8 min (right) did not originate from nucleotide-derivatives as could be concluded from the absorption spectrum (not shown).

Table 2. Influence of 5-FU on the incorporation of sugars into glycoconjugates of H35 hepatoma cells.
Cells were incubated with [^{14}C]leucine and a [^{3}H]sugar during 4 h, in the presence of various concentrations of 5FU. All values are the mean of two incubations and expressed as ^{3}H/^{14}C ratios.

sugar	secreted (glyco)-proteins			cellular + matrix proteins and glycoconjugates		
5FU (µM):	0	17	50	0	17	50
exp. A:						
N-acetyl-[^{3}H]mannosamine	6.3	6.0	7.6	0.32	0.28	0.27
[^{3}H]fucose	6.8	6.6	7.4	0.34	0.38	0.41
exp. B *:						
[^{3}H]glucosamine	5.6	7.5	6.2	0.43	0.41	0.42
[^{3}H]fucose	7.3	8.9	7.9	0.47	0.53	0.62

* preincubated with 5FU for 2 h.

performed with monolayers in Waymouth medium, serum-supplemented, after a long preincubation. We observed considerable differences in 5FU-metabolism between hepatocytes in suspension and as monolayer (Peters et al., unpublished results). This might also explain why Sommadossi et al. did not find FU-anabolic compounds.

The incorporation of [^{3}H]sugars into glycoconjugates of H35 hepatoma cells relative to [^{14}C]leucine incorporation, is shown in Table 2. The effects of 5FU are small, the maximal difference is a 30% increase of [^{3}H]fucose incorporation in cellular + matrix glycoconjugates (exp. B). In hepatocytes the differences were even smaller (results not shown). It should be emphasized that analysis of the nucleotide sugars, as immediate precursors of oligosaccharide synthesis, is necessary for the interpretation of sugar incorporation. We therefore measured the radioactivity incorporated into the sugar moiety of the nucleotide sugars. At 50 µM 5FU in the medium the radioactivity in the (F)UDP-[^{3}H]HexNAc peak increased to 160%, compared with the control, accompanied by an increase in UV-absorption to 175%. We did not determine the sra's of UDP-HexNAc and of FUDP-HexNAc separately, but these might be different. Although 5FU-derivatives are substrates for various enzymes (8), no information is available about the utilization of FUDP-HexNAc by N-acetylhexosaminyltransferases, and about the passage of FUDP-HexNAc through ER membranes. Membrane passage is a prerequisite for the incorporation of most HexNAc residues into oligosaccharides. Nevertheless, from the above mentioned increase in radioactivity and UV-absorbtion, and from the observation that the concentrations of normal nucleotides were not noticeable affected by 5FU, we concluded that the specific radioactivity of the HexNAc-donor had not been changed in the presence of 5FU. Therefore, the results presented in Table 2 for the incorporation of [^{3}H]glucosamine do not need correction.

The incorporation of [^{3}H]fucose was used as a pyrimidine-independent reference for glycoconjugate synthesis, with the purine nucleotide GDP-fucose as sugar donor. A 30% increase in the incorporation was observed in cellular and matrix glycoconjugates, but a similar increase in radioactivity of the GDP-fucose peak was found. Determination of the exact sra of GDP-fucose was not possible, because its concentration was below the detectable UV-level. We think that the enhancement of [^{3}H]fucose incorporation is the result of the increase of GDP-[^{3}H]fucose radioactivity. Combination of GDP-[^{3}H]fucose values with the incorporation of fucose into secreted glycoproteins do not result in a measurable effect of 5FU-metabolism.

N-Acetylmannosamine was used as a specific precursor of sialic acid resulting in labeled CMP-neuraminate as sugar donor. It appeared that N-Acetylmannosamine was metabolized to a large extent in H35 hepatoma cells, in contrast to hepatocytes. All ribose-containing nucleotides contained radioactivity. Therefore, the radioactivity incorporated in glycoconjugates is the result of the incorporation of various radioactive sugars and is dependent on many metabolic processes.

The results of these analysis of nucleotide sugars in combination with the small variations observed in incorporation of the radioactive sugars into glycoconjugates lead us to the conclusion that a 4-6 h incubation with 5FU does not result in a measurable alteration of the overall glycosylation in hepatoma cells.

Acknowledgements

We thank E.Laurensse for his skillful technical assistance. G.J.Peters is a recipient of a senior research fellowship of the Royal Netherlands Academy of Arts and Sciences.

REFERENCES

1. Pinedo,H.M. & Peters,G.J. (1988) 5-Fluorouracil: biochemistry and pharmacology. J.Clin.Oncol. in press
2. Heidelberger,C., Danenberg,P.V. & Moran,R.G. (1983) "Fluorinated pyrimidines and their nucleosides" in Advances in Enzymology, Vol. 54:57 (Meister, A., ed.) John Wiley & Sons, New York.
3. Ho.D.H., Townsend,L., Luna,M.A. & Bodey,G.P. (1986) Distribution and inhibition of dihydrouracil dehydrogenase activities in human tissues using 5-fluorouracil as a substrate. Anticancer Res. 6:782
4. Weber,G. (1985) Biochemical Strategy of cancer cells and the design of chemotherapy: GHA Glowes memorial lecture. Cancer Res. 43:3466
5. Nadal,J.C., Van Groeningen,C.J., Pinedo,H.M. & Peters,G.J. (1988) In vivo potentiation of 5-fluorouracil by leucovorin in murine colon carcinoma. Biomed.Pharmacother., in press
6. Peters,G.J., Van Dijk,J., Laurensse,E., Van Groeningen,C.J., Lankelma,J., Leyva,A., Nadal,J.C. & Pinedo,H.M. (1988) In vitro biochemical and in vivo biological studies of the uridine "rescue" of 5-fluorouracil. Brit.J.Cancer 57:259
7. Sharon,N. (1984) Glycoproteins. Trends Biochem.Sci. 9:198
8. Weckbecker,G. & Keppler,D.O.R. (1984) Substrate properties of 5-fluorouridine diphospho sugars detected in hepatoma cells. Biochem.Pharmacol. 33:2291
9. McDowell, W., Weckbecker,G. & Schwarz,R.T. (1986) Uridine sugar nucleotides modified in their nucleoside moieties and their effect on lipid-linked saccharide formation in vitro, Bioscience Rep. 6:435
10. Peters,G.J., Laurensse,E., Lankelma,J., Leyva,A. & Pinedo,

H.M. (1984) Separation of several 5-fluorouracil metabol ites in various melanoma cell lines. Evidence for the synthesis of 5-fluorouracil-nucleotide sugars. Eur.J.Cancer Clin.Oncol. 20:1425
11. Peters,G.J., Laurensse,E., Leyva,A. & Pinedo,H.M. (1987) Purine nucleosides as cell-specific modulators of 5-fluorouracil metabolism and cytotoxicity. Eur.J.Cancer Clin.Oncol. 23:1869
12. Peters,G.J., Laurensse,E., Van Groeningen,C.J., Meijer,S. & Pinedo,H.M. (1988) In vitro and in vivo inhibition of thymidylate synthase of human colon cancer by 5-fluorouracil. These proceedings.
13. Peters,G.J., Laurensse,E., Leyva,A., Lankelma,J. & Pinedo, H.M. (1986) Sensitivity of human, murine and rat cells to 5-fluorouracil and 5`deoxy-5-fluorouridine in relation to drug metabolizing enzymes. Cancer Res. 46:20
14. Pels Rijcken W.R., Blok,C.M. & Ferwerda,W. (1987) Effect of alteration in pyrimidine nucleotide metabolism on the biosynthesis of sugar nucleotides in rat hepatocytes. Klin. Wochenschrift 65, suppl.X:38
15. Berry,M.N. & Friend,D.S. (1969) High yield preparation of isolated rat liver parenchymal cells. J.Cell Biol. 43:506
16. Van Dijk,W., Boers,W., Sala,M., Lasthuis,A-M. & Mookerjea,S. (1986) Activity and secretion of sialyltransferase in primary monolayer cultures of rat hepatocytes cultured with and without dexamethasone. Biochem.Cell Biol. 64:79
17. Khym,J.X. (1975) An analytical system for rapid separation of tissue nucleotides at low pressures on conventionel anion exchangers. Clin.Chem. 21:1245
18. Sommadossi,J-P., Gewirtz,D.A., Cross,D.S., Goldman,I.D., Cano,J-P. & Diasio,R.B. (1985) Modulation of 5-fluorouracil catabolism in isolated rat hepatocytes with enhancement of 5-fluorouracil glucuronide formation. Cancer Res. 45:116

TRANSPORT AND INTRACELLULAR METABOLISM OF FLUORINATED PYRIMIDINES IN CULTURED CELL LINES

Michihiko Uchida, Dah Hsi W.Ho*, Ken-ichi Kamiya, Teruo Yoshimura, Kin-ya Sasaki, Hiroshi Tsutani and Toru Nakamura

The First Department of Internal Medicine, Fukui Medical School.23 Shimo-aizuki, Yoshida-gun, Matsuoko, Fukui 910-11 Japan *Department of Chem. Res. Box 52, The University of Texas System Cancer Center, M.D.Anderson Hospital and Tumor Institute,Texas Medical Center,6723 Bertner Avenue,Houston, Texas,77030 USA

INTRODUCTION

5-fluorouracil (FU) is a potent anticancer drug and used widely for the treatment of solid cancer. Extensive clinical trials have been performed on this study by many researchers, and results showed that some tumors were responsive to the drug. They included the breast cancer (response rate: 33%), colon and rectum cancer (response rate: 16%) and cervical cancer (response rate: 29%)(1). Although efforts were made to increase the response rate (2,3,4,5), the results have not been satisfactory to date.

FU displays the anticancer activity following the conversion to FU nucleotides, 5-fluorouridine-5'-monophosphate (FUMP) and 5-fluoro-2'-deoxyuridine-5'-monophosphate (FdUMP). FUMP is incorporated into the various RNA species by which some RNA functions such as translation by messenger RNA and the post-transcriptional methylation of ribosomal RNA were affected. FdUMP is a preferential substrate for TMP synthetase and forms ternary complex with 5,10-methylene tetrahydrofolate, leading to TMP depletion in the cells. Several pathways have been suggested by many biochemical investigations, by which FU can be activated: Orotic acid phosphoribosyl transferase catalyzes the interaction of FU and PRPP to form FUMP. Another pathway involves conversion from FU to FUMP in two steps in which fluorouridine (FUR) is synthesized by interaction of uridine-deoxyuridine phosphorylase and FU with ribose-1-phosphate, then converted to FUMP via action of uridine kinase. FdUMP can be formed from FUMP through the serial conversion of FUDP and FdUDP. There is also an alternate pathway for the conversion of FU to FdUMP: Thymidine phosphorylase is capable of catalyzing the formation of 5-fluoro-2'-deoxyuridine (FUdR) from FU utilizing deoxyribose-1-phosphate and thus thymidine kinase completes the conversion of FU to FdUMP. The present study was carried out in an attemt to investigate the transport mode and intracellular conversion to FU nucleotides of fluorinated pyrimidines in cultured murine leukemia cell and three cultured human cancer cell lines to provide some clues for more clinical effectiveness.

MATERIALS AND METHODS

Cells

Cell lines, L-1210, HL-60, CCRF-CEM have been maintained by serial culture with RPMI 1640 medium containing 10% FBS in our laboratory. The cells were obtained at their exponential growth phase. HeLa-S3 cells were obtained from the ascitic fluid of BALB/c-nu/nu mice to which 5×10^7 cells were intraperitoneally inoculated 11 days before. Human RBCs were obtained from a healthy man by vein puncture and prepared by washing twice with a cold 0.9% NaCl solution.

Chemicals

5-[^{3}H]-fluorouracil (14.3 Ci/mmol), 5-[^{3}H]-fluorouridine (16.4 Ci/mmol) and 5-[^{3}H]-fluorodeoxyuridine (14.3 Ci/mmol) were purchased from New England Nuclear (Boston, Mass). 5-fluorouracil and 5-fluorodeoxyuridine were obtained from Nakarai Chem. LTD. (Japan). 5-fluorouridine was obtained from Fluka AG. Chem. Fabrik (Buchs, Switzerland) and FdUMP from Sigma Chem. Co. (St. Louis, Mo.). RPMI 1640 medium and fetal bovine serum were purchased from GIBCO Laboratories (Santa Clara, Calif.). HB 101 medium was purchased from Hana Biologics, Inc. (Berkeley, Calif.). Polygram sil TLC plastic sheet was obtained from Macherey-Nagel (Postfach, Germany). All other reagents were of analytical grade.

Transport Assay

The leukemia cells were suspended in the HB 101 medium at the concentration of up to 2×10^7 cells/ml and incubated at 37°C in the presence of labeled fluorinated compound. After the designated time, 1 ml of cell suspension was transferred to a test tube containing 0.5 ml of silicone oil (specific gravity: 1.034), followed by centrifugation at 12,000Xg for 7 seconds. The supernatant and silicone oil were aspirated and the cell pellet was then transferred to a counting vial. After 2 ml of NCS tissue solubilizer (Amersham Corp. Arlington Heights, IL.) was added, the mixture was kept at room temperature overnight to allow the cell to lyse. The radioactivity of the mixture was measured by counting in toluene-triton based scintillator. The same procedure was used for assaying RBC except that the concentration was set at 15×10^7/ml and the cells were lysed by freezing and thawing twice before counting.

Intracellular Metabolism

After incubation at 37°C, 1 ml of cell suspension was transferred to the centrifuge tube containing 0.5 ml of silicone oil followed by centrifugation at 12,000Xg for 7 seconds. The supernatant was aspirated and the test tube wall was washing with water. Silicone oil and water were aspirated and then 0.5 ml of 0.3M TCA was added. The mixture was vortexed and left for standing at 4°C for 30 minutes. Following centrifugation at 15,600Xg for 20 seconds, 0.3 ml of acid-soluble extract obtained was transferred to a new centrifuge tube and mixed with 0.6 ml of cold freon containing 0.5M tri-n-octylamine. The mixture was centrifuged again at 15,600Xg for 20 seconds. 200 µl of supernatant was applied onto a thin layer of silica gel and developed with ethylacetate, formic acid and water (60:5:35, v/v) along with authentic markers, FdUMP, FUR, FUdR and FU.

RESULT

FU, FUR and FUdR Transport by Human RBC

The incorporation of FU into human RBC increased as the extracellular concentration of FU increased from 1 μM to 100 μM. The transport profiles of the 3 compounds were almost the same with the radioactivity showing the peak after 5-minute incubation and gradually dropping for 30 minutes. By incubating human RBC with 10 μM of each compound for 1 minute, FU, FUR and FUdR were incorporated into 15×10^7 RBCs by 212 pmol, 211 pmol and 186 pmol, respectively, and they were not significantly different from each other. All were not converted metabolically to any other forms in RBCs.

Table 1

	p mol/2×10^7 cells		
	FU	FUR	FUdR
L-1210	292	1700	180
HL-60	17	6470	142
CCRF-CEM	184	1190	218
HeLa-S3	67	148	130

FU, FUR and FUdR Transport by Cancer Cells

The incorporation profiles of the three compounds in cancer cells were different from those in RBCs. When 10 μM of the compounds were added, their intracellular radioactivity increased for up to 15 minutes. Table 1 shows the transport amounts after 15-minute incubation.

The result showed that FUR was transported in greater amounts that FU and FUdR in L-1210, HL-60 and CCRF-CEM cells but not in HeLa-S3 cells.

Intracellular Metabolism

The incorporated compounds and intracellular metabolites were identified by the way their radioactive spots coincided with each authentic marker on the TLC plate. The intracellular content was calculated by the following formula;

$$\frac{\text{total transport amount} \times \text{radioactivity (dpm) of each spot}}{\text{total radioactivity (dpm) of applied sample}}$$

In L-1210 and HL-60 cells, a little or no amount of FU and FUdR were converted to FU nucleotide. In contrast, most FUR incorported into the cells was converted to FU nucleotide effectively. In CCRF-CEM cells, FUR and FUdR were converted to FU and this conversion was not observed in the above 2 cell lines. FUR was converted to FU nucleotide in a large amount. HeLa-S3 cells also showed the conversion of FUR and FUdR to FU with little production of FU nucleotide. Tabel 2 summarizes the data.

DISCUSSION

It is of the great importance to know which pathway of FU to FU nucleotide is predominant in the cancer cells. We designed an experiment in order to provide some clues to this question. First, the transports of FU, FUR and FUdR by human RBCs were measured at 1-minute incubation.

It was supposed that the apparent initial velocity would by given in this system, since all the compounds are metabolically inactive in human RBCs and therefore the velocity is not affected by intracellular metabolism. The result showed that the incorporations of FU, FUR and FUdR into human RBCs were not different each other. On the other

Table 2

cells	product	compound added (10μM) 5-FU	5-FUR	5-FUdR
L-1210	5-FU	127.9(a)	N.D.(b)	N.D.
	5-FUR	N.D.	118.0	N.D.
	5-FUdR	N.D.	N.D.	96.6
	FU nucleotide	27.4	1966.3	5.1
HL-60	5-FU	12.9	N.D.	N.D.
	5-FUR	N.D.	N.D.	N.D.
	5-FUdR	N.D.	N.D.	76.2
	FU nucleotide	N.D.	6412.2	52.0
CCRF-CEM	5-FU	20.9	194.9	101.7
	5-FUR	N.D.	N.D.	N.D.
	5-FUdR	N.D.	N.D.	16.6
	FU nucleotide	24.8	988.6	75.6
HeLa-S3	5-FU	19.2	43.2	71.2
	5-FUR	N.D.	N.D.	N.D.
	5-FUdR	N.D.	N.D.	23.9
	FU nucleotide	20.4	43.2	23.2

(a) Product amount was expressed as pmol/15min/2X10^7 cells.
(b) N.D.: not detected

hand, the incorporation of FUR was higher than those of FU and FUdR in L-1210, HL-60 and CCRF-CEM cells but not in HeLa-S3 cells. The increased incorporation of FUR was apparently related to the production of a large amount of FU nucleotide, since FU and FUdR were not well converted in any cell line tested and a little amount of FU nucleotide was produced from FUR in HeLa-S3 cells although the apparent initial velocity of transport of FUR was not different from FU and FUdR. This showed that the conversion of FUR to FU nucleotide contributed to the intracellular retention of the compound. The intracellular metabolisms of FU, FUR and FUdR vary with the cells tested. FU and FUdR showed a little conversion to FU nucleotide in L-1210 cells. On the other hand, FUR was effectively converted to FU nucleotide without allowing the presence of FU inbetween, since little FU could be converted to FU nucleotide in this cell line. In HL-60 cells, FU was not converted to FU nucleotide and a part of FUdR was directly converted to FU nucleotide. In contrast, all the FUR incorporated was converted to FU nucleotide. In CCRF-CEM and HeLa-S3 cells, a part of FUdR seemed to be converted in two steps; that is, it was once converted to FU, and then to FU nucleotide. Although FUR and FUdR may be cleaved to FU from which FU nucleotide can be produced by the action of OPRTase, the two-step conversion theory can not be confirmed since, the pathway of direct conversion to FU nucleotide was not inhibited in this system and our chromatography system could not separate FdUMP from FUMP. The main pathway of FU nucleotide production was the direct conversion from FUR in CCRF-CEM cells since the amount of FU nucleotide was relatively higher than that expected in the pathway of conversion from FU. FUR or FUdR production from FU was not observed in any cell line.

FU is the substrate for dihydrouracil dehydrogenase which catalyzes the formation of 5-fluorodihydrouracil and 5-fluoro-ureidopropionic acid from FU (6). FU might be cleaved by this enzyme in cells, resulting in apparently decreased incorporation of FU. In case of FUR and FUdR, cleavage to FU is, however, needed before it is the substrate for this enzyme. The FU content was higher in CCRF-CEM cells treated with FUR than in those treated with FU and FUdR so that the difference in incorporation between FUR and FUdR could not be explained by this mechanism.

Keefer et al reported that there was a temporal change in uridine kinase activity over ten generations in the Novikoff Hepatoma cells (7). Dihydrouracil dehydrogenase activity also decreased in accordance with the increase in hepatoma growth rate (8). The low conversion rate of 5-FUR to FU nucleotide in HeLa-S3 cells might be a temporal phenomenon and due to the change in enzyme activity caused by different growth rate because only HeLa-S3 cells were maintained in the ascitic fluid of mice.

The varied fluoropyrimidine metabolism in the different human cancer cells should be emphasized for cancer therapy in the clinical setting.

ACKNOWLEDGMENTS

This investigation was supported by Grant-in-Aid for Scientific Research from the Ministry of Education, Science and Culture, and the Ministry of Health and Welfare. We wish to thank Miss Toshie Kawabata and Miss Tomoko Kanoh for their assistance in editing our manuscript.

REFERENCES

(1) Ansfield,F.J., Schroeder,J.M., and Curreri,A.R. Five years clinical experience with 5-fluorouracil. J.A.M.A., 181: 295-299 (1962).

(2) Kanzawa,F., Hoshi,A., and Kuretani,K. Improvement of therapeutic effect of 5-fluorouracil by orotic acid. J.Pharm.Dyn., 2: 257-259 (1979).

(3) Byfield,J.E., Barone,R., Mendelsohn,J., Frankel,S., Quinol,L., Sharp,T., and Seagren,S. Infusional 5-fluorouracil and x-ray therapy for non-resectable esophageal cancer. Cancer, 45: 703-708 (1980).

(4) Speyer,J.L., Collins,J.M., Dedrick,R.L., Brennan,M.F., Buckpitt,A.R., Londer,H., Devita,Jr.,V.T., and Myers,C.E. Phase I and pharmacological studies of 5-fluorouracil administered intraperitoneally. Cancer Res., 40: 567-572 (1980).

(5) Gewirtz,A.M., and Canman,E. Preliminary Report on the efficacy of sequential methotrexate and 5-fluorouracil in advanced breast cancer. Cancer, 47: 2552-2555 (1981).

(6) Ho,D.H., Townsend,L., Luna,M.A., and Bodey,G.P. Distribution and inhibition of dihydrouracil dehydrogenase activities in human tissues using 5-fluorouracil as a substrate. Anticancer Res., 6: 781-784 (1986).

(7) Keefer,R.C., McNamara,D.J., Schumm,D.E., Billmire,D.F., and Webb,T.E. Early temporal changes in the uridine kinase isozyme profile of the Novikoff hepatoma in response to 5-azacytidine treatment. Biochem. Pharmacol. 24: 1287-1290 (1975).
(8) Queener,S.F., Morris,H.P., and Weber,G. Dihydrouracil dehydrogenase activity in normal, differentiating, and regenerating liver and in hepatomas. Cancer Res., 31: 1004-1009 (1971).

FLUOROPYRIMIDINE METABOLISM IN HUMAN HEAD AND NECK CANCER XENOGRAFTS AND MURINE COLON TUMORS

Emile J. Laurensse, Boudewijn J.M. Braakhuis*, Herbert M. Pinedo and Godefridus J. Peters#

Departments of Oncology and *Otolaryngology
Free University Hospital
PO Box 7057, 1007 MB Amsterdam, the Netherlands

INTRODUCTION

Fluoropyrimidines are widely used for the treatment of solid tumors, such as breast-, colorectal- and head & neck cancer [1-3]. A better therapeutic efficacy than for 5-fluorouracil (5FU) has been claimed for the analog 5'-deoxy-5-fluorouridine [1-6] (5'dFUR, doxifluridine) which cannot be converted directly to the nucleotide level due to the presence of a 5-deoxy-ribose moiety and requires conversion to 5FU (Fig. 1) for its activation [5,7,8].

The human xenograft model has a potential unique value in screening and selecting new drugs for clinical trials [9]. We have developed a panel of head and neck cancer xenograft (HNX) tumor lines in order to select compounds for Phase II trials [10,11]. Table 1 summarizes the sensitivities of several HNX lines and 2 murine colon tumor lines in mice for 5FU and 5'dFUR. Up to now no biochemical evaluation in relation to sensitivity to fluoropyrimidines of HNX lines has been performed.

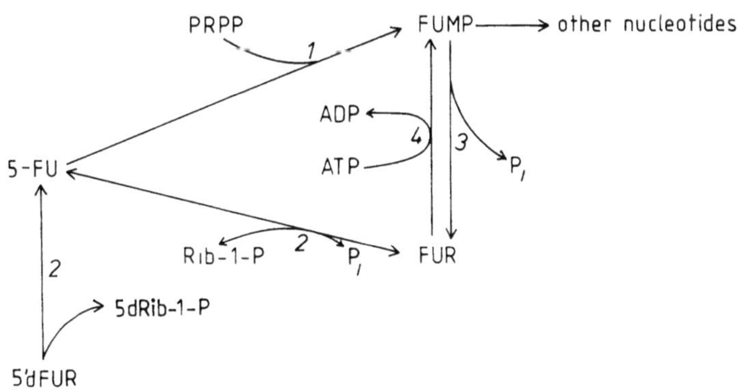

Fig. 1. Metabolism of 5FU and 5'dFUR; The enzymes catalyzing these reactions are: 1, orotate phosphoribosyltransferase (OPRT); 2, pyrimidine nucleoside phosphorylases; 3, 5'nucleotidase and phosphatases; 4, uridine kinase. PRPP, 5-phosphoribosyl-1-pyrophosphate; 5dRib-1-P, 5-deoxy-Rib-1-P; FUR, fluorouridine.

Table 1. SENSITIVITY OF 6 TUMOR LINES TO 5FU AND 5'DFUR

Tumor	Origin	Drug (mg/kg)	Dose	Sensitivity
HNX-DU	Hypopharynx	5FU	50	-
		5'dFUR	1000	-
HNX-KE	Larynx	5FU	50	-
		5'dFUR	1000	++
HNX-G	Oral cavity	5FU	50	-/+
		5'dFUR	400	-
HNX-E	Skin	5FU	50	-
		5'dFUR	1000	+
Colon 26	Murine colon	5FU	100	-/+
		5'dFUR	2000	+++++
Colon 38	Murine colon	5FU	100	+++++
		5'dFUR	1000	+++++

Drugs were administered i.p. weekly.

The conversion of 5'dFUR to 5FU is catalyzed by a pyrimidine nucleoside phosphorylase. It has been suggested that activation of 5'dFUR to 5FU occurred selectively in tumor cells compared to normal cells [12]. The activity of pyrimidine nucleoside phosphorylase appears to be related to the toxicity of 5'dFUR [5,8,12,13], however, a strict correlation has not been demonstrated [5,14]. We postulated that although sufficient conversion of 5'dFUR to 5FU is essential for a cell to be sensitive to 5'dFUR, this is not the only critical factor which determines the activity of 5'dFUR [14]. Further activation of the 5FU formed from 5'dFUR appeared to be essential. A cell line in which the direct conversion of 5FU to FUMP plays an important role, appeared to be very sensitive to 5'dFUR [14]. In this cell line 5'dFUR was also able to decrease the levels of PRPP [15]. 5FU can also be converted to FUMP via FUR. In the presence of sufficient dRib-1-P as co-substrate, 5FU can be directly converted to 5-fluoro-deoxy-uridine (FUdR). However, the concentration of dRib-1-P is neglectable under physiological concentrations [16,17], but enhanced at administration of e.g. deoxyinosine.

In order to know whether 5FU resistance could be overcome by administration of 5'dFUR we attempted to correlate the sensitivity of several tumor lines to fluoropyrimidines with the biochemical pharmacology of the drugs in each line; four human tumor xenografts and two murine colon carcinomas. In these tumors we measured the activities of enzymes involved in the activation of 5'dFUR and 5FU. The results demonstrated that antitumor activity of 5'dFUR not only depends on the rate of conversion to 5FU but also on the further metabolism of 5FU.

MATERIALS AND METHODS

5FU and 5'dFUR were obtained from Hoffman-La Roche, Mijdrecht, the Netherlands. All other chemicals were of the highest quality commercially available. The head and neck tumor lines were established from tumors from untreated patients (Table 1) and maintained in female B10.LP/Cpb nude mice as described previously [10]. The murine colon tumors Colon 26 and Colon 38 were maintained in female Balb/c mice and C57Bl/6 mice, respectively, as described previously [18]. Tumors (ranging in size between 200-2000 mm^3) were obtained from non-treated mice. Tumors were removed immediately, and stored directly either in liquid nitrogen or at -70^0C. Frozen tissues were pulverized using a micro-dismembrator as described [19]. The powder was suspended in assay buffer (50 mM Tris-1 mM EDTA, pH

7.4), (one gram tissue per 4 ml buffer). After centrifugation the supernatant was immediately used for determination of enzyme activities. Enzyme assays were performed at 37°C. Assays with 5FU as substrate were performed as described previously [14,19]. Products were separated from the substrate 5FU using thin-layer chromatography [20]. The reaction mixture for the pyrimidine nucleoside phosphorylase assay contained 0.3-60 μg protein, 5 mM $MgCl_2$ and the cofactors Rib-1-P or dRib-1-P at 2.5 mM final concentration. For the measurement of FUMP and FdUMP synthesis (via FUR or FUdR, respectively), more protein (200-650 μg) was present in the assay. In order to prevent breakdown of nucleotides by phosphatases we added 15 mM 2-glycerol-phosphate to these assays. ATP and (d)Rib-1-P were present at 2.5 mM final concentration. For the measurement of the direct conversion of 5FU to FUMP catalyzed by OPRT the pentose phosphates were substituted by 2 mM PRPP; 0.6 mM α,β-methylene-ADP was present to inhibit 5'-nucleotidases; 200-650 μg protein was present. The reactions were initiated by addition of radiolabeled 5FU (final concentration 0.27 mM [6-^{14}C]-5FU]. The phosphorolysis of 5'dFUR to 5FU was measured using a recently developed HPLC assay [21]. The assay mixture was performed with the 11,000 g supernatant. The assay mixture contained 5-100 μg protein, 40 mM KH_2PO_4 and was initiated by addition of 5'dFUR (final concentrations 0.25-2 mM).

RESULTS

Pyrimidine nucleoside phosphorylase was assayed using a recently described HPLC assay [21] with 5'dFUR as substrate. A high Km was observed in all tumors varying from 0.6-2 mM. The maximal activity with 5'dFUR varied considerably between the several tumor lines (Table 2). The activity in the Colon 26 was about 50 times higher than in HNX-DU and HNX-KE. The activity in the other two HNX tumor was intermediate.

Pyrimidine nucleoside phosphorylases also catalyzes the further anabolism of the product 5FU with Rib-1-P or dRib-1-P as co-substrates (Fig. 2). The activity with dRib-1-P was higher than with Rib-1-P in all tumors, in the HNX tumors at least 10-fold, in both colon tumors 3-fold. With Rib-1-P the highest activity was found in the Colon 26 and the lowest activity in the HNX-KE and Colon 38. With dRib-1-P the pattern was different, the highest activity was present in Colon 26 and the lowest in Colon 38. However, the activities in the HNX tumors were all higher than in Colon 38 and the difference of HNX-E with Colon 26 was less than with Rib-1-P as co-substrate.

TABLE 2. ACTIVITY OF PYRIMIDINE NUCLEOSIDE PHOSPHORYLASE WITH 5'dFUR AS SUBSTRATE.

Tumor line	Vmax (nmol/hr per mg protein)	Km (mM)
HNX-E	2300 ± 460	1.16 ± 0.30
HNX-DU	273 ± 28	0.80 ± 0.11
HNX-KE	366 ± 54	1.16 ± 0.09
HNX-G	2565 ± 293	1.02 ± 0.34
Colon 26	9892 ± 2087	1.89 ± 0.54
Colon 38	671 ± 238	2.19 ± 1.56

Values are means ± SE of 3-5 different tumors.

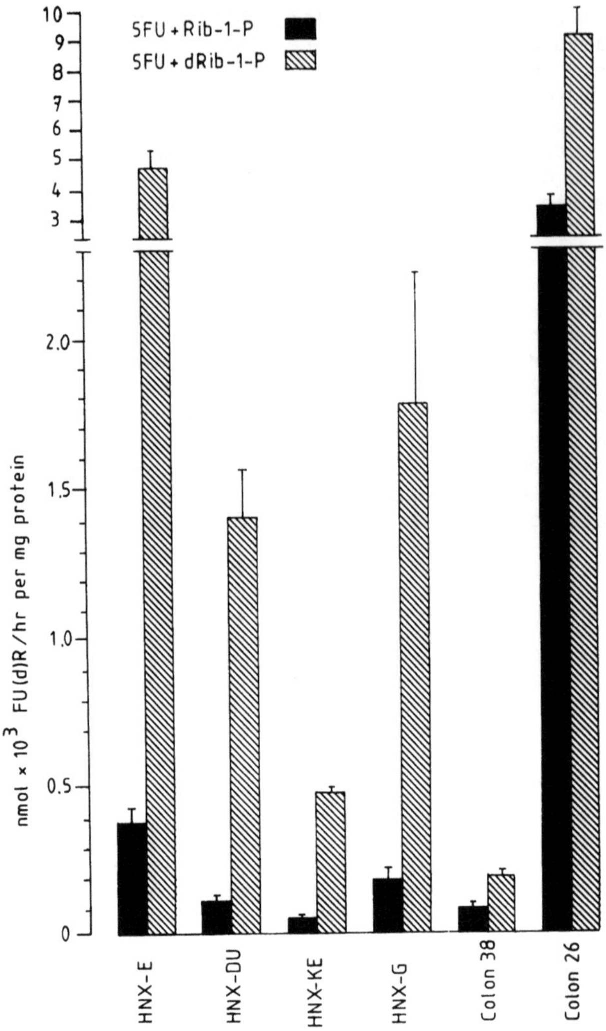

Fig. 2. Activity of pyrimidine nucleoside phosphorylase with 5FU as substrate and Rib-1-P or dRib-1-P as co-substrate. Values represent means + SE of 3-5 different tumors.

Conversion of 5FU to either FUMP or FdUMP via FUR or FUdR, respectively, was measured by supplying for the ATP at a physiological concentration in the assay for pyrimidine nucleoside phosphorylase. Under these conditions an estimate of the conversion of 5FU to the nucleotide will be obtained [14]. This conversion was termed "chanelling". The rate of chanelling was measured in the presence of a phosphatase inhibitor to prevent the degradation of newly formed nucleotides. The rate of chanelling of 5FU to FUMP was highest in the Colon 26 and very low in Colon 38 (Fig. 3). The activity in all HNX tumor was at least 10 times higher than in Colon 38, but also less than 5% of that in Colon 26. The rate of chanelling of 5FU to FdUMP showed a completely different pattern. The activity was highest in Colon 26 but the difference with the other tumors was less, the activity in HNX-DU and KE was even 50% of that in Colon 26. The rate of chanelling in the other 3 tumor lines was much lower. The direct conversion of 5FU catalyzed by the pyrimidine de novo enzyme OPRT was highest in Colon 26 and Colon 38 (Fig. 4); for the HNX tumors the activity was highest in the HNX-KE.

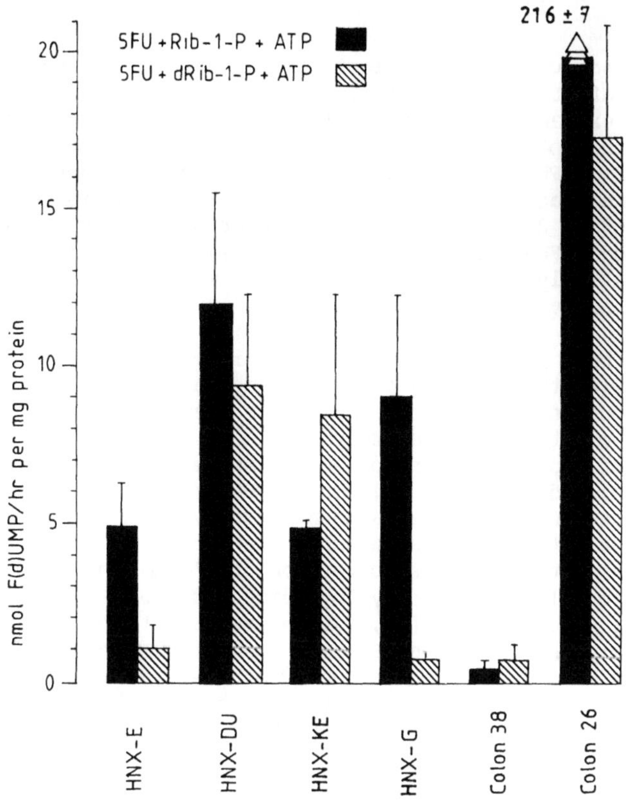

Fig. 3. Synthesis of FUMP and FdUMP from 5FU via FUR or FUdR, respectively. The "channeling" reaction was measured with Rib-1-P or dRib-1-P as co-substrates in the presence of ATP. Values represent means + SE of 3-5 different tumors.

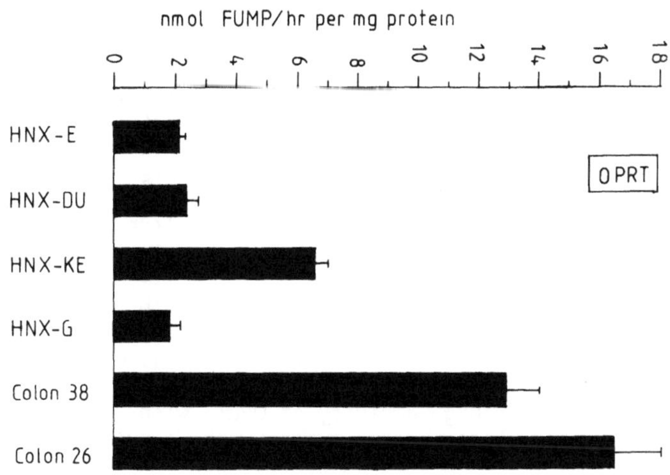

Fig. 4. Activity of OPRT with 5FU as substrate and PRPP as co-substrate. Values represent means + SE of 3-5 different tumors.

DISCUSSION

An attractive way to select active anticancer agents is the use of human xenograft tumor lines [9,10]. The present study tries to give a biochemical explanation for the fact that in human xenografts resistant to 5FU the therapeutic efficacy might be increased by treatment with the 5FU prodrug 5'dFUR. The improved therapeutic efficacy might be related to the rate of activation of 5'dFUR to 5FU which is very high in the sensitive Colon 26, but the HNX-KE tumor and Colon 38, which are also sensitive to 5'dFUR, have rather low activities. All sensitive tumors have a relatively high activity of OPRT, which might play a key role in the further activation of 5U.

It has been postulated that the sensitivity of tumors to 5'dFUR is correlated with the activity of pyrimidine nucleoside phosphorylase [5,8, 12,13] although no strict correlation has been demonstrated [5,14]. This discrepancy might be related to the role of uridine phosphorylase in the activation of 5'dFUR. Uridine phosphorylase might be responsible for the conversion of 5'dFUR to 5FU [8] but 5'dFUR is also a substrate for thymidine phosphorylase [22]. The pattern of activity of pyrimidine nucleoside phosphorylase with 5'dFUR as a substrate correlated with that of 5FU and dRib-1-P as substrates. This conversion of 5FU to FudR and the cleavage of FUdR are mainly catalyzed by thymidine phosphorylase and/or a uridine-deoxyuridine phosphorylase [14,23]. Enzyme kinetics of pyrimidine nucleoside phosphorylase with 5'dFUR as a substrate were comparable to other studies. The Km for 5'dFUR observed by others was about 1 mM [7,24,25]. The Km for 5'dFUR is much higher than that for uridine [26] or thymidine [27] it may be concluded that 5'dFUR has a low affinity for pyrimidine phosphorylases. It might be substrate for the various phosphorylases which have been reported thusfar [23,24].

From the data presented it can be concluded that for this panel of tumors the activity of uridine phosphorylase is not related to the sensitivity of the tumor to 5'dFUR. The Colon 38 is most sensitive to 5'dFUR but the rate of 5'dFUR phosphorolysis is very low. Colon 26, which is also sensitive to 5'dFUR has a very high rate of 5'dFUR phosphorolysis, but the HNX-KE which is also sensitive has low activity, lower than HNX-E and HNX-G. This means that other factors than the rate of phosphorolysis of 5'dFUR play an important role in the sensitivity to 5'dFUR. Activation to FdUMP via FUdR can be excluded because of the low endogenous levels of the co-substrate dRib-1-P. The levels of Rib-1-P are usually high enough for activation of 5FU to FUMP via FUR, but the rate of this pathway was very low in the 5'dFUR sensitive Colon 38, high in the 5'dFUR resistant HNX-DU, and very high in Colon 26. So, this pathway might only be important for Colon 26. However, previously we demonstrated for cell lines that this way of phosphorylation of 5FU formed from 5'dFUR is unlikely [14]. This could be partially due to; a) the presence of the modified pentose phosphate 5-deoxy-ribose-1-phosphate which might interact with the substrate Rib-1-P and b) an interference of 5'dFUR itself with pyrimidine nucleoside phosphorylase. The only alternative for phosphorylation of 5FU will be direct conversion to FUMP. WiDr cells, appeared to be most sensitive to 5'dFUR of the panel of cell lines which were tested [14,15] and had a relatively high activity of OPRT both with the analog substrate 5FU [14] and the natural substrate orotic acid [15]. With the present tumor lines it appears that the most sensitive lines Colon 26, Colon 38 and HNX-KE have a relatively high activity of OPRT with 5FU as a substrate. This means that only those human tumors with a sufficient capacity to convert 5'dFUR to 5FU and subsequently 5FU to nucleotides may be sensitive to 5'dFUR.

It may be that 5FU resistant tumors do not use efficiently the pathway catalyzed by OPRT, for the activation of 5FU, possibly due to a low availability of PRPP or a high activity of pyrimidine nucleoside phosphorylase. Thus inhibition of pyrimidine nucleoside phosphorylase may lead to a more efficient use of the OPRT pathway, and therapy of 5FU resistant tumors might be improved by an efficient use of the OPRT pathway.

Recently we demonstrated that several human colon tumor have a relatively high activity of OPRT compared to adjacent normal mucosal tissue [19]. This might lead to the conclusion that colorectal cancer is an attractive tumor type to be treated with 5'dFUR, although results up to now only showed a limited therapeutic effect. Tumors with a high OPRT activity might also be treated better with a combination of 5FU and a modulator which selectively employs the activation of 5FU via the direct phosphorylation pathway [28].

In conclusion, the enhanced sensitivity to 5'dFUR might not strictly be related to the rate of conversion of 5'dFUR to 5FU. Although a certain amount of activity is essential to convert 5'dFUR to 5FU, further metabolism to nucleotides might be limiting. This might lead to an improved therapeutic efficacy of 5'dFUR compared to 5FU in tumors with a high OPRT activity such as the HNX-KE and the Colon 26. These findings indicate that further studies attempting at reducing the severe clinical toxicity observed with 5'dFUR are warranted.

ACKNOWLEDGEMENTS

This work was supported by the Netherlands Cancer Foundation (Koningin Wilhelmina Fonds) by grants IKA 83-16 and AUKC-VU 82-11, and by Hoffmann-La Roche, Mijdrecht, the Netherlands. We also thank Drs. J. van Dijk, Mrs. M. Bagnay and Mr. E.J. Schoevers for their contributions. Dr. G.J. Peters is a senior research fellow of the Royal Netherlands Academy of Sciences.

REFERENCES

1. Abele, R., Alberto, P., Kaplan, S., & 6 others (1983). J. Clin. Oncol. 1, 750.
2. Abele R., Kaplan E., Grossenbacher, R., Schmidt, H.J. & Cavalli, F. (1984). Eur. J. Cancer Clin Oncol., 20, 333.
3. Alberto, P., Rozencweig, M., Clavel, M. & 6 others (1986). Cancer Chemother. Pharmacol., 16, 78.
4. Bollag W. & Hartmann, H.R. (1980). Eur. J. Cancer, 16, 427.
5. Hartmann, H.R. & Matter, A. (1982). Cancer Res., 42, 2412.
6. Hurteloup, P., Armand, J.P., Cappelaere, P., & 16 others (1986). Cancer Treatm. Rep., 70, 731.
7. Armstrong, R.D. & Diasio, R.B. (1980). Cancer Res., 40, 3333.
8. Ishitsuka, H., Miwa, M., Takemoto., K, Fukuoka, K., Itoga, A., & Maruyama, H.B. (1980). Gann 71, 112.
9. Winograd, B., Boven, E., Lobbezoo, M.W. & Pinedo, H.M. (1987). In Vivo, 1, 1.
10. Braakhuis, B.J.M., Schoevers, E.J., Heinerman, E.C.M., Sneeuwloper, G. & Snow G.B. (1983). Br. J. Cancer, 48, 711.
11. Braakhuis, B.J.M. & Snow, G.B. (1988). ESO monographs, Human xeno grafts in anticancer drug development. (Eds.: Winograd, B., Peckham, M.J. & Pinedo, H.M.). Springer Verlag, Heidelberg, pp. 37-39.
12. Armstrong, R.D. & Diasio, R.B. (1981). Cancer Res., 41, 4891.

13. Armstrong, R.D., Gesmonde, J., Wu, T. & Cadman, E. (1983). Cancer Treatm. Rep., 67, 541.
14. Peters, G.J., Laurensse, E., Leyva, A., Lankelma, J. & Pinedo, H.M. (1986). Cancer Res., 46, 20.
15. Peters, G.J., Laurensse, E., Leyva, A. & Pinedo, H.M. (1985). Int. J. Biochem., 17, 95.
16. Barankiewicz, J. & Henderson, J.F. (1977). Biochem. Med., 17, 45.
17. Peters, G.J., Laurensse, E., Leyva, A. & Pinedo, H.M. (1987). Eur. J. Cancer Clin. Oncol., 23, 1869.
18. Peters, G.J., Van Dijk, J., Nadal, J.C., Van Groeningen, C.J. Lankelma, J. & Pinedo, H.M. (1987). In Vivo, 1, 113.
19. Peters, G.J., Laurensse, E., Leyva, A. & Pinedo, H.M. (1986). Clin. Chim. Acta., 158, 193.
20. Peters, G.J., Laurensse, E., Lankelma, J., Leyva, A. & Pinedo, H.M. (1984). Eur. J. Cancer Clin. Oncol., 20, 1425.
21. Peters, G.J., Laurensse, E., Leyva, A. & Pinedo H.M. (1987). Klin. Wochenschr., 65 (Suppl X), 18.
22. Siegel, S.A. & Lin, T.-S. (1986). Drugs Future, 11, 961.
23. Woodman, P.W., Saril, A.M. & Heidelberger, C. (1980). Cancer Res., 40, 507.
24. Choong, Y.S. & Lee, S.P. (1985). Clin. Chim. Acta, 149, 175.
25. Miwa, M., Nakamura, J. & Ishitsuka, H. (1981). Gann 72, 965.
26. Leyva, A., Kraal, I., Lankelma, J., Delemarre, J.F.M. & Pinedo, H.M. (1983). AntiCancer Res., 3, 227.
27. Wataya, Y., Santi, D.V. (1981). Anal. Biochem 112, 96.
28. Pinedo, H.M. & Peters, G.J. (1988). J. Clin. Oncol., in press.

REDUCTION OF ARA-C CYTOTOXICITY IN HL 60 CELLS BY ADDITION OF DEOXYCYTIDINE, CYTIDINE OR INCREASED LEVEL OF CYTIDINE DEAMINASE

Johannes Mejer and Børge Thing Mortensen

The Finsen Institute
The Laboratory of the Medical Department
Copenhagen, Denmark

INTRODUCTION

Ara-C is one of the main drugs in the treatment of acute myeloid leukemia. It has a structural resemblance to the natural nucleosides, deoxycytidine and cytidine. The metabolism of Ara-C follow the pathway of deoxycytidine. Necessary for the cytotoxicity is phosphorylation to Ara-CTP initially and rate limiting catalyzed by the enzyme deoxycytidine kinase (1). Ara-CTP is a potent inhibitor of DNA polymerase (2). The inhibition is competetive with respect to dCTP (3). Degradation of Ara-C is catalyzed by cytidine deaminase (4). Deoxycytidine and cytidine share this degradative pathway. Ara-U is only toxic in very high concentrations (5). Previously, we have investigated this initial metabolism of Ara-C in myeloid leukemia cells and in normal cells (6,7). No correlations were found between the activities of deoxycytidine kinase or cytidine deaminase, measured in extracts of the cells, and response to a subsequent treatment including Ara-C. However, we observed that the maximal velocity for Ara-C phosphorylation in myeloid leukemic cells was the double of that for deoxycytidine, while in normal cells the maximal velocity with the two substrates was equal. There was no difference between the ratio of Ara-C versus deoxycytidine degradation in leukemic and normal cells. We suggested, based on this specificity changes of deoxycytidine kinase in leukemic cells in comparison with normal cells, that deoxycytidine could possibly rescue normal cells more efficiently than malignant cells against the cytotoxicity of Ara-C. In the present study we therefore investigated the rescue of Ara-C cytotoxicity.

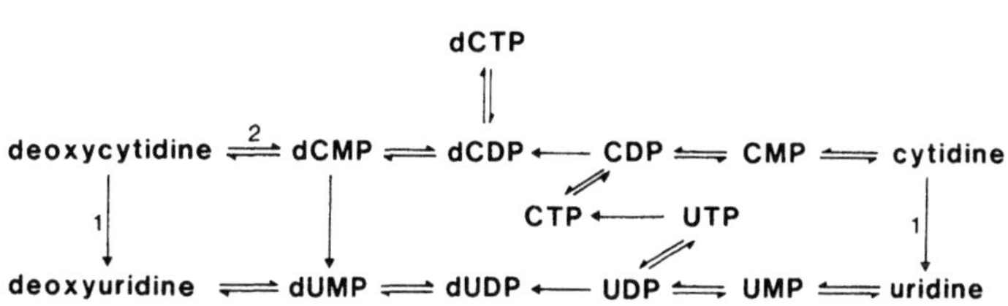

Fig. 1. 1: cytidine deaminase, 2 : deoxycytidine kinase.

METHODS

Reagents: Fine chemicals were obtained from Sigma Chemicals Co. St. Louis, U.S.A. Tetrahydrouridine was obtained from Calb. Chem. Boehringer Corp., La Jolla, Calif. U.S.A.. 1,25 dihydroxy D_3 was a gift from Roche, Basel, Switzerland. 2-^{14}C cytidine was obtained from Amershan, U.K.. 5-^{3}H deocycytidine was obtained from CEA, GIf-sur-Yvette, France.

Cells: our method for the culture of HL 60 cells has been described before (8). The cell number was reduced twice a week to 10^5 cells/ml to keep the cells in exponential growth. Cell counts were performed using a coulter counter (Coulter Electric Ltd., Dunstable, U.K.). The Cell line was regularly tested for mycoplasma infection and discarded if positive.

The influence of 1,25 dihydroxy D_3 on the cytidine deaminase activity was measured after addition of various concentrations of 1,25 dihydroxy D_3.

The combined effect of 1,25 dihydroxy D_3, retinoic acid or tetrahydrouridine and Ara-C was estimated as the change of growth 72 hours after addition of 8 nM Ara-C to the suspension cultures of HL 60 cells grown in multiwell plates (1 ml/well, 2 x 10^5 cells/ml). The Ara-C addition was postponed 24 hours to permit the metablic changes to take place.

The influence of deoxycytidine and cytidine on the Ara-C cytotoxicity was estimated 72 hourts after the concomitantly incubation with Ara-C (5 nM)

Enzymes: Deoxycytidine kinase and cytidine deaminase were measures in a radiochemical assay as previously described (6, 9).

RESULTS

Previously we have observed, that 1,25 dihydroxy D_3 induced an increase in the cytidine deaminase activity of HL 60 cells, whereas another differention inducer, retinoic acid, was without effect.

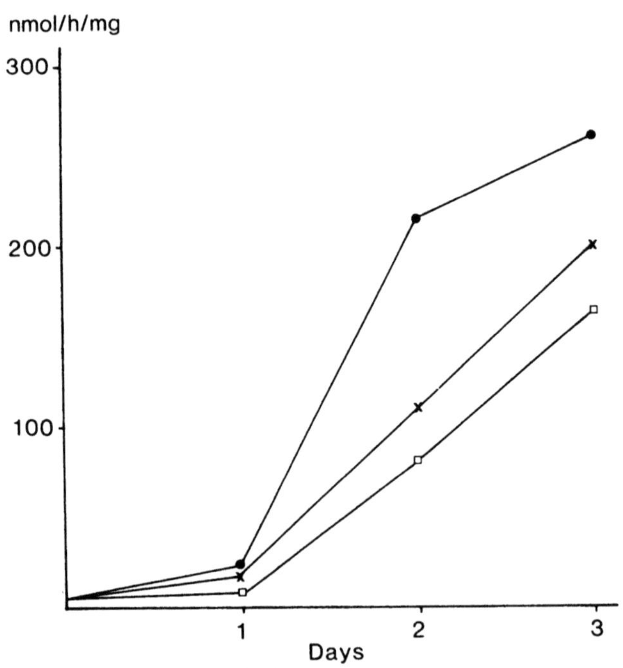

Fig. 2. The activities of cytidine deaminase after addition of 1,25 dihydroxy D_3 to HL 60 cells. 10^{-6} M●, 10^{-7} M×, 10^{-8} M□.

Figure 2 demonstrate that the effect of 1,25 dihydroxy D_3 on the activity of cytidine deaminase in HL 60 cells is dependent on the concentration of 1,25 dihydroxy D_3. It also shows, that even low (10 nM) concentrations of 1,25 dihydroxy D_3 were able to induce increase in the activity of cytidine deaminase and that the effect was postponed 24 hours before it was measurable.

Because Ara-C is a S phase specific drug, and because differentiation arrest the cells in G_0 phase of the cell cycle, the effect of both 1,25 dihydroxy D_3 and retinoic acid on the Ara-C cytotoxicity was investigated. However, flow cytometric evaluation revealed, that uniform and minor changes were the results of the low dosages of both 1,25 dihydroxy D_3 and retinoic acid used.

Table 1. Effect of induction or inhibition of cytidine deaminase on the Ara-C cytotoxicity in HL 60 cells.

The effect of changes in the cytidine deaminase activity on the Ara-C induced inhibition of growth in suspension cultures of HL 60 cells were investigated. After exposure to the mediator of the changes in the cytidine deaminase activity for 24 hours the cells were additionally exposed to Ara-C (8 nM) for 72 hours. Mean $\pm$ SEM of three experiments.

% of control growth	Additional treatment
60 $\pm$ 1	1,25 dihydroxy D_3 (60 nM)
52 $\pm$ 2	0
36 $\pm$ 4	Retinoic acid (20 nM)
24 $\pm$ 1	Tetrahydrouridine (100 μM)

The reduction in the increement induced by Ara-C was less, when the cells were pretreated with 1,25 dihydroxy D_3. Tetrahydrouridine, on the other hand, increased the reduction of the cell counts extered by Ara-C and so did retinoid acid.

Rescue of Ara-C cytotoxicity could possibly also be obtained by addition of nucleosides, which could be converted to dCTP. The effect of cytidine and deoxycytidine was therefore investigated.

Table 2. The effect of natural nucleosides similar to Ara-C on the cytotoxicity of Ara-C in HL 60 cells.

HL 60 cells were grown in suspension culture in the continously presence of 5 nM Ara-C + additional treatments in 72 hours. Mean $\pm$ SEM of three experiments.

% control growth	Additional treatment
44 $\pm$ 2	0
62 $\pm$ 2	Cytidine 100 μM
49 $\pm$ 2	Cytidine 100 μM 1 mM
62 $\pm$ 4	Deoxycytidine 1 μM
32 $\pm$ 4	Deoxycytidine 1 μM + Tetrahydrouridine 1 mM

Cytidine added concomitantly with Ara-C inhibited the cytotoxicity of Ara-C. This effect could be abolished by the further addition of uridine. The same reduction of Ara-C cytotoxicity could be obtained by deoxycytidine at a concentration 100 times less than the concentration of cytidine. This effect could be balanced by the otherwise atoxic tetrahydrouridine.

To secure that the change in deoxycytidine kinase also was present in HL 60 cells as in other leukemic cells (6, 7) the activities with the two substrates, deoxycytidine and Ara-C, were measured in HL 60 cells and bone marrow cells from healthy volunteers.

Table 3. Ara-C and deoxycytidine phosphorylation
Normal bone marrow cells were isolated by density cut separation (density 1.077), and deoxycytidine kinase was measured as previously described (6, 9).

	Ara-C	Deoxycytidine
HL 60 cells V_{max}	5.3 ± 1.0	2.5 ± 0.5
Normal marrow cells V_{max}	1.4 ± 0.2	1.1 ± 0.1

Values are nmol/h/mg. N = 3.

DISCUSSION

The presented results demonstrates that the Ara-C degradating enzyme, cytidine deaminase, has an impact on the cytotoxicity of Ara-C, although we did not find a correlation between the clinical outcome and the activity of cytidine deaminase in the leukemic cells from patients (6). Increasing the activity of cytidine deaminase inhibits cytotoxicity of Ara-C and inhibition of the enzyme increases the cytoxicity (Table 1).

Together with the cytidine deaminase inducer, 1,25 dihydroxy D_3, also cytidine and deoxycytidine were able to inhibit the Ara-C cytotoxicity (Table 2). The necessary concentration of cytidine to excert the same effect as deoxycytidine on the Ara-C cytotoxicity was 100 times that of deoxycytidine. This is in agreement with the observation, that deoxycytidine is more efficiently converted to dCTP than cytidine (10).

Uridine was able to abolish the cytidine rescue on the Ara-C cytotoxicity (Table 2). This means (Fig. 1), that phosporylation of cytidine catalyzed by uridine/cytidine kinase is necessary for the cytidine rescue effect

Although deoxycytidine and Ara-C have in common the degradative pathway inhibition of the degradation with tetrahydrouridine resulted in a reversal of the deoxycytidine prevention of Ara-C cytotoxicity in HL 60 cells (Table 2). The maximal velocity for Ara-C phosphorylation in HL 60 cells was higher than that for deoxycytidine. This is a potential explanation of the reversal effect of tetrahydrouridine on the deoxycytidine rescue of the Ara-C cytotoxicity. An explanation of especial interest, because normal marrow cells exhibited the same activity with the two substrates, Ara-C and deoxycytidine (Table 3). Higher maximal velocity with Ara-C compared to deoxycytidine phosphorylation has also been found in leukemic cells from patients with acute myeloid leukemia (6) chronic myeloid leukemia (7), and in lymphoblasts from patients (11).

The combination of high dose Ara-C and deoxycytidine has recently been reported to exert selective cytotoxicity in acute myloid leukemia cells, while sparing normal myeloid cells (12). The present observations rise the possibility of using lower more clinical acceptable dosages of both Ara-C and deoxycytidine if combined with tetrahydrouridine.

ACKNOWLEDGEMENTS

Supported by grants from Jacob Madsen and Olga Madsen Foundation and from the Lundbeck Foundation.
I.Kobbernagel is thanked for excellent technical assistance.

REFERENCES

1. Momparler, R. L. and Fischer, G.A. Mammalian deoxynucleoside kinases. I. Deoxycytidine kinase: Purification, properties, and kinetic studies with cytosine arabinoside. J. biol. Chem., 243: 4289-4304, 1968.
2. Furth, J. S., and Cohen, S. S. Inhibition of mammalian DNA polymerase by the 5'-triphosphate of 1-B-D-arabinofuranosylcytosine and 5'-triphosphate of 1-β-D-arabinofuranosyladenine. Cancer Res., 23: 2061-2067, 1968.
3. Liliemark, J. O., and Plumket, W. Regulation of 1-β-D-arabinofuranosylcytosine 5'-triphosphate accumulation in human leukemia cells by deoxycytidine 5'-triphosphate. Cancer Res., 46: 1079-1083, 1986.
4. Camiener, G. W., and Smith, C. G. Studies on the enzymatic deamination of cytosine arabinoside - I. Enzyme distribution and species specificity. Biochem. Pharmacol., 14: 1405-1416, 1965.
5. Müller, W. E. G., and Zahn, R. K. Metabolism of 1-β-D-arabinofuranosyluracil in Mouse L5178Y cells. Cancer Res., 39: 1102-1107, 1979.
6. Mejer, J., and Nygaard, P. Cytosine arabinoside phosphorylation and meamination in acute myeloblastic leukemia cells. Leukemia Res., 2: 127-131, 1978.
7. Mejer, J. Enzymatic studies on possible improvement of cytosine arabinoside treatment. Scand. J. clin. Lab. Invest., 42: 401-406, 1982.
8. Mortensen, B.T., Hartmann, N. R., Christensen, I. J., Larsen, J. K., Kristensen, T., Wieslander, S. B., and Nissen, N. I. Synchronization of the human promyelocytic cell line HL60 by thymidine. Cell Tissue Kinct. 19: 351-364, 1986.
9. Mejer, J., and Mortensen, B. T. Changes in the activities of cytidine deaminase during differentiation of HL 60 cells induced by 1,25 dihydroxy D_3. Leukemia Res., 12: 405-409, 1988.
10. Hunting, D. Hordern, J. and Henderson, F. Quantitative analysis of purine and pyrimidine metabolism in chinese hamster ovary cells. Can. J. Biochem., 29: 838-847, 1981.
11. Wiley, J. S., Taupin, J., Jamieson, G. P., Snook, M., Sawyer, W. H., and Finch, L. R. Cytosine arabinoside transport and metabolism in acute leukemias and T cell lymphoblastic leukemia. J. Clin. Invest., 75: 632-642, 1985.
12. Bhalla, K., MacLaughlin, W., Cole, J., Arlin, Z., Baker, M., Graham, G., and Grant, S. Deoxycytidine preferentially protects normal versus leukemic myeloid progenitor cells from cytosine arabinoside-mediated cytotoxicity. Blood, 70: 568-571, 1987.

INTRACELLULAR PHARMACOKINETICS OF N^4-BEHENOYL-1-β-D-ARABINOFURANOSYLCYTOSINE

Masamune Higashigawa, Hiroki Hori, Hidee Ochiai
Toshi Ohkubo, Hajime Kawasaki, Tsutomu Nobori
Hitoshi Kamiya and Minoru Sakurai

Department of Pediatrics, Mie University
School of Medicine, Edobashi, Tsu 514, Japan

INTRODUCTION

Ara-C is one of the most important antitumor agents in the treatment of acute human myelogenous leukemia[1]. However the optimal therapeutic effects of ara-C can not often be obtained, mainly because ara-C is rapidly inactivated by the enzyme cytidine deaminase into 1-β-D-arabinofuranosyluracil(ara-U)[2]. To overcome this deamination, a highly lipophilic group was introduced at the 4-amino position of cytosine[3]. Among a series of N^4-acyl derivatives of ara-C ,BHAC showed the greatest activity against the mouse leukemia L1210[4]. BHAC is also known as a masked, depot type compound of ara-C,and remains in the body for a long time because of its lipophilicity[5]. But it is not certain whether BHAC is phosphorylated directly to produce N^4-behenoyl-ara-CTP or whether,once converted to ara-C, it subsequently produces ara-CTP, the active metabolite of the drug.The experiments were carried out to clarify how BHAC is incorporated into DNA.

MATERIALS AND METHODS

Chemicals

[Methyl-^{3}H]thymidine(specific activity,6.7Ci/mmol) was purchased from ICN Radiochemicals. [5-^{3}H]ara-C(specific activity,27 Ci/mmol) was purchased from Amersham Japan Ltd. [cytosine-2-^{14}C]BHAC(8.26 mCi/mmol) and [acyl-1-^{14}C]BHAC (10.2 mCi/mmol) were synthesized by Daiichi Pure Chemicals Co., Ltd,. Tokyo,Japan, and were kindly supplied by Asahi chemicals Co., Ltd., Tokyo,Japan. Nuclease P_1, proteinase K,sodium dodecyl sulfate (SDS), standard samples of deoxycytidine 5'- monophosphate (d-CMP), deoxyadenosine 5'-monophosphate(d-AMP), deoxythymidine 5'-monophosphate(d-TMP),deoxyguanosine 5'-monophosphate(d-GMP) and RNase were purchased from Sigma Chemical Co., St. Louis,MO. 1,1,2-trichlorotrifloro-ethane(freon) was purchased from E Merk

AG.,Darmstadt,West Germany and tri-n-octylamine(alamine) was purchased from Tokyo Chemical Industry Co., Ltd.,Tokyo,Japan. Ethanol, phenol, chloroform, isoamyl alcohol, and potassium dihydrogenphosphate(KH_2PO_4) were purchased from Hayashi Pure Chemical Industry Co.,Ltd.,Osaka,Japan. Disodium hydrogen phosphate(Na_2HPO_4) was purchased from Nakarai Chemicals,Co.,Ltd.,Kyoto,Japan.

Tumor cell lines

P388 murine leukemic cells were maintained by serial in vitro culture in RPMI 1640 medium supplemented with 10% FCS and 20 M 2-mercaptoethanol. $3.0X10^7$ cells in logarithmic growth phase were used in each experiment at a concentration of 6.0×10^6 cells/ml.

Determination of incorporation of [5-^{3}H]thymidine into DNA and recovery with phenol method

P388 cells($3.0X10^7$ cells) in duplicate Costar 3025 tissue culture flasks containing RPMI 1640 supplemented with 10% FCS were incubated with [5-^{3}H]thymidine(0.2μCi/ml of medium) at 37 °C for 3 hours under 5% CO_2,95% air. After incubation,the cells were washed three times with ice cold 0.9% saline.Then, using one flask cells,the acid soluble fraction was extracted from the cells with 0.6 N TCA. Acid insoluble fraction was lysed with 0.4 N KOH. At the same time using another flask cells, DNA was extracted with phenol, and was lysed with 0.4 N KOH. The acid soluble fraction,the acid insoluble fraction and the lysate of phenol extracted DNA were transferred to scintillation vials and radioactivity was determined by liquid scintillation counter.

Sample preparation for HPLC

DNA was extracted from cells using standard techniques. The extracted DNA was dissolved in distilled water (1mg/ml),and was heated at 100 °C for 5 minutes,and then cooled rapidly in ice bath.The heat denatured DNA solution was mixed with the same volume of nuclease P_1 solution (0.1 mg/ml of 1/35 M acetate buffer containing $2\ 10^{-4}$M $ZnCl_2$,pH 5.3). And the mixture was incubated at 50 °C for 1 hour. The DNA was completely digested to deoxyribonucleosides 5'-monophosphate,and the hydrolysate was applied to HPLC.

Ribonucleotide pools were extracted with TCA by modified method of Khym[6] and Santi[7]. Briefly, after the cells were incubated with BHAC, the ribonucleotide pools were extracted with 200μl of 0.6 N TCA.The acid soluble fraction was neutralized with 300 μl of 0.5M freon-alamine,and applied to HPLC.

HPLC

The separation of deoxyribonucleosides 5'-monophosphate from digested DNA was achieved using a Model 6000A solvent delivery system equipped with a Model U6K injector,a Model 440 absorbance detector at 254nm(Waters Associates, Inc.,Milford,Mass.) and a model C-R2A peak integrator (Shimazu Co., Japan). An isocratic separation was performed using a TSK gel DEAE 2SW column(Tosoh,Ltd.,Japan) with 0.01

M Na_2HPO_4 with 10% CH_3CN,pH 3.0. The flow rate was 0.7 ml/min,and the fractions were collected every 1 min. The separation of the acid soluble ribonucleosides di- and triphosphates was achieved using a Waters ALC/GPC 204 HPLC equipped with two Model 510 solvent delivery systems, a Model 490 programmable multiwavelength detector at 254 nm, a Model 680 automated gradient controller ,a Model 710B sample processor and a Model 7000B chromatogram processor(System Instruments Co.,Japan).The eluents were:A,0.05 M KH_2PO_4 (pH3.35);B,0.5 M KH_2PO_4(pH4.7). A linear gradient separation was performed using a Partisil 10-SAX column(Whatman Ltd.) from 0% eluent B at 0 min to 75% at 20 min and equilibration for an additional 40 min. The flow rate was 1 ml/min, and fractions were collected every 1 min.

RESULTS

Recovery of DNA with phenol method

The recovery of DNA was calculated as follows;(Mean radioactivity of phenol extracted DNA / Mean radioactivity of acid insoluble fraction)X100, and the result was 73.4%(Table 1).

HPLC profile of the deoxyribonucleosides 5'-monophosphate and ara-CMP standard mixture

The mixture of standard samples(d-CMP,ara-CMP,d-AMP,d-TMP and d-GMP) was applied to HPLC. Deoxyribonucleosides 5'-monophosphate were clearly separated on this condition. Peak of ara-CMP was appeared just behind the peak of d-CMP(Fig.1).

Incorporation of [cytosine-2-^{14}C]BHAC and [acyl 1 ^{14}C]BHAC

When cells were incubated with [cytosine-2-^{14}C],the radioactivities of the acid soluble fraction,the freon alamine fraction,the acid insoluble fraction and the phenol extracted DNA were 10.8%, 2.2%, 87.0% and 23.5% of the total count, respectively. On the other hand, when cells were incubated with [acyl-1-^{14}C]BHAC, the radioactivities of the acid soluble fraction,the freon alamine fraction,the acid insoluble fraction and the phenol extracted DNA were 0.5%, 2.5%, 97.0% and 0.2% of the total count, respectively. [acyl-1-^{14}C]BHAC was mainly taken into cells as acid insoluble fraction, but the radioactivities of the acid soluble fraction and the phenol extracted DNA were hardly detected(Fig.2).

Table 1.Recovery of DNA by the phenol extraction

Radioactivity (CPM/1.0 X 107 P388 cells)		
Acid soluble fraction	Acid insoluble fraction	Phenol extracted DNA
11102.5+1620.0	335381.0+4703.0	247320+31893.3
		73.4% *

* The extent of recovery is defined as the radioactivity of the phenol extracted DNA divided by that of the acid insoluble fraction.

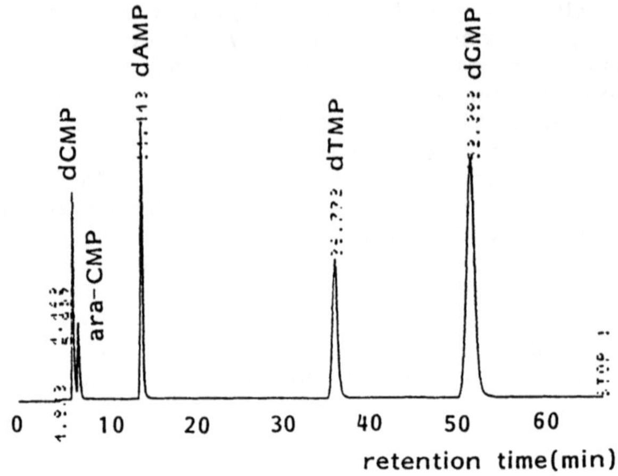

column; TSKgel DEAE-2SW(4mmX25cm)
eluent; 10% CH_3CN in 0.01 M Na_2HPO_4

Fig.1. The HPLC profile of the mixture of the standard samples

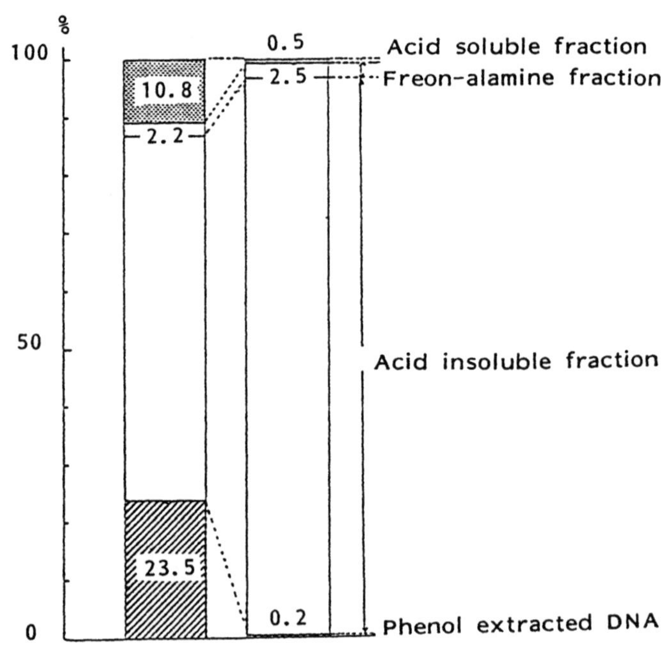

Fig.2. Incorporation of (cytosine-2-^{14}C)BHAC and (acy-1-^{14}C)BHAC into DNA by P388 murine leukemic cells

Analysis of acid soluble fraction

HPLC profile of the acid soluble fraction, which was extracted from the cells incubated with [cytosine-2-^{14}C]BHAC, is shown in Fig.3. As the peak of ara-CTP appeared between CTP and ATP on this condition[8], the main peak radioactive fraction was determined as ara-CTP.

HPLC profile of the hydrolyzed DNA by nuclease P_1 and its peak radioactivity

Fig.4 shows the HPLC profile of the hydrolyzed DNA, from the cells incubated with [5-^{3}H]ara-C. The main peak radioactive fraction appears just behind the peak of the d-CMP. Therefore the radioactive fractions are determined as ara-CMP. Fig.5 shows the HPLC profile of the hydrolyzed DNA, from the cells incubated with [cytosine-2-^{14}C]BHAC. The position of the peak radioactive fraction is the same as that of the hydrolyzed DNA, from the cells incubated with [5-^{3}H]ara-C. This fact means that BHAC is incorporated into DNA as ara-C.

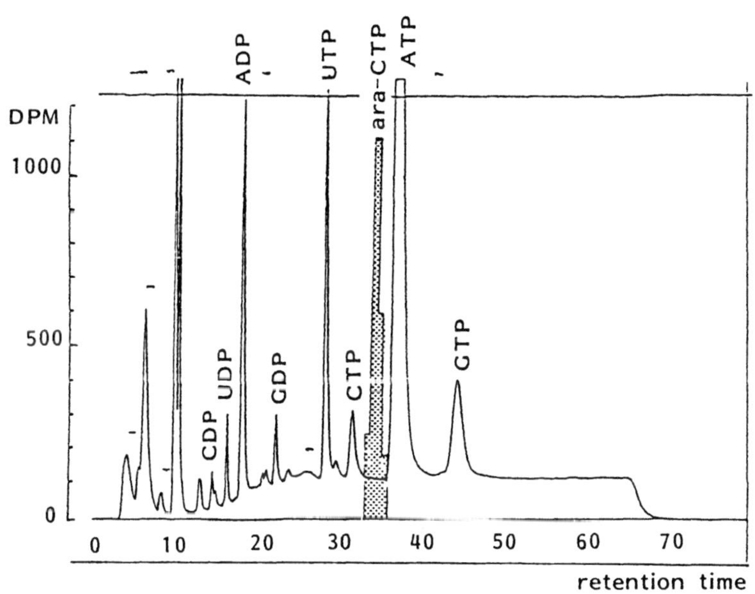

Fig.3. The HPLC profile of the acid soluble fraction and the radioactive fractions

eluent: (A) 0.05 M KH_2PO_4, pH3.35; (B) 0.5 M KH_2PO_4, pH4.7

gradient: A linear gradient separation was performed from 0% eluent B at 0 min to 75% at 20 min and equilibration for an aditional 40 min.

The radioactive fractions were determined as ara-CTP.

DISCUSSION

BHAC is a newly synthesized lipophilic derivative of ara-C, and has been widely used in the clinical treatment of acute myelogenous leukemia in Japan[9,10]. But the mechanism of action of this interesting drug still remains unexplained at present. The purpose of this study is to clarify how BHAC is metabolized and acts on DNA. For analysis of incorporation of ara-C or BHAC into DNA , we used a unique enzyme, nuclease P_1[11]. This enzyme is highly and specifically active towards RNA and single stranded DNA. One mg of the enzyme can hydrolyze 200 mg of single stranded nucleic acids to nucleosides 5'-monophosphate. Incorporation of a small amount of drugs into DNA could be detected by using a combination of this method and isotopic labeling.

When the cells were incubated with [cytosine-2-^{14}C]BHAC, the radioactivity of DNA was 23.5% of the total count and that of hydrolyzed DNA was mainly determined as ara-CMP by using the HPLC. And the radioactivity of acid soluble fraction was 10.8% of the total count and mainly detected as ara-CTP.

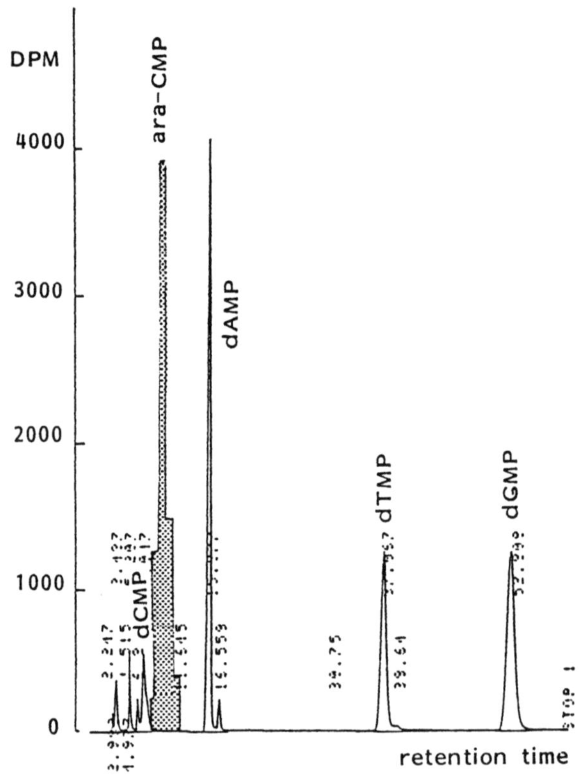

Fig.4. The HPLC profile of the hydrolyzed DNA,from the cells incubated with (5-^{3}H)ara-C

The main radioactive fraction was determined as ara-CMP.

On the other hand, when cells were incubated with [acyl-1-^{14}C]BHAC,the radioactivity of acid insoluble fraction was 97.0% of the total count, but the radioactivities of acid soluble fraction and the phenol extracted DNA were 0.5% and 0.2%,respectively.The lipophilic side chain of BHAC is stable during the phenol extraction of DNA. If the carbon atom at the C-1 position of behenoyl moiety of BHAC remained uncleaved,the radioactivity of the phenol extracted DNA had to be detected, but it was hardly recovered. Ueda et al.[12] demonstrated that BHAC was accumulated markedly in the membrane, mitochondria and microsome fractions in L1210 cells. Our results may support that BHAC is once bound to membranes without releasing the acyl group. Oh-ishi et al.[13] reported the existence of N^4-succinyl-ara-C, a novel metabolite of BHAC, in the urine and bile of mice. Although the radioactivity of the freon alamine fraction (up to 2.5% of the total count) may indicate the possibility of the existence of phosphorylated BHAC or other N^4-acyl ara-C, our results suggest that BHAC is not phosphorylated directly to produce N^4-behenoyl-ara-CTP, but mainly once converted to ara-C and it subsequently produces ara-CTP, the active metabolite of the drug, and then incorporated into DNA.

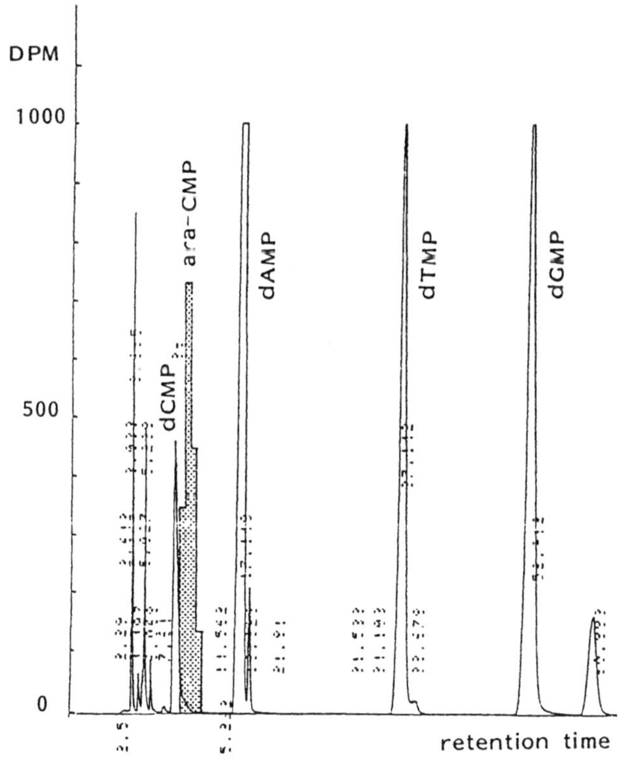

Fig.5. The HPLC profile of the hydrolyzed DNA,from the cells incubated with (cytosine-2-^{14}C)BHAC The main radioactive fraction was determined as ara-CMP.

REFERENCES

1. Ellison R.R., Holland J.F., Weil M. Jacquillat C., Boiron M., Bernrd J., Sawitsky A., Rosner F., Gussoff B., Silver R.T., Karanas A., Cuttner J., Spurr C.L., Hayes D.M., Blom J., Leone L.A., Haurani F., Kyle R., Hutchison J.L., Forcier R.J. and Moon J.H: Arabinosyl cytosine: a useful agent in the treatment of acute leukemia in adults. Blood 32, 507(1968)
2. Ho D.H.W. and Frei E.III: Clinical pharmacology of 1-β-D-arabinofuranosylcytosine. Clin.Phamacol.Ther. 12, 944 (1971).
3. Akiyama M., Ohishi J., Shirai T., Akashi K., Yoshida K., Nishikido J., Hayashi H., Usubuchi Y., Nishimura D., Itoh H., Shibuya C. and Ishida T: The synthesis of new derivatives of 1-β-D-arabinofuranosylcytosine. Chem.Pharm. Bull. 26, 981(1978)
4. Aoshima M., Tsukagoshi S., Sakurai Y., Ohishi J., Ishida T. and Kobayashi H: N^4-behenoyl-1-β-D-arabinofuranosyl-cytosine as a potential new antitumor agent. Cancer Res. 37, 2481(1977)
5. Ueda T., Nakamura T., Ando S., Kagawa D., Sasada M., Uchino M., Johno I. and Akiyama Y: Pharmacokinetics of N^4-behenoyl-1-β-D-arabinofuranosylcytosine in patients with acute leukemia. Cacer Res. 43, 3412(1983)
6. Khym J.X: An analytical system for rapid separation of tissue nucleotides at low pressures of conventional anion exchagers. Clin.Chem. 21, 1245(1975)
7. Garrett C. and Santi D.V: A rapid and sensitive high pressure liquid chromatography assay for deoxyribonucleoside triphosphates on cell extracts. Anal. Biochem. 99, 268(1979)
8. Ohkubo T: Biochemical modulation of 5-fluorouracil and cytosine arabinoside by sequential administration of methotrexate. Mie Med. J. 35, 15(1985)
9. Kimura K., Yamada K., Uzuka Y., Maekawa Y., Takaku M., Shimoyama M., Ogawa M., Amaki I., Osamura S., Ito M., Sakai Y., Oguro M., Hattori K., Hoshino A., Hirota Y., Ohta K., Nakamura T., Masaoka T., Kimura I. and Ichimura M: Phase I-II study of N^4-behenoyl-1-β-D-arabinofuranosylcytosine. Recent Results. Cancer Res. 76, 232(1981)
10. Kimura K., Ohno R., Amaki I., Hattori K., Hirota Y., Hoshino A., Ichimura M., Ito M., Kimura I., Maekawa T., Masaoka T., Nakamura T., Ogawa M., Oguro M., Ohta K., Osamura S., Shimoyama M., Tkaku F., Uzuka Y. and Yamada K: Treatment of acute myelogenous leukemia in adults with N^4-behenoyl-1- -D-arabinofuranosylcytosine. Cancer 56, 1913(1983)
11. Ueda T., Nakamura T., Kagawa D., Yamamoto K., Uchino M., Sasada M., Johno I. and Akiyama Y: Intracellular distribution of N^4-behenoyl-1-β-D-arabinofuranosylcytosine in blood cells. Gann 74, 445(1983)
12. Kaneko T., Katoh K., Fujimoto M., Kumagai M., Tamaoka J. and Fujimura Y.K: Detection of the nucleotide composition of a deoxyribonucleic acid by high-performance liquid chromatography of its enzymatic hydrolysate: a review. J. Microbiol.Methods 4, 229(1986)
13. Oh-ishi J., Kataoka T., Tsukagoshi S., Sakurai Y., Shibukawa M. and Kobayashi H: Production of N^4-succinyl-1-β-D-arabinofuranosylcytosine, in mice and its biological significance. Cancer Res. 41, 2501(1981)

SEQUENTIAL COMBINATION OF METHOTREXATE AND 1-β-D-ARABINOFURANOSYLCYTOSINE SHOWS SYNERGISTIC EFFECT ON THE GENERATION OF DNA STRAND BREAKS IN A HUMAN PROMYELOCYTIC LEUKEMIA CELL LINE

Akihiko Tanizawa, Masaru Kubota, Tetsuya Takimoto, Toshiyuki Kito, Yuichi Akiyama, Yukio Kiriyama and Haruki Mikawa

Department of Pediatrics, Kyoto University
Kyoto, Japan

For cancer chemotherapy, it is important to use anti-cancer drugs which interact one another to enhance tumor cell toxicity with minimum adverse effects. For example, the sequential combination of methotrexate (MTX) and 1-β-D-arabinofuranosylcytosine (Ara-C) has been demonstrated to have synergistic cytotoxic effects [1,2]. Although the biochemical basis for this synergism is not clearly elucidated, MTX is believed to modulate intracellular pyrimidine pools and ultimately to alter intracellular 1-β-D-arabinofuranosylcytosine 5'-triphosphate (Ara-CTP) pools.

Recently Li and Kaminskas found the accumulation of DNA strand breaks after MTX treatment[3]. Ara-C, on the other hand, is an inhibitor of DNA repair synthesis or replication. Therefore, we investigated the formation of DNA strand breaks by the combination of MTX and Ara-C. In this report, we showed that this combination worked synergistically on the generation of DNA damage in a human promyelocytic leukemia cell line.

MATERIALS AND METHODS

Cell Culture and Drug Treatment

A human promyelocytic leukemia cell line, HL-60, was maintained in RPMI 1640 supplemented with 10% heat-inactivated fetal bovine serum (FBS), 100 units/ml penicillin, 100 μg/ml streptomycin and 2mM L-glutamine at 37°C in a 95% air/5% CO_2 atmosphere. The cultures were checked routinely for mycoplasma. MTX and leucovorin were gifts from Japan Lederle Company (Tokyo Japan) and Ara-C from Nippon Shinyaku Company (Kyoto Japan). [^{3}H]Ara-C (27 Curies/mmole) was purchased from Amersham Japan (Tokyo Japan) and other agents from Sigma.

Cells in the late logarithmic phase were washed twice with cold phosphate buffered saline (PBS) and resuspended in fresh RPMI 1640 with 10% FBS. Cells were then exposed to the first drug(s) for 2 hr. Subsequently the second drug was added followed by further incubation for 2 hr. The cells were washed in cold PBS twice, then DNA strand breaks, Ara-CTP levels and the incorporation of Ara-C into DNA were analyzed.

DNA Strand Break Analysis

DNA strand breaks were measured by the "fluorometric analysis of DNA unwinding" technique described by Birnboim and Jevcak[4]. The basis of this method is that during alkaline treatment, DNA unwinding occurs as a function of the amount of strand breaks and/or alkali-labile sites. The number of strand breaks present can be estimated by correlation with the radiation dose necessary to produce the same change in DNA unwinding. For dose-effect curves the cells were X-ray irradiated on ice and immediately analyzed fluorometrically. Therefore, the amount of DNA strand breaks were expressed in rad equivalents. Increment of DNA strand breaks by the combination was caluculated as follows;

$$\text{Increment of SB} = SB_{(MTX+Ara-C)} - (SB_{(MTX)} + SB_{(Ara-C)})$$

where SB is the amount of DNA strand breaks expressed in rad equivalents.

Measurement of Ara-CTP

After drug exposure and two washings with cold PBS, the cells were extracted with 0.4N perchloric acid and neutralized with Alamine/Freone. Ara-CTP was separated and quantified by high performance liquid chromatography on a Partisil 10-SAX anion exchange column eluted with 0.5M KCL/0.25M KH_2PO_4, pH 3.4 at 1.0 ml/min[5].

Analysis of Incorporation of Ara-C into DNA

The amount of incorporated Ara-C into DNA was measured by the method described by Fram et al with modifications[6]. Briefly, cells were exposed to $[^3H]$Ara-C with or without pretreatment of MTX. After two washings with cold PBS, the cells(10^6) were first digested with proteinase K(50μg/ml) and extracted with phenol. Aqueous phase was treated with RNase(50μg/ml) followed by the second extraction with phenol and chroloform. After third extraction with chroloform, the DNA was precipitated with NaCl and ethanol. Then the DNA was filterd on Whatman GF/A filters, and its radioactivity was measured.

Measurement of Cytotoxicity

After treatment of various combinations of drugs, cells were washed twice with regular media and resuspended at a density of 1.5×10^5/ml. After three days, viable cells were determined by trypan blue dye exclusion test.

RESULTS

First we examined whether the sequence of MTX and Ara-C was an essential factor for synergism in causing DNA strand breaks. As shown in Fig. 1, the simultaneous incubation with MTX(1μM) and Ara-C (10μM) for 2 hr produced the small increment of strand breaks approximately 20 rad equivalents. When Ara-C was added 2 hr prior to MTX, the increase of strand breaks was also small(approximately 40rad equivalents). On the contrary, the enhancement by the sequential combination of MTX then Ara-C was more than 100rad equivalents. These findings postulated that the order of the two drugs is one critical determinant for the synergistic formation of DNA strand breaks.

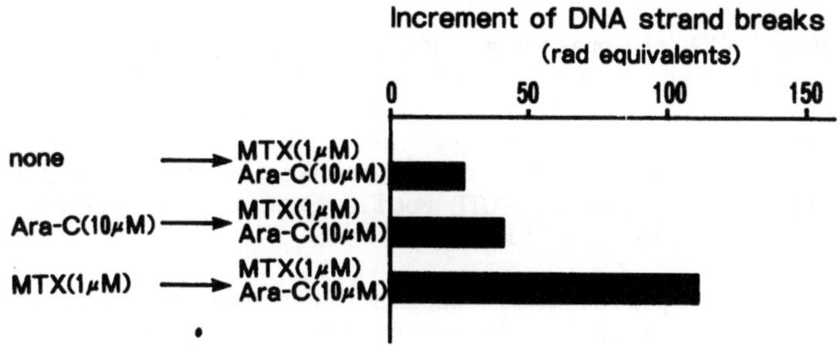

Fig. 1. Amount of DNA strand breaks by various combinations of MTX(1μM) and Ara-C(10μM) was measured. The results were expressed as the augmentation of strand breaks calculated as in Materials and Methods. The mean of more than 5 experiments is presented.

Next we investigated whether the synergistic effect on DNA damage depended on the drug concentrations. Cells were first exposed to varying concentrations of MTX (0.001 to 10μM). Two hours later, Ara-C was added to achieve a concentration of 10μM, and the cells were further incubated for 2 hr. As illustrated in Fig. 2, the sequential treatment of MTX and Ara-C increased the generation of DNA strand breaks as a function of MTX concentrations. Secondly we evaluated DNA strand breaks under the same drug sequence by changing Ara-C concentrations. When the concentration of MTX was 1μM, the addition of Ara-C (more than 1μM) augmented the production of DNA strand breaks significantly(data not shown).

The cytotoxicity of Ara-C is dependent on the phosphorylation of Ara-C to an active metabolite, Ara-CTP. It exerts cell killing through the

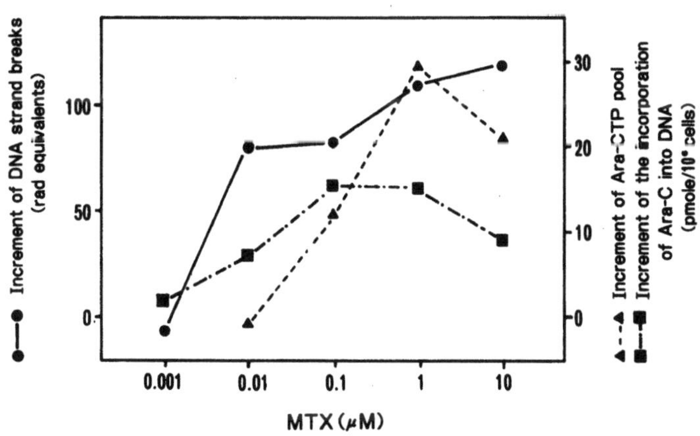

Fig. 2. Enhancement of DNA strand breaks(●—●), Ara-CTP formation (▲---▲) and the incorporation of Ara-C into DNA (■-·-■) with MTX pretreatment is shown. 10μM Ara-C for 2 hr accumulated 74.8±4.8 (pmole/10^6 cells) Ara-CTP and incorporated 11.9±1.7 (pmole/10^6 cells) of Ara-C into DNA. Means of more than 5 experiments are shown.

inhibition of DNA polymerase α and/or the incorporation into DNA. Therefore, we evaluated the synthesis of Ara-CTP after MTX pretreatment and found the significant enhancement of Ara-CTP generation as reported in previous works[1,2] (Fig. 2). However, there was a discrepancy between augmentation of Ara-CTP and DNA strand breaks. Namely, MTX pretreatment at 0.01μM could not increase Ara-CTP pool, while it augmented the formation of DNA strand breaks significantly.

Kufe et al reported that the Ara-C residue was incorporated into DNA and behaved as a relative DNA chain terminator. Furthermore, they demonstrated that the formation of (Ara-C)DNA was a highly predictive parameter of lethal cellular events[7]. Therefore, the experiments were undertaken to determine the relationship between DNA strand breaks and the incorporation of Ara-C into DNA. As illustrated in Fig. 2, preincubation with MTX enhanced radioactivity of Ara-C in extracted DNA, when the concentration of MTX was more than 0.01μM. But with 10μM MTX, this enhancement effect decreased like that in Ara-CTP pool synthesis.

Next we investigated the reversal of this synergism when the biochemical effects of MTX were modulated with thymidine (TdR) and/or hypoxanthine (Hx), or leucovorin. Each of TdR (10μM), Hx (100μM), TdR(10μM) +Hx (100μM) or leucovorin (10μM) did not induce significant DNA strand breaks after 4 hours of treatment. Fig. 3 shows that only when TdR and Hx was simultaneously added to MTX, they almost completely prevented the synergistic increase in DNA strand breaks. On the contrary, leucovorin or TdR could partially reverse this synergism. Yoshioka et al proposed that deoxyribonucleoside triphosphate imbalance induced by 5-fluorodeoxyuridine triggers the production of endonuclease, which is inhibited by cycloheximide[8]. Since treatment with MTX affected deoxyribonucleotide pools, we investigated the ability of cycloheximide (10μg/ml) to modulate this synergism and found that it perfectly inhibited the synergistic induction of strand breaks. Furthermore, the addition of TdR and Hx reduced the enhancement of the incorporation of Ara-C into DNA to 4.4pmole/10^6 cells. On the other hand, cycloheximide was not effective enough to inhibit the enhancement (Fig.3).

The relationship between the amount of DNA damage and the cytotoxicity was determined finally. Table 1 clearly shows that the combination of MTX and Ara-C has significant synergy also in the biological activity. Moreover, the simultaneous addition of TdR and Hx reversed the

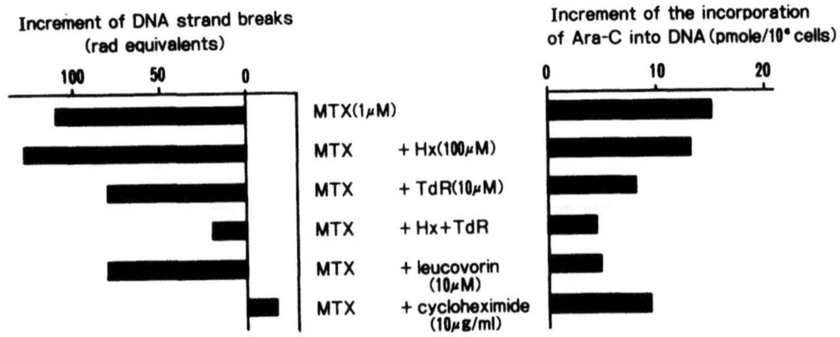

Fig. 3. First cells were exposed to MTX(1μM) with or without modulators. Two hours later, Ara-C was added to achieve 10μM and further incubated for 2hr. Then DNA strand breaks and the incorporation of Ara-C into DNA were analyzed. The mean of more than 5 experiments is shown.

Table 1. Effects of modulators on the cytotoxicity by the sequential treatment of MTX(1μM) and Ara-C(10μM)

first drug(s)	second drug	viable cells ($x10^5$/ml)
none	none(control)	9.1±0.8
MTX	none	6.0±0.7
none	Ara-C	3.2±0.4
MTX	Ara-C	1.2±0.3
MTX+Hx(100μM)	Ara-C	1.4±0.4
MTX+TdR(10μM)	Ara-C	2.1±0.3
MTX+Hx+TdR	Ara-C	3.1±0.5
MTX+leucovorin (10μM)	Ara-C	2.6±0.4
MTX+cycloheximide (10μg/ml)	Ara-C	2.4±0.2

Cells were exposed to the first drug(s) for 2hr and Ara-C was added without washing. Two hour later, the cells were washed twice with regular media and seeded at the density of $1.5x10^5$/ml. Viable cells were counted at day 3. None of Hx, TdR, Hx+TdR and leucovorin alone showed any toxic effects. Cycloheximide has some cytotoxic effects(4.3± $0.3x10^5$/ml). Values represent the means ± SD of more than three experiments in duplicates.

cytotoxicity to that of Ara-C alone, 3.1±$0.5x10^5$, 3.2±$0.4x10^5$/ml respectively. Interestingly, cycloheximide partially prevented the cytotoxicity by the combination of MTX and Ara-C.

DISCUSSION

The biochemical mechanism for the combination of MTX and Ara-C has been investigated extensively. The changes of intracellular Ara-CTP pool and/or the incorporation of Ara-C into DNA have been reported as essential factors for synergistic cytotoxicity. In the present study, we show clearly the interaction of MTX and Ara-C in terms of DNA strand break formation. However, the amount of strand breaks dose not necessarily alter proportionally with the Ara-CTP synthesis and the incorporation into DNA. This discrepancy suggests that one or more factors other than the changes of the Ara-CTP synthesis and the (Ara-C)DNA formation are involved in the synergism leading to increased DNA damage.

At present, we do not know the underlying mechanism for this synergistic effect on DNA damage. However, there are two possibilities; i) MTX induces DNA strand breaks which are quickly sealed under regular conditions and Ara-C may inhibit this repair process. ii) MTX enhances Ara-C activity, which is a main factor causing breaks, through incresing the Ara-CTP synthesis and/or depleting dCTP pools. Our data favors the former hypothesis because of following reasons; i) There is the apparent discrepancy among the changes of DNA strand breaks, Ara-CTP synthesis and the incorporation of Ara-C into DNA, when the concentration of MTX is changed. ii) Aphidicolin, another inhibitor of DNA polymerase α, also shows the synergistic effect with MTX on DNA damage(data not shown). iii) The addition of cycloheximide can abolish the synergistic formation of DNA strand breaks and protect cells from cytotoxic effects of the combination treatment, though it can not prevent the increase of Ara-C incorporation into DNA.

In regard to the relationship between DNA strand breaks and the cytotoxicity of anti-cancer drugs, DNA strand breaks may promote the suicidal activation of poly(ADP-ribosy)lation in very large amount[9], or inhibit DNA replication, and ultimately they can induce cell death. Whatever the underlying mechanism of this synergism is, our present data provide one possible biochemical basis for sequential MTX and Ara-C therapy.

REFERENCES

1, M.L. Hoovis, and M.Y. Chu, Enhancement of the antiproliferative action of 1-β-D-arabinofuranosylcytosine by methotrexate in murine leukemic cells(L5178Y), Cancer Res. 33:521 (1973).
2, E. Cadman, and F. Eiferman, Mechanism of synergistic cell killing when methotrexate precedes cytosine arabinoside; study of L1210 and human leukemic cells, J. Clin. Invest. 64:788 (1979).
3, J.C. Li, and E. Kaminskas, Accumulation of DNA strand breaks and methotrexate cytotoxicity, Proc. Natl. Acad. Sci. USA. 81:5694 (1984).
4, H.C. Birnboim, and J.J. Jevcak, Fluorometric method for rapid detection of DNA strand breaks in human white blood cells produced by low doses of radiation. Cancer Res. 41:1889 (1981).
5, M. Kubota, T. Takimoto, A. Tanizawa, Y. Akiyama, and H. Mikawa, Differential modulation of 1-β-D-arabinofuranosylcytosine metabolism by hydroxyurea in human leukemic cell lines, Biochem. Pharmacol. 37:1745 (1988).
6, R.J. Fram, N. Robichaud, S.D. Bishov, and J.M. Wilson, Interactions of cis-diamminedichloroplatinum(II) with 1-β-D-arabinofuranosylcytosine in LoVo colon carcinoma cells, Cancer Res. 47:3360 (1987).
7, D.W. Kufe, D. Munroe, D. Herrick, E. Egan, and D. Spriggs, Effects of 1-β-D-arabinofuranosylcytosine incorporation on eukaryotic DNA template function, Mol. Pharmacol. 26:128 (1984).
8, A. Yoshioka, S. Tanaka, O. Hiraoka, Y. Koyama, Y. Hirota, D. Ayusawa, T.Seno, C. Garrett, and Y. Wataya, Deoxyribonucleoside triphosphate imbalance; 5-fluorodeoxyuridine-induced DNA double strand breaks in mouse FM3A cells and the mechanism of cell death, J. Biol. Chem. 262:8235 (1987).
9, A. Tanizawa, M. Kubota, T. Takimoto, Y. Akiyama, S. Seto, Y. Kiriyama, and H. Mikawa, Prevention of adriamycin-induced interphase death by 3-aminobenzamide and nicotinamide in a human promyelocytic leukemia cell line, Biochem. Biophys. Res. Commun. 144:1031 (1987).

SYNERGISTIC INTERACTION BETWEEN ETOPOSIDE AND 1-β-D-ARABINOFURANOSYLCYTOSINE

Toshiki Ohkubo, Hiroki Hori, Masamune Higashigawa, Hajime Kawasaki, Hitoshi Kamiya, Minoru Sakurai, Yoshiyuki Kagawa*, Eiji kakito*, Katsumi Sumida*, and Kazuya Ooi**

Department of Pediatrics,*Department of Pharmacy, Mie university, School of Medicine, 2-174,Edobashi, Tsu, Mie, 514, **Department of Pharmacy, Social Insurance Hazu Hospital, Yokkaichi,Mie,513, Japan

ABSTRACT

The sequence-dependency of the antitumor effect of etoposide(VP-16) and 1-β-D-arabinofuranosylcytosine(ara-C) and its mechanisms were investigated in L1210 ascites tumor. Treatment with VP-16(15 mg/kg) and ara-C(25 mg/kg) was administered intraperitoneally on days 1, 4, and 7 after tumor inoculation. Three-hour pretreatment with VP-16 followed by ara-C produced 70% of cure rate, but only 20% of cure rate was obtained with the reverse sequence. Simultaneous administration led to the worst therapeutic result. To clarify this sequence-dependent interaction of the both drugs, the effect of VP-16 on incorporation of ara-C into DNA was investigated in vivo. On day 3 after intraperitoneal transplantion of $1x10^6$ L1210 cells, 15 mg/kg of VP-16 and 1μCi of (^{3}H)ara-C was administered intraperitoneally with or without time interval. The effect of VP-16 on ara-C incorporation was also strikingly schedule-dependent and compatible with the schedule dependency of the antitumor effect. At 1 hr after injection of ara-C, simultaneous administration of 15 mg/kg of VP-16 reduced ara-C incorporation by 67% of ara-C injection alone. But in sharp contrast to simultaneous-administration, 3 hour preadministration of VP-16 increased ara-C incorporation up to 247%. From 1 to 6 hour after intraperitoneal administration of 15 mg/kg of VP-16, the sedimentation rate of L1210 DNA was increasing on alkaline sucrose gradients. This repairing period from VP-16 induced DNA damage might be the vulnerable time to ara-C through an increase in ara-C incorporation into DNA.

INTRODUCTION

The combination chemotherapy of VP-16 and ara-C has been shown to be effective for treatment of human leukemia(1,2).

VP-16 produces topoisomerase II-mediated DNA strand breaks, and a correlation has been observed between the degree of drug-induced DNA breaks and cytotoxicity(3-6). On the other hand VP-16 inhibits nucleoside transport reversibly. Yalowich and Goldman(7) reported that VP-16 inhibits thymidine transport in Ehrlich cells in vitro. ara-C is one of the most potent anticancer agents for remission induction in acute nonlymphocytic leukemia. After transport into the cell, ara-C is phosphorylated to ara-CTP which inhibits DNA polymerase. ara-C is also incorporated into DNA, and the amount of ara-C incorporation correlates closely with cytotoxicity in L1210 cells in vitro(8-11). Since ara-C is an analogue of deoxycytidine, and the effect of VP-16 on the membrane transport is reversible, there is a possibility that sequential administration of VP-16 and ara-C is more potent than simultaneous administration. This work describes the effect of time interval of drug administration on the antitumor effect and on incorporation of ara-C in the combination of VP-16 and ara-C.

MATERIALS AND METHODS

Mice and Tumor

Male BDF1 mice weighing from 20 to 22g each were kept at Mie university animal center under constant conditions (12hr light:dark regimen, Oriental Chow pellet food, and water ad libitum). The strain of leukemia L1210 used was obtained from Shionogi Pharmacological Laboratory, Osaka and has been maintained in BDF1 mice by weekly intraperitoneal (i.p.) inoculation of $1x10^6$ cells. Mice which were to be used in experiments received $1x10^6$ ascites tumor cells i.p. in 0.2ml of 0.9% NaCl solution. After transplantation, mice were randomized into groups of 10 mice.

Drug and Drug Administration

VP-16 was supplied from Bristol Myers Co., Tokyo, Japan in vials containing 100mg VP-16, and ara-C from Nippon Shin'yaku Co., Kyoto, Japan. (^{3}H)ara-C(10 Ci/mmol) was obtained from Amersham, Tokyo. The drug was prepared just before administration and diluted with isotonic NaCl solution until the dose per gram of mice was contained in 0.01 ml.

antitumor effect of VP-16 and ara-C

On days 1,4,and 7 after tumor inoculation mice received the drug treatment intraperitoneally. The doses of VP-16 and ara-C were 15 mg/kg and 25 mg/kg respectively. When VP-16 and ara-C were given in combination, the drugs were injected either simultaneously or with a time interval of 3 and 6hr in between. Mice in the control group were given 0.2ml isotonic NaCl solution in the same schedule.Antitumor effect of the drugs was evaluated by comparing cure rate and prolongation of the postinoculation life span. One hundred and twenty-day survivors from inoculation were defined as "cures". The data were analyzed by t-test for life prolongation of uncured mice without inclusion of cured mice, and by Chi-square test for cure rate.

incorporation of ara-C into DNA

Incorporation of ara-C into DNA was quantified by determining the radioactivity from (^{3}H)ara-C(specific activity 10 Ci/mmol) incorporated into acid-insoluble fraction. Using CsSO4 gradient method, Kufe et al. documented that ara-C was incorporated only into DNA(11). On day 3 after tumor inoculation, 15 mg/kg of VP-16 and 1 μCi of (^{3}H)ara-C was administered intraperitoneally in the schedules described under results. The mice were sacrificed by cervical dislocation of the spine, and the cells were removed by flushing the abdominal cavity with 1 ml of ice-cold PBS. Ten million cells were washed twice in ice-cold PBS. After centrifugation(1500 rpm for 5 min at 4°C), 1 ml of ice-cold 0.6N trichloroacetic acid was added and mixed well. The mixture was kept for 10 min on the ice bath, and centrifuged at 2500 rpm for 10 min at 4°C. The precipitant was washed twice in ice-cold 0.6 N trichloroacetic acid, and solubilized by heating with 0.5 ml of 0.4 N KOH at 60°C for 15 min. Radioactivity was determined after addition of 10 ml of acidified Aquasol II(New England Nuclear) in a LKB 1219 liquid scintillation counter.

VP-16-induced strand breaks in L1210 DNA

VP-16-induced DNA damage was evaluated using alkaline sucrose density gradient method. (methyl-5-^{3}H)Thymidine (specific activity, 8.7 Ci/mmol) was diluted with 4 volume of sterile 0.9% NaCl solution to yield a final concentration of 200 μCi/ml. On day 3 after inoculation of 1×10^6 L1210 cells i.p., 40 μCi of (methyl-5-^{3}H)thymidine was administered i.p. to each mouse. After 12 to 16 hr, the mice were given i.p. injection of VP-16. The mice were sacrificed by cervical dislocation of the spine,and the cells were removed by flushing the abdominal cavity with 1 ml of 0.9% NaCl solution containing 0.01 M EDTA. The suspension of ascites cells were diluted to 5×10^5 cells/ml, and 1 ml of suspension were slowly pipetted onto a 1 ml layer of 0.45 M NaOH, 0.55M NaCl, and 0.01 M EDTA overlying 28 ml of a 5-20% sucrose gradient in 0.1 M NaOH, 0.9 M NaCl, and 0.01 M EDTA. A 5 ml cushion of 60% sucrose was present underneath the gradient. After layering the cells onto the precooled solutions, the tubes were allowed to stand at 4°C for 2 hr before centrifugation. Centrifugation was at 4°C for 2 hr at 25,000 rpm in a Beckman SW27 rotor. Gradients were collected from the bottom in 1.2-ml fractions. Fifty μl of a solution of DNA(10 mg/ml) were added to coprecipitate with the tritiated L1210 DNA in each fraction. After adding 100μl of 60% trichloroacetic acid in 1.2 M HCl and allowing the tube to stand for 10 min in an ice bath, the precipitant were collected by centrifugation. The supernatant fluids were replaced with 200 μl of 5% trichloroacetic acid,and the DNA was hydrolyzed by heating in a boiling water bath for 15 min. Radioactivity was counted after addition of 10ml of Aquasol II(New England Nuclear) in a LKB 1219 Liquid Scintillation counter.

RESULTS

Antitumor effect of VP-16 and ara-C

As compared with 7.20 days of the control group, the

life span of the mice treated with 15 mg/kg of VP-16 was cextended significantly to 19.80 days(175%ILS, $P<0.01$ by t-test, Fig.1). Treatment with 25 mg/kg of ara-C also expanded the postinoculation life span to 13.55 days(88%ILS, $P<0.01$). Cures were not obtained by single drug treatment. Of 50 mice treated with 15mg/kg of VP-16 and 25mg/kg of ara-C, 26 died with tumor and 24 were cured. The mean life span of uncured mice ranged from 18.50(157%ILS) to 41.50(476%ILS)in 5 treatment groups. Although the mean life span of all combination group was significantly longer($p<0.01$) than that of the group treated with 25 mg/kg of ara-C(13.55 days), only when VP-16 preceded ara-C(41.50 and 27.67 days) and when ara-C preceded VP-16 by 6 hr(24.33 days) did survival of the mice significantly exceed that of VP-16 alone(19.80 days, $p<0.01$). Cure rate of mice treated with VP-16-6hr-ara-C, VP-16-3hr-ara-C, VP-16+ara-C, ara-C-3hr-VP-16, and ara-C-6hr-VP-16 was 80, 70, 0, 20, 70% respectively. Simultaneous administration of 25 mg/kg of ara-C showed the worst antitumor effect, and 6 hr and 3 hr pretreatment with VP-16 and 6 hr pretreatment with ara-C showed synergistic antitumor effect (statistically different from simultaneous treatment, $P<0.01$, and 3 hr pretreatment with ara-C, $P<0.05$).

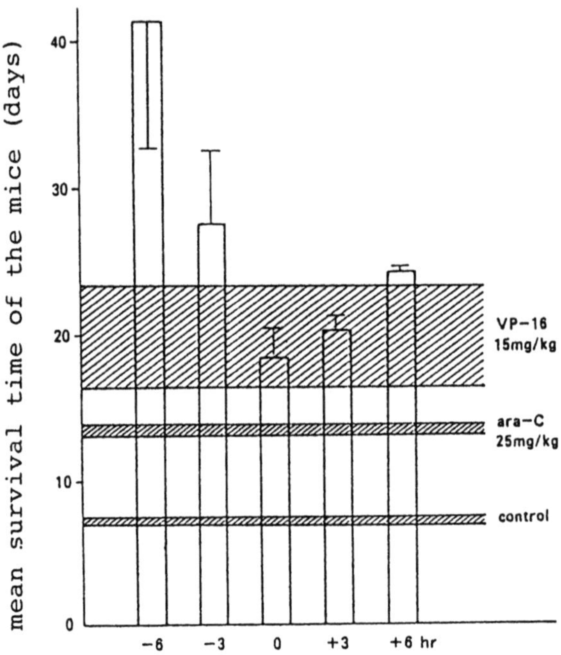

Fig.1. Effect of drug scheduling of VP-16 and ara-C on survival time of mice bearing L1210 leukemia. Groups of 10 mice were inoculated i.p. with 10^6 L1210 cells. Treatment as indicated in the table was given on days 1,4, and 7.

Effect of VP-16 on incorporation of ara-C into L1210 DNA

To clarify the mechanism of sequence-dependent antitumor effect of VP-16 and ara-C, the effect of timing of administration of VP-16 on ara-C incorporation was investigated in vivo. The effect of VP-16 on ara-C incorporation was strikingly schedule-dependent and compatible with the schedule dependency of antitumor effect. At 1 hr after injection of 1μCi of ara-C, 0.14 pmol/10^7 cells of ara-C was incorporated into L1210 DNA, and simultaneous administration of 15 mg/kg of VP-16 reduced ara-C incorporation by 67% as compared to that of ara-C injection alone. In contrast to simultaneous administration, 3 hour preadministration of VP-16 increased ara-C incorporation up to 247% (Fig.2-a). The effect of VP-16 on ara-C incorporation at 6 hr after injection of ara-C, was quite similar to that at 1 hr after injection of ara-C. Simultaneous administration of VP-16 reduced ara-C incorporation by 33% ,and 3 hr preadministration enhanced it up to 192%(Fig.2-b).

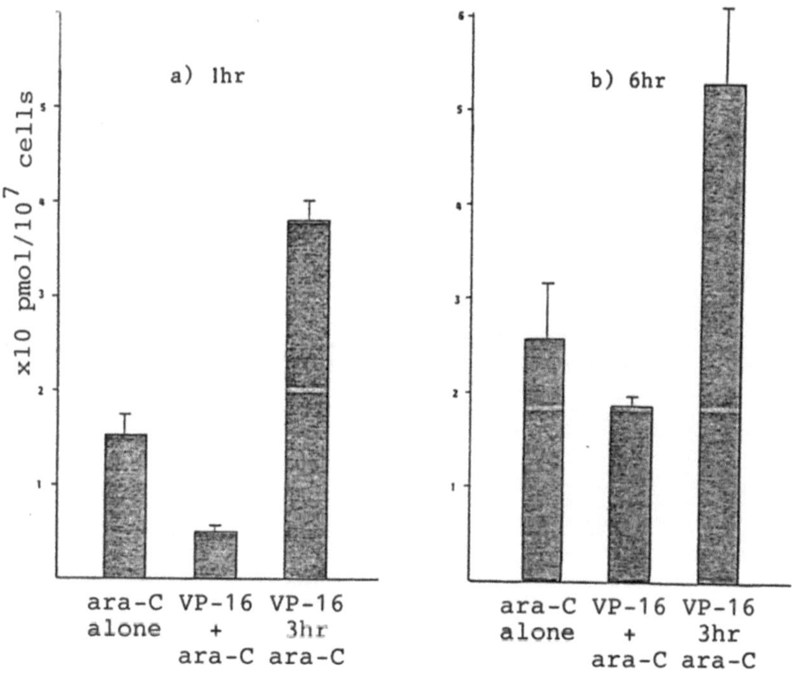

Fig.2. Effect of timing of administration of VP-16 on ara-C incorporation into L1210 DNA. On day 3 after tumor inoculation, 1 μCi of (^{3}H)ara-C and 15 mg/kg of ara-C was injected i.p. Incorporation of ara-C was quantified by determining the radioactivity from (^{3}H) ara-C incorporated into acid-insoluble fraction.

Strand breaks in L1210 DNA after treatment with VP-16

The L1210 DNA from untreated mice sedimented around fraction 8 of the gradient (Fig.3-a). On the other hand, at 10 min after i.p. injection of 15 mg/kg of VP-16, the maximum concentration of DNA was observed in near the top of gradients (Fig.3-a). Strand breaks in DNA occurred within 10 min after i.p. injection of VP-16, but the DNA sedimented more rapidly and the maximum concentration of DNA was observed in fraction 20 when cells were removed at 1 hr after injection of VP-16 (Fig.3-b). As the time interval between injection of VP-16 and removal of the cells from experimental mice was elongated to 3 hr and then 6 hr, the rate of sedimentation of the DNA further increased, suggesting the continuous repairing of DNA (Fig.3-c,d). At 6 hr after injection of VP-16, the DNA sedimented in a major fraction of No.10 and a minor fraction of No.16.

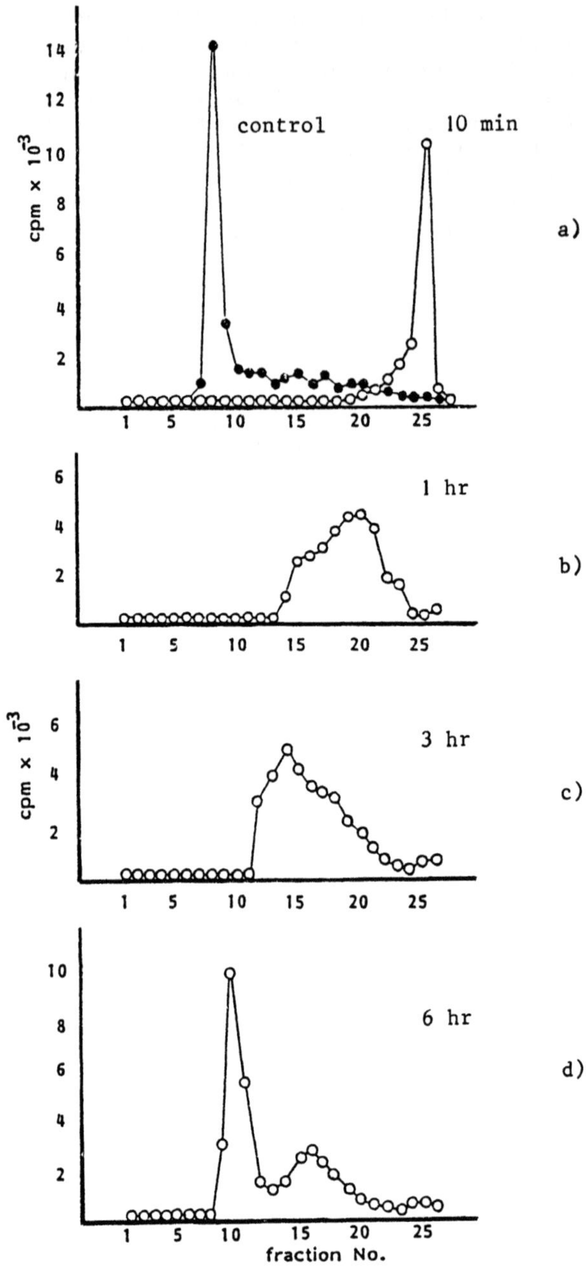

Fig.3. Sedimentation of DNA from L1210 cells at 10 min, 1, 3, and 6 hr after treatment in vivo with 15 mg/kg of VP-16. At 12 to 16 hr before treatment with VP-16, mice bearing L1210 leukemia received 40μCi of (methyl-5-^{3}H) thymidine i.p. After treatment with VP-16, $5x10^5$ of the L1210 ascites cells were removed and placed on top of a 5 to 20% alkaline sucrose gradient.

DISCUSSION

The present studies demonstrated that the antitumor effect of VP-16 and ara-C was strikingly sequence-dependent. Although the combination chemotherapy of VP-16 and ara-C has been shown to promising in recent clinical studies, little attention has been paid to the timing of administration of the both drugs. In the present study in L1210 ascites tumor, pretreatment with VP-16 produced the best therapeutic results, producing an impressive increase of cure rate, and simultaneous administration of 25 mg/kg of ara-C produced the worst therapeutic results.

Our studies also demonstrated the sequence-dependency of the effect of VP-16 on ara-C incorporation, which is compatible with the sequence-dependency of the antitumor effect. Simultaneous administration of VP-16, which produced the worst therapeutic results decreased ara-C incorporation to one third of the control. It was of note that pretreatment with VP-16 increased the incorporation of ara-C into L1210 DNA up to about 2.5-fold of the control. This effect of VP-16 on ara-C incorporation may partially explain the sequence-dependency of the antitumor effect of the both drugs. The effect of VP-16 in diminishing the membrane transport of nucleoside was investigated in vitro and reported already. In the early investigation, Loike and Horwitz(12) found that 30μM VP-16 inhibits thymidine transport by 50% in HeLa cells and that the inhibition is completely reversible. Yalowich and Goldman(7) showed that VP-16 inhibits both transport and accumulation of thymidine at low concentrations in Ehrlich ascites cells. White et al.(13) reported the effects of VP-16 on ara-C transport and accumulation in freshly isolated human leukemic blasts. Since the intraperitoneal concentration of VP-16 decreases rapidly after i.p.injection by physiological clearance, this reversible effect of VP-16 in diminishing the membrane transport of ara-C may explain why simultaneous administration of VP-16 decreased the incorporation of ara-C and pretreatment with VP-16 did not decrease the incorporation of ara-C. The enhancement of ara-C incorporation by pretreatment with VP-16 may result from a number of factors, including the cytokinetic effect, DNA damage induced by VP-16 and following repair of DNA. VP-16 causes slowing of cell transit through late S and G2 phase of the cell cycle, producing an increase in the percentage of S phase cells (14). Since ara-C is a cell cycle phase- specific drug and its cytotoxicity and incorporation into DNA depend on the percentage of the cells in S phase, increase in ara-C incorporation by pretreatment with VP-16 can be partially explained by the cytokinetic effects of VP-16. The cytotoxic effect of ara-C depends not only on the percentage of cells in S phase but also on the rate of DNA synthesis(15). From 1 to 6 hour after intraperitoneal administration of 15 mg/kg of VP-16, the sedimentation rate of L1210 DNA was increasing on alkaline sucrose gradients. Although the rate of DNA synthesis for proliferation would be reduced strongly in this period, DNA repair from VP-16-induced DNA damage might be the vulnerable step to ara-C through an increase in ara-C incorporation into DNA.

REFERENCES

1. Sauter C.,Fehr J.,Frick P.,Gmuer J.,Honegger H., and Martz G.,1982, Acute myelogeneous leukemia:successful treatment of relapse with cytosine arabinoside, VP-16-213, vincristine, and vinblastine(A-Triple-V). Eur.J.Cancer Clin. Oncol.,18: 733-737.
2. Creutzig U.,Ritter J.,Budde M. et al.,1986, Current results of cooperative AML therapy studies in children: BFM-78 and 83. Klin. Padiatr.,198: 183-190.
3. Loike J.D., and Horwitz S.B.,1976, Effect of VP-16-213 on the intracellular degradation of DNA in HeLa cells. Biochemistry,15:5443-5448.
4. Wozniak A.J., and Ross W.E.,1983, DNA damage as a basis for 4'-demethylepipodophyllotoxin-9-(4,6-O-ethylidene-β-D-glucopyranoside)(etoposide) cytotoxicity. Cancer Res.,43: 120-124.
5. Long B.H.,Musial S.T.,and Brattain M.G.,1986, DNA breakage in human lung carcinoma cells and nuclei that are naturally sensitive and resistant to etoposide and teniposide. Cancer Res.,46:3809-3816.
6. Long B.H., Musial S.T., and Brattain M.G.,1984, Comparison of cytotoxicity and DNA breakage activity of congeners of podophyllotoxin including VP16-213 and VM-26: a quantitative structure-activity relationship. Biochemistry,23:1183-1188.
7. Yalowich J.C., and Goldman I.D.,1984, Analysis of the inhibitory effects of VP-16-213(etoposide) and podophyllotoxin on thymidine transport and metabolism in Ehrlich ascites tumor cells in vitro. Cancer Res.,44:984-989.
8. Karon M., and Shirakawa S.,1969, The locus of action of 1-β-D-arabinofuranosylcytosine in the cell cycle. Cancer Res.,29:687-696.
9. Graham F.L., and Whitmore G.F.,1970, Studies in mouse L-cells on the incorporation of 1-β-D-arabinofuranosylcytosine into DNA and on inhibition of DNA polymerase by 1-β-D-arabinofuranosylcytidine 5'-triphosphate. Cancer Res.,30,: 2636-2644.
10. Major P.P., Egan E.M., Beardsley G.P., Minden M.E., and Kufe D.W.,1981, Lethality of human myeloblasts correlates with the incorporation of arabinofuranosylcytosine into DNA. Proc. Natl. Acad. Sci. U.S.A.,78:3235-3239.
11. Kufe D.W.,Major P.P., Egan E.M.,and Beardsley G.P.,1980, Correlation of cytotoxicity with incorporation of ara-C into DNA. J.Biol.Chem.,255:8997-9000.
12. Loike J.D. and Horwitz S.B.,1976, Effects of podophyllotoxin and VP-16-213 on microtubule assembly in vitro and nucleoside transport in HeLa cells. Biochemistry,15:5435-5443.
13. White J.C.,Hines L.H.,and Rathmell J.P.,1985, Inhibition of 1-β-D-arabinofuranosylcytosine transport and net accumulation by teniposide and etoposide in Ehrlich ascites cells and human leukemic blasts. Cancer Res.,45:3070-3075.
14. Misna N.C. and Roberts D.,1975,Inhibition by 4'-demethyl-epipodophyllotoxin-9-(4,6-O-2-thenlidene-β-D-glucopyranoside) of human lymphoblast cultures in G2 phase of the cell cycle. Cancer Res.,35:99-105.
15. Shackney S.E., Ford S.S., Occhipinti S.J., Ritch P.S., Riccardi R., and Erickson B.W.Jr.,1982, Schedule optimization of hydroxyurea and 1-β-D-arabinofuranosylcytosine in sarcoma 180 in vitro. Cancer Res.,42:4339-4347.

ARA-CTP METABOLISM FOLLOWING HYDROXYUREA OR METHOTREXATE TREATMENT IN HUMAN LEUKEMIA CELL LINES

Masaru Kubota, Tetsuya Takimoto, Toshiyuki Kitoh, Akihiko Tanizawa, Yuichi Akiyama, Yukio Kiriyama and Haruki Mikawa

Department of Pediatrics,
Kyoto University, Kyoto, Japan

INTRODUCTION

Cytosine arabinoside (Ara-C) is one of the most effective drugs in the treatment of acute non-lymphocytic leukemia in childhood. Ara-C is used as a single agent; however, several drugs such as 5-fluorouracil, thymidine, methotrexate (MTX), hydroxyurea (HU) and L-asparaginase are reported to act synergistically with Ara-C [1]. In the present study, we investigated the biochemical basis for the synergism between HU or MTX and Ara-C, using human leukemia cell lines.

MATERIALS AND METHODS

Cell lines

Two human leukemia cell lines, the T lymphoblastoid Molt 4 and the the promyelocytic HL-60, were the main ones analyzed in this study. Other cell lines included were; CCRF-CEM (T lymphoblastoid), Raji and BALL-1 (B lymphoblastoid) and U937 (Histiocytic lymphoma). All cells were grown in suspension cultures in RPMI-1640 medium supplemented with 10% heat-inactivated fetal bovine serum, 2mM L-glutamine, penicillin (100U/ml) and streptomycin (100μg/ml). The cultures were kept at 37°C in a humidified atmosphere of 95% air/5% CO_2 and were checked routinely for mycoplasma.

Measurement of Ara-CTP synthesis

Logarithmically growing cells were preincubated with or without HU (1mM) or MTX (1μM) for 2 hr followed by further incubation with Ara-C for 2 hr. After that, the cells were extracted with 0.4M perchloric acid and neutralized with KOH. Ara-CTP was separated and quantified by HPLC on a Partisil 10-SAX column [2].

Ara-C cytotoxicity

The cells were pretreated with or without HU or MTX followed by further incubation with various concentrations of Ara-C. Then they were washed twice and seeded at a density of $1x10^5$/ml. After three days viable cells were enumerated by trypan blue dye exclusion and were expressed as a percent growth of control cultures.

Determination of dCTP and TTP pools

For the determination of dCTP and TTP, ribonucleotides in the cell extracts were destroyed by sodium periodate as described by Garrett and Santi [3]. The samples were analyzed by the method described previously [2].

Determination of deoxycytidine (CdR) excretion

The cells were resuspended in the medium containing dialyzed fetal bovine serum at a density of $3\text{-}4 \times 10^6$/ml. After 4 hr the supernatants were obtained and were deprotenized then neutralized. They were analyzed by HPLC on a C_{18} μ-Bondapak eluted with 1.0% acetonitrile either directly or after SEP-PAK C_{18} treatment as reported earlier [4].

RESULTS

Effects of HU or MTX treatment on Ara-CTP generation

In HL-60, BALL-1 and Raji cells, HU significantly augmented Ara-CTP production. However, in Molt 4, CCRF-CEM and U937 cells, HU did not have any effect (Table 1). In our examinations, all B cell lines belonged to the former group (termed B cell type, hereafter), whereas T cell lines belonged to the latter one (T cell type). On the other hand, MTX significantly increased Ara-CTP synthesis in both cell types, although to a lesser degree.

Effects of HU or MTX on the cytotoxic activity of Ara-C

Next, we examined the effects of HU or MTX on Ara-C induced cytotoxicity. In HL-60 cells of the B cell type, the addition of HU increased Ara-C toxicity about 20-fold, but no significant effect was observed in Molt 4 cells. In contrast, MTX showed synergy with Ara-C in both cell types, although MTX treatment itself was significantly cytotoxic (Fig.1). Thus, the differential biochemical modulation described above seemed to be correlated with the biological activity.

Table 1. Effects of HU or MTX on Ara-CTP generation

	Ara-CTP (pmol/10^6cells)		
Treatment	None	HU (1mM)	MTX (1μM)
HL-60	94	222 (236)	129 (137)
BALL-1	216	471 (218)	290 (134)
Raji	196	1130 (577)	315 (161)
Molt 4	348	410 (118)	436 (125)
CCRF-CEM	264	267 (101)	358 (135)
U937	284	257 (90)	363 (128)

The data represent the mean of more than five experiments. The numbers in parentheses show the percentage of controls.

There are several possible mechanisms for different response(s) to HU/MTX. First, the sensitivity of the target enzyme, ribonucleotide reductase, to HU is different. Although we have not done studies with the purified enzyme, it is unlikely for two reasons. [a] a 72-hr cytotoxicity study against HU revealed that both cell types were equally sentitive to the drug, the ID_{50} being around 0.05mM; [b] 1mM HU completely inhibited [^{3}H]-thymidine incorporation into DNA in both cells. Second, the effect at the level of drug transport was also negative. In fact, we measured [^{3}H]-Ara-C entry into the cells for 30-60 seconds by the silicon-oil technique and pretreatment had no effect. Third, the activation of deoxycytidine kinase (dCK) by HU or MTX in intact cells was investigated further.

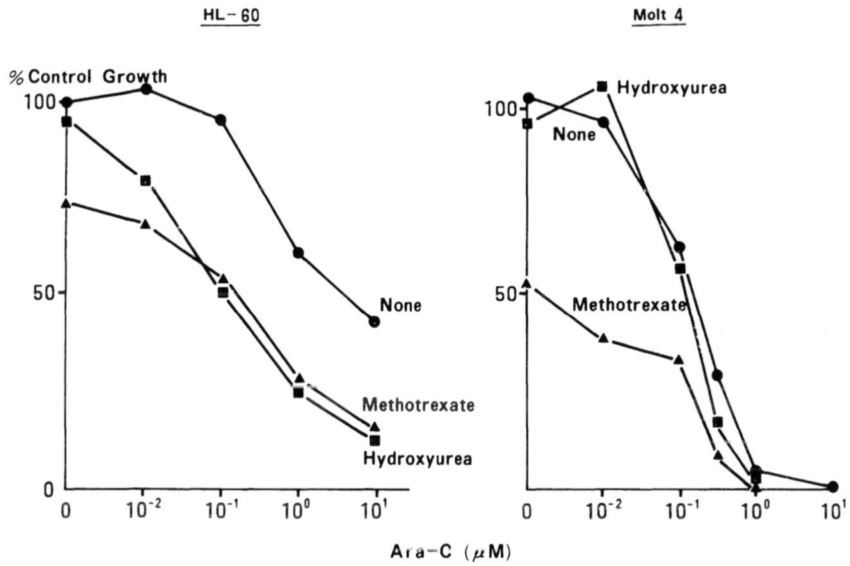

Figure 1. Effect of hydroxyurea or methotrexate on Ara-C toxicity

Changes of dCTP or TTP following drug treatment

dCK, the rate limiting enzyme of the phosphorylation of Ara-C to Ara-CTP, is known to be regulated by various factors in intact cells: i.e., dCTP interferes with the enzyme by feedback inhibition [5] and, on the contrary, TTP activates dCK either directly or through the depletion of dCTP [6]. Therefore, we examined the changes in the deoxynucleotide pools after HU or MTX exposure. HU caused a decrease of dCTP pools and an increase of TTP pools which were the same in HL-60 and Molt 4 cells. On the other hand, MTX treatment resulted in a decrease of TTP in both cell types (Table 2). These results taken together argue against the possibility that the changes of pyrimidine dNTP levels play a major role in regulating dCK during HU or MTX exposure.

Table 2. Effects of HU or MTX on dCTP and TTP pools

Cell line	Treatment	dCTP	TTP
		(pmol/10^6cells)	
HL-60	None	6.0	16.0
	HU(1mM)	2.6	12.4 (21.5)
	MTX(1μM)	7.1	3.9
Molt 4	None	7.2	42.5
	HU(1mM)	2.4	43.4 (62.1)
	MTX(1μM)	9.5	16.6

Logarithmically growing cells ($1x10^6$/ml) were incubated in the presence or absence of the drug for 4hr and the nucleotides were extracted. Data show the mean of more than five experiments. The number in parenthesis represent TTP levels at 2hr incubation.

CdR excretion

CdR has a higher affinity for dCK than Ara-C. When CdR and Ara-C were added simultaneously to HL-60 cells, Ara-CTP synthesis was inhibited in a dose-dependent manner and the cells were rescued from the cytotoxic effects of Ara-C (data not shown). So, we determined the endogenous production of CdR under incubation with HU or MTX. Since the direct measurement of CdR generation is difficult, we evaluated CdR excretion into the culture medium on the assumption that it reflects de novo CdR synthesis. Under normal culture conditions, about 3-fold higher synthetic rates of the nucleoside were observed in HL-60 cells than in Molt 4 cells (Table 3). However, in the presence of HU, the production of CdR decreased drastically in HL-60 cells at a velocity comparable to that of the T cell line. MTX did not have any significant effect on CdR synthesis (preliminary data).

DISCUSSION

Ara-CTP is an active metabolite of Ara-C. It exerts its anti-tumor effect by inhibiting either DNA polymerase α or DNA chain elongation after incorporation into DNA [7]. Thus, the synthesis and degradation of Ara-CTP is a crucial factor in the response of leukemic cells to the drug. In the present study, we evaluated the biochemical mechanism for the synergy between HU or MTX and Ara-C focusing on Ara-CTP synthesis. We have clearly shown that HU increased Ara-CTP generation in cells of B cell origin, but not in those of T cell origin, in correlation with the biological activity.

This effect seemed to be through in vivo activation of dCK, since deoxyadenosine or Ara-A were also more effectively phosphorylated by the addition of HU only in B cell type [2]. As a possible mechanism, we propose an important role of de novo CdR production instead of changes in dCTP or TTP pools. The main enzymes responsible for CdR production are ribonucleotide reductase and deoxy-5'-nucleotidase. Since deoxy-5'-nucleotidase activity is generally higher in B lymphoblasts than in T lymphoblasts [8], the production of CdR is accelerated in the former cells. Thus, under normal culture conditions, the accumulation of CdR competes with the phosphorylation of Ara-C at the level of dCK, which is more specific for the cells of B cell

Table 3. Effect of HU on deoxycytidine excretion

	Deoxycytidine excretion (pmol/hr/10^7cells)	
	Hydroxyurea (1mM)	
Cell line	(-)	(+)
HL-60 wild type	28.9±2.4	<10.0
HL-60 Ara-CR1	36.0±3.2	<10.0
HL-60 Ara-CR2	42.0±2.1	11.3±0.5
Molt 4 wild type	<10.0	not done

HL-60 Ara-CR1-2 are the mutants deficient in dCK selected in our laboraroty [2]. dCK activity is less than 1% of wild type cells. The data show the mean±SD of more than three experiments.

origin. When HU is added to inhibit ribonucleotide reductase, the inhibition of CdR synthesis is more remarkable in B cells, which results in the functional activation of dCK. In the previous meeting, Carson et al. presented two mutants which had elevated activity of either ribonucleotide reductase or deoxy-5'-nucleotidase selected on the basis of their resistance to deoxyadenosine [4,9]. These cells produced more deoxynucleosides than wild type cells, thus exhibiting cross resistance to Ara-C. The mechanism by which MTX enhances Ara-CTP generation cannot be explained by this hypothesis and needs further investigation.

REFERENCES

1. L.E.Damon and E.C.Cadman, The metabolic basis for combination chemotherapy, Pharmac.Ther. 38;73 (1988)
2. M.Kubota, T.Takimoto, A.Tanizawa, Y.Akiyama and H.Mikawa, Differential modulation of 1-beta-D arabinofuranosylcytosine metabolism by hydroxyurea in human leukemic cell lines, Biochem.Pharmac. 37;1745 (1988)
3. C.Garrett and D.V.Santi, A rapid and sensitive high pressure liquid chromatography assay for deoxyribonucleoside triphosphates in cell extracts, Analyt.Biochem. 99;268 (1979)
4. M.Kubota, C.J.Carrera, D.B.Wasson and D.A.Carson, Deoxynucleoside overproduction in deoxyadenosine-resistant, adenosine deaminase-deficient human histiocytic lymphoma cells, Biochim.Biophys.Acta. 804;37 (1984)
5. D.H.Ives and J.P.Durham, Deoxycytidine kinase III, Kinetics and allosteric regulation of the calf thymus enzyme, J.Biol.Chem. 245;2285 (1970)
6. M.S.Blumenreich, T.C.Chou, M.Andreeff, K.Vale, B.D.Clarkson and C.W.Young Thymidine as a kinetic and biochemical modulator of 1-beta-D arabinofuranosylcytosine in human acute nonlymphocytic leukemia, Cancer Res. 44;825 (1984)
7. D.W.Kufe, P.P.Major, E.M.Egan and G.P.Beardsley, Correlation of cytotoxicity and incorporation of ara-C into DNA, J.Biol.Chem. 255;8997 (1980)
8. D.A.Carson, J.Kaye and D.B.Wasson, The potential importance of soluble deoxynucleotidase activity in mediating deoxyadenosine toxicity in human lymphoblasts, J.Immunol. 126;348 (1981)
9. D.A.Carson, C.J.Carrera, M.Kubota, D.B.Wasson and T.Iizasa, Genetic analysis of deoxyadenosine toxicity in dividing human lymphoblasts, in " Purine and Pyrimidine Metabolism in Man V" W.L.Nyhan et al.ed., Plenum Press, New York, Part B-207 (1986)

RELATIONSHIP BETWEEN INTRAELLULAR dCTP/ARA-CTP RATIO AND CYTOTOXIC EFFECT OF ARA-C

Hajime Kawasaki, Masamune Higashigawa, Toshiki Ohkubo, Hitoshi Kamiya, and Minoru Sakurai

Mie Univ. School of Med. Dep.of Ped.
Tsu city, Mie pref. Japan

INTRODUCTION

Actually, in order for ara-C to exert its cytotoxic effect, ara-C must first be transported across the cell membrane and activated to ara-CTP by deoxycytidine kinase, whose activity was regulated by the intracellular dCTP level. On other side, the incorporation of ara-C into DNA is one of the cytotoxic mechanism of it(1,2). Ara-C incorporation into DNA is related to endogenous dCTP pool at the level of DNA polymerase. So the cytotoxic activity of ara-C was ultimately dependent on the intracellular dCTP level. However intracellular dCTP levels were varied on cell type, and also fluctuated under phase of cell cycle(3). So the precise mechanism of ara-C cytotoxicity is unknown. To clarify of the mechanism of ara-C cytotoxicity, we studied relations between cell growths and ratios of dCTP/ara-CTP in R.1.1. mice splenic cells, CCRF-CEM human T cell leukemic cells and ML-1 human myelogenous leukemic cells using by HPLC methods(4).

MATERIAL AND METHOD

Standard nucleotides of Sigma Co., tri-n-octylamine (Amine) of Tokyo Kasei Co., and 1,1,2-trichlorotrifluoroethan (Freon) of MERCK Co. were used. Cell growth was analyzed by counting viable cell number incubated in varying concentration of ara-C. Intracellular acidsoluble nucleotide were extracted with TCA, then neutralized with Freon and Amine according to the method of Khym et al(5). After periodate oxidation(6), deoxyribonucleotide were obtained. Intracellular ara-CTP and deoxyribonucleotide pools were analyzed by the HPLC method(Fig.1). The HPLC of ALC/GPC type, Waters Co., and TSK gel DEAE-2SW column were used.

RESULT

(1)Cell growth

Growth of R.1.1. cells was inhibited at concentration up to 10 uM of ara-C(Fig.2). Inhibitory effects to the growth of CCRF-CEM human T cell leukemic cells was shown 1 uM of ara-C or more(Fig.3). The growth of ML-1 cells was not significantly effected at concentration of ara-C up to 0.1 uM. Slightly inhibitory effects to the growth of ML-1 cells was shown 1 uM of ara-C(Fig.4).

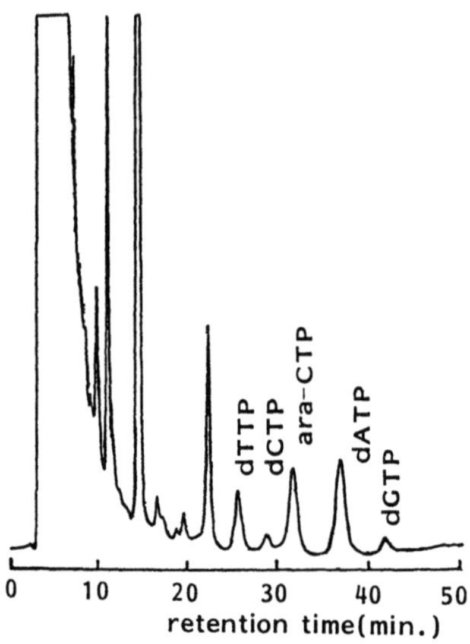

Fig.1 Ara-CTP and dNTPs profiles
instrument:Waters ALC/GPC 204
column: TSK gel DEAE-2SW(4mmx30cm)
eluent: 20% CH_3CN in 0.06M Na_2HPO_4(pH6.8)
detecter: UV 254nm

(2)Intracellular ara-CTP and dCTP levels

Intracellular dCTP and ara-CTP level in the R.1.1. cells exposure to 1 uM of ara-C were 56.0 and 48.0 pmol/10^6 cells, respectively, whereas exposure to 10 uM of ara-C, intracellular dCTP and ara-CTP level were 44.0, 351.0 pmol/10^6cells, respectively(Fig.5). Intracellular dCTP and ara-CTP level in the CCRF-CEM cells exposured to 0.1 uM of ara-C were 22.1 and 20.6 pmol/10^6cells, respectively. After incubation of ara-C 1 uM, at which concentration the inhibitory effect to the cell growth was shown, the intracellular ara-CTP level was 103.8 pmol/10^6cells, though the level of dCTP was 25.5 pmol/10^6cells(Fig.6). Same relation was shown in ML-1 cells. With ara-C concentration between 0.1 uM and 1 uM, a ratio of dCTP/ara-CTP was reversed to less than 1(Fig.7).

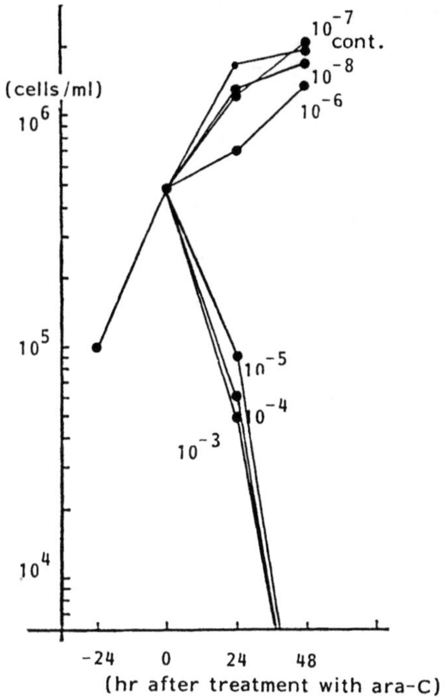

Fig.2 Effect of ara-C on growthof R.1.1. cells.

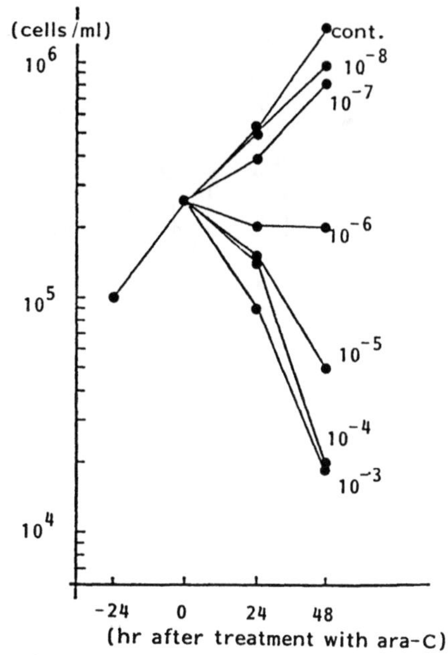

Fig.3 Effect of ara-C on growth of CCRF CEM cells.

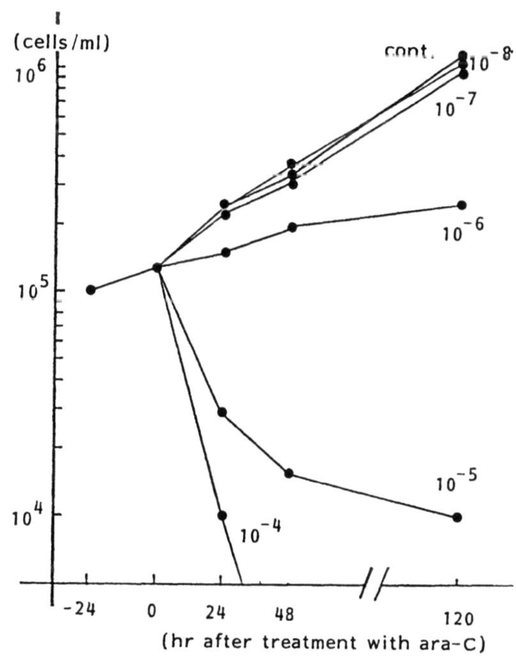

Fig.4 Effect of ara-C on growth of ML-1 cells.

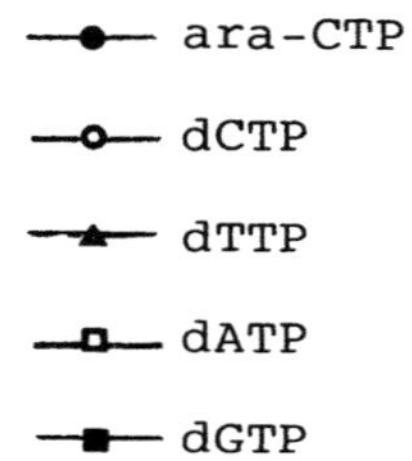

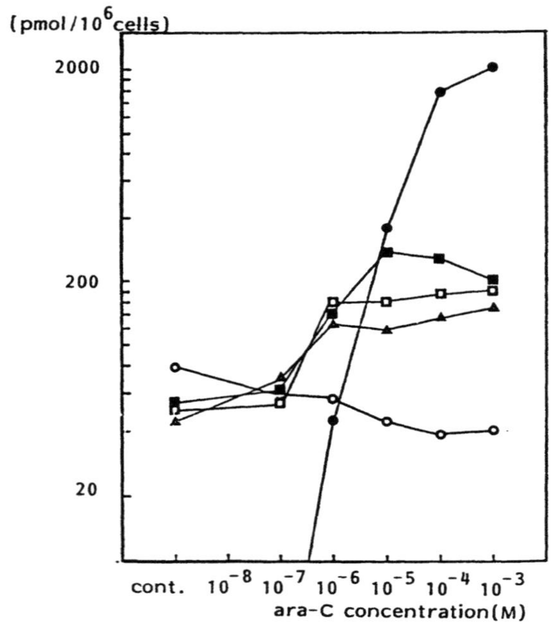

Fig.5 Relations of ara-CTP and dCTP pools in R.1.1. cells.

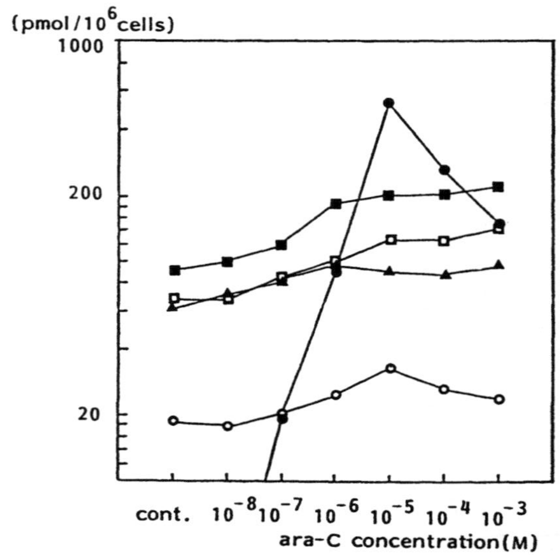

Fig.6 Relations of ara-CTP and dCTP pools in CCRF CEM cells.

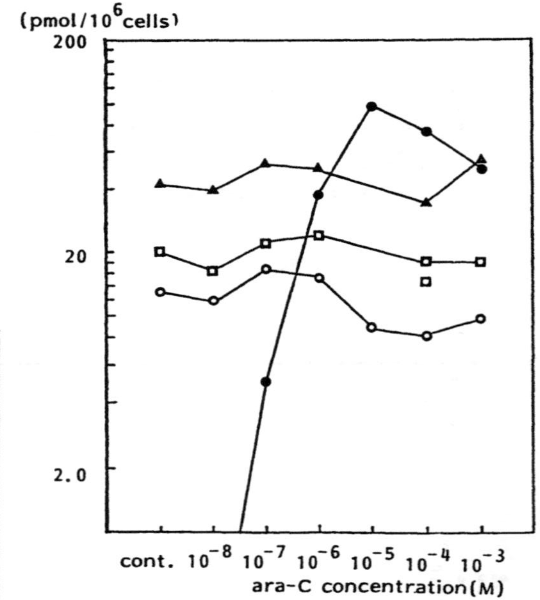

Fig.7 Relations of ara-CTP and dCTP pools in R.1.1. cells.

DISCUSSION

In MTX and ara-C sequential therapy, pretreatment of MTX causes the intracellular reduction of dTTP and dCTP pools. The reduction of dCTP level cause the activation of deoxycytidine kinase, which phosphorylate both deoxycytidine and ara-C. And thus ara-C will be more phospholylated after treatment with sequential MTX and ara-C[7,8,9,10,11]. One of the mechanism of synergistic interaction between ara-C and MTX might therefore be the stimulated phosphorylation of ara-C with MTX activated deoxycytidine kinase[12,13,14,15].

Farther more, using by HPLC, we analyzed each concentration of ara-CTP in human leukemic cells after incubation with ara-C. Ara-CTP production in relapsed cells were lower than in untreated cells[16]. Especially in relapsed AML cells, which were usually treated on the induction therapy included ara-C, were definitely lower than in untreated AML cells. In such a patient with low ara-CTP production in blast cells, low activity of intracellular deoxycytidine kinase was suspected. Hence in patients with low ara-CTP production, high-dose ara-C therapy was done. But they needed up to 10 times as conventional dose of ara-C to get effective ara-CTP level. In such a case, MTX-Ara-C sequential therapy was effective. Because the reduction of dCTP by MTX occurs an activation of deoxycytidine kinase. From these clinical experience suggested that cytotoxicity of ara-C was influenced by intracellular dCTP/ara-CTP ratio.

In this paper, we studied relations between cell growth and ratio of dCTP/ara-CTP in R.1.1. mice splenic cells, CCRF-CEM human T cell leukemic cells and ML-1 human myelogenous leukemic cells. Growth of R.1.1. cells was inhibited at concentration up to 10 uM of ara-C. Intracellular dCTP and ara-CTP level in the R.1.1. cells exposure to 1 uM of ara-C were 56.0 and 48.0 pmol/10^6cells, respectively, whereas exposure to 10 uM of ara-C, intracellular dCTP and ara-CTP level were 44.0, 351.0 pmol/10^6cells, respectively. So, at the higher level of ara-CTP than of dCTP, ara-C significantly effected on the growth of R.1.1. cells. Same relations were shown in CCRF-CEM cells and ML-1 cells. These data suggested that ara-C had significant effects on the growth of cells at the higher level of ara-CTP than of dCTP. So the dCTP/ara-CTP ratio is very important factor in ara-C cytotoxicity.

REFERENCE

(1) D.W.Kufe, D.Spriggs, M.Egan, D.Munroe,Relationships among ara-CTP pools,formation of (ara-C)DNA, and cytotoxicity of human leukemic cells, Blood,64:54 (1984).

(2) D.W.Kufe, D.Munroe, D.Herrick, M.Egan and D.Spriggs, Effects of 1-B-D-arabinofuranosylcytosine Incorporation on Eukryotic DNA Template Function, Molecular Pharamaco., 26:128 (1984).

(3) J.O.Liliemark, W. Plumkett, Regulation of 1-B-D-arabinofuranosylcytosine-5'-triphosphate accumulation in human leukemic cells by deoxycytidine-5'-triphosphates, Cancer Res.,46:1079 (1986).

(4) H.Kawasaki, T.Ohkubo, T.Nobori, H.Kamiya, M.Sakurai, Analysis of intracellular deoxyribonucleoic acid pool using HPLC equipped with DEAE-2SW column, Igaku no Ayumi 137:1019 (1986).
(5) J.X.Khym, An analytical system for rapid separation of tissue nucleotides at low pressure on conventional anion exchangers, Clini.Chem. 21:124 (1975).
(6) A.L.Pogolotti and D.V.Santi, High-pressure liquid chromatography-ultraviolet analysis of intracellular nucleotides, Anal.Biochem. 126,335 (1982).
(7) T.Ohkubo, Biochemical modulation of 5-fluorouracil and cytosine arabinoside by sequential administration of methotrexate, Mie Med.J. 35:155 (1985).
(8) H.Kawasaki, T.Nobori, H.Kamiya, M.Sakurai, Biochemical modulation in cancer chemotherapy, Jpn.J. Cancer Chemother. 12:815 (1985)
(9) T.Nobori, H.Ochiai, N.Shimizu, H.Kawasaki, Enhanced production of intracellular ara-CTP after sequential use of methotrexate and 1-B-D-arabinofuranosyl-cytosine, Leukemia Res. 9:1263 (1985).
(10) J.R.Bertino, Clinical use of methotrexate with emphasis on use of high doses, Cancer Treat.Rep. 65:131 (1981).
(11) T.L.Avery and D.Robert, Dose-related synergism of cytosine arabinoside and methotrexate against murine leukemia L1210, Eu.J.Cancer 10:425 (1974).
(12) H.Kawasaki, T.Ohkubo, T.Nobori, H.Kamiya, M.Sakurai, The effect of Methotrexate on Intracellular Nucleotide pools, Jpn.J. Cancer Chemother. 12:815 (1985).
(13) T.Nobori, H.Ochiai, N.Shimizu, H.Kawasaki, Effects of methotrexate on the intracellular nucleotide pools and its role, in combination chemotherapy, In: "Purine and purine metabolism in man V," W.L.Nyhan, L.F.Thompson and R.W.E. Watts, ed., Plenum Publishing Corp, New York (1986).
(14) R.C.Jackson and R.J.Harkrader, Synergistic and antagonistic interactions of methotrexate and 1-B-D-arabinofuranosylcytosine in hepatoma cells, Biochem. Pharmacol. 30:223 (1981).
(15) E.Cadman and F.Eifeman, Mechanism of synergistic cell killing when methotrexate precedes cytosine arabinoside, J.Clin.Invest. 64:788(1981).
(16) T.Nobori, H.Kamiya, Mechanism of resistance to anticancer drugs and trials to overcome their resistance, Jpn.J.Clin. Hematol, 27:1460 (1986).

THE RELATIONSHIP BETWEEN DIHYDROOROTIC ACID DEHYDROGENASE AND IN VITRO AND IN VIVO CYTOSTATIC EFFECTS OF BREQUINAR SODIUM (DUP-785; NSC 368390)

Godefridus J. Peters, Emile Laurensse, Erik de Kant
Jorge C. Nadal, and Herbert M. Pinedo

Department of Oncology, Free University Hospital
PO Box 7057, 1007 MB Amsterdam, the Netherlands

INTRODUCTION

Dihydroorotic acid dehydrogenase (DHO-DH) is a critical enzyme in the de novo pyrimidine nucleotide synthesis (Fig. 1). It is the fourth enzyme in this pathway, and is located on the outer site of the inner membrane of the mitochondrion [1,2]. Brequinar Sodium (BS; DUP-785; NSC 368390) is a novel potent inhibitor of DHO-DH [3,4] with a Ki varying between 10 and 100 nM. Treatment of cells led to an accumulation in the S phase [4]. The growth-inhibitory effects of BS could be prevented and reversed by addition of uridine or cytidine to the cell culture medium [4,5], but not by thymidine or deoxycytidine. So, a selective depletion of one of the pyrimidine deoxynucleotides is not likely to be responsible for the growth-inhibitory effects of BS. Depletion of pyrimidine deoxyribonucleotides apppeared to be proportional to that of the ribonucleotides [5], and could be reversed by uridine. Inhibition of UMP synthesis appears to be crucial for the in vitro action of BS. To get more insight in the role of DHO-DH in the synthesis of pyrimidine nucleotides and in the action of BS we measured the activity of DHO-DH and the inhibition by BS in several cell lines from different histological origin and determined their sensitivity to BS.

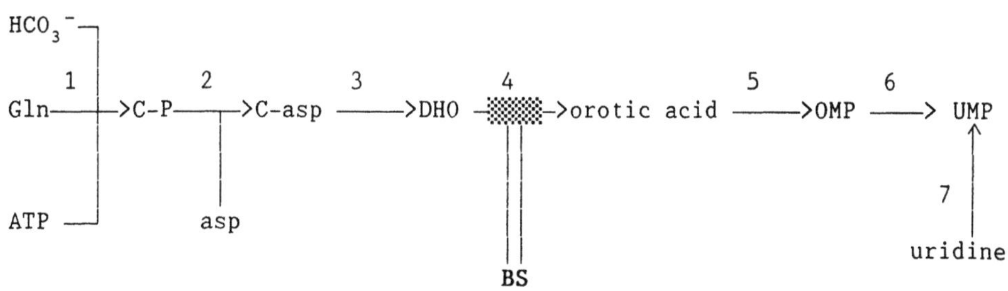

Fig. 1: Pyrimidine de novo nucleotide synthesis. The enzymes are; 1, carbamyl-P synthetase II; 2, aspartate transcarbamylase (ATC); 3, DHOase; 4, DHO-DH; 5, orotate phosphoribosyltransferase (OPRT); 6, OMP decarboxylase; 7, uridine kinase. Gln, glutamine; C-P, carbamyl-P; asp, aspartate; C-asp, carbamyl-asp; DHO, dihydroorotic acid. The bar represents inhibition by BS.

In vitro we demonstrated that a short term exposure of cells to BS followed by culture in drug-free medium did not affect growth rate of the cells [4,5]. This appeared to be related to the retention of inhibition of DHO-DH [4] and depletion of pyrimidine nucleotides [5]. A prolonged presence of BS was essential to maintain both enzyme inhibition and nucleotide depletion. In vitro it is easily possible to manipulate the levels of possible rescue compounds in the medium, e.g. by the use of dialyzed serum. Cell lines cultured in dialyzed serum were more sensitive to BS than when cultured in non-dialyzed serum [4]. In vivo uridine is present at relatively high concentrations in tissues, compared to e.g. plasma [6,7], which might affect the antitumor activity of BS. In order to study the in vivo antipyrimidine effects of BS we used two murine colon tumors as models in which we measured the retention of the inhibition of DHO-DH and the effect of BS on nucleotide pools.

MATERIALS AND METHODS

BS was synthesized and obtained from the Medicinal Chemistry Section, DuPont Pharmaceuticals, Wilmington, Delaware, USA. Pyrimidine compounds were obtained from Sigma, Ohio, while other compounds were of analytical grade quality. The origin of cell culture media and sera has been described previously [4,5,8]. Female C57Bl/6 and Balb-c mice were obtained from the animal breeding station "Proefdieren-bedrijf TNO", Zeist, the Netherlands and were kept in an area with standardized light-dark cycle for at least 10-14 days prior to the beginning of an experiment. Mice had excess to food and water ad libitum. The sources, growth characteristics and evaluation of antitumor activity of Colon 26 and Colon 38 (both murine colon adenocarcinomas) have been described previously [9, 10]. Colon 26 was maintained in Balb-c and Colon 38 in C57Bl/6 mice, both subcutaneously in the thoracic region.

Cell lines were maintained as described previously in 10% heat-inactivated non-dialyzed fetal bovine serum [4,5,8]. Growth inhibition was determined by comparison of cell number at the start and the end of culture. Only cells in logarithmic phase were used for enzyme assays. Cells were harvested, counted, centrifuged and frozen at -70°C. The cell pellets were suspended in 0.1 M Tris-HCl (pH 8.0) and lysed by sonication, then DHO-DH was assayed as described using a sensitive HPLC method [12]. The enzyme was assayed at 158 μM L-DHO and inhibition was measured at 1.3 μM BS. The antitumor activity of BS was determined by i.p. injec-

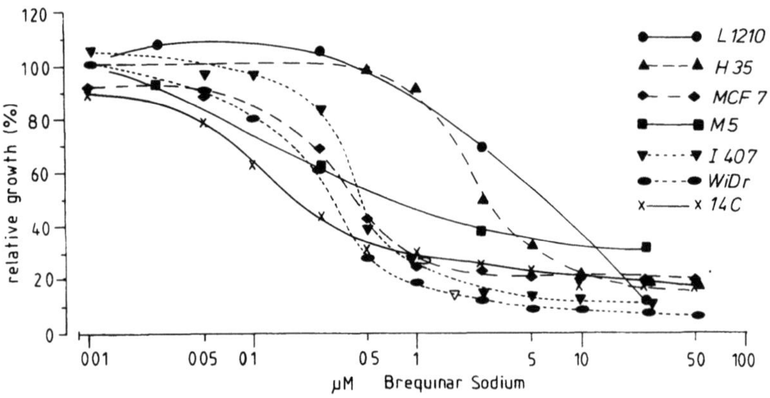

Fig. 2. Growth-inhibition of 7 cell lines by BS. Cells were continuously exposed to BS for 48 hr and relative growth was determined as described previously [7]. Mean values of 3-4 separate experiments are shown. Values of M5 and L1210 are from [4].

tion of BS; mouse weight and tumor size was measured twice weekly; tumor size and doubling times were calculated relative to that of day 0 and used for evaluation of antitumor effect by means of the growth delay factor (GDF) as described [9,10].

The effects of BS on the activity of DHO-DH and the levels of pyrimidine nucleotides were determined after a single i.p. injection of BS at 50 mg/kg. Mice were killed by cervical dislocation and tumors were removed immediately. Mitochondria were prepared [11] and used for the measurement of DHO-DH essentially as described above for cell lines. Other tumors were immediately frozen in liquid nitrogen and used for the measurement of pyrimidine nucleotides with anion-exchange HPLC as described [6] after pulverization using a microdismembrator [6,12].

RESULTS

In vitro sensitivity and DHO-DH activity

Seven cell lines from different histological origin were tested for their sensitivity to BS (Fig. 2). Cells were continuously exposed to BS, since we had demonstrated previously that at short term exposure the growth-inhibition was not retained [4,5]. 14C was the most sensitive and L1210 and H35 were the most resistant cell lines with a more than 14-fold difference in sensitivity. However, in none of the cell lines a complete growth-inhibition was observed at higher concentration, even at 50 μM BS some cell growth was observed. A relationship was observed between growth rate and sensitivity. The most sensitive line, 14C, was also the most slowly growing line, while the most resistant line, L1210, was the fastest growing cell line (Fig. 3).

DHO-DH activity was measured at a saturating substrate concentration and the inhibition of 1.3 μM BS was determined. The activity was lowest in 14C, intermediate in M5, and about 2-times higher in the other cell lines. The extent of inhibition by BS varied considerably among the tested cell lines. The inhibition was highest in 14C and MCF7 cells, intermediate in WiDr and Intestine 407, and moderate in H35 and L1210 cells. It appeared that the most sensitive cell line also displayed the lowest activity of DHO-DH. However, the other cell lines had a comparable activity although there was more than ten-fold difference in sensitivity. A good correlation was observed between sensitivity to BS and the remaining DHO-DH activity in the presence of BS. For 14C the highest enzyme inhibition was observed, while H35 and L1210 showed a slight inhibition.

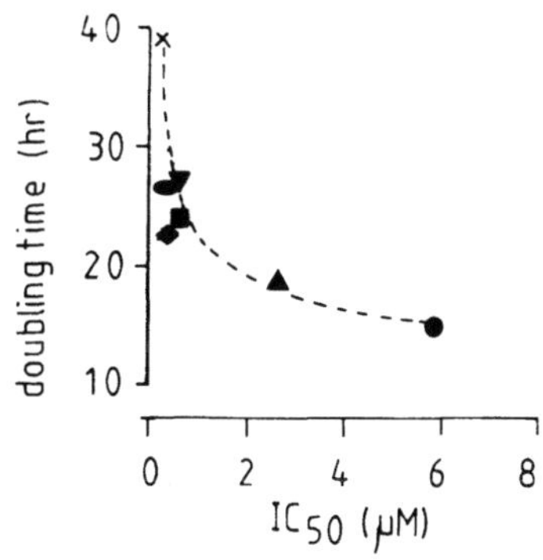

Fig. 3. Relationship between growth rate of the cell lines and sensitivity to BS. Symbols are the same as in Fig. 2.

Table 1. IC50 FOR BS AND ACTIVITY OF DHO-DH IN CELL LINES.

Cell line	Origin	IC50 (μM)	DHO-DH activity - BS	DHO-DH activity + BS
14C	squamous cell ca.	0.19 ± 0.07	7.6 ± 0.5	0.2 ± 0.1
WiDr	colon ca.	0.32 ± 0.06	17.9 ± 2.5	2.2 ± 0.5
M5	melanoma	0.39 ± 0.06	10.0; 8.2	0.5
MCF7	breast ca.	0.42 ± 0.09	16.5 ± 2.7	0.4 ± 0.1
I407	tr.epithelial intestine	0.45 ± 0.09	16.1 ± 2.1	2.0 ± 1.1
H35	rat hepatoma	2.53 ± 0.29	15.0 ± 1.4	7.1 ± 1.9
L1210	murine leukemia	5.81 ± 2.10	16.3 ± 1.2	9.5

Enzyme activity is given as nmol/hr per 10^6 cells. Values are means ± SE of 3-5 separate experiements. Inhibition of DHO-DH was determined in the presence of 1.3 μM BS. Ca., carcinoma; tr, transformed; I407, Intestine 407. All cell lines were from human origin, unless otherwise indicated.

Table 2. ANTITUMOR EFFECT OF BS ON COLON 26 AND COLON 38, IN RELATION TO DHO-DH ACTIVITY AND INHIBITION BY BS

Tumor	Doubling time (days)	GDF 5xD	GDF Dx5	DHO-DH activity control	DHO-DH activity + 1.3 μM BS
Colon 26	2.7	0.04	0.69	551 ± 71	28.5 ± 3.4
Colon 38	5.1	0.62	2.12	123 ± 24	12.2 ± 3.8

BS was injected i.p. at 50 mg/kg at day 0, 4, 8 and 12 (5xD) or daily at 25 mg/kg at day 0-5 (Dx5); GDF < 1, insensitive; > 1, sensitive. DHO-DH activity (nmol/hr per mg mitochondrial protein) was determined in untreated tumors; values on enzyme activity are means ± SE of 4 tumors.

In vivo sensitivity and antipyrimidine effects of BS

The *in vivo* effects of BS were studied in two murine colon carcinomas. A better antitumor effect was observed at at daily administration in both tumor lines (Table 2). Colon 26 was not sensitive to BS and had a high basal activity of DHO-DH, while inhibition of DHO-DH was more than 80%. Colon 38 was more sensitive to BS than Colon 26 and showed a lower basal activity of DHO-DH. In these tumors we also studied the retention of antipyrimidine effects, i.e. retention of inhibition of DHO-DH and the effect of BS on pyrimidine nucleotide pools in the tumors. In both tumor lines a single injection of 50 mg BS/kg resulted in an almost complete inhibition of DHO-DH after 4 hr (Fig. 4), but was recovered considerably after 1 day in both tumor lines. In Colon 38 DHO-DH activity returned to basal levels after 4 days. In Colon 26 DHO-DH activity remained partially inhibited and did not return to pretreatment levels within 4 days, although the absolute activity in treated tumors was never lower than in treated Colon 38 at the same day.

The *in vivo* antipyrimidine effects of BS were also studied by measurement of the pyrimidine nucleotide levels at several time-points after one injection of BS. Only total pyrimidine nucleotides (UΣP and CΣP) were measured, by degradation of UTP, UDP and UDP-sugars to UMP and CTP, CDP and cytidine nucleotide sugars to CMP [6]. CΣP were comparable in Colon

26 and Colon 38 and amounted to about 200 pmol/g wet weight. BS had not effect on CΣP in both tumor lines (data not shown). However, in Colon 38 BS caused a depletion of UΣP after 1 day, followed by a rebound after 2 and 4 days (Fig. 5). In Colon 26 an increase in UΣP was observed after 2 and 4 hr, after 1 day UΣP levels decreased to about 80% of control levels followed by a rebound after 2 and 4 days.

DISCUSSION

Seven different cell lines from human and animal origin differed widely in their sensitivity to BS. It appeared that the activity of DHO-DH is not an important determinant of the sensitivity to BS. The most sensitive cell line had the lowest DHO-DH, while the other cell lines showed a higher enzyme activity, but still displayed a large variation in

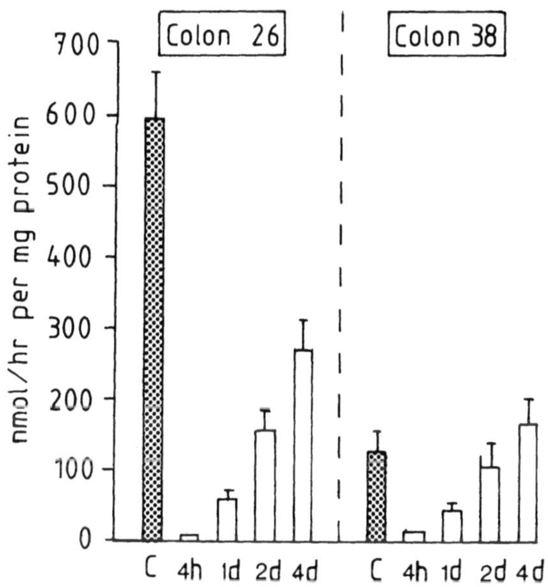

Fig. 4. Effect of one injection of BS at 50 mg/kg on the DHO-DH activity of Colon 26 and Colon 38 tumors. Values represent means ± SE of 3-7 tumors.

sensitivity. It might be that only very low or very high activities are determinants for sensitivity or resistance, respectively. However, the latter could not be investigated in the present study, because of the relatively low variation in DHO-DH activity. For another pyrimidine antimetabolite, N-phosphonacetyl-L-aspartate (PALA) an inhibitor of ATC, a relationship between enzyme activity and sensitivity could be demonstrated [13,14]. Induced resistance against PALA appeared to be associated with increased enzyme activity [15,16]. A better correlation was observed between the activity of DHO-DH in the presence of BS and the sensitivity of these cell lines to BS. The most sensitive cell line, 14C, had the lowest remaining activity. Also in vivo the enzyme activity and its inhibition correlated with the antitumor activity, although only data of two tumor lines are presented. Another determinant for the sensitivity to BS appears to be the proliferation rate of the cells. This was also observed for the two tumor lines. A less clear relationship between proliferation rate and sensitivity was observed for the effect of PALA

[13,14]. We recently demonstrated that BS caused an accumulation of cells in the S-phase [5], but this block was less for the fast growing L1210 cells. So more cells might pass through the cycle and (partly) explain the lower sensitivity.

In cell lines treatment with BS resulted in a marked reduction of pyrimidine nucleotides [3,5]. Retention of depletion of UTP and CTP and of DHO-DH inhibition [4] appeared to be determinants of the extent and duration of growth-inhibition. Data shown in Fig. 5 suggest that in vivo nucleotide depletion is related to the antitumor effect of BS. In Colon 38, which is more sensitive than Colon 26, UΣP decreased. This decrease, however, was followed by a rebound in nucleotide pools. Also in Colon 26 the pyrimidine nucleotide pool was higher after 4 days. The effects of BS

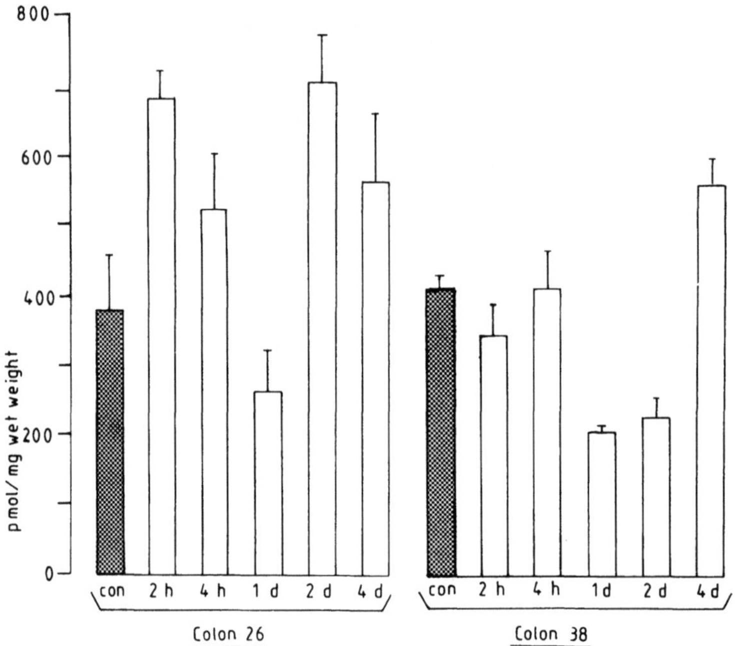

Fig. 5. Effect of BS on the levels of UΣP in Colon 26 and Colon 38. Values are means ± SE of 3-14 tumors.

are unlike to those observed with PALA [16-19] or pyrazofurin, an inhibitor of OPRT [20]. Extent of depletion, observed with BS was less, while the rebound was not observed with PALA. In most studies pools were only measured at 24 hr. However, we observed in B16 cells that both UTP and CTP remained depleted for 5 days [19]. Furthermore, in Lewis lung carcinoma one injection of PALA resulted in inhibition of ATC which was retained for 6 days in a sensitive line [16]. Retention of DHO-DH inhibition was only a few days in the murine tumors and parallelled the effects on UΣP. It should be noted that although the extent of inhibition in the Colon 26 was higher than in Colon 38, that the absolute DHO-DH activity in Colon 26 was never lower than in Colon 38. So, it might be concluded that also in vivo the absolute activity in the presence of or after BS is important. Recently we demonstrated that in lymphocytes of patients treated with BS that DHO-DH was inhibited for up to one week [21]. The extent of antipyrimidine effects appeared to be related to toxicity.

Sensitivity of cells and tumors to BS might also be influenced by other factors than activity of DHO-DH. High pools of intracellular or intratumoral salvage precursors and/or an effective salvage pathway might result in a bypass of pyrimidine de novo nucleotide inhibition. In cell lines the pool of precursors can be regulated by the use of dialyzed serum. However, in vivo it has been demonstrated that tissues can have high uridine pools [6,7]. In Colon 38 we measured a uridine pool of about 100 µmol/kg and in Colon 26 of about 10 µmol/kg, which are both higher than the plasma level. Although this might imply an effective rescue in Colon 38, this is apparently not the case, since this tumor is more sensitive than Colon 26. This might be related to the difference in uridine metabolizing enzymes. Elsewhere in these proceedings [22] we demonstrate that the activities of these enzymes in Colon 26 are several times higher than in Colon 38.

Despite the limited antipyrimidine effects (at least in comparison to PALA or pyrazofurin) BS shows a broad antitumor activity against both experimental tumors and human xenografts [23]. The role of DHO-DH in cellular function is not clear. The activity of DHO-DH in the M5 and WiDr is higher than that of OPRT [24]. During hepatocarcinogenesis DHO-DH is decreased in activity [25], while cells with defective DHO-DH did not require exogenous uridine for growth [26]. DHO-DH is an unique enzyme in pyrimidine pathway since it is located in the mitochondrion [1,2] and it is coupled to the electron transport chain. Thus, it has been demonstrated that under hypoxic conditions DHO-DH activity is decreased [27], while hypoxic conditions also appeared to be synergistic with another inhibitor of DHO-DH. Since most tumors are more or less hypoxic this might be an important determinant of the antitumor activity of BS. E.g. the Colon 38 is a very necrotic tumor and thus hypoxic. DHO-DH might represent an unique target for cancer chemotherapy.

In conclusion, the sensitivity to BS is related to the extent of DHO-DH inhibition in vitro. In vivo retention of DHO-DH inhibition parallelled the effects on UΣP. However, the observed antipyrimidine effects might only be a marker but not a determinant of the effects of BS.

ACKNOWLEDGEMENTS

This work was supported by the Netherlands Cancer Foundation (Koningin Wilhelmina Fonds) by grant IKA 83-16 and by DuPont de Nemours & Co, Geneva, Switzerland and Wilmington, DE, USA. Dr. G.J. Peters is a recipient of a senior fellowship of the Netherlands Academy of Sciences.

REFERENCES

1. Jones ME (1980). Ann Rev Biochem 49, 253-279.
2. Peters GJ, Veerkamp JH (1984). Adv Exp Med Biol 165A, 531-534
3. Chen SF, Ruben RL, Dexter DL (1986). Cancer Res 46, 5014-5019
4. Peters GJ, Sharma SL, Laurensse E, Pinedo HM (1987). Invest New Drugs 5, 235-244
5. Schwartsmann G, Peters GJ, Laurensse E, De Waal FC, Loonen AH, Leyva A, Pinedo HM (1988). Biochem Pharmacol, in press
6. Peters GJ, Van Groeningen CJ, Laurensse EJ, Lankelma J, Leyva A, Pinedo HM (1987). Cancer Chemother Pharmacol 20, 101-108
7. Darnowsky JW, Handschumacher RE (1986). Cancer Res 46, 3490-3494

8. Peters GJ, Laurensse E, Leyva A, Lankelma J, Pinedo HM (1986). Cancer Res 46, 20-28
9. Peters GJ, Van Dijk J, Nadal JC, Van Groeningen CJ, Lankelma J, Pinedo HM (1987). In Vivo 1, 113-118
10. Peters GJ, Van Dijk J, Van Groeningen CJ, Laurensse EJ, Leyva A, Lankelma J, Nadal JC, Pinedo HM (1988). Brit J Cancer 57, 259-265
11. Peters GJ, Laurensse E, Leyva A, Pinedo HM (1987). Anal Biochem 161, 32-38
12. Peters GJ, Laurensse E, Leyva A, Pinedo HM (1986). Clin Chim Acta 158, 193-198
13. Leyva A, Appel H, Smith P, Lankelma J, Pinedo HM (1981). Cancer Lett 12, 169-173
14. Johnson RK, Swyryd EA, Stark GR (1978). Cancer Res 38, 371-378.
15. Kempe TD, Swyryd EA, Bruist M, Stark GR (1976). Cell 9, 541-550
16. Kensler TW, Mutter G, Hankerson JG, Reck LJ, Harley C, Han N, Ardalan B, Cysyk RL, Johnson RK, Jayaram HN, Cooney DA (1981). Cancer Res 41, 894-904
17. Moyer JD, Handschumacher RE (1979). Cancer Res 39, 3089-3094.
18. Martin DS, Stolfi RL, Sawyer RC, Spiegelman S, Casper ES, Young CW (1983). Cancer Res 43, 2317-2321
19. Lankelma J, Peters GJ, Laurensse EJ, Leyva A, Pinedo HM (1983). Proc ECCO 2, Abstract 02-37
20. Brockman RW, Shaddix SC, Rose LM (1977). Cancer (Phila.) 40, 2681-2691.
21. Peters GJ, Nadal JC, Schwartsmann G, De Kant E, Laurensse EJ, Pinedo HM (1988). Proc AACR 29, 350 (Abstract 1392)
22. Laurensse EJ, Braakhuis BJM, Pinedo HM, Peters GJ (1988). Adv Exp Med Biol, these proceedings
23. Dexter DL, Hesson HP, Ardecky RJ, Rao GV, Tippett DL, Dusak BA, Paull KD, Plowman J, DeLarco BM, Narayanan, VL, Forbes M (1985). Cancer Res 45, 5563-5568
24. Peters GJ, Laurensse E, Leyva A, Pinedo HM (1985). Int J Biochem 17, 45-49
25. Elliott WL, Sawick DP, DeFrees SA, Heinstein PF, Cassady JM, Morré DJ (1984). Biochim Biophys Acta 800, 194-201
26. Stamato TD, Patterson D (1979). J Cell Physiol 98, 459-468
27. Löffler M (1980). Eur J Biochem 107, 207-215

ESTIMATION OF THE IN VITRO AND IN VIVO INHIBITORY EFFECTS OF ANTIFOLATES UPON THYMIDYLATE SYNTHASE (TS) IN WHOLE CELLS

Gordon A. Taylor, Ann L. Jackman, Kathryn Balmanno
Leslie R. Hughes#, and A. Hilary Calvert

Inst. Cancer Res. Drug Development Section, Sutton, Surrey
and #ICI Pharmaceuticals, Macclesfield, Cheshire, UK

INTRODUCTION

Thymidylate synthase (TS) is the final enzyme in the de novo pathway of thymidine nucleotide synthesis catalysing the reductive transfer of a methyl group from 5,10-methylene tetrahydrofolate to deoxyuridylate (dUMP) to produce thymidylate (TMP). This enzyme may be inhibited by analogues of either the substrate dUMP (e.g. 5 Fluoro dUMP) or the folate co-factor (e.g. CB3717 [10-propargyl- 5,8-dideazafolic acid], structure 6, Table 1). Antifolates tend to be more specific in their locus of action than pyrimidine based inhibitors, as they cannot form metabolites which may be incorporated into nucleic acids. Many folate antimetabolites and natural folates are substrates for folylpolyglutamate synthetase (FPGS), an enzyme which catalyses the addition of glutamate residues to form a negatively charged polyglutamate chain. It is thought that the formation of polyglutamates serves primarily to retain natural folates within the cells (analogous to the retention of nucleosides by phosphorylation to nucleotides), the polyglutamated forms being the prefered substrates for enzymes using folate cofactors. Antimetabolites which are substrates for FPGS would, by analogy, be expected to be retained within cells, as has been observed with methotrexate[1]. The polyglutamated forms of CB3717[2,3] and some 5,8-dideazafolate (DAF) analogues[4] have been shown to have a higher affinity for mammalian TS than their non polyglutamated counterparts.

The evaluation of antimetabolites affecting thymidine nucleotide synthesis using rodent tumour models is complicated by species differences in the circulating levels of deoxy-pyrimidine nucleosides. Indeed the concentration of thymidine in mouse plasma has been shown to be greater than that required to circumvent the cytotoxicity of CB3717 to L1210 cells in tissue culture[5]. This is thought to be at least in part responsible for the lack of antitumour activity seen with CB3717 in most rodent tumour screens. In man the circulating level of thymidine is approximately 10 fold lower than in mice[6] and is therefore less readily available for the circumvention of inhibition of TS.

The selection of TS as a suitable antineoplastic locus has been vindicated by the demonstrable clinical efficacy of CB3717[7-8]. Unfortunately unpredictable liver and kidney toxicities were responsible for the withdrawal of CB3717 from clinical evaluation. Part of our search for less toxic analogues of CB3717 has involved the synthesis of 2-desamino-2-substituted-5,8-dideazafolic acid analogues with various substitutions at the N^{10} position[9-11] (Table 1). In order to evaluate these

Table 1. Structures of 5,8-dideazafolic acid analogues

R (C^2)	R' (N^{10})	
Methyl	H	1
Methyl	Methyl	2
Hydroxymethyl	Methyl	3
Methyl	Ethyl	4
Methyl	Hydroxymethyl	5
Amino	Propargyl	6
Methyl	Hydroxypropyl	7
Desamino	Propargyl	8
Methyl	Propargyl	9
Fluoromethyl	Propargyl	10
Ethyl	Propargyl	11
Hydroxymethyl	Ethyl	12
Hydroxymethyl	Ethyl	13
Methoxy	Propargyl	14

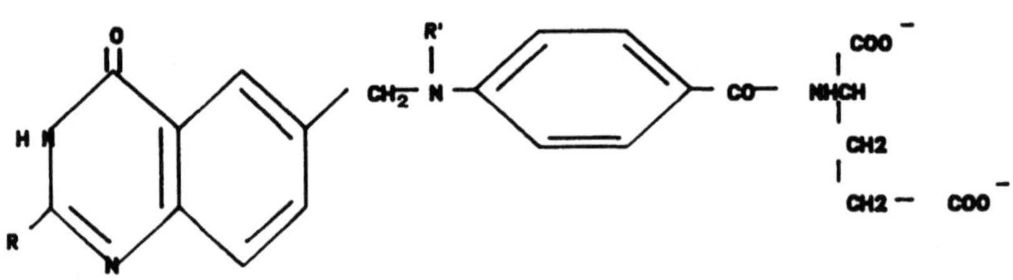

analogues it was necessary to develop methods which would allow the activity of the TS to be monitored in intact cells. Radio-labelling using nucleic acid precursors would be the obvious method but standard procedures are not valid if the specific activity of the substrate varies, e.g. due to pool size fluctuations, following exposure to an antimetabolite. The determination of TS activity in intact cells by the method of Yalowich and Kalman[13] was not used as this did not lend itself to the more detailed biochemical studies we propose to undertake with selected analogues of interest. The approach of Nicander et al[14] has been adopted as this permits an in depth investigation of the relevant pathways as well as individual enzymes. The TS activity in intact cells was estimated by monitoring ^3H_2O, released during the reaction catalysed by TS, in cells exposed to [5-^{3}H] UdR. The rate of ^3H_2O release was determined following the attainment of a state of dynamic equilibrium. As the specific activity of dUMP remained essentially constant during the period in which the rate of ^3H_2O release was determined the activity of folate based TS inhibitors can be compared without the necessity for complex specific activity determinations.

We report here some of our preliminary structure activity studies where analogues (Table 1) were compared with respect to their ability to inhibit TS in intact L1210 cells in vitro at approximately equitoxic doses. In addition the in vivo inhibition of TS in an L1210 tumour line (NCI L1210) which is totally refractory to CB3717, is compared for 10-propargyl DAF analogues modified at the 2 position.

MATERIALS AND METHODS

The [5-^{3}H] UdR was purified and the specific activity determined by HPLC immediately prior to each experiment. Initially [5-^{3}H] UdR was not readily available[1] and was obtained using alkaline phosphatase (Bovine Intestine, SIGMA, UK) to dephosphorylate [5-^{3}H] dUMP (ammonium salt 34.4 mCi/mg, Amersham, UK). Up to 5mls (5 mCi) of [5-^{3}H] dUMP in ethanol water 1:1 v/v was dried under reduced pressure at 45°C, resuspended in 100μl of 0.3M ammonium carbonate containing 6 units of alkaline phosphatase, and incubated at 37°C for 1hr. After drying under reduced pressure at 45°C, the label was dissolved in 200μl of water and injected onto a 10cm x 4.6mm id Apex 3μ C18 column (Jones Chromatography, UK) running isocratically in 0.05M ammonium acetate pH 4.7 at 2ml/min. The UdR peak, with a retention time of -250 seconds, was collected and diluted with unlabelled UdR to a concentration of 30μM.

In vitro studies

The dose required to reduce the cell count to 50% of control after 48 hours (IC_{50}) was determined for suspension cultures of L1210 cells in RPMI1640 medium containing 10% horse serum. Duplicate suspension cultures of L1210 cells were exposed for 4 hours to the DAF analogues at 10x IC_{50} The cells were then incubated in the presence of 0.3μM [5-^{3}H] UdR either directly in the presence of the analogues or immediately following the washing of the cells to remove exogenous inhibitors. Triplicate 400μl aliquots were taken at various times, between 15 and 90 minutes, into microfuge tubes containing 400μl of ice cold 1M PCA. A further 400μl sample was diluted with 9.6mls of Isoton and the cell concentration of each cell suspension determined using a Coulter counter model ZM. After -20 minutes at 0°C 500μl of a suspension of activated charcoal (200mgs/ml activated charcoal, 10mgs/ml Dextran T-70 in water) was added to each tube. Fifteen minutes later the tubes were spun for 4 mins at 13400g to sediment the charcoal together with the adsorbed UdR. The ^3H_2O released from the cells was estimated by counting aliquots (500μl) of the supernatant in 5mls of scintillant using a Packard Tricarb 2000 scintillation counter.

In vivo studies

L1210/NCI tumour cells were maintained in DBA2 mice and passaged (i.p.) into male BDF1 hybrid mice such that the mice had a tumour burden of -5x10^7 cells on the day of treatment. The mice were injected with the agents under study via the tail vein after brief (-20 seconds) local hyperthermia in water (47°C). At the times indicated the tumour cells were harvested by flushing the peritoneal cavity with warm (37°C) medium (RPMI1640) and centrifuged for 90 seconds at 1200g. The cells were resuspended to a concentration of -10^6 cells/ml in warm medium and 6 ml dispensed into a tube containing 60μl of [5-^{3}H] UdR. Samples were taken and manipulated as described above for the in vitro studies.

Data manipulation

Data from the HPLC separations were analysed using a Trilab 2000 datasytem (Trivector Scientific Ltd., Sandy, UK). Disintegrations per minute (dpm), calculated by the scintillation counter, were stored on floppy disc and subsequently coordinated with appropriate cell counts and sample times in a sreadsheet (Symphony, Lotus Development Corpn.). After conversion to an appropriate format the sample data (times and dpm/10^6 cells) were imported into Statgraphics (Statistical Graphics Corpn. Maryland, USA) and subjected to linear regression

---

1 Recently [5-^{3}H]UdR has become available (Sigma Radiochemicals, UK) which required only the final HPLC purification described above.

analysis to determine the rate of ^3H_2O release for each mouse/flask. Care was taken to ensure that estimations were made from linear portions of the curves i.e. during a period when equilibrium had been established.

RESULTS AND DISCUSSION

In all studies a final concentration of 0.3μM [5-^{3}H] UdR (2-5μCi/ml) was used in the incubation medium as this did not unduly perturb the intracellular dUMP pool size nor the concentration of exogenous label (>=0.285μM at the end of the incubation period) data not shown.

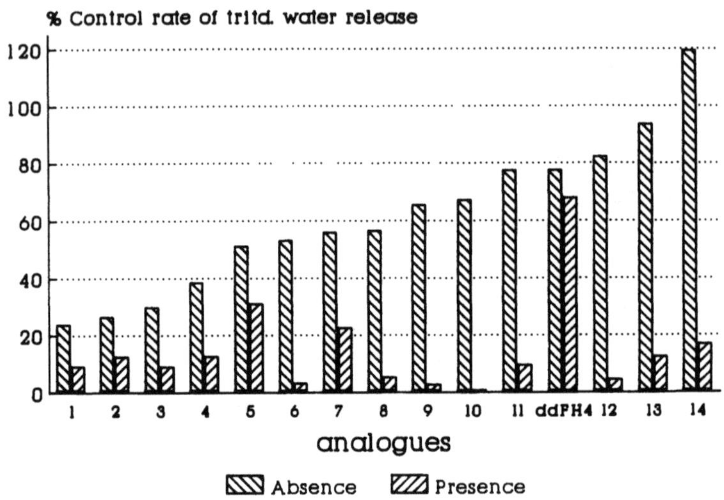

Figure 1. TS activity (as measured by ^3H_2O release) after 4 hours exposure to 5,8-dideazafolic acid analogues at 10 times IC_{50}

In vitro studies

The in vitro data from a structure activity study of DAF analogues are illustrated in Fig 1. L1210 cells were exposed for 4 hours to -equitoxic doses (10 x IC_{50}) of the analogues, with variations at positions 2 and 10 (structures shown in Table 1). TS activity as measured by the rate of ^3H_2O release was then determined in the presence or absence of exogenous compound. TS was effectively inhibited in the presence of the DAF analogues, but a spectrum of activities were observed after resuspension in fresh medium. For some compounds resuspension in fresh medium resulted in the rate of ^3H_2O release rapidly returned to control levels, suggesting non retention of the compounds. With other analogues the inhibition of TS was maintained for several hours which would suggest that either the parent compound or an inhibitory metabolite was being retained.

By comparison of the results obtained in the presence or absence of exogenous inhibitors it is possible to investigate the ability of these agents to generate non-effluxable (presumably polyglutamated) inhibitory forms. DAF analogues ranged from well retained (e.g. 2-methyl, 10-H) through intermediate (e.g. CB3717) to unretained (e.g. 2-methoxy, 10-propargyl). 5,10-Dideaza tetrahydrofolate[2] (ddFH4) is an antifolate which does not have TS as its cytotoxic locus and was included as a negative control. The reduced rate of ^3H_2O release following exposure to this agent was presumably due to the effects of this inhibitor of de novo purine synthesis rather than its weak TS inhibition[14].

Modifications at the 2 position dramatically affected the retention of the analogues, which was additionally modulated by variations of the 10 substituent. Retention of inhibitory activity within cells did not correlate with the inhibition of isolated TS under these conditions.

In vivo studies

Due to the presence of high circulating levels of thymidine it is probable that tumour cells in vivo may be able to continue to replicate and to synthesise new TS, even in the face of effective inhibition of this enzyme. Prolonged exposure is essential to induce measurable antitumour response. The ability of a folate based inhibitor of TS to generate polyglutamates is, therefore, an important determinant of its activity particularly in vivo where polyglutamation may extend the exposure of cells to inhibitors beyond the period of the availability of the parent compound in the plasma. It may also serve to increase potency and/or offer selectivity towards tissues in which the toxicity of antifolates is dependant on their FPGS activity.

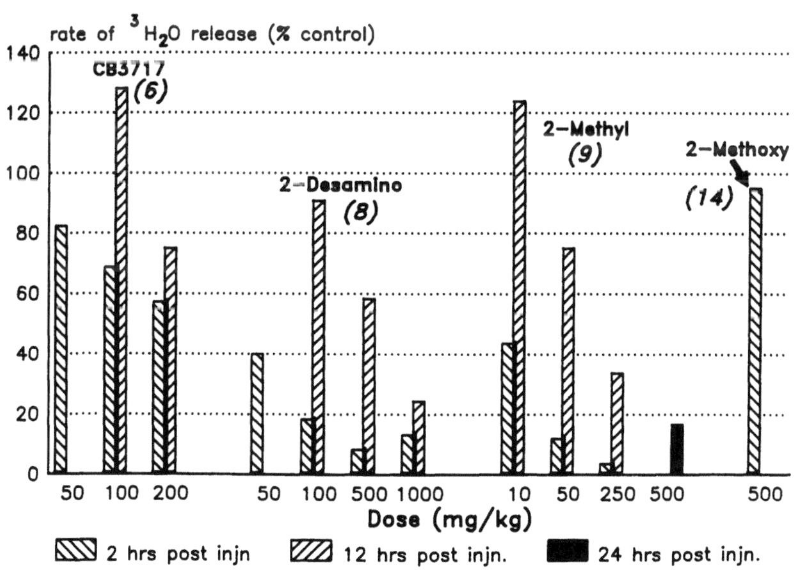

Figure 2. TS activity in L1210 cells after in vivo exposure to 5,8-dideazafolate analogues

---

2 ddFH$_4$ was supplied by Dr. G.B. Grindey, Lilly Research Laboratories Indiana, USA

The inhibition of TS by non effluxable metabolites was estimated in L1210/NCI ascitic tumour cells which were totally refractory to CB3717 and some of its analouges at maximally tolerated doses (unpublished data). In this way the additional factors associated with host metabolism and pharmacokinetics can be assessed with respect to the ability of an analogue to inhibit the target enzyme.

Dose dependant inhibition of the release of ^3H_2O was seen in vivo (Fig 4) following the intravenous (i.v.) administration of analogues which had been shown to be retained in vitro. The 2-methoxy compound (14) which was not retained in vitro did not exhibit TS inhibition even at 2 hours after a high dose (500 mg/kg) in vivo. CB3717 exhibited only weak inhibitory activity with only 43% reduction in the rate of ^3H_2O release 2 hours after the administration of the MTD (200 mgs/kg). The 2-modified analogues were more rapidly cleared than CB3717 from the plasma of the mouse such that five fold higher doses of the 2-desamino or 2-methyl analogues would be required to obtain a similar area under the concentration vs time curve (AUC) compared to CB3717[15]. Despite this the 2-desamino analogue was considerably more potent in that all the doses in the range tested (50-1000 mgs/kg) reduced the rate of ^3H_2O release more than the MTD of CB3717. The 2-methyl analogue (2) was in turn even more potent. Twelve hours after treatment the rate of ^3H_2O release had increased compared, to 2 hours, and again appeared to be dose dependant. This apparent recovery of TS activity is probably as a result of dilution effects due to cell division and synthesis of new enzyme during a period when the cytotoxic locus (TS) is effectively circumvented by the salvage of exogenous thymidine. Even 24 hours after a single i.v. injection of 500 mgs/kg of the 2-methyl analogue extensive inhibition of ^3H_2O release was observed.

We conclude that in accord with the in vitro studies the modifications of the DAF molecule also modulate the in vivo potency of these analogues with respect to in situ TS inhibition.

EFFECTS OF 3'-AZIDO-3'-DEOXYTHYMIDINE ON THE DEOXYNUCLEOSIDE TRIPHOSPHATE POOLS OF CULTURED HUMAN CELLS

Lloyd W. Frick and Donald J. Nelson

Department of Experimental Therapy, The Wellcome Research Laboratories, Research Triangle Park, North Carolina 27709

INTRODUCTION

3'-azido-3'-deoxythymidine (AZT, zidovudine, Retrovir®, 1-ß-D-(3'-azido-2',3'-dideoxy)ribofuranosylthymine, BW A509U) has potent antiviral action against the human immunodeficiency virus (HIV) in vitro (1). Although AZT is clinically useful for the treatment of HIV infections, it is associated with megaloblastic anemias in some patients (2). In vitro, AZT interferes with the maturation of human bone marrow cells (3). Relatively high concentrations of AZT (50-200 μM), have been shown to alter the deoxynucleoside metabolism of cultured human cells (4, 5). These alterations consist primarily of declines in the dTTP and dGTP pools, as well as an increase in the dCTP pool. The decline of dTTP in AZT-treated cells has been attributed to inhibition of thymidylate kinase by AZT-5'-MP (4). The changes seen in the dGTP and dCTP pools are consistent with models of the allosteric regulation of ribonucleotide reductase (6, 7). Neither AZT nor its metabolites directly affect ribonucleotide reductase (8). An initial report of a decline in the dCTP pool in H-9 cells treated with AZT (4) has not been confirmed after further investigation (5).

We have studied the effects of 24 hr exposures to various concentrations of AZT on the deoxynucleotide metabolism of four human cell lines. We have used H-9 and RPMI-8402 cells, which are derived from human T-cell leukemias (9, 10). We have also used HL-60 and K-562 cells, which represent stages in the maturation of granulocyte/macrophages and erythrocytes, respectively (11, 12). We employed radiolabeled thymidine to investigate the possibility of thymidylate kinase inhibition in cells incubated with AZT.

MATERIALS AND METHODS

Cells and Cell Culture. HL-60 and K-562 cells were obtained from the Tissue Culture Facility of the Lineberger Cancer Center at the University of North Carolina at Chapel Hill. H-9 cells were obtained from B. Hampar (National Cancer Institute, Frederick Cancer Center). RPMI-8402 cells were from B. Ullman (Portland, Oregon). Cells were cultured in RPMI-1640 (Gibco) supplemented with penicillin (50 U/ml), streptomycin (50 μg/ml), and 10% fetal calf serum (Gibco). Cell numbers were determined by counting in a hemocytometer, using Trypan Blue as an indicator of viability. Cultures were free from mycoplasma (Gen-Probe, San Diego, California). Cell volumes of log-phase cells were determined using tritiated water and [^{14}C]sucrose, obtained from New England Nuclear and dibutyl phthalate from Fisher Chemical (13).

Preparation of Extracts. Cell cultures were split the day before drug treatment. The cell density at the time of AZT addition was 150-450x10^3 cells/ml. Cells were harvested by centrifugation and metabolites extracted overnight with 5 ml of 60% methanol at -20°C. After centrifugation to remove cell debris, the solvent was removed under reduced pressure and the dried extracts were dissolved in 200 µl of deionized water. Samples were stored at -20°C before use.

Samples of the culture media obtained after harvest of the cells were immediately ultrafiltered (Centrifree Micropartition System, Amicon) and stored frozen at -20°C.

Quantification of Metabolites . Deoxynucleoside triphosphates and AZT-5'-MP were quantified by HPLC, using UV detectors set at 254 and 280 nm. Extracts were loaded on a SAX column (Whatman Partisil 4.6x250 mm, 10µ) and eluted with a gradient of ammonium phosphate (pH 3.5) in 7% ethanol. The gradient for AZT-5'-MP was 10 to 700 mM over 50 min and that for dNTPs was 300 to 700 mM over 50 min. AZT-5'-MP appeared as a major peak at about 11 minutes. Deoxynucleoside triphosphates were measured using a modified periodate procedure (14, 15). The extract of 10 to 50 million cells was used to determine the dNTPs and one fifteenth of that amount to determine AZT-5'-MP. Cells from duplicate cultures were analyzed separately and the average values reported. The peak areas of metabolites were digitized and integrated by computer (Digital Specialties, Chapel Hill, NC) and the concentrations determined by comparison to standards.

Because of the large amount of coeluting substances, it was not possible to measure the pools of either dTMP or dTDP using UV data. Pool sizes of dTMP and dTDP were therefore estimated using tritiated thymidine (New England Nuclear) to determine the specific activity of the pool of dTTP. It was assumed that the specific activities of the pools of dTMP and dTDP were equal to the specific activity of the dTTP pool (16). The specific activity of the dTTP pool was measured directly by comparison of the radioactivity associated with the dTTP peak to the amount of dTTP in the peak as determined using the UV data. This value for the specific activity was then used to quantify the amounts of dTMP and dTDP based on the radioactivity eluting at the appropriate retention times.

The degree of inhibition of thymidylate kinase was estimated using the rate equation and values of 8.6 µM as the K_i of the enzyme for dTMP and 3 µM as the K_m of the enzyme for dTMP (4).

RESULTS

The magnitude of the effect of 24 hr incubations with different AZT concentrations varied widely among the cell lines. The dNTP pools of HL-60 cells were relatively unaffected by AZT (Fig. 1a). Even at 200 µM AZT, dTTP was 70 % of control values. Pools of dCTP dropped to 60% of control at intermediate levels of AZT. Pools of dGTP were almost unaffected, although dATP pools increased incrementally with increasing AZT.

In H-9 cells (Fig. 1b), pools of dTTP declined to 50 % of control at 100 to 200 µM AZT. Pools of dGTP also decreased in these cells, but were still equal to 60 % of control at 200 µM AZT. Elevation of the dCTP pool in H-9 cells was maximal (2.5 times control) at 100 µM AZT.

AZT caused more dramatic changes in the dNTP pools of K-562 cells (Fig. 1c). At 10 µM AZT, dTTP pools increased to 30 % above control, whereas at 200 µM AZT, pools of dTTP had declined to only 30 % of control values. Pools of dCTP in K-562 cells were relatively unchanged after 24 hr of incubation in media containing 10 µM AZT. However, at higher AZT concentrations, dCTP increased sharply, reaching levels nearly 7 times greater than those of control cultures. In 200 µM AZT, pools of dGTP were decreased 50 % relative to control. Pools of dATP increased steadily in K-562 cells as the AZT concentration was raised. At the highest level of drug, dATP pools were twice those of the control cultures.

The dNTP metabolism of RPMI-8402 cells was markedly affected by AZT also (Fig 1d). In these cells incubated in 200 µM AZT, dTTP pools declined to 40 % of control and dGTP pools declined to 20 % of control. As in the K-562 cells, pools of dCTP were increased sharply at higher AZT concentrations, reaching more than 6 times control values at 200 µM AZT. The dATP pool of RPMI-8402 cells was decreased by incubation with AZT. At concentrations of AZT above 30 µM, dATP pools were 60 % of control.

The relationship between the concentration of AZT in the culture medium and the intracellular concentration of AZT-5'-MP after 24 hrs of incubation is shown in Figure 2. Incremental increases in the intracellular AZT-5'-MP concentration are apparent, even at relatively high concentrations of AZT in the medium. The slopes of the lines describing the relationships between AZT-5'-MP and AZT differ among the four cell lines, with K-562 cells having a noticeably steeper slope than the others.

Pools of dTMP were elevated in cells exposed to 10 μM AZT for four hours (Table 1). The increase was similar in the different cell lines, averaging slightly more than three-fold. An independent technique involving isolation of dTMP by SAX HPLC followed by alkaline phosphatase treatment and reverse-phase HPLC gave similar results for the expansion of the dTMP pool in the HL-60, H-9 and K-562 cells (not shown). The technique

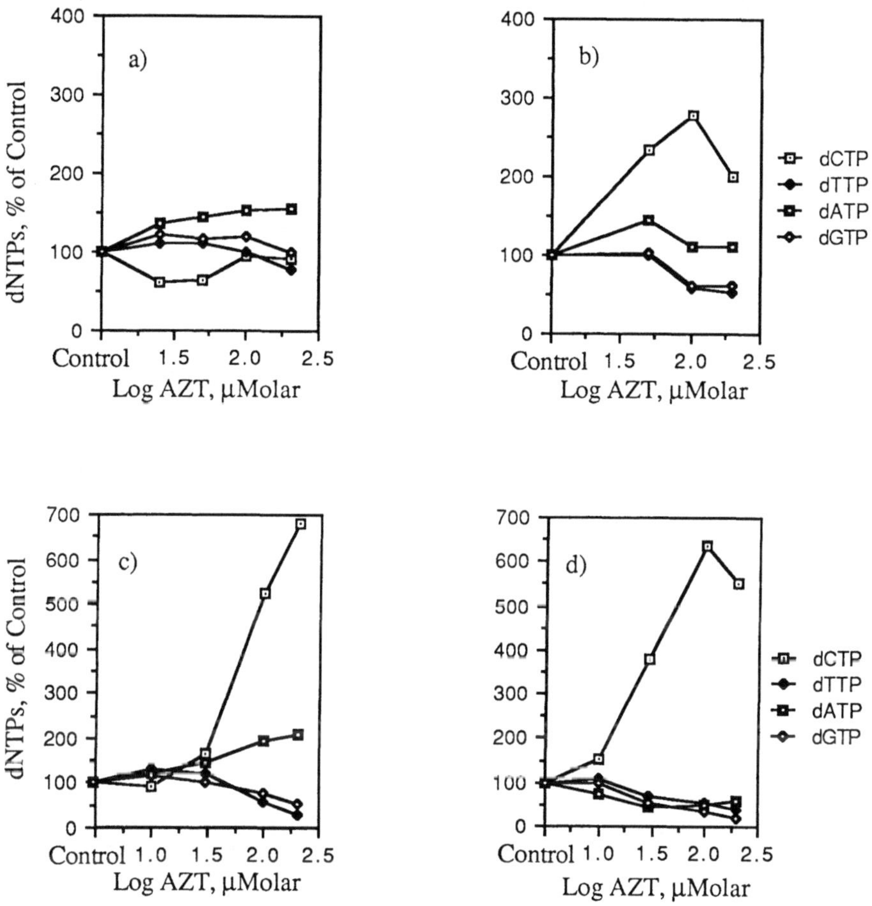

Figure 1. AZT-induced fluctuations in the deoxynucleoside triphosphate pools of four human cell lines after 24 hr incubations at different AZT concentrations. a) HL-60 cells, incubated at 25, 50 , 100, and 200 μM AZT.b) H-9 cells, incubated at 50, 100, and 200 μM AZT, c) K-562, incubated at 10, 30, 100, and 200 μM AZT, and d) RPMI-8402 cells, incubated at 10, 30, 100, and 200 μM AZT. Values are the means of two determinations and are presented as % of the values of control cultures lacking AZT. Note differences in scale.

used to measure dTMP and dTDP described here depends on the rapid establishment of isotopic equilibrium between the pools of dTTP, dTDP and dTMP, so that the specific activities are the same. However, it became obvious that this assumption does not hold in RPMI-8402 cells. Still, after directly determining the specific activities of the dTMP and dTDP pools using SAX HPLC, alkaline phosphatase digestion, and reverse-phase HPLC, the dTMP pool appeared to be elevated 2-3 fold in these cells, whereas the dTDP pool was essentially unchanged (not shown).

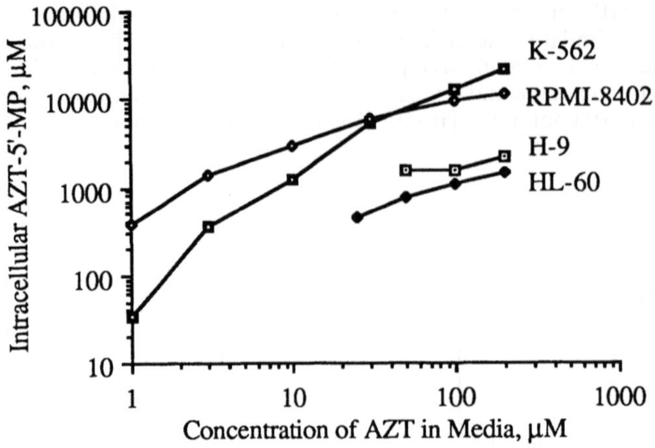

Figure 2. Accumulation of intracellular AZT-5'-MP after 24 hrs as a function of AZT concentration in the medium. Values are the averages of two determinations.

Table 1. Effects of AZT on dTMP, dTDP, and dTTP pools of cultured human cells

	dTMP	dTDP (μM)	dTTP	AZT-MP	Calculated inhibition of thymidylate kinase (%)[a]
HL-60	5	5	50		
HL-60 + AZT	15	5	45	300	85 %
H-9	65	55	140		
H-9 + AZT	120	50	115	580	70 %
K-562	7	25	80		
K-562 + AZT	30	5	35	2960	95 %

Concentrations are in μM. Cells were incubated in 10 μM AZT for 3.5 hrs, tritiated thymidine (2 μCi/ml) was added, and the incubation continued for a further 30 min.
[a] The inhibition of thymidylate kinase was calculated using the Michealis-Menten rate equation as described in the Methods section.

DISCUSSION

An earlier investigation of the time-dependent effects of AZT on deoxynucleoside metabolism has shown that the perturbations induced by AZT are short lived at low concentrations of drug (5). Therefore, at the low drug concentrations in Figure 1, most pools had returned to normal by 24 hrs. Despite this limitation, the dependence of the alterations in the dNTP pools on AZT concentration is apparent. The relatively modest effects of AZT on the dNTP pools of HL-60 and H-9 cells may be due to the fact that comparatively little AZT-5'-MP accumulates in these cells, due in part to a greater rate of degradation (4, 5).

That relatively high levels of AZT-5'-MP accumulate in human cells when they are exposed to AZT in the culture medium has been noted earlier (4, 5, 17). AZT-5'-MP, because of its ability to act as a competitive inhibitor of thymidylate kinase (K_i 8.6 μM), is thought to play a central role in the dNTP pool changes caused by AZT (4). The relationship between intracellular AZT-5'-MP and AZT in the medium is therefore of considerable interest. Figure 2 shows that this relationship differs among the human cell lines studied here. Particularly interesting is the fact that raising the AZT concentration to high levels continues to result in still greater levels of intracellular AZT-5'-MP, even though thymidine kinase activity, which has a K_m of 3 µM for AZT, should have been nearly saturated. Also intriguing is the difference in the slopes of the AZT/AZT-5'-MP relationship, with K-562 cells having a steeper slope than the other three lines. It is thus interesting that the dNTP pools of K-562 cells are affected more strongly than those of RPMI-8402 cells at higher AZT concentrations, but less strongly at low concentrations (Figure 1). This difference may be due to the fact that K-562 cells accumulate less AZT-5'-MP than do the RPMI-8402 cells at low AZT concentrations, but accumulate more at high AZT concentrations The differences in the slope of the AZT/AZT-5'-MP lines are probably not due to transport phenomena, since AZT enters cells by passive diffusion through the cell membrane (18). In part, the apparently low affinity of thymidine kinase for AZT may be due to competition with thymidine in the media. Thymidine is known to increase in the media of cells treated with AZT, reaching levels as high as 10 µM (5). Another explanation of these data may reside in the allosteric regulation of thymidine kinase (19). As dTTP pools drop, thymidine kinase activity should rise, which would generate still more AZT-5'-MP. Thus, the incremental change in the cellular concentration of AZT-5'-MP as the AZT concentration is raised may be determined by a complex interplay of enzyme activities and dTTP levels, and not be interpretable solely on the basis of Michaelis-Menten kinetics.

Cells are able to overcome the imbalances induced in the dNTP pools by AZT, especially at lower concentrations of drug (5). If the dNTP imbalances were caused by the inhibition of thymidylate kinase by AZT-5'-MP, then one way in which cells could restore their pools to a more or less normal state would be to increase the pool of the natural substrate for this enzyme, dTMP. Table 1 shows that dTMP pools were in fact increased substantially in cells cultured in 10 µM AZT. In spite of this increase, the calculated % inhibition of thymidylate kinase was still high. The lack of a big drop in the dTDP pool for all but the K-562 cells, where the enzyme was especially strongly inhibited, suggests that only a portion of the thymidylate kinase activity is adequate to meet most demand for dTTP.

Pharmacological levels of AZT in the plasma of patients undergoing AZT therapy are typically 3-5 µM (21) and clonogenic assays of human bone marrow stem cells are affected by even lower concentrations of AZT (3). The fact that changes in the dNTP pools did not persist at low concentrations of AZT suggests that the bone marrow-suppressive action of AZT may not be due to its effects on deoxynucleotide metabolism. However, the data in Figure 1 show that there are quantitative differences in the responses of each cell line to AZT. It is therefore at least possible that bone marrow stem cells experience dNTP pool perturbations similar to those seen here, although at unusually low AZT concentrations, and that such imbalances could play a role in the anemias associated with AZT therapy.

REFERENCES

1. Yarchoan, R., Klecker, R.W., Weinhold, K.J., Markham, P.D., Lyerly, H.K., Durack, D.T., Gelman, E., Lehrman, S.N., Blum, R.,M.,. Barry, D.W., Shearer, G.M., Fischl, M.A., Mitsuya, H., Gallo, R.C., Collins, J.M., Bolognesi, D.P., Myers, C.E., and Broder, S. (1986) Lancet i, 575-581.
2. Richman, D.D., Fischl, M.A., Grieco, M.H., Gottlieb, M.S., Volberding, P.A., Laskin, O.L., Leedom, J.M., Groopman, J.E., Mildvan, D., Hirsch, M.S., Jackson, G.G., Durack, D.T., Nusinoff-Lehrman, S., et al. (1987) New Engl. J. Med. 317, 192-197.
3. Sommadosi, J.P. and Carlisle, R. (1987) Antimicrob. Agents Chemother. 31, 452-454.
4. Furman, P.A., Fyfe, J.A., St. Clair, M., Weinhold, K., Rideout, J.L., Freeman, G.A., Lehrman, S.N., Bolognesi, D.P. Broder, S., Mitsuya, H., and Barry, D.W. (1986) Proc. Nat. Acad. Sci. USA 83, 8333-8337.
5. Frick, L.W., Nelson, D.J., St. Clair, M.H., Furman, P.A., and Krenitsky, T.A. Biochem. Biophys. Res. Commun. (1988) in press
6. Thelander, L. and Reichard, P. (1979) Annu. Rev. Biochem. 48,133-158.
7. Reichard, P. (1987) Biochemistry 26, 3245-3248.

8. Harrington, J.A., Miller, W.H. and Spector, T. (1987) Biochem. Pharmacol. 36, 3757-3761.
9. Popovic, M., Sarngadharan, M.G., Reed E., and Gallo, R.C. (1984) Science 224, 497-500.
10. Huang, C.C., Hou, Y., Woods, L.K., Moore, G.E., & Minowada, J. (1974) J. Nat. Cancer Instit. 53, 655-660.
11. Collins, S.J., Gallo, R.C., and Gallagher, R.E. (1977) Nature 270, 347-349.
12. Andersson, L.C., Jokinen, M., and Gamberg, C.G. (1979) Nature 278, 364-365.
13. Wohlhueter, R.M., Marz, R., Graff, J.C., and Plagemann, P.G.W. (1978) Methods Cell. Biol. 20, 211-236.
14. Garret, C. and Santi, D.V. (1979) Anal. Biochem. 99, 268-273.
15. Furman, P.A., de Miranda, P., St. Clair, M.H., and Elion, G.B. (1981) Antimicrob. Agents Chemother. 18,741-745.
16. Nicander, B. and Reichard, P. (1983) Proc. Nat. Acad. Sci. USA 80, 1347-1351.
17. Balzarini, J., Pauwels, R., Baba, M., Herdewijn, P, De Clerq, E., Broder, S., and Johns D.G. (1988) Biochem. Pharmacol. 37, 897-903.
18. Zimmerman T. P., Mahony, W.B., and Prus, K.L. (1987) J. Biol. Chem. 262,
20. Maley, F., and Maley, G.F. (1962) Biochemistry 1, 847-851.
21. Klecker, R.W., Collins, J.M., Yarchoan, R., Thomas, R., Jenkins, J.F., Broder, S. and Meyers, C.E. (1987) Clin. Pharmacol. Ther. 41, 407-412.

2-HALO-2',3'-DIDEOXYADENOSINES: METABOLICALLY STABLE DIDEOXYNUCLEOSIDES WITH ACTIVITY AGAINST THE HUMAN IMMUNODEFICIENCY VIRUS (HIV)

Dennis A. Carson, Thomas Haertle, Carlos J. Carrera, Erik H. Willis, D. Bruce Wasson, and Douglas D. Richman*

Department of Basic and Clinical Research
Research Institute of Scripps Clinic
10666 North Torrey Pines Road, La Jolla, California
*University of California, San Diego/Veterans Administration Medical Center, San Diego, California

INTRODUCTION

The acquired immune deficiency syndrome (AIDS) is caused by a retrovirus that has been termed human immunodeficiency virus-1 (HIV). Retrovirus infections in animals persist for prolonged periods, usually for life (1). This finding is consistent with the integration of the proviral genome into the host DNA. It has been estimated that nearly all subjects with antibodies to HIV are infected with the virus and are at risk for the development of AIDS (2). Millions of people have been infected with HIV to date, and from 1-2% of these subjects may develop AIDS during each subsequent year. Clearly, means for eradicating or controlling the illness are sorely needed.

HIV selectively binds to and infects cells that bear the CD4 (T4) membrane complex. In humans, CD4 has been detected on "helper" T lymphocytes, thymocytes, some monocyte/macrophages, and certain B lymphoblastoid cell lines (3). Perhaps some neuronal cells probably are also targets for HIV infection *in vivo*.

Mitsuya and Broder found that the 2',3'-dideoxynucleoside derivatives of adenosine, guanosine, inosine, cytidine, and thymidine inhibited the infectivity and cytopathic effect of HIV *in vitro* at concentrations from 10-20-fold lower than those that blocked the proliferation of uninfected T cells (4). Both 2',3'-dideoxycytidine and 2',3'-dideoxyadenosine are phosphorylated by T lymphocytes, and converted to the 5'-triphosphate derivatives. The 2',3'-dideoxynucleoside-5'-triphosphates are established substrates for retroviral RNA-dependent DNA polymerases (5). Indeed, this knowledge forms the basis of the direct mRNA sequencing technique that is widely used in molecular biology laboratories. Like 3'-azido-3'-deoxythymidine, the 2',3'-dideoxynucleosides lack the 3'-hydroxyl group that is necessary to form a 3'-5'-phosphodiester bond. Hence, the 2',3'-dideoxynucleoside 5'-triphosphates terminate DNA chain elongation. Almost certainly, this is why they inhibit HIV infection.

2',3'-dideoxycytidine has been administered to several patients with AIDS (6). Although the drug appeared to have anti-HIV activity, it displayed unexpected neurotoxicity. This has prompted the development of many other dideoxynucleoside analogs as potential anti-HIV agents.

Metabolism of 2',3'-dideoxyadenosine

2',3'-dideoxyadenosine has been shown to have potent anti-HIV activity *in vitro*, with minimal toxicity, despite the fact that adenosine deaminase rapidly converts the compound to 2',3'-dideoxyinosine (4). Surprisingly, 2',3'-dideoxyadenosine and 2',3'-dideoxyinosine exert equivalent anti-HIV activity.

We analyzed the metabolism of 2',3'-dideoxyadenosine and 2',3'-dideoxyinosine in human CEM T lymphoblasts, and in CEM mutants deficient in adenosine kinase or deoxycytidine kinase (7). The 2',3'-dideoxyinosine was phosphorylated by the T cells, and metabolized to 2',3'-dideoxyadenosine 5'-monophosphate. CEM T cells that were incubated for 4 hours with either 2 µM 2',3'-dideoxyadenosine or 2 µM 2',3'-dideoxyinosine formed equal amounts of ddATP (about 1 pmol/10^8 cells in each case).

In an effort to potentiate the uptake and anti-HIV activity of 2',3'-dideoxyadenosine, the medium was supplemented with deoxycoformycin, a tight binding inhibitor of adenosine deaminase. Despite the inclusion deoxycoformycin in the culture medium, the CEM T cells still deaminated exogenously added 2',3'-dideoxyadenosine. The deamination was six-fold higher in normal human monocytes than in T lymphoblasts. The mutational loss of adenosine kinase in T cells did not prevent 2',3'-dideoxyinosine formation from 2',3'-dideoxyadenosine in deoxycoformycin-supplemented medium, while the absence of both adenosine kinase and deoxycytidine kinase reduced 2',3'-dideoxyinosine production by 70-75%. The deamination of 2',3'-dideoxyadenosine was not related to residual deoxycoformycin-resistant adenosine deaminase activity, since enzymatic activity was negligible in extracts of deoxycoformycin-supplemented T cell cultures.

At a concentration of 50 µM, coformycin has been reported to inhibit AMP deaminase as well as adenosine deaminase. We found that coformycin blocked nearly completely the degradation of 2',3'-dideoxyadenosine by both lymphocyte and monocyte cultures. These results suggest that T lymphocytes and monocytes can convert 2',3'-dideoxyadenosine to 2',3'-dideoxyinosine via a pathway that is independent of adenosine deaminase. We presume that the transformation is mediated by AMP deaminase, since the mutational loss of deoxycytidine kinase reduced deamination by 75% in deoxycoformycin-supplemented medium. Furthermore, coformycin, at concentrations reported to inhibit both AMP deaminase and adenosine deaminase, prevented 2',3'-dideoxyadenosine degradation in both T cells and monocytes. Based upon the respective rates of 2',3'-dideoxyadenosine 5'-triphosphate accumulation (about 4 pmols/10^8 cells/hr) and 2',3'-dideoxyadenosine deamination (800-1000 pmols/10^8 cells/hr) in deoxycoformycin-supplemented cultures, it is apparent that more than 99% of the 2',3'-dideoxyadenosine 5'-monophosphate that is formed in T cells is degraded by AMP deaminase to the IMP derivative and subsequently dephosphorylated and released into the medium as 2',3'-dideoxyinosine. Collectively, these findings explain why 2',3'-dideoxyadenosine and 2',3'-dideoxyinosine have equivalent anti-HIV activity.

Synthesis of 2-halo-2',3'-dideoxyadenosine derivatives

As summarized above, 2',3'-dideoxyadenosine is catabolized to 2',3'-dideoxyinosine, which has been shown to be a substrate for purine nucleoside phosphorylase (8). Thus, these deoxynucleosides could be converted by human cells to hypoxanthine and eventually to uric acid. In previous studies, we have shown that 2-chloro-2'-deoxyadenosine is not catabolized significantly by human cells and is phosphorylated efficiently by T lymphocytes (7). The 2-chloro-2'-deoxyadenosine has also demonstrated potent anti-lymphocyte activity in humans at dosages that cause minimal bone marrow depression (9). Therefore, it was of interest to prepare the 2-chloro and related 2-halo derivatives of 2',3'-dideoxyadenosine and to analyze their metabolism and anti-HIV activity in T lymphoblasts.

Initially, the 2-chloro, 2-fluoro, and 2-bromo derivatives of 2',3'-dideoxyadenosine were generated from the corresponding 2-halo-2'-deoxynucleosides by Barton elimination of the 3'-OH group. However, we subsequently discovered that the 2-halo-2',3'-dideoxyadenosine derivatives, as well as many other 2',3'-dideoxynucleosides, could be prepared simply by enzymatic trans-glycosylation (9). In this reaction, a partially purified nucleoside deoxyribosyltransferase from *Lactobacillus helveticus* is incubated with 2',3'-dideoxycytidine or 2',3'-dideoxythymidine as a dideoxyribosyl donor, and with any one of several purine or pyrimidine bases as an acceptor. At equilibrium, the products are purified by preparative reverse phase high performance liquid chromatography (HPLC). Although the yields of 2',3'-dideoxynucleosides are lower following trans-glycosylation than after chemical synthesis, the simplicity of the method makes it attractive. It is particularly useful for the preparation of radiolabelled 2',3'-dideoxynucleosides.

Metabolism and Anti-HIV Activity of 2-halo-2',3'-dideoxyadenosine Derivatives

The 2-chloro, 2-fluoro and 2-bromo derivatives of 2',3'-dideoxyadenosine were extremely poor substrates for adenosine deaminase (10). The relative efficiencies of deamination, estimated by the ratio V_{max}/K_m, were two orders of magnitude smaller than the calculated value for 2'-deoxyadenosine.

The metabolism of 2-chloro-2',3'-dideoxyadenosine in viable T lymphoblasts was compared to that of 2',3'-dideoxyadenosine. Extracts of wild type CEM cells that had been incubated with 3 μM [8-^{3}H]2-chloro-2',3'-dideoxyadenosine for 4 hr were fractionated by ion exchange HPLC. Detected radioactive peaks corresponded to the free 2',3'-dideoxynucleoside and to the 5'-mono-, di-, and triphosphates, respectively. The total intracellular 2-chloro-2',3'-dideoxynucleotide content was 30 pmols/10^8 cells. The deoxycytidine kinase-deficient CEM variant accumulated only 20% as much 2-chloro-2',3'-dideoxyadenosine phosphates as well type CEM cells, under the same culture conditions. In contrast, when assayed under the same conditions 2',3'-dideoxyadenosine uptake was similar in the wild type T cells and the deoxycytidine kinase-deficient variant (approximately 20-30 pmols/10^8 cells in each case).

The anti-HIV activities of the 2-halo-2',3'-dideoxyadenosine derivatives were tested in MT-2 T lymphoblasts, as this system readily permitted a differentiation between anti-viral and anti-proliferative effects (10). Exposure of the MT-2 T cell line to cell-free HIV led to the formation of syncytia within 4-5 days and to a parallel decline in cell viability. The addition to the medium of 2-fluoro, 2-chloro, or 2-bromo-2',3'-dideoxyadenosine (each at 10 μM) blocked syncytia formation. In contrast, none of the corresponding 2-halo-2',3'-dideoxyinosine derivatives displayed anti-HIV activity.

The 2-halo-2',3'-dideoxynucleosides inhibited cell proliferation by 50% at concentrations approximately 2-fold higher than those that blocked HIV-induced syncytia formation. However, cultures that were supplemented with the 2-halo-2',3'-dideoxynucleosides were protected from virus-induced cell killing, as determined visually.

The deoxycytidine kinase-deficient mutant of the CEM cell line was at least 10-20-fold more resistant than parental cells to growth inhibition by the 2-halo-2',3'-dideoxyadenosine derivatives. Furthermore, as shown by p24 (*gag*) antigen detection, 2-chloro-2',3'-dideoxyadenosine did not protect the deoxycytidine kinase deficient variant from HIV infection. In contrast, 2',3'-dideoxyadenosine exerted equivalent anti-HIV activity in wild type CEM cells and the deoxycytidine kinase-deficient mutant.

These results show that 2',3'-dideoxyadenosine is a more specific inhibitor of HIV infection of T lymphoblasts than 2-chloro-2',3'-dideoxyadenosine, in spite of the resistance of the latter compound to deamination by cellular enzymes. It is also remarkable that 2',3'-dideoxyadenosine displays anti-HIV activity even in cells that lack both adenosine kinase and deoxycytidine kinase. In the future it will be of interest to compare the metabolism and effectiveness of these 2',3'-dideoxynucleosides in animal models of retroviral infection.

ACKNOWLEDGMENT

This research was supported, in part, by National Institutes of Health Grants AI24466 and GM23200, and grants from the Veterans Administration, the State of California, as allocated by the recommendation of the University Task Force on AIDS, and by the American Foundation for AIDS Research. This is publication number 5466BCR from the Research Institute of Scripps Clinic, La Jolla, California 92037.

REFERENCES

1. A.T. Haase, Pathogenesis of lentivirus infections, Nature 322:130 (1986).
2. D.D. Ho, R.J. Pomerantz, and J.C. Kaplan, Pathogenesis of infection with human immunodeficiency virus, New Engl. J. Med. 317:278 (1987).
3. J.A. Levy, J. Shimabukruo, T. McHugh, C. Casavant, D. Stites, and L. Oshiro, AIDS-associated retroviruses (RAV) can productively infect other cells besides human T helper cells, Virol. 147:441 (1985).
4. H. Mitsuya, and S. Broder, Inhibition of the *in vitro* infectivity and cytopathic effect of human T-lymphotropic virus type III/lymphadenopathy associated virus (HTLV-III/LAV) by 2',3'-dideoxynucleosides, Proc. Natl. Acad. Sci. USA 83:1911 (1986).
5. Y.-C. Cheng, G.E. Dutschman, K.F. Bastone, M.G. Sarngadharan, and Ting, R.Y.C., Human immunodeficiency virus reverse transcriptase, J. Biol. Chem. 262:2187 (1987).
6. R. Yarchoan, C.F. Perno, R.V. Thomas, R.W. Klecker, J.-P.- Allain, R.J. Wills, N. McAtee, M.A. Fischl, R. Dubinsky, M.C. McNeely, H. Mitsuya, J.M. Pluda, T.J. Lawley, M. Leuther, B. Safai, J.M. Collins, C.E. Myers, and S. Broder, Phase I studes of 2',3'-dideoxycytidine in severe human immunodeficiency virus infection as a single agent and alternating with zidovudine (AZT), Lancet 1:76 (1988).
7. D.A. Carson, T. Haertle, D.B. Wasson, and D.D. Richman, Biochemical genetic analysis of 2',3'-dideoxyadenosine metabolism in human T lymphocytes, Biochem. Biophys. Res. Comm. 151:788 (1988).
8. D.A. Cooney, G. Ahluwalia, M. Hiroaki, A. Fridland, M. Johnson, H. Zhang, M. Dalal, J. Balzarini, S. Broder, and D.G. Johns, Initial studies on the cellular pharmacology of 2',3'-dideoxyadenosine, an inhibitor of HTLV-III infectivity, Biochem. Pharmacol. 36:1765 (1987).
9. D.A. Carson, and D.B. Wasson, Synthesis of 2',3'-dideoxynucleosides by enzymatic trans-glycosylation, submitted for publication.
10. T. Haertle, C.J. Carrera, D.B. Wasson, L.C. Sowers, D.D. Richman, and D.A. Carson, Metabolism and anti-human immunodeficiency virus-1 activity of 2-halo-2',3'-dideoxyadenosine derivatives, J. Biol. Chem. 263:5870 (1988).

3'-AZIDO-3'-DEOXYTHYMIDINE AND ACYCLOVIR: ANTIVIRAL NUCLEOSIDE ANALOGUES WITH UNUSUAL CELL MEMBRANE PERMEATION PROPERTIES

Thomas P. Zimmerman, Karen L. Prus, William B. Mahony and Barbara A. Domin

Experimental Therapy Department, Wellcome Research Laboratories, Research Triangle Park, N.C. 27709

INTRODUCTION

Most of the clinically effective antiviral drugs developed to date are nucleoside analogues. The earlier systemic antiherpetic agent vidarabine (VIRA-A®, 9-β-D-arabinofuranosyladenine) has been superseded largely by acyclovir (ACV, ZOVIRAX®, 9-[(2-hydroxyethoxy)methyl]guanine), while vidarabine, acyclovir, idoxuridine (5-iodo-2'-deoxyuridine) and trifluridine (VIROPTIC®, 5-trifluoromethyl-2'-deoxyuridine) are all used to treat ocular herpes infections. Ribavirin (VIRAZOLE®, 1-β-D-ribofuranosyl-1,2,4-triazole-3-carboxamide) has been shown to have broad spectrum antiviral activity and has been approved in the United States for treatment of respiratory syncytial virus in infants. Zidovudine (AZT, RETROVIR®, 3'-azido-3'-deoxythymidine) has shown efficacy against human immunodeficiency virus. Ganciclovir (9-[1,3-dihydroxy-2-propoxymethyl]guanine) exhibits activity against all of the herpesviruses and has demonstrated usefulness in the management of cytomegalovirus infections in immunocompromised patients.[1-3]

Since all of the above nucleoside analogues must first enter the cytoplasm of an infected cell before being able to exert their antiviral effect, it is of interest to understand the mechanism by which each of these agents permeates cell membranes. The membrane transport of drugs is fundamental to their cellular pharmacology and can, in general, occur by either of two mechanisms: facilitated (i.e., carrier-mediated) or nonfacilitated (i.e., simple diffusion through the lipid bilayer) diffusion. Earlier reports have indicated that vidarabine[4-6], idoxuridine[7], trifluridine[8] and ribavirin[8,9] enter cells by means of a broad specificity nucleoside transport system, while AZT[10] surprisingly permeates human erythrocytes and lymphocytes by nonfacilitated diffusion. The present report, by describing the transport properties of 2',3'-dideoxythymidine (ddThd), provides insight concerning the molecular features of AZT which contribute to its unusual membrane permeability behavior. In addition, ACV is shown to enter human erythrocytes neither by the nucleoside transporter nor by nonfacilitated diffusion but via a purine nucleobase carrier. The chemical structures of these permeants are shown in Fig. 1.

MATERIALS AND METHODS

Materials

[*Methyl*-^{3}H]ddThd (6.0 Ci/mmol), [*methyl*-^{3}H]thymidine (dThd, 40 Ci/mmol) and [8-^{14}C]guanine (56 mCi/mmol) were purchased from Moravek Biochemicals. [*Side chain*-2-^{3}H]ACV (18.6 Ci/mmol) and [^{3}H(G)]hypoxanthine (8-12 Ci/mmol) were obtained from Dupont-New England Nuclear. [8-^{3}H]Adenine (26 Ci/mmol) was from Amersham Corp. The radiochemical purity of these permeants was assessed by reversed-phase high performance liquid chromatography and, when necessary, these reagents were purified by reversed-phase chromatography to increase their radiochemical purity to >95%. Other chemicals were from Sigma Chemical Co. or from sources identified previously.[7,10-12]

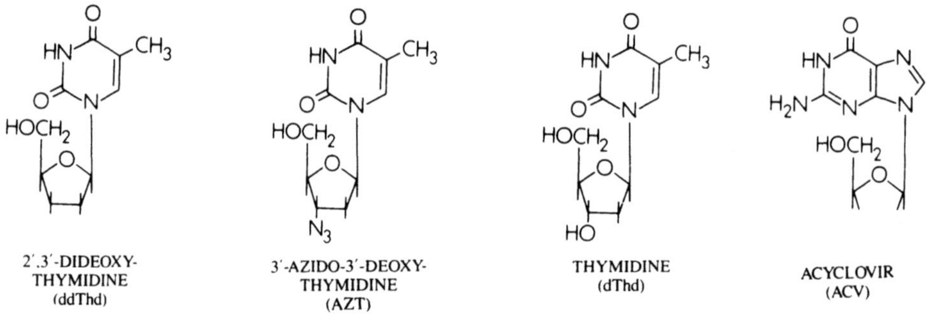

Fig. 1. Chemical structures of permeants.

Preparation of Human Erythrocytes and Lymphocytes

Blood was withdrawn from healthy human volunteers into evacuated tubes coated with sodium heparin, and erythrocytes and lymphocytes were prepared as described previously.[10]

Kinetics of Drug Influx into Erythrocytes

Influx assays were performed in 1.5-ml polypropylene microcentrifuge tubes. Each assay was performed by two persons and was timed audibly with an electric metronome set at 120 tones/min. Phosphate-buffered saline was the medium used for transport studies with AZT, ddThd, dThd and ACV, while 10 mM 4-(2-hydroxyethyl)-1-piperazineethanesulfonic acid (pH 7.3) in physiological saline was used as the buffer for nucleobase transport assays.

AZT Influx Kinetics. The influx of [^{3}H]AZT into human erythrocytes (20°C) and lymphocytes (37°C) was measured with an "oil-stop" assay described previously.[10]

dThd Influx Kinetics. [^{3}H]dThd influx into human erythrocytes (20°C) and lymphocytes (37°C) was measured with an "inhibitor-stop" assay using dilazep as a quench agent.[7,10]

ddThd Influx Kinetics. The influx of [^{3}H]ddThd into human erythrocytes was measured at 20°C with an "oil-stop" assay similar to that for AZT.[10]

ACV Influx Kinetics. [^{3}H]ACV influx into human erythrocytes was assessed at 37°C with an "inhibitor-stop" assay using an equal volume of 3.0 mM papaverine as a quench agent.[11]

Nucleobase Influx Kinetics. The influx of [^{3}H]adenine, [^{3}H]hypoxanthine and [^{14}C]guanine into human erythrocytes was measured at 37°C with an "inhibitor-stop" assay using seven assay volumes of cold 19 mM papaverine as a quench agent.[12]

Initial velocities of permeant influx were determined by linear regression analysis of the plots of cell-associated radioisotope *versus* assay time, using data obtained during the linear phase of influx (0-1.0 s for dThd, 0-1.5 s for the nucleobases, 4.0-8.5 s for AZT and ddThd, 0-8.0 s for ACV). These velocities are expressed in terms of moles of permeant taken up per second per 10 µl of packed cells. Kinetic constants were determined by directly fitting the data to a hyperbola according to the method of Wilkinson[13] and the computer program of Cleland.[14]

Statistics. Statistical comparisons of influx rates were determined with the PROC GLM computer program from SAS Institute, Inc. (Cary, NC).

RESULTS AND DISCUSSION

Structural Features of AZT Contributing to Its Ability to Permeate Cell Membranes by Nonfacilitated Diffusion

Recognition of the unusual ability of AZT to permeate cell membranes by nonfacilitated diffusion, rather than via a nucleoside transporter, prompted a study of how the closely related nucleoside ddThd enters cells. The initial velocity of ddThd influx into human erythrocytes at 20°C was shown to be linear for at least 8.5 s (Fig. 2). As was found with AZT[10], the initial velocity of ddThd influx into these cells was a linear function of ddThd concentration (0.05 to 12 mM) (Fig. 3). By contrast, the initial velocity of dThd influx was clearly saturable by 0.5 mM dThd. Nitrobenzylthioinosine (NBMPR), dipyridamole and dilazep, potent inhibitors of the erythrocyte nucleoside transporter, had no effect upon the rates of influx of either ddThd or AZT under conditions where dThd influx was totally inhibited (Table 1). Known permeants of the erythrocyte nucleoside transporter (dThd, uridine, 2-chloroadenosine) also had no effect upon the influx of either ddThd or AZT, although these permeants markedly inhibited the influx of dThd (Table 1). Similarly, adenine and acyclovir, which are demonstrated permeants of the erythrocyte nucleobase transporter[11,12], were without effect on the influx of ddThd and AZT into these cells (Table 1). The erythrocyte uptake of ddThd, like that of AZT and dThd[10], was shown to be nonconcentrative and to occur in the absence of permeant metabolism. On the basis of these results, it is concluded that ddThd, like AZT, enters human erythrocytes chiefly by nonfacilitated diffusion. This unusual cell membrane permeation property of ddThd and AZT, relative to that of dThd, is attributed to the lipophilicity imparted to both of these nucleosides by the elimination of the 3'-hydroxyl moiety. The presence of the 3'-azido moiety in AZT results in both a larger partition coefficient and a greater rate of nonfacilitated diffusion relative to ddThd (Table 2). Lack of utilization of both ddThd and AZT by the human erythrocyte nucleoside transporter indicates the importance of the 3'-hydroxyl moiety for transport of thymine-containing nucleosides by this carrier, as suggested previously.[15]

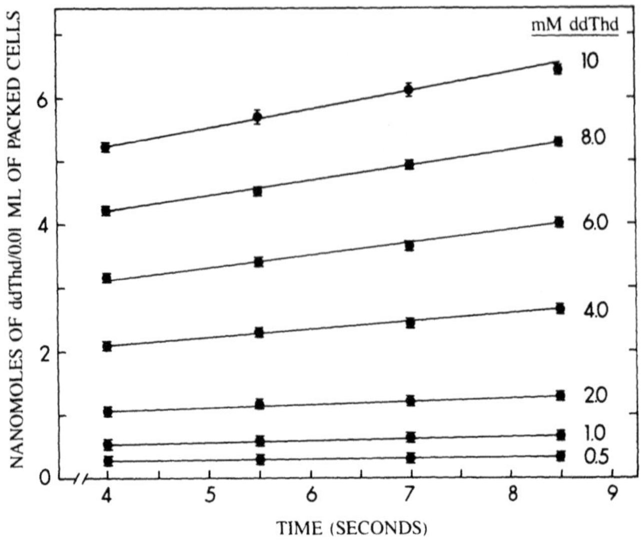

Fig. 2. Time dependence of ddThd influx into human erythrocytes. [3H]ddThd (1.6 mCi/mmol) was added to the different assays at the specified concentrations. Assay temperature, 20°C. Each point represents the mean ± S.E. for three assays.

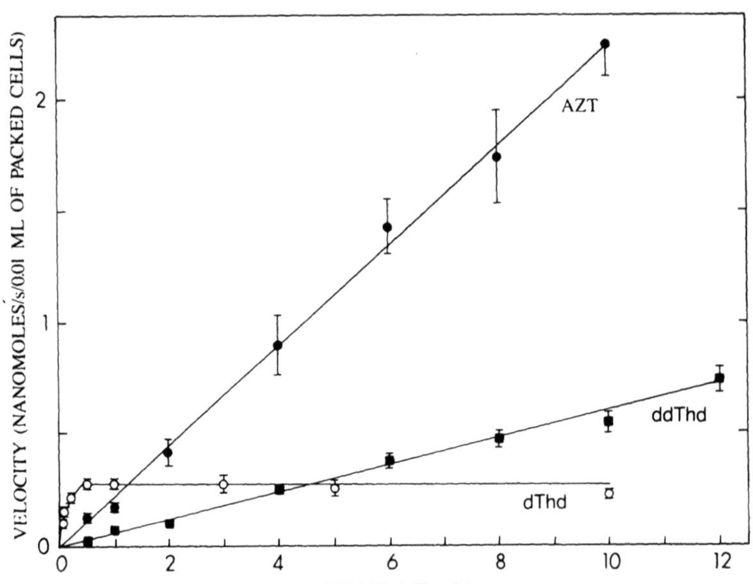

Fig. 3. Plots of initial velocity of influx into human erythrocytes (20°C) *versus* permeant concentration for ddThd, AZT and dThd. Initial velocities of ddThd influx were obtained from time courses like those shown in Fig. 2, while those for AZT and dThd are from Reference 10.

Table 1. Effects of Inhibitors of Nucleoside Transport, Nucleosides, Adenine and ACV on the Rates of Influx of [^{3}H]ddThd, [^{3}H]AZT and [^{3}H]dThd into Human Erythrocytes at 20°C

Additive	Permeant (1.0 μM)		
	[^{3}H]ddThd[a]	[^{3}H]AZT[b]	[^{3}H]dThd[c]
		percent inhibition of influx	
NBMPR (2.5 μM)	8	0	100
Dipyridamole (2.5 μM)	0	0	100
Dilazep (2.5 μM)	3	0	100
Uridine (1.0 mM)	6	9	92
2-Chloroadenosine (1.0 mM)	4	7	99
dThd (1.0 mM)	6	0	95
AZT (1.0 mM)	ND[d]	0	62
ddThd (1.0 mM)	6	0	17
Adenine (1.0 mM)	8	0	17
ACV (1.0 mM)	8	0	4

[a]Control rate was 84 ± 8 fmol/s/10 μl of packed cells.
[b]Control rate was 198 ± 22 fmol/s/10 μl of packed cells.
[c]Control rate was 4530 ± 260 fmol/s/10 μl of packed cells.
[d]ND, not determined.

With human erythrocytes and 1 μM permeant, the rates of nonfacilitated influx of ddThd and AZT at 20°C were only 2 and 5%, respectively, of the rate of carrier-mediated dThd influx (Fig. 3 and Table 1). However, with human peripheral blood lymphocytes, which are relatively deficient in nucleoside transport sites[16,17], and 1-2 μM permeant, the rate of nonfacilitated influx of AZT at 37°C exceeded that of facilitated dThd influx.[10]

Table 2. Rates of Nonfacilitated Diffusion and Partition Coefficients of dThd, ddThd and AZT at 20°C

Permeant	Nonfacilitated Diffusion Influx Rate (fMol/s/μM/10 μl Packed Cells)	Partition Coefficient[a]
dThd	(Undetectable)	0.06
ddThd	60 ± 2	0.18
AZT	220 ± 22	0.86

[a]Determined in 1-octanol:phosphate buffer, pH 7.4.

ACV Transport Characteristics

The initial velocity of ACV influx into human erythrocytes at 37°C was found to be linear for at least 8 s (Fig. 4). A plot of initial velocity of ACV influx *versus* ACV concentration (Fig. 5) provided clear evidence that ACV influx into erythrocytes is a saturable process and yielded a K_m of 260 ± 20 μM (n=8).[11] Potent inhibitors of nucleoside transport (NBMPR, dilazep, dipyridamole), as well as several permeants of the nucleoside transporter, were without effect upon ACV influx.[11] However, several purine nucleobases (adenine, hypoxanthine, guanine) which are co-permeants of the human erythrocyte nucleobase transporter[12] inhibited ACV influx into these cells in a competitive manner with K_i values similar to their respective K_m values for influx.[11] Moreover, ACV was shown to inhibit the influx of adenine, hypoxanthine and guanine into erythrocytes in a competitive manner with K_i values (240-280 μM) similar to its K_m for influx.[11] These results indicate that ACV permeates human erythrocytes predominantly by means of the purine nucleobase carrier.

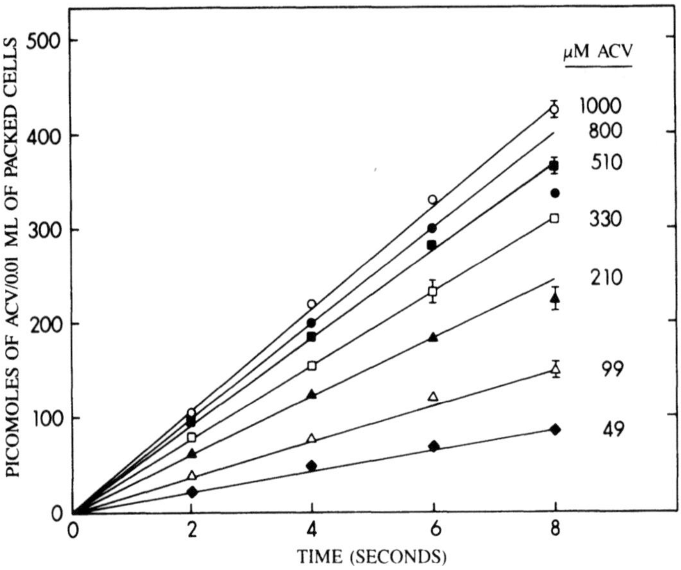

Fig. 4. Time dependence of ACV influx into human erythrocytes. [^{3}H]ACV (8.9 mCi/mmol) was added to the different assays at the specified concentrations. Assay temperature, 37°C. Each point represents the mean ± the average deviation of duplicate assays; error bars were omitted when they did not extend beyond the symbol boundaries.

Therapeutic Implications of AZT and ACV Influx Mechanisms

Both AZT[18] and ACV[1,3] are phosphorylated intracellularly by a thymidine kinase (host- or virus-specified, respectively). Thus, intracellular dThd can compete with both of these antiviral nucleosides for phosphorylation and can thereby reduce their extent of anabolism to their biologically active triphosphate metabolites. Since both AZT and ACV can enter cells by a means other than the nucleoside transporter, whereas dThd appears to be totally dependent upon this carrier for cell membrane permeation, selective inhibitors of the nucleoside transporter can prevent the cellular influx of dThd without affecting the ability of

AZT and ACV to enter cells. These biochemical circumstances appear to be responsible for the abilities of NBMPR, dilazep and dipyridamole to potentiate the antiviral effects of AZT[10] and ACV[11] *in vitro*. Indeed, the ability of purified dThd phosphorylase to also potentiate the antiviral activities of these two nucleoside analogues[10,11] supports the conclusion that this effect of the nucleoside transport inhibitors is due to their prevention of cellular uptake of dThd from the medium. In this regard, it is of interest that, with human erythrocytes at 37°C, the V_{max} for ACV influx (66 pmol/s/10 µl cells)[11] is only 5% that for dThd influx (~1450 pmol/s/10 µl cells).[10] These results suggest that inhibitors of nucleoside transport may have therapeutic value when used in combination with either AZT or ACV.

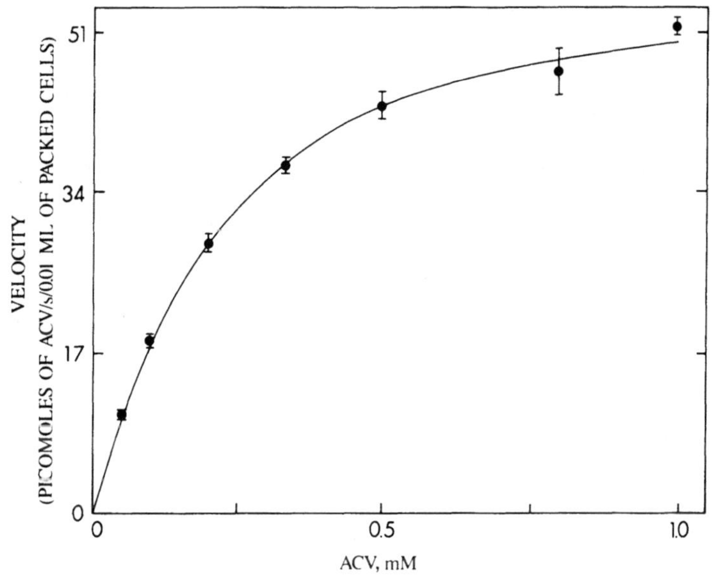

Fig. 5. Plot of initial velocity of ACV influx *versus* ACV concentration. Initial velocities of ACV influx were obtained with erythrocytes at 37°C from time courses like those shown in Fig. 4.

REFERENCES

1. M. A. Sande and G. L. Mandell, Antimicrobial Agents - Antifungal and Antiviral Agents, in: The Pharmacological Basis of Therapeutics, A. G. Gilman, L. S. Goodman, T. W. Rall, and F. Murad, eds., p. 1219, Macmillan Publishing Co., New York (1985).
2. M. M. Mansuri and J. C. Martin, Antiviral Agents, in: Annual Reports in Medicinal Chemistry, D. M. Bailey, ed., Vol. 22, p. 147, Academic Press, Orlando (1987).
3. E. D. Reines and P. A. Gross, Antiviral Agents, Med. Clin. N. Amer. 72:691 (1988).
4. C. E. Cass and A. R. P. Paterson, Mediated Transport of Nucleosides by Human Erythrocytes: Specificity toward Purine Nucleosides as Permeants, Biochim. Biophys. Acta 291:734 (1973).

5. C. E. Cass and A. R. P. Paterson, Inhibition by Nitrobenzylthioinosine of Uptake of Adenosine, 2'-Deoxyadenosine and 9-β-D-Arabinofuranosyladenine by Human and Mouse Erythrocytes, Biochem. Pharmacol. 24:1989 (1975).
6. C. E. Cass, N. Kolassa, Y. Uehara, E. Dahlig-Harley, E. R. Harley, and A. R. P. Paterson, Absence of Binding Sites for the Transport Inhibitor Nitrobenzylthioinosine on Nucleoside Transport-Deficient Mouse Lymphoma Cells, Biochim. Biophys. Acta 649:769 (1981).
7. W. B. Mahony and T. P. Zimmerman, An Assay for Inhibitors of Nucleoside Transport Based upon the Use of 5-[^{125}I]Iodo-2'-deoxyuridine as Permeant, Anal. Biochem. 154:235 (1986).
8. A. R. P. Paterson, S.-E. Yang, E. Y. Lau, and C. E. Cass, Low Specificity of the Nucleoside Transport Mechanism of RPMI 6410 Cells, Mol. Pharmacol. 16:900 (1979).
9. A. Sherwood and V. Stollar, Evidence that Ribavirin is Transported in *Aedes albopictus* Cells by the Nucleoside Transport System, Antiviral Res. 4:317 (1984).
10. T. P. Zimmerman, W. B. Mahony, and K. L. Prus, 3'-Azido-3'-deoxythymidine: An Unusual Nucleoside Analogue that Permeates the Membrane of Human Erythrocytes and Lymphocytes by Nonfacilitated Diffusion, J. Biol. Chem. 262:5748 (1987).
11. W. B. Mahony, B. A. Domin, R. T. McConnell, and T. P. Zimmerman, Acyclovir Transport into Human Erythrocytes, J. Biol. Chem. 263:9285 (1988).
12. B. A. Domin, W. B. Mahony, and T. P. Zimmerman, Purine Nucleobase Transport in Human Erythrocytes: Reinvestigation with a Novel "Inhibitor-Stop" Assay, J. Biol. Chem. 263:9276 (1988).
13. G. N. Wilkinson, Statistical Estimations in Enzyme Kinetics, Biochem. J. 80:324 (1961).
14. W. W. Cleland, Computer Programmes for Processing Enzyme Kinetic Data, Nature 198:463 (1963).
15. W. P. Gati, H. K. Misra, E. E. Knaus, and L. I. Wiebe, Structural Modifications at the 2'- and 3'-Positions of Some Pyrimidine Nucleosides as Determinants of Their Interaction with the Mouse Erythrocyte Nucleoside Transporter, Biochem. Pharmacol. 33:3325 (1984).
16. J. S. Wiley, S. P. Jones, W. H. Sawyer, and A. R. P. Paterson, Cytosine Arabinoside Influx and Nucleoside Transport Sites in Acute Leukemia, J. Clin. Invest. 69:479 (1982).
17. C. Woffendin and P. G. W. Plagemann, Interaction of [^{3}H]Dipyridamole with the Nucleoside Transporters of Human Erythrocytes and Cultured Animal Cells, J. Membrane Biol. 98:89 (1987).
18. P. A. Furman, J. A. Fyfe, M. H. St. Clair, K. Weinhold, J. L. Rideout, G. A. Freeman, S. Nusinoff Lehrman, D. P. Bolognesi, S. Broder, H. Mitsuya, and D. W. Barry, Phosphorylation of 3'-Azido-3'-deoxythymidine and Selective Interaction of the 5'-Triphosphate with Human Immunodeficiency Viral Reverse Transcriptase, Proc. Natl. Acad. Sci. U.S.A. 83:8333 (1986).

THE ANTIRETROVIRAL AND CYTOSTATIC ACTIVITY, AND METABOLISM OF 3'-AZIDO-2',3'-DIDEOXYTHYMIDINE, 3'-FLUORO-2',3'-DIDEOXYTHYMIDINE AND 2',3'-DIDEOXYCYTIDINE ARE HIGHLY CELL TYPE-DEPENDENT

Jan Balzarini[1], Eckart Matthes[2], Pascal Meeus[2], David G. Johns[3] and Erik De Clercq[1]

[1]Rega Institute for Medical Research, Katholieke Universiteit Leuven, B-3000 Leuven, Belgium, [2]Forschungszentrum fur Molekularbiologie und Medizin, Akademie der Wissenschaften, D-1115 Berlin, DDR, [3]National Cancer Institute, National Institutes of Health, Bethesda, Maryland 20892, U.S.A.

INTRODUCTION

2',3'-Dideoxyribonucleosides (ddNs) are known to inhibit the replication of human immunodeficiency virus (HIV), the causative agent of the acquired immunodeficiency syndrome (AIDS) (1). In addition to the 2',3'-dideoxyribonucleosides, various derivatives thereof, i.e. 2',3'-didehydro-2',3'-dideoxy- and 3'-azido- and 3'-fluoro-substituted 2',3'-dideoxyribonucleosides have proven to be potent and selective inhibitors of HIV replication (for an overview, see refs. 2,3). The 2',3'-dideoxyribonucleoside analogues are assumed to act at the virus-specified reverse transcriptase. To be able to interact with this target enzyme, the 2',3'-dideoxynucleoside analogues must be phosphorylated intracellularly to their 5'-triphosphate (ddNTP) metabolites. As has been clearly demonstrated for 3'-azido-2',3'-dideoxythymidine (AzddThd, AZT, Zidovudine) (4), 2',3'-dideoxycytidine (ddCyd) (5) and 3'-fluoro-2',3'-dideoxythymidine (FddThd) (6), their anabolism to the 5'-mono-, 5'-di- and 5'-triphosphate metabolites is similar in uninfected and HIV-infected cells. These observations suggest that the phosphorylation steps are catalysed by cellular enzymes (i.e. nucleoside and nucleotide kinases). Indeed, the thymidine (dThd) kinase from H9 cells efficiently converts AzddThd to its 5'-monophosphate (AzddTMP) (4), and thymidylate (dTMP) kinase further phosphorylates AzddTMP to AzddTDP (4). The K_m value of AzddThd for dThd kinase was 3 µM (as compared to 2.9 µM for dThd) and the V_{max} of phosphorylation of AzddThd was 60 % that of dThd (4). A K_i/K_m value of 0.66 was found with AzddThd for the dThd kinase of Molt/4F cells (7,8). The apparent K_m value of AzddTMP for dTMP kinase was 8.6 µM (as compared to 4.1 µM for dTMP), but the V_{max} of phosphorylation of AzddTMP was only 0.3 % that of dTMP (4). With Ehrlich ascites carcinoma cell dThd kinase a K_m of 11.2 µM has been found for FddThd while the corresponding value for dThd was 3.2 µM; the V_{max} of FddThd was 25 % that of dThd (9). Also, a K_i/K_m value of 3.4 for FddThd with Molt/4F dThd kinase has been noted (8). Thus, FddThd is a good although slightly less efficient substrate for dThd kinase than is AzddThd.

It is evident that the extent and rate by which the 2',3'-dideoxynucleoside analogues are phosphorylated to the corresponding 5'-triphosphates and the affinity of the ddNTPs for HIV-specified reverse transcriptase (relative to their affinity for the cellular DNA polymerases) determines both their potency and selectivity as anti-retrovirus agents. As has been directly demonstrated with AzddTTP, FddTTP and ddCTP, HIV reverse transcriptase is more readily inhibited by these ddNTP's than is DNA polymerase α (4,10,11), and this may explain the selectivity of these compounds against HIV. We have now compared the antiretroviral and cytostatic activity of AzddThd, FddThd and ddCyd and their metabolism in different human T-lymphocyte cell lines *in vitro*. Striking differences were found in the anti-retrovirus and cytostatic activity, and the intracellular metabolism of the test compounds, depending on the nature of the cell line.

MATERIALS AND METHODS

Cells. The origin and growth characteristics of the human H9, ATH8 and MT4 cells used in our study have been described elsewhere (12-14).

Viruses. Human immunodeficiency virus (HIV) was obtained from the culture supernatant of a H9 cell line persistently infected with human T-cell lymphotropic virus type III_B (14). Moloney murine sarcoma virus (MSV) was prepared from tumors induced following intramuscular inoculation of 3 day-old NMRI mice (15).

Compounds. 3'-Azido-2',3'-dideoxythymidine (AzddThd) and 3'-fluoro-2',3'-dideoxythymidine (FddThd) were synthesized as described previously (16). 2',3'-Dideoxycytidine (ddCyd) was obtained from Pharmacia PL-Biochemicals (Piscataway, NJ, USA). The structural formulae of AzddThd, FddThd and ddCyd are depicted in Fig. 1.

Radiochemicals. [^{3}H]AzddThd (specific radioactivity 10 Ci/mmol) was synthesized by Moravek Biochemicals Inc. (Brea, California, USA); [^{3}H]-FddThd (specific radioactivity : 4 Ci/mmol) was synthesized by von Janta-Lipinski (Akademie der Wissenschaften, Berlin, DDR).

Antiviral Assays. The procedures for measuring anti-HIV activity in ATH8 and MT-4 cells, and anti-MSV activity in C3H cells, have been described previously (13,17,18).

Cytostatic Assays. Cytostatic effects of the compounds were assessed by measuring the inhibition of cell proliferation. The experimental procedures have been described previously (17). Briefly, exponentially growing MT-4 and H9 cells were seeded in 200 µl microplate wells and incubated in the presence of varying concentrations of the test compound (∿ 50,000 cells/well). After 3, 4 and 5 days of incubation the cell number was determined in a Coulter counter. ATH8 cells were seeded at 10^4 cells/culture tube (2 ml/tube) in the presence of the test compound, and after 7 days of incubation the number of living cells was determined in a blood cell counting chamber using the trypan blue dye exclusion technique.

Metabolism of [^{3}H]AzddThd and [^{3}H]FddThd in human MT-4 and H9 cells. The metabolism of the radiolabeled compounds was monitored according to previously established procedures (19). Briefly, the cells were seeded at 2-4 x 10^5 cells/ml and incubated with 1 µM of the radiolabeled compounds (AzddThd or FddThd). At different time intervals, cells were centrifuged, washed and precipitated with cold TCA (10 %). After centrifugation, the supernatants were neutralized with tri-n-octylamine in freon. HPLC analysis of the neutralized cell extracts was carried out using a Partisil-SAX radial compression column. A linear gradient of 0.01 M ammonium phosphate pH 3.6 to 0.6 M ammonium phosphate pH 3.8 was used. The different fractions of the eluate were assayed for radioactivity in a toluene-based scintillant.

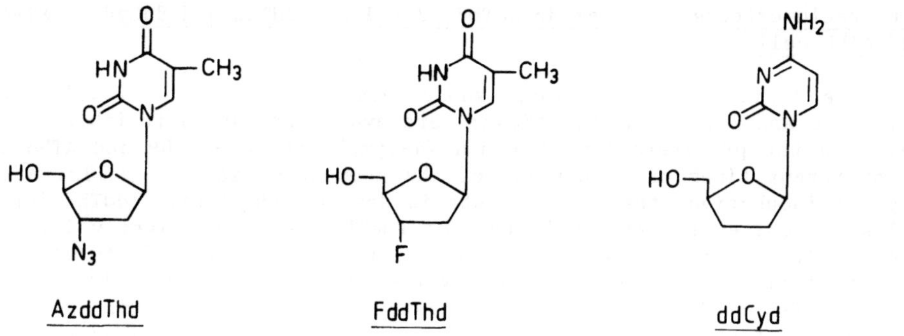

Fig. 1. Structural formulae of 3'-azido-2',3'-dideoxythymidine (AzddThd), 3'-fluoro-2',3'-dideoxythymidine (FddThd) and 2',3'-dideoxycytidine (ddCyd).

RESULTS

Inhibitory effects of 3'-azido-ddThd, 3'-fluoro-ddThd and ddCyd on HIV replication in MT-4, ATH8 and H9 cells

Remarkable differences were noted in the inhibitory effects of AzddThd and FddThd on HIV replication, depending on the cell lines in which HIV replication was examined (Table 1). AzddThd and FddThd proved about 1000-fold more inhibitory to HIV-induced cytopathogenicity in MT-4 cells than in ATH8 cells, while ddCyd was almost equally effective against MT-4 and ATH8 cells. Anti-HIV activity of the compounds was slightly more pronounced in H9 cells than in ATH8 cells (Table 1).

Table 1. Anti-HIV activity of AzddThd, FddThd and ddCyd in MT-4, ATH8 and H9 cells

Compound	50 % Effective dose (μM)		
	MT-4	ATH8[a]	H9[b]
AzddThd	0.004	2.4	0.4
FddThd	0.001	1.4	-
ddCyd	0.5	0.2	0.1

[a]Data taken from ref. 16.
[b]Data taken from refs. 1, 20.

Cytostatic effects of 3'-azido-ddThd, 3'-fluoro-ddThd and ddCyd on MT-4, H9 and ATH8 cells

The 3'-azido- and 3'-fluoro-substituted derivatives of 2',3'-dideoxythymidine and 2',3'-dideoxycytidine were evaluated for their inhibitory effects on the proliferation of human T4-lymphocyte MT-4, H9 and ATH8 cells at different times of incubation of the compounds with the cells. After 3 days of incubation, the 50 % cytostatic dose of ddCyd and AzddThd for MT-4 cells was quite similar, while that of FddThd was only slightly lower (Table 2). After 4 or 5 days of incubation, the cytostatic effects of AzddThd and ddCyd had not markedly changed, while the cytostatic activity of FddThd against MT-4 cells increased significantly (20- and 120-fold, as compared to the cytostatic activity noted after 3 days). Neither AzddThd nor FddThd proved inhibitory to H9 cell proliferation at a concentration as high as 1000 µM. When the cytostatic activity of the compounds was evaluated against ATH8 cells (after 7 days of incubation), the ID_{50} values were comparable to those found with MT-4 cells after a 3-day incubation period (Table 2).

Table 2. Cytostatic effects of AzddThd, FddThd and ddCyd on MT-4, ATH8 and H9 cells

Compound	Cell line	50 % Cytostatic dose (µM)			
		day 3	day 4	day 5	day 7[a]
AzddThd	MT-4	53	42	36	-
	ATH8	-	-	-	45
	H9	> 1000	> 1000	> 1000	-
FddThd	MT-4	12	0.6	0.10	-
	ATH8	-	-	-	15
	H9	> 1000	> 1000	> 1000	-
ddCyd	MT-4	40	-	35	-
	ATH8	-	-	-	35

[a]Data taken from ref. 16. Due to the long time the ATH8 cells required for proliferation, the inhibitory effects of the compounds were measured only after 7 days of incubation.

Phosphorylation of [^{3}H]AzddThd and [^{3}H]FddThd in MT-4 and H9 cells

The phosphorylation of 1 µM AzddThd and 1 µM FddThd to their corresponding 5'-mono-, 5'-di- and 5'-triphosphates has been examined in MT-4 and H9 cells after 24 hrs and 48 hrs of incubation (Table 3). In both cell lines AzddThd mainly accumulated as its 5'-monophosphate derivative, and little AzddTMP was further metabolized to its 5'-di and 5'-triphosphate. In contrast with AzddThd, FddThd was readily metabolized to its 5'-triphosphate. FddThd also accumulated as its 5'-monophosphate, but, irrespective of the cell line studied, FddTMP levels were lower than AzddTMP levels. There were no striking differences between MT-4 and H9 cells in the phosphorylation pattern of AzddThd and FddThd (Table 3).

Table 3. Phosphorylation of 1 µM [^{3}H]FddThd and 1 µM [^{3}H]AzddThd in MT-4 and H9 cells

Compound	5'-phosphorylated products (pmol/10^6 cells)		
	5'-monophosphate	5'-diphosphate	5'-triphosphate
AzddThd in MT-4 cells			
24 hrs	92	2.1	1.2
48 hrs	96	1.7	1.2
AzddThd in H9 cells			
24 hrs	75	0.8	0.8
48 hrs	44	0.4	0.5
FddThd in MT-4 cells			
24 hrs	18	2.3	7.3
48 hrs	17	1.7	4.8
FddThd in H9 cells			
24 hrs	27	2.2	6.5
48 hrs	28	1.7	5.4

DISCUSSION

The cytostatic and antiretroviral properties of AzddThd, FddThd and ddCyd, and intracellular metabolism have been measured in several human T4-lymphocyte cell lines, including ATH8, MT-4 and H9 cells. Striking differences in antiretroviral and cytostatic action, and intracellular metabolism of the compounds were noted between the different cell lines examined. The cytostatic activity of FddThd not only differed from one cell line to another but it also varied in function of the incubation time : i.e. the antiproliferative effects of FddThd increased substantially with longer incubation times for MT-4 cells, but not H9 or ATH8 cells. In contrast, the cytostatic effects of AzddThd and also ddCyd remained virtually unchanged during the whole incubation period of MT-4, ATH8 or H9 cells. The antiretroviral effects of the compounds also varied depending on the cell system in which they were examined : MT-4 cells proved exquisitely sensitive to the inhibitory effects of AzddThd and FddThd on HIV replication. In ATH8 cells they were about 1000-fold less active against HIV than in MT-4 cells. In contrast, the anti-HIV activity of ddCyd did not markedly differ from one cell line to another.

These and previous observations indicate that MT-4 cells are highly sensitive to the anti-HIV effects of several ddThd analogues, while cytosine and purine (i.e. adenine, guanine, 2,6-diaminopurine) analogues are equally effective against HIV in MT-4, H9, ATH8 and HUT-78 cells (2,16,21). These findings caution against premature conclusions based on the antiviral and antimetabolic data obtained in only one cell line, without having extended the exploration of the compounds to other suitable cell lines.

The reason(s) for the exquisite sensitivity of MT-4 cells to the anti-HIV effects of ddThd analogues such as AzddThd and FddThd are still unclear. We found that the affinities of AzddThd and FddThd for MT-4 dThd kinase and H9 dThd kinase are almost similar and that comparable levels of dThd kinase are present in both cell lines (data not shown). Also, AzddThd nor FddThd are not more extensively phosphorylated to their 5'-triphosphates in MT-4 than in H9 cells (Table 3). In addition, we also ascertained that intracellular levels of dTTP in MT-4 cells are not substantially lower than dTTP pool levels in H9 cells (data not shown). Consequently, AzddTTP and FddTTP should not benefit of a greater competitional advantage to dTTP at the HIV reverse transcriptase level in MT-4 cells than in H9 cells.

The 5'-triphosphate levels of FddThd obtained in MT-4 and H9 cells were quite similar (Table 3). This contrasts with the remarkable difference in the cytostatic effect of FddThd on MT-4 and H9 cells, as well as its differential anti-HIV activity in these cell lines. A striking difference is noted between the inhibitory effects of the 5'-triphosphate derivatives of AzddThd and FddThd on DNA polymerases α and β. Furman reported that both DNA polymerase α and DNA polymerase β are almost insensitive to the inhibitory effects of AzddTTP (4). In contrast, Matthes et al. (10) recently found that FddTTP is a potent inhibitor of DNA polymerase β but not of DNA polymerase α. To what extent these findings may be related to the differential toxicity of both AzddThd and FddThd in different cell lines is now under investigation.

REFERENCES

1. H. Mitsuya and S. Broder, Inhibition of the in vitro infectivity and cytopathic effect of human T-lymphotropic virus type III/lymphadenopathy-associated virus (HTLV-III/LAV) by 2',3'-dideoxynucleosides, Proc. Natl. Acad. Sci. USA 83:1911 (1986).
2. E. De Clercq, Perspectives for the chemotherapy of AIDS, Anticancer Res. 7:1023 (1987).
3. E. De Clercq, New selective antiviral agents active against the AIDS virus, Trends Pharmacol. Sci. (TIPS) 8:339 (1987).
4. P.A. Furman, J.A. Fyfe, M.H. St. Clair, K. Weinhold, J.L. Rideout, G.A. Freeman, S. Nusinoff-Lehrman, D.P. Bolognesi, S. Broder, H. Mitsuya and D.W. Barry, Phosphorylation of 3'-azido-3'-deoxythymidine and selective interaction of the 5'-triphosphate with human immunodeficiency virus reverse transcriptase, Proc. Natl. Acad. Sci. USA 83:8333 (1986).
5. D.A. Cooney, M. Dalal, H. Mitsuya, J.B. McMahon, M. Nadkarni, J. Balzarini, S. Broder and D.G. Johns, Initial studies on the cellular pharmacology of 2',3'-dideoxycytidine, an inhibitor of HTLV-III infectivity, Biochem. Pharmacol. 35:2065 (1986).
6. E. Matthes, Ch. Lehmann, D. Scholz, H.A. Rosenthal and P. Langen, Phosphorylation, anti-HIV activity and cytotoxicity of 3'-fluorothymidine, submitted for publication (1988).
7. J. Balzarini and S. Broder, Principles of antiretroviral therapy for AIDS and related diseases, in: "Clinical Use of Antiviral Drugs", E. De Clercq, ed., Martinus Nijhoff Publ., Boston, Dordrecht, Lancaster, p. 361 (1988).
8. J. Balzarini, M. Baba, R. Pauwels, P. Herdewijn and E. De Clercq, Anti-retrovirus activity of 3'-fluoro- and 3'-azido-substituted pyrimidine 2',3'-dideoxynucleoside analogues, Biochem. Pharmacol.: in press (1988).
9. P. Langen, G. Kowollik, G. Etzold, H. Venner, and H. Reinert, The phosphorylation of 3'-deoxy-3'-fluorothymidine and its incorporation into DNA in a cellfree system from tumor cells, Acta biol. med. germ. 29:483 (1972).
10. E. Matthes, Ch. Lehmann, B. Drescher, W. Büttner, and P. Langen, 3'-Deoxy-3'-fluorothymidinetriphosphate: inhibitor and terminator of DNA synthesis catalysed by DNA polymerase β, terminal deoxynucleotidyl transferase and DNA polymerase I, Biomed. Biochim. Acta 44:K63 (1985).

11. M.C. Starnes and Y.-C. Cheng, Cellular metabolism of 2',3'-dideoxycytidine, a compound active against human immunodeficiency virus in vitro, J. Biol. Chem. 262:988 (1987).
12. H. Mitsuya, H.-G. Guo, J. Cossman, M. Megson, M.S. Reitz Jr. and S. Broder, Functional properties of antigen-specific T cells infected by human T-cell leukemia-lymphoma virus (HTLV-I), Science 225:1484 (1984).
13. R. Pauwels, E. De Clercq, J. Desmyter, J. Balzarini, P. Goubau, P. Herdewijn, H. Vanderhaeghe and M. Vandeputte, Sensitive and rapid assay on MT-4 cells for detection of antiviral compounds against the AIDS virus, J. Virol. Methods 16:171 (1987).
14. M. Popovic, M.G. Sarngadharan, E. Read and R.C. Gallo, Detection, isolation, and continuous production of cytopathic retroviruses (HTLV-III) from patients with AIDS and pre-AIDS. Science 224:497 (1984).
15. E. De Clercq and T. Merigan, Moloney sarcoma virus-induced tumors in mice: inhibition or stimulation by (poly rI).(poly rC), Proc. Soc. Exp. Biol. Med. 137:590 (1971).
16. P. Herdewijn, J. Balzarini, E. De Clercq, R. Pauwels, M. Baba, S. Broder and H. Vanderhaeghe, 3'-Substituted 2',3'-dideoxynucleoside analogues as potential anti-HIV (HTLV-III/LAV) agents. J. Med. Chem. 30:1270 (1987).
17. J. Balzarini, G.-J. Kang, M. Dalal, P. Herdewijn, E. De Clercq, S. Broder and D.G. Johns, The anti-HTLV-III (anti-HIV) and cytotoxic activity of 2',3'-didehydro-2',3'-dideoxyribonucleosides: a comparison with their parental 2',3'-dideoxyribonucleosides, Mol. Pharmacol. 32:162 (1987).
18. J. Balzarini, R. Pauwels, M. Baba, M.J. Robins, R. Zou, P. Herdewijn and E. De Clercq, The 2',3'-dideoxyriboside of 2,6-diaminopurine selectively inhibits human immunodeficiency virus (HIV) replication in vitro, Biochem. Biophys. Res. Commun. 145:269 (1987).
19. J. Balzarini, R. Pauwels, M. Baba, P. Herdewijn, E. De Clercq, S. Broder and D.G. Johns, The in vitro and in vivo anti-retrovirus activity, and intracellular metabolism of 3'-azido-2',3'-dideoxythymidine and 2',3'-dideoxycytidine are highly dependent on the cell species, Biochem. Pharmacol. 37:897 (1988).
20. H. Mitsuya, K.J. Weinhold, P.A. Furman, M.H. St. Clair, S. Nusinoff Lehrman, R.C. Gallo, D. Bolognesi, D.W. Barry and S. Broder, 3'-Azido-3'-deoxythymidine (BW A509U): an antiviral agent that inhibits the infectivity and cytopathic effect of human T-lymphotropic virus type III/lymphadenopathy-associated virus in vitro. Proc. Natl. Acad. Sci. USA 82:7096 (1985).
21. J. Balzarini, M. Baba, R. Pauwels, P. Herdewijn, S.C. Wood, M.J. Robins and E. De Clercq, Potent and selective activity of 3'-azido-2,6-diaminopurine-2',3'-dideoxyriboside, 3'-fluoro-2,6-diaminopurine-2',3'-dideoxyriboside, and 3'-fluoro-2',3'-dideoxyguanosine against human immunodeficiency virus, Mol. Pharmacol. 33:243 (1988).

DIDEOXYCYTIDINE METABOLISM IN WILD TYPE AND MUTANT CEM CELLS DEFICIENT IN NUCLEOSIDE TRANSPORT OR DEOXYCYTIDINE KINASE

Buddy Ullman

Department of Biochemistry
The Oregon Health Sciences University
Portland, Oregon 97201

ABSTRACT

The growth inhibitory effects and metabolism of 2',3'-dideoxycytidine (ddC) were examined in wild type human CEM T lymphoblasts and in mutant populations of CEM cells that were genetically deficient in either nucleoside transport or deoxycytidine kinase activity. Whereas ddC at a concentration of 4 uM inhibited growth of the wild type CEM parental strain by 50%, two nucleoside transport-deficient clones were four-fold resistant to the pyrimidine analog. The deoxycytidine kinase-deficient cell line was virtually completely resistant to growth inhibition by the dideoxynucleoside (ddN) at a concentration or 1024 uM. An 80% diminished rate of [^{3}H]ddC influx into the two nucleoside transport-deficient lines could account for their resistance to the ddN, while the resistance of the deoxycytidine kinase deficient cells to ddC toxicity could be explained by a virtually complete failure to incorporate [^{3}H]ddC <u>in situ</u>. Two potent inhibitors of mammalian nucleoside transport, 4-nitrobenzylthioinosine and dipyridamole, mimicked the effects of a genetic deficiency in nucleoside transport with respect to ddC toxicity and incorporation. These data indicate that the intracellular metabolism of ddC in CEM cells is initiated by the nucleoside transport system and the cellular deoxycytidine kinase activity.

INTRODUCTION

The human immunodeficiency virus (HIV), is a lymphotropic retrovirus and the etiologic agent of acquired immunodeficiency syndrome (AIDS) and related diseases. Although considerable information is now available on both the molecular biology of HIV and the clinical treatment of patients suffering from AIDS, there are still no effective chemotherapeutic regimens for the treatment of this insidious disease. The design of chemotherapeutic agents for the treatment of AIDS is essential, because the disease is almost invariably fatal within several years after onset of clinical symptoms, is infectious, and the incidence and prevalence of AIDS continues to increase exponentially. 3'-Azido-3'-deoxythymidine is one such candidate antiretroviral drug (1), one which gets phosphorylated to the triphosphate level and then acts as a terminator of DNA synthesis (2). Recently, Mitsuya and Broder have demonstrated that a class of compounds known as 2',3'-dideoxynucleosides (ddNs) can inhibit the <u>in vitro</u> infectivity and cytopathic effect of HIV toward human T cells (3). Mitsuya et al. subsequently showed that all four ddNTPs were incorporated into elongating DNA chains by the HIV reverse transcriptase, leading to inhibition of both viral DNA synthesis and subsequent messenger RNA expression (4).

A rational strategy for the exploitation of ddN in the treatment of AIDS depends on a fundamental understanding of the metabolic pathways by which these compounds are converted to their active metabolites. In this manuscript, we describe the cytotoxic effects and the metabolism of one ddN, 2',3'-dideoxycytidine (ddC), in a human T lymphoblast cell line (CCRF-CEM) and several clonal derivatives which are genetically deficient in either their ability to transport nucleosides or in deoxycytidine kinase activity.

MATERIALS AND METHODS

Chemicals and Reagents - All chemicals, reagents, and materials were of the highest quality commercially available.

Cell Culture and Origin of Strains - CCRF-CEM cells are a human T lymphoblast cell line originally derived from a patient with acute lymphocytic leukemia. The S49 cell line is a murine T cell lymphoma that was originally derived after mineral oil injection into a Balb/c mouse. Both the human CEM cells and the mouse S49 cells were propagated in Dulbecco's Modified Eagle medium supplemented with 10% horse serum that had been previously heat inactivated at 56° for 30 min in a humidified 10% CO_2 atmosphere at 37° as previously described. The ARAC-8D cell line is a clonal derivative of an hypoxanthine-guanine phosphoribosyltransferase (HGPRTase)-deficient parent CEM cell line originally isolated in semi-solid agarose containing 8 uM arabinosylcytosine. The ARAC-8D cell line was originally derived to analyze the pathways by which purine 2'-deoxynucleosides were metabolized in human T cells. The HGPRTase deficiency was valuable for these previous metabolic studies and continues to serve as a useful marker during the long term culture of these cell lines in the laboratory. For the purposes of these investigations, the absence or presence of HGPRTase, a purine base metabolizing enzyme, is not pertinent to the metabolism or cytotoxicity of ddC, a pyrimidine ddN, and therefore the HGPRTase-deficient parent is the equivalent of a wild type CEM cell line. The two nucleoside transport-deficient cell lines employed in these studies, ARAC-8C and TUB-4D, were also isolated from an HGPRTase-deficient derivative of wild type CEM cells (10). The ARAC-8C and TUB-4D clones were isolated by virtue of their resistance to 8 uM arabinosylcytosine and 4 uM tubercidin, respectively.

Growth Rate Determinations - The ability of wild type and mutant CEM cells to grow in ddC was assessed by previously described methods (7). In order to determine the effects of NBMPR or DPA, two potent inhibitors of nucleoside transport in mammalian cells (8, 9), on the growth of wild type cells in the presence of ddC, cells were grown as above except in the presence of a 10 uM concentration of each inhibitor of nucleoside transport. After 96 hr, the number of cells in each well was enumerated using a Coulter Counter Model ZBM. The initial cell density was subtracted from the final cell density, and the resulting difference was plotted as a percentage of the difference in cell density in control wells as a function of the exogenous ddC concentration. Typically the control cell density at enumeration had reached a late stage of exponential growth, 1.6-2.5 X 10^6 cell per ml, approximately 4-4.5 cell divisions.

Dideoxycytidine Incorporation into Cellular Nucleotides - The ability of the wild type, ARAC-8C, TUB-4D, and ARAC-8D cells to incorporate [^{3}H]ddC into phosphorylated metabolites was carried out as described (10). Cells were exposed to various concentrations (0.05uM - 200uM) of radiolabelled ddC for several time intervals (1-8 hr), harvested by centrifugation after the addition of an excess of ice cold phosphate buffered saline, the diluted radioactive culture medium decanted, and the cells lysed in 100ul of 1%

Triton X-100. The lysed cell extracts were then adsorbed onto 1.5 X 1.5 cm DEAE-Whatman filter discs and allowed to dry. The discs were pooled and washed four times with 1 liter of 1 mM ammonium formate until no more radioactive ddN could be eluted from the discs. The dried DEAE-Whatman paper discs were incubated in 0.1N HCl containing 0.3M KCl for 1 hr at 37o and counted by liquid scintillation spectrometry.

RESULTS

Effects of ddC on Growth of Wild Type and Deoxycytidine Kinase-Deficient Cells - The effects of increasing concentrations of ddC on the growth of wild type, nucleoside transport-deficient, and deoxycytidine kinase-deficient CEM cells were compared. As illustrated in Figure 1, the effective concentration of ddC which inhibited the growth of wild type CEM cells by 50% (EC_{50}) value was approximately 4 uM. The EC_{50} values of the nucleoside transport-deficient CEM cells for the ddN were four-fold greater than that for parental cells, whereas concentrations of ddC as high as 1 mM had virtually no growth inhibitory or cytotoxic effects on the deoxycytidine kinase-deficient ARAC-8D cells (Figure 1). The effects of a genetic deficiency in nucleoside transport on the growth sensitivity of wild type cells to ddC could be simulated by 10 uM concentrations of either NBMPR or DPA, two tight binding inhibitors of nucleoside transport in mammalian cells (8, 9), either of which increased the EC_{50} value of wild type cells for the pyrimidine analog four-fold (data not shown). Interestingly, the murine S49 T cell line was completely insensitive to growth inhibition by 400 uM ddC (data not shown).

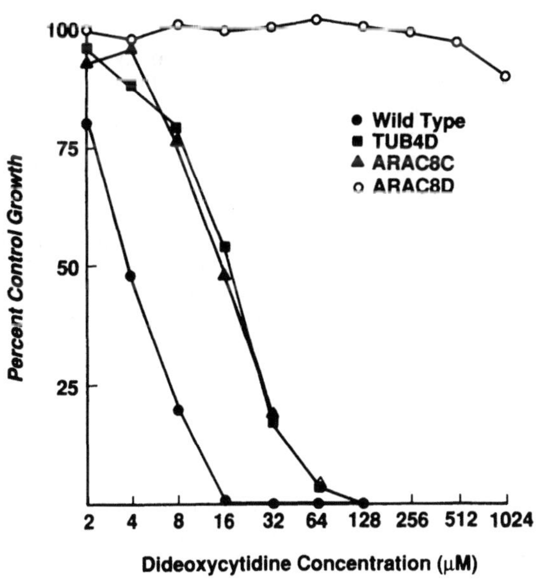

Figure 1. Growth Sensitivities of Wild Type, ARAC-8C, TUB-4D, and ARAC-8D cells to ddC. The sensitivities of wild type, ARAC-8C, TUB-4D and ARAC-8D cells to growth inhibition by ddC were ascertained as described in the Materials and Methods section. This experiment is one of approximately ten similar experiments that gave virtually identical results.

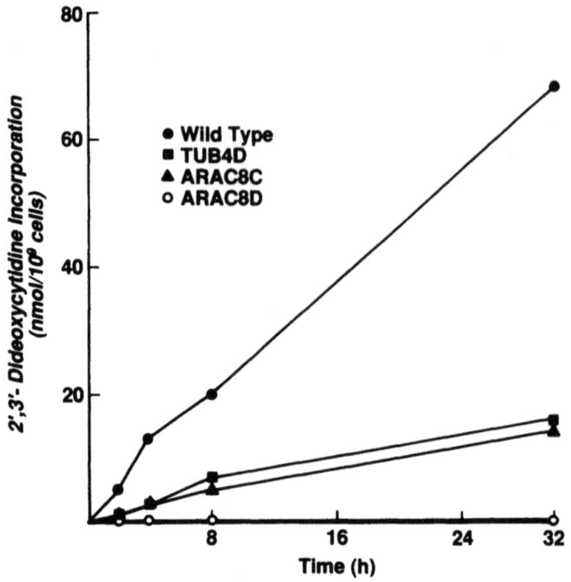

Figure 2. Rate of Incorporation of [^{3}H]ddC into Phosphorylated Metabolites. The abilities of wild type, ARAC-8C, TUB-4D, and ARAC-8D cells to incorporate 5 uM [^{3}H]ddC (20 mCi/mmol) into phosphorylated derivatives were determined after 1, 2, 4, 8, and 32 hr incubations with the ddN. The anionic ddC metabolites were isolated on DEAE-Whatman filter discs and processed as described in the Materials and Methods section.

<u>Uptake of ddC into Wild Type and Mutant CEM Cells</u> - One of the mechanisms by which cells become resistant to the cytotoxic effects of nucleosides or nucleoside analogs is by their failure to incorporate these compounds into cellular nucleotides. Therefore, the abilities of the wild type and mutant CEM cell lines to incorporate 5 uM [^{3}H]ddC (20 mCi/mmol) into radioactive phosphorylated derivatives were evaluated. As demonstrated in Figure 2, wild type CEM cells could incorporate [^{3}H]ddC into phosphorylated metabolites in a relatively linear fashion for 32 hr, whereas the ARAC-8D cells were essentially completely incapable of phosphorylating ddC <u>in situ</u>. The two nucleoside transport-deficient cell lines incorporated [^{3}H]ddC into phosphorylated derivatives at about 20% the rate of its wild type parent. Either 10 uM NBMPR or 10 uM DPA inhibited [^{3}H]ddC incorporation into wild type cells by about 70-80%, whereas the two inhibitors of nucleoside transport had no effect on [^{3}H]ddC metabolism in the nucleoside transport-deficient ARAC-8C cell line (data not shown).

DISCUSSION

The important cellular parameters involved in both the entry and metabolism of ddC have been defined in human CEM T lymphoblasts, a cell line which can be infected by HIV (11). Studies with nucleoside transport-deficient mutants and with inhibitors of nucleoside transport indicated that ddC enters CEM cells both by nucleoside transporter-dependent and nucleoside transporter-independent routes. Similarly, Cooney et al. (12) demonstrated that 20 uM NBMPR inhibits ddC uptake by 80% in P388 cells, a non-thymic murine cell line. Although nucleoside transport-deficient or NBMPR-treated wild type S49 cells do not possess sizable passive diffusion components for ribonucleoside entry (13), that ddNs cross the lipophilic plasma membrane more readily is not that surprising, since the sugar moiety is much less hydrophilic. The biochemical genetic analysis of ddC transport into human CEM cells has important implications for the rational utilization of the pyrimidine nucleoside in anti-AIDS chemotherapeutic protocols. The existence of two routes of ddN entry into T cells, one that is amenable to pharmacologic manipulation by NBMPR and DPA, offers an approach for the modulation and targeting of the drug in AIDS patients.

These biochemical and genetic studies have also confirmed that deoxycytidine kinase initiates the metabolism and subsequent cytotoxic effects of ddC toward this human T lymphoblast cell line. The conclusion that deoxycytidine kinase is responsible for the initiation of ddC metabolism is consistent with the studies of Cooney et. al who demonstrated that deoxycytidine could abolish ddC phosphorylation in human ATH8 cells, that a deoxycytidine kinase-deficient derivative of the nonthymic P388 cell line did not effectively metabolize ddC, and that the ddN was not an effective substrate for either cytidine deaminase or pyrimidine nucleoside phosphorylase activities (12). The demonstration that the former route is responsible for ddC incorporation also has some important chemotherapeutic implications, since the flux of the ddN through the deoxycytidine kinase is amenable to pharmacological perturbation. For instance, deoxycytidine kinase in human T cells has a broad substrate specificity and is capable of recognizing deoxyadenosine and deoxyguanosine, as well as its preferred substrate deoxycytidine, and the enzyme is sensitive to feedback inhibition by a variety of deoxynucleoside triphosphates, especially deoxyCTP. The cellular deoxyCTP pools can be manipulated exogneously since deoxynucleoside triphosphates cause profound allosteric alterations in the cellular ribonucleotide reductase, the enzyme that synthesizes the deoxyribonucleotide precursors required for DNA synthesis. For instance, incubation of mammalian T cells with deoxyadenosine, deoxyguanosine, or thymidine results in a virtually complete depletion of deoxyCTP. Recently, Balzarini et al. have shown that thymidine can stimulate the formation of phosphorylated ddC metablites in murine L1210 cells and enhances its protective effect toward HIV-infected ATH8 cells (14).

ACKNOWLEDGEMENTS

This work was supported by Grant DK 38809 from the National Institutes of Health. B.U. is supported by a Research Career Development Award from the National Institutes of Health.

REFERENCES

1. Mitsuya, H., Weinhold, K.J., Furman, P.A., St. Clair, M.H., Nusinoff Lehrman, S., Gallo, R.C., Bolognesi, D., Barry, D.W., & Broder, S. (1985) Proc. Natl. Acad. Sci. USA 82, 7096-7100

2. Furman, P.A., Fyfe, J.A., St. Clair, M.H., Weinhold, K., Rideout, J.L., Freeman, G.A., Nusinoff Lehrman, S., Bolognesi, D., Broder, S. Mitsuya, H., & Barry, D.W. (1986) Proc. Natl. Acad. Sci. USA 83, 8333-8337

3. Mitsuya, H. & Broder, S. (1986) Proc. Natl. Acad. Sci. USA 83, 1911-1915

4. Mitsuya, H., Jarrett, R.F., Matsukura, M., Di Marzo Veronese, F., DeVico, A.L., Sarngaharan, M.G., Johns, D.G., Reitz, M.S., & Broder, S. (1987) Proc. Natl. Acad. Sci. USA 84, 2033-2037

5. Foley, G.E., Lazarus, H., Farber, S., Geren Uzman, B., Boone, B.A., & McCarthy, R.E. (1965) Cancer 18, 522-529

6. Hershfield, M.S., Fetter, J.E., Small, W.C., Bagnara, A.S., Williams, S.R., Ullman, B., Martin, D.W. Jr., Wasson, D.B., and Carson, D.A. (1982) J. Biol. Chem. 257, 6380-6386

7. Ullman, B., Cohen, A., & Martin, D.W. Jr. (1976) Cell 9, 205-211

8. Paterson, A.R.P. & Oliver, J.M. (1971) Can. J. Biochem. 49, 271-274

9. Scholtissek, C. (1968) Biochim. Biophys. Acta 158, 435-447

10. Ullman, B., Coons, T., McCartan, K., and Rockwell, S. J. Biol. Chem., in press.

11. Dalgleish, A.G., Beverley, P.C.L., Clapham, P.R., Grawford, D.H., Greaves, M.F., & Weiss, R.A. (1984) Nature 312, 763-767

12. Cooney, D.A., Dala, M., Mitsuya, H., McMahon, J.B., Nadkarni, M., Balzarni, J., Broder, S., & Johns, D.G. (1986) Biochem. Pharmacol. 35, 2065-2068

13. Berlin, R.D. & Oliver, J.M. (1975) Int. Rev. Cytol. 42, 287-336

14. Balzarini, J., Cooney, D.A., Dalal, M., Kang, G.-J., Cupp, J.E., De Clercq, E., Broder, S., & Johns, D.G. (1987) Molec. Pharmacol. 32, 798-806

COORDINATE REGULATION OF ADENOSINE A1 AND A2 RECEPTORS

Roberta M. Palmour, Anna-Marie Babey and Andrew J. K. Smith

Departments of Psychiatry, Biology and Centre for Human Genetics
McGill University
Montreal PQ Canada

INTRODUCTION

Adenosine (ADO) has potent receptor-mediated modulatory effects on brain and behavior and on many physiological integrative processes[1-3], such as hormone release and interaction of immune cells. In many instances, this modulation is thought to be mediated through stimulation or inhibition of adenylate cyclase. A2 ADO receptors are positively coupled to cyclase via the guanine nucleotide stimulatory protein, while A1 receptors are negatively coupled. Thus stimulation of A1 receptors generates a biochemical signal in opposition to that produced by stimulation of A2 receptors.

Pharmacological definition of A1 vs A2 receptors has been accomplished in two ways. First, there is differentiation on the basis of agonist concentration. A1 response requires nanomolar concentrations of ADO or ADO analogs, while A2 receptors are activated by submicromolar to micromolar concentrations of agonist. Second, some compounds are selectively potent at one receptor. The generally agreed order of potency is NECA > PIA > CHA > CPA at A2 receptors and CPA > CHA > PIA > NECA at A1 receptors[4-6].

At the tissue level, activation of A1 receptors often initiates processes which oppose those actions stimulated by A2 receptor agonism. For example, in the guinea pig trachea, A1 agonism induces contraction, while A2 activation stimulates relaxation[7]. In kidney, renin secretion is stimulated by A2-appropriate agonists, but inhibited by A1 doses or analogs[8]. In brain, A1 activation diminishes basal acetylcholine output, while A2 stimulation enhances electrically-stimulated acetylcholine release[9].

In the intact animal, A2 stimulation decreases blood pressure and increases heart rate, while reverse effects are seen after A1 activation[10]. Similarly, locomotion and agonistic behaviors are increased by the administration of A1-selective drugs and reduced by A2-selective drugs or doses[11,12]. To appreciate the relative contributions of A1 and A2 receptor actions <u>in vivo</u> and especially to understand the organismal effects of ADO receptor antagonists (most of which are equally potent at A1 and A2 sites), it is necessary to study a preparation comprising both receptor types.

We have used cultured murine neuroblastoma (NB) cells to define (1) the individual actions of ADO agonists and antagonists on A1 and A2 receptors, (2) the net effect of analogs in cells possessing both receptor subtypes, and (3) the coordinate effect on receptor regulation of acute and chronic administration of theophylline (a potent ADO antagonist) and of ethanol.

METHODS

Neuroblastoma cells Clonal NB cells (N4 Scl 3) were grown in DME, supplemented with 10% FCS, Pen-Strep and Fungizone. For whole cell studies, 35 mm plates (log phase, 250-300 ug protein/dish) were washed free of serum-containing medium and preincubated 10' in DME + 300 uM Ro 20-1724. Adenylate cyclase activity was stimulated with graded concentrations of ADO analog ± PGE_1 and/or theophylline (THEO). After 15' incubation, reactions were terminated with boiling Na acetate buffer (50 mM, pH 4). Cyclic AMP was determined according to the method of Gilman, and expressed as pmol/mg protei (Lowry), or as % maximal stimulation.

Binding of $[^3H]$-NECA to either A1 or A2 ADO receptors was studied in membranes freshly prepared from NB cells, as described previously[13]. A1 receptors were labelled with 4 nM $[^3H]$-NECA and defined by 100 nM CPA. A2 receptors were labelled with 8 nM $[^3H]$-NECA in the presence of 20 nM CPA and were defined by 100 uM unlabelled NECA. Binding data were analyzed by non- linear curve fitting programs derived from LIGAND (Munson and Rodbard.

Pharmacological pretreatments NB cells were pretreated either overnight or chronically with 30 uM THEO, then washed with 1 additional change of mediu prior to whole cell or binding studies. In one set of experiments, 30 plates were preincubated overnight with 2 uM cycloheximide administered before THEO; parallel sets of plates contained THEO alone, cycloheximide alone or vehicle. For chronic studes of both A1 and A2 receptors, cells were treated with THEO-supplemented medium for a period of 2 weeks prior to whole cell and binding studies. For ethanol experiments, a similar protocol was followed using 0.5% ethanol acutely and for periods of 2 or 3 weeks.

RESULTS

A1 and A2 receptors on clonal neuroblastoma cells Although it has beer known for many years that various murine NB lines have A2 ADO receptors[14,15], we have recently presented data[13] supporting the contention that clonal N4 NB cells also have A1 receptors. Stimulation of whole cells with submicromolar doses of ADO analogs alone increased the intracellular levels of cAMP (Table 1), while treatment with nanomolar concentrations of ADO analogs reduced PGE_1-stimulated cAMP accumulation (Table 2). The order of potency of ADO analogs at the A2 NB

Table 1. ADENOSINE ANALOGS INCREASE cAMP LEVELS IN N4 NEUROBLASTOMA CELLS

Analog	EC_{50} (uM)	Efficacy (-fold increase cAMP)
NECA	1.3 ± 0.2	14.8 ± 3.6
CADO	2.5 ± 0.5	12.8 ± 4.2
R-PIA	3.8 ± 0.7	11.2 ± 3.6
ADO	4.0 ± 0.9	12.8 ± 4.1
CPA	13.5 ± 4.6	5.6 ± 2.3
CHA	20 ± 5.8	10.4 ± 3.2
S-PIA	23 ± 8	3.7 ± 1.5

All values given as mean ± SD.

Table 2. REDUCTION OF PGE_1-STIMULATED cAMP ACCUMULATION IN N4 CELLS

Analog	cAMP, pmol/mg 100 nM PGE1	% reduction, 100 nM PGE1 + 2 nM analog	% reduction, 100 nM PGE1 + 4 nM analog	% reduction, 100 nM PGE1 + 10 nM analog
CPA	1235 ± 41	4 ± 2	17 ± 6	44 ± 13
NECA	1410 ± 33	28 ± 4	42 ± 7	37 ± 6
R-PIA	1126 ± 31	32 ± 3	36 ± 4	27 ± 4

All values given as mean ± SD.

receptor was NECA > 2CADO > PIA > CHA > CPA. At the A1 receptor, this order of potency was reversed. Interestingly, there were marked differences in the efficacy of analogs at the A2, but not the A1, receptor. A2 concentration-effect curves were shifted to the right by THEO; 120 uM THEO approximately doubled the EC_{50} values for NECA, PIA and CADO. Similarly, THEO dose-dependently reduced the ability of ADO analogs to inhibit PGE_1-stimulated cAMP accumulation.

Two classes of $[^3H]$-NECA binding sites can be distinguished in N4 NB cells when a fixed concentration of radioligand (4 nM) is displaced by increasing concentrations of unlabelled NECA, PIA or CPA. Computerized non-linear curve fitting analysis (adapted from LIGAND) of 6 replications suggested K_d and B_{max} values of 3.15 ± 0.35 nM and 62.4 ± 10 fmol/mg protein for A1 receptors and 46 ± 16 nM and 188 ± 14 fmole/mg protein for A2 receptors in N4 NB cells.

Regulation of ADO receptors by the antagonist THEO Green and Stanbury[15] and Green[16] showed that overnight pretreatment with THEO stimulated ADO-sensitive cAMP accumulation in N2 NB cells, and several investigators[17-19] have documented the effects upon brain A1 binding sites of chronic caffeine administered *in vivo* to rats. To date, there have been n investigations of antagonist regulation of *both* A1 and A2 receptors in a single tissue or cell preparation.

As shown in Figure 1, 16 hour pretreatment of NB cells, as described in methods, increases the extent to which a given dose of NECA stimulates cAMP production. Parallel binding studies (Figure 2B) demonstrate an increase in the number of binding sites (B_{max}), but no change in K_d). This effect is dose- and time-dependent, being effectively produced by doses of antagonis from 10-300 uM and by treatment times of 8 hr or longer. As shown in Figure 2C, the "up-regulation" of A2 receptors is effectively blocked by cycloheximide pretreatment (Figure 2C), suggesting that synthesis of new protein is required.

The effects of chronic THEO treatment are more complex. At the A1 receptor (Figure 3A), ADO analogs are significantly less effective in reducin PGE_1-stimulated cAMP accumulation after chronic THEO treatment. The profile for $[^3H]$-NECA binding to A1 sites (3B) is substantially altered, with a change in both affinity and number for the low-affinity state, and reduced numbers of high-affinity receptors. At the A2 receptor (Figure 3C), similar amounts of cAMP are produced in respose to a given dose of NECA in

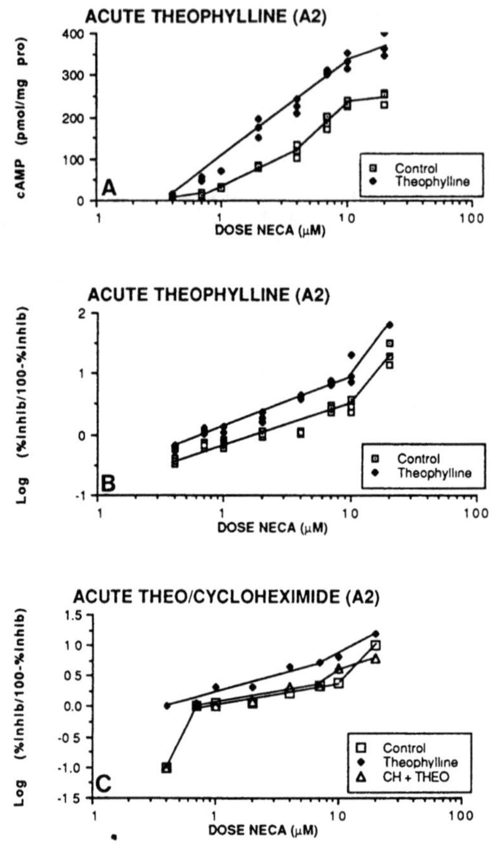

Figure 1. Regulation of A2 ADO receptors by acute pretreatment with THEO, a described in the text.

A. The effects of overnight pretreatment upon A2 specific cAMP accumulation.

B. The effects of overnight pretreatment upon A2 specific $[^3H]$-NECA binding.

C. Reduction of the effects of acute THEO by concommitant cycloheximid treatment.

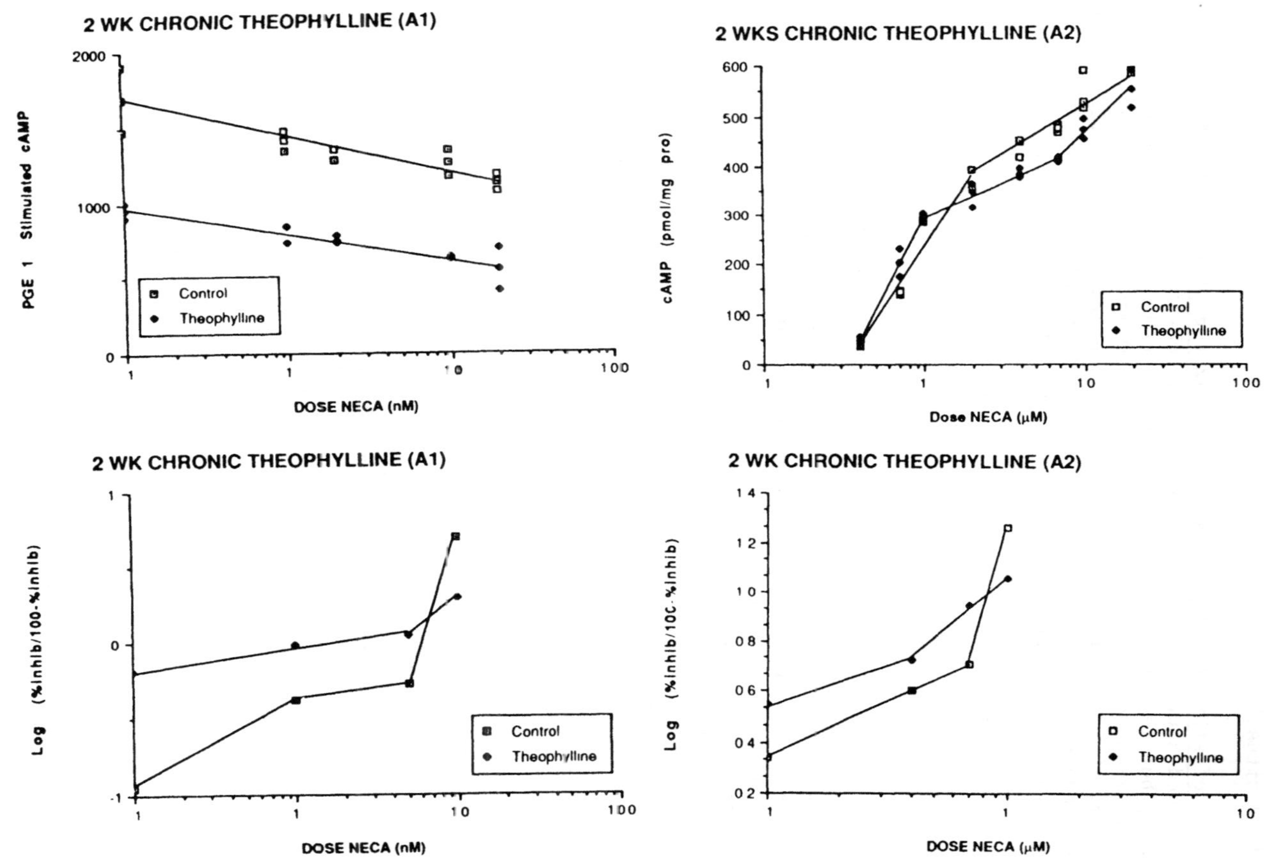

Figure 2. The effects of chronic THEO administration upon A1 and A2 receptors in murine NB cells, as described.
Top row: The effects on adenylate cyclase activity at A1 (left) and A2 (right) sites.
Bottom row: The effects upon A1 (left) and A2 (right) specific binding of [3H]-NECA, as described in the text.

control cells and in cells grown for 2 weeks in the continuous presence of 3 uM THEO. Although there is a small shift in the profile of [^{3}H]-NECA binding (3D), this difference is not significant. In fact, preliminary experiments suggest that THEO shifts the relative proportions of high- and low-affinity coupling states for both A1 and A2 receptors, but additional studies are required to evaluate this possibility. The net effect of acute THEO in N4 NB cells is to favor actions at the A1 receptor, by increasing th ratio of A1:A2 sites.

Effects of ethyl alcohol on ADO A1 and A2 receptors Ethanol (0.2%-1.0%) added acutely to NB cells dose-dependently increases the basal level of cAMP, but decreases the extent to which cAMP accumulation is stimulated by A2-appropriate concentrations of ADO analogs (Figure 4A). Concommitantly, the dose-effect curve for A1 receptors is shifted left 4-fol so that the range of concentrations over which A1 stimulation occurs is increased by nearly an order of magnitude (4B). This does not simply reflec initial values, as cAMP levels achieved at 100 nM PGE_1 are nearly identical in the presence or the absence of added ETOH. ETOH, at concentrations between 0.2 and 1%, non-specifically facilitates binding of [^{3}H]-NECA and many other neurotransmitter radioligands. Consequently, binding studies were not done in parallel.

NB cells grow readily in the presence of ETOH concentrations between 0. and 0.8%, and can be gradually adapted to grow well at higher concentrations As shown in Figure 4C, after 2 weeks of growth in 0.5% ETOH, there is a dramatic reduction in the quantity of cAMP accumulated in the presence of uM concentrations of NECA and a flattening of the A2 dose-effect curve. At 3 weeks, the pattern is nearer that of the control cells. Although it is tempting to speculate that these changes reflect the development of toleranc and addiction[20], additional time points will be required to verify this trend. At A1-appropriate concentrations (4D), there is little change in the absolute reduction of PGE_1-stimuated cAMP levels, but the dose-effect curve is again extended. Preliminary binding studies suggest an increase in A1 specific sites at the expense of A2 sites.

DISCUSSION

In summary, we have shown that clonal N4 NB cells possess both A1 and A receptors, while in other cellular aggregates or tissues, one cell type has receptors (eg., B lymphocytes) and another cell type (such as T lymphocytes) has A2 receptors. In addition, we report here two instances in which acute chronic treatment with a common, widely available psychotropic agent elicits change in the relative effects of adenosine analogs at both A1 and A2 receptors. In general, acute effects are straightforward sensitizations or desensitizations, while chronic effects are more complex and probably involv the elaboration of a new level of homeostasis.

In a number of instances, there is now very strong evidence for balance opposition of physiological effects mediated by A1 and A2 receptors, respectively[7-9, 22, 23]. For an integrated tissue with both A1 and A2 receptors, the sum of actions at these receptors will reflect: (a) the relative abundance of A1 vs A2 receptors; (b) the concentrations of endogeno or exogenous agonists; (c) the presence or absence of other modifying factor such as antagonists or other psychotropic drugs. Because ADO is a very general modulator, and has effects upon many other transmitter and hormone systems, it is not possible to arrive at a mathematical summation of the net basal steady state. It may, however, be possible to document the specific changes which accompany the development of a new steady state induced by a receptor antagonist or other drug.

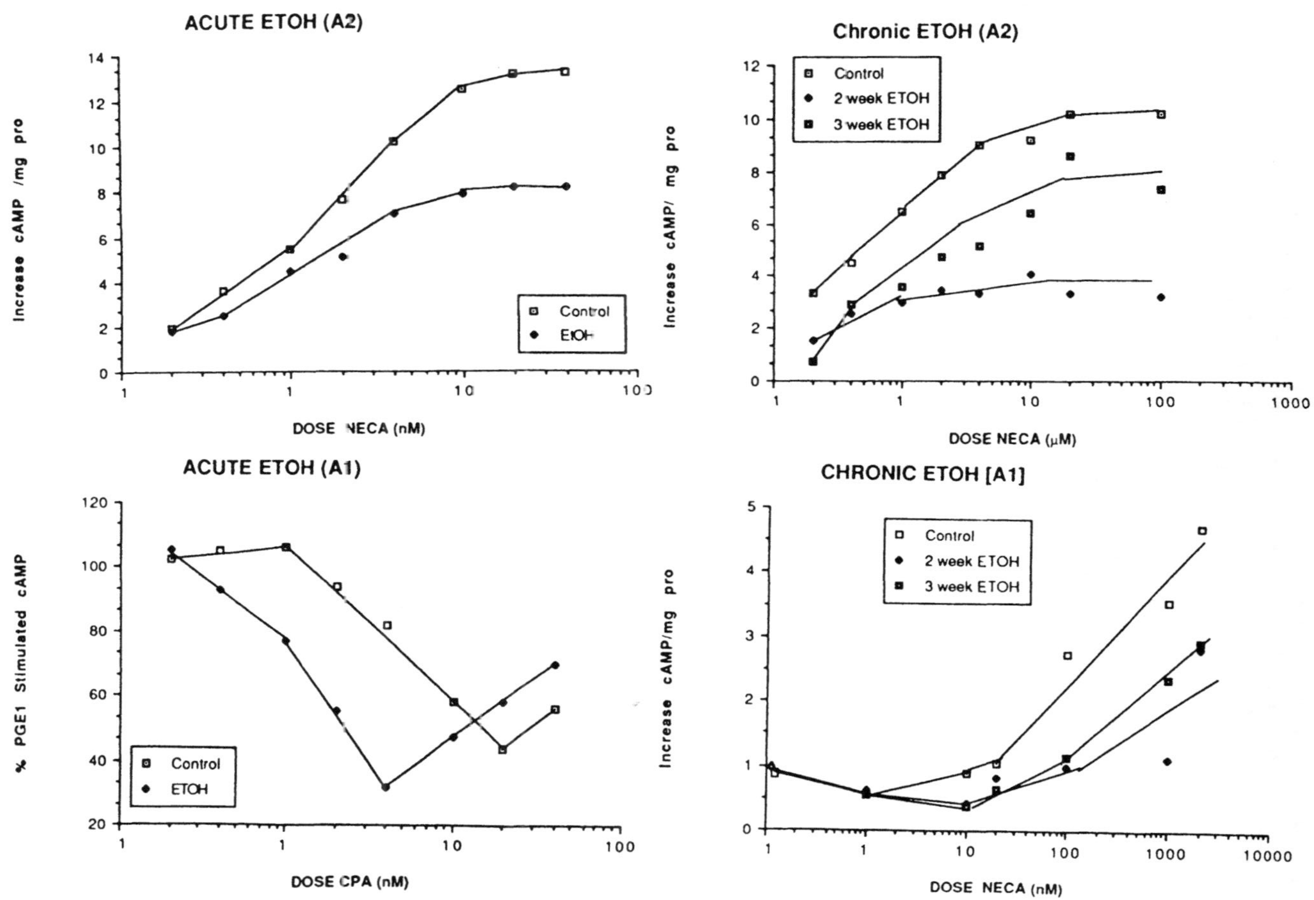

Figure 3. The effects of acute and chronic ethanol (ETOH) pretreatment on A1 and A2 receptors in murine NB cells.
Left: The effects of acute ETOH on A2 and A1 receptor interactions with adenylate cyclase, as described in the text.
Right: The effects of chronic ETOH on A2 and A1 receptor interactions with adenylate cyclase, as described in the text.

REFERENCES

[1]Snyder, S. H., Adenosine as a neuromodulator, Ann. Rev. Neurosci 8: 103 (1985).

[2]Dunwiddie, T. V., The phsyiological role of adenosine in the central nervous system, Intl. Rev. Neurobiol. 27, 63 (1985).

[3]Williams, M., Adenosine receptors, in Psychopharmacology: The Third Generation of Progress Raven Press, New York (1987).

[4]Londos, C., Cooper, D. M. F., Wolff, J., Subclasses of external adenosine receptor, Proc. Natl. Acad. Aci. USA 77: 2551 (1980).

[5]Stone, T. W., Summary of a symposium discussion on purine receptor nomenclature, In: Purines: Pharmacology and Physiological Roles, ed. T. W. Stone (VCH Publications, Deerfield Beach, Florida) p. 1-4 (1985).

[6]Bruns, R. F., G. H. Lu and T. A. Pugsley, Characterization of the A2 adenosine receptor labeled by [^{3}H] NECA in rat striatal membranes, Mol. Pharmacol. 29, 331 (1986).

[7]Ghai, G., Zimmerman, M. B. and Hopkins, M.F., Evidence for Al and A2 adenosine receptors in guinea pig trachea, Life Sci. 41: 1215 (1987).

[8]Churchill, P.C. and M. C. Churchill, Al and A2 Adenosine receptor activation inhibits and stimulates renin secretion of rat renal cortical slices. J. Pharmacol. Exp. Therap. 232, 589 (1984).

[9]Spignoli G., F. Pedata and G. Pepeu, Al and A2 adenosine receptors modulate acetylcholine release from brain slices, Eur. J. Pharmacol. 97, 341 (1984).

[10]Coffin, V. L. and Spealman, R. D., Behavioral and cardiovascular effects of analogs of adenosine in cynomolgus monkeys J Pharmacol Exp Therap 241: 76 (1987).

[11]Spealman R. D. and Coffin, V.L., Behavioral effects of adenosine analogs in squirrel monkeys: relation to adenosine A2 receptors, Psychopharmacol. 90: 419 (1986).

[12]Palmour, R.M., C.J. Lipowski, C.K. Simon and F.R. Ervin. 1988. Adenosine analogs inhibit fighting in isolated male mice. Life Sci.

[13]Palmour, R. M. and Oblin, C., Coexistence of adenosine Al and A2 receptors on murine neuroblastoma cells, submitted (1988).

[14]Blume A. J. and C. J. Foster, Neuroblastoma adenylate cyclase, J. Biol. Chem. 251, 3399 (1976).

[15]Green, R. D. and Stanberry, L. R., Elevation of cyclic AMP in C-1300 murine neuroblastoma by adenosine and related compounds of the antagonism of this response by methylxanthines, Biochem. Parmacol. 26: 37 (1977).

[16]Green, R. D., Release of adenosine by C1300 neuroblastoma cells in tissue culture, J. Supramolec. Struc. 13: 175 (1980).

[17]Fredholm, B.B., Adenosine actions and adenosine receptors after 1 week treatment with caffeine. Acta Physiol. Scand. 115:283 (1982).

[18]Boulenger J.-P., Patel J., Post R. M., Parma A. M. and Marangos P. J., Chronic caffeine consumption increases the number of brain adenosine receptors, Life Sci 32: 1135 (1983).

[19]Green R. M. and Stiles G. L., Chronic caffeine ingestion sensitizes the Al adenosine receptor-adenylate cyclase system in rat cerebral cortex, J. Clin Investig 77: 222 (1986).

[20]Sharma, S. K., Klee, W. A. and Nirenberg, M., Dual regulation of adenylate cyclase accounts for narcotic dependence and tolerance, Proc. Natl. Acad. Sci. USA 72:3092 (1975).

[21]Proll, M. A., R. B. Clark and R. W. Butcher, Al and A2 adenosine receptors regulate adenylate cyclase in cultured human lung fibroblasts Molec. Cell. Endocrinol. 44: 211 (1986).

[22]Garcia-Sainz, J. A. and Torner, M. L., Rat fat-cells have three types of adenosine receptors (Ra, Ri and P), Biochem. J. 232, 439 (1985).

A_1 AND A_2 ADENOSINE RECEPTORS IN SYNOVIAL CELLS FROM PATIENTS WITH RHEUMATIC DISEASES

Claudia Martini, Gino Giannaccini, Antonio Lucacchini, Laura Bazzichi*, Annalia Soletti* and Maria L. Ciompi*

Istituto Policattedra di Discipline Biologiche
Istituto di Patologia Speciale Medica*
University of Pisa, Pisa (Italy)

INTRODUCTION

Adenosine mediates physiological or pharmacological effects via two subtypes of membrane-localized site receptors at cell surface. Differences in the rank order of potency observed for stimulation and inhibition by several adenosine analogues have led to the classification of these receptors as inhibiting (A_1) and activating (A_2) adenylate cyclase activity (1,2). The physiological responses are often inhibitory in nature and can be observed in smooth and cardiac muscle, platelets, lymphocytes, adipose tissue. Adenosine evokes an inhibitory response on human neutrophyl function as the inhibition of release of toxic oxigen species (3). Moreover purine metabolism plays a pivotal role in the control of immune functions. Inborn errors in this pathway have been shown to lead to specific immune deficiency states (4). In addition it has been demonstrated that physiological concentrations of the purine adenosine play an important role in the modulation of immune responses in vitro (5, 6, 7). Recently it has been possible to label adenosine receptors with adenosine analogues such as ^{3}H-N^6-cyclohexyladenosine (CHA) for A_1 receptors and ^{3}H-5′-N-ethylcarboxamideadenosine (NECA) for A_2 receptors. In the present report we described the presence of A_1 and A_2 receptors in cells from synovial fluid from rheumatic patients and time-, protein-, temperature-dependence characteristics of these ^{3}H-NECA binding sites.

MATERIALS AND METHODS

The synovial fluids were obtained from patients with rheumatic diseases (rheumatoid arthritis, psoriatic arthritis, ostheoartrosis, serum negative arthritis) by knee punctions. The cells, which are known to be essentially neutrophils (80-90%), obtained by centrifugation were washed with physiological solution, centrifuged at 50000 g for 20 min at 4° C and homogenized in 20 vol. of 20 mM Tris-HCl pH 7.4 containing protease inhibitors (bacitracin 200 µg/ml, benzamidine 1mM, soybean trypsin inhibitor 20 μg/ml) then centrifuged at 50000 g for 20 min. The resulting pellet was treated with adenosine deaminase to remove endogenous adenosine, centrifuged and utilized for ^{3}H-CHA and ^{3}H-NECA binding assays. A_1 binding assay was performed as described in (8) incubating membrane preparations (0.6-0.8 mg proteins) with 10 nM ^{3}H-CHA at 25° C for 45 min. and filtering on Whatman GF/B glass fiber filters. The specific binding was determined in the presence of L-PIA (100 μM). A_2 binding assay was performed incubating membrane preparations (0.6-0.8 mg proteins) with 10 nM ^{3}H-NECA at 0° C for 120 min. in 40 mM Tris-HCL (pH 7.5) containing 1 mM EDTA. The samples were filtered into Whatman GF/B glass fiber filters under vacuum. After washing twince with 3X5 ml of Tris buffer, the radioactivity was counted in 8 ml of Ready-Solv HP Beckman scintillation cocktail containing 0.4 ml 0.01 N KOH. The specific binding was obtained in the presence of adenosine 5 mM.

RESULTS

The specificities of ^{3}H-CHA and ^{3}H-NECA bindings to membrane of synovial fluid cells (essentially neutrophyls 80-90%) are shown in table I and II and demonstrated an association with A_1 and A_2 adenosine binding sites. However the ^{3}H-CHA specific binding is very low in the membranes ranging from 2 to 4 fmol/mg of protein while ^{3}H-NECA specific binding ranges from 790 to 2545 fmol/mg of protein.

Table I

^{3}H-NECA binding specificity

Compounds	IC_{50} (μM)
NECA	0.25
2-Chloroadenosine	1.65
Adenosine	35
Theophylline	580
L-Phenylisopropyladenosine	1

Table II

^{3}H-CHA binding specificity

Compounds	IC_{50} (μM)
NECA	0.080
2-Chloroadenosine	0.021
CHA	0.003
Theophylline	19.50
L-Phenylisopropyladenosine	0.0008

Binding characteristic studies show that the binding of ^{3}H-NECA increased linearly with increasing protein concentrations (0.135-1.5 mg proteins). The amount of specifically bound ^{3}H-NECA was temperature-dependent and the highest value was reached at 0°C. On the contrary, the amount of the non specifically bound ^{3}H-NECA remained constant at all incubation temperatures tested (0C, 4C, 10C, 25C, 37C). The incubation time selected for the binding assay was 120 min., since at that time equilibrium had been reached between a ligand and the binding sites. In order to test the saturability of ^{3}H-NECA specific binding sites, membrane preparations were incubated with increasing concentrations of ^{3}H-NECA (5-450 nM) as described under Methods. Specific ^{3}H-NECA binding was saturable; in contrast non specific ^{3}H-NECA binding increased linearly. Non specific binding amounted to only 4-10% of total binding at 5 nM and 450 nM ^{3}H-NECA respectively. Plotting the data of specific ^{3}H-NECA binding according to Scatchard method a single straight line was obtained as shown in the figure 1, indicating that a single population of specific binding sites and a Kd of 172 nM.
In table III the binding of ^{3}H-NECA and ^{3}H-CHA to membrane preparations of synovial cells from patients with rheumatic diseases is reported.

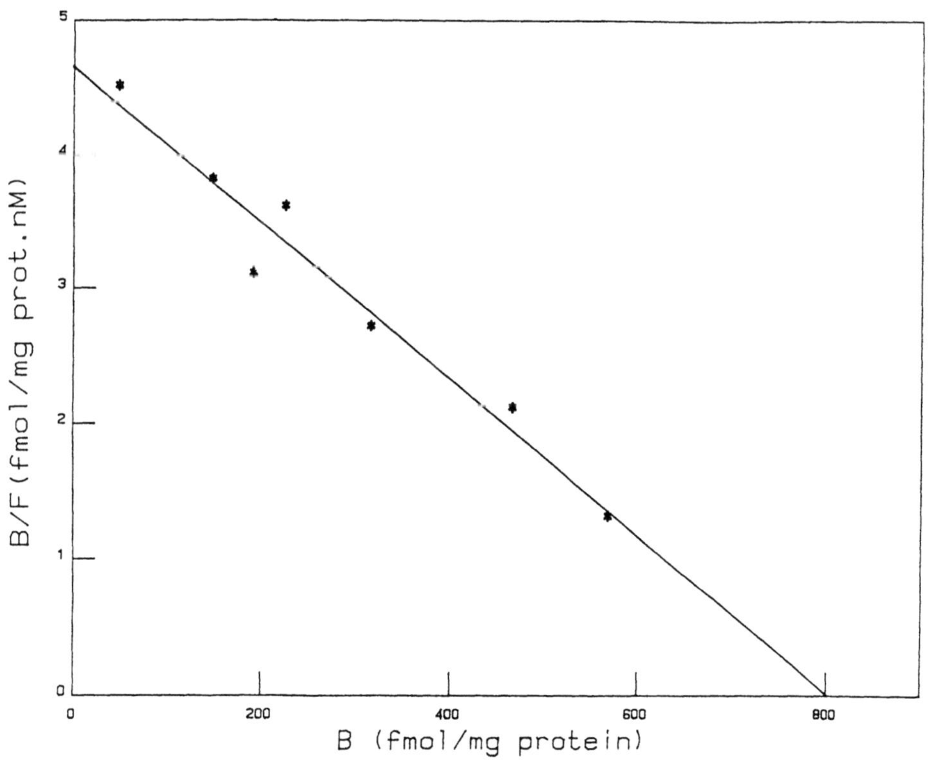

Fig.1. Scatchard plot of ^{3}H-NECA binding to membrane preparations. The membranes were obtained from synovial fluid cells (neutrophils). The assay was carried out as described under Materials and Methods.

TABLE III

Bmax values for A_1 and A_2 receptors in some rheumatic diseases

Disease	A_1 fmol/mg protein	A_2 fmol/mg protein
Rheumatoid arthritis	3.38±0.25	1376±436
Psoriatic arthritis	1.50±0.10	599±250
Ostheoartrosis	1.60±0.15	777±150
Serum negative arthritis	3.77±0.20	821±305
Gout	0.50±0.02	2242±200

DISCUSSION

The results of the radioligand binding studies demonstrate the presence of ^{3}H-CHA and ^{3}H-NECA binding sites in human synovial fluid cell membranes.

The ^{3}H-CHA specific binding (A_1 receptors) is very low in the membranes ranging from 2 to 4 fmol/mg of protein. The ^{3}H-NECA specific binding (A_2 receptors) ranges from 790 to 2545 fmol/mg of protein.

The pharmacological characteristics of adenosine site present in the synovial fluid cell support the hypothesis that the binding is essentially to A_2 adenosine receptors. The binding satisfies essential criteria for the identification of membrane bound receptors since it was found to be rapid, rapidly reversible and dependent on protein concentration and temperature.

Variations in the number of binding sites associated with disease were observed. Studies are in progress to separate the different cell types of synovial fluid, to determine the relative distribution of A_1 and A_2 receptors correlating to disease activity. In fact in gout where the inflammatory response is preminent the number of A_2 receptor sites essentially present in neutrophils is higher than in the chronic arthritis.

REFERENCES

1. D.van Calker, M.Muller,B.Hamprecht, Adenosine regulates via two different types of receptors the accumulation of cyclic AMP in cultured brain cells, J.Neurochem. 33: 999 (1979)
2. H.S.Snyder, Neurotransmitter receptor binding and drug discovery, J.Med.Chem. 26: 1677 (1983)
3. B.N.Cronstein, S.M.Kubersky, G.W.Weissmann, R.Hirschhorn, Clin.Immunol.Immunopathol. 42: 76 (1987)
4. B.S.Mitchell, W.N.Kelly, Purinogenic immunodeficiency diseases: clinical features and moleculaar mechanism, Am.Inter.Med. 92: 826 (1980)
5. G.Marone, M.Plant, L.M.Lichtenstein, Characterization of a specific adenosine receptor on human lymphocytes, J.Immun. 11: 2153 (1978)
6. G.Marone, S.R.Findlay, L.M.Lichtenstein, Adenosine receptor on human basophils: modulation of histamine release, J.Immunol. 123: 1473 (1979)

7. G.Marone, R.Petracca, M.Condorelli, Adenosine receptors on human infiammatory cells, in "Purine Metabolism in Man", Bruyn, Simmonds, Muller eds., Plenum Publishing Corp., New York (1984)
8. C.Martini, E.Pennacchi, M.G.Poli, A.Lucacchini, Solubilization of adenosine A_1 binding sites from sheep cortex, Neurochem.Int. 7: 1017 (1985)

USE OF PURINE NUCLEOTIDE AND NUCLEOSIDE METABOLISING ENZYMES AS TOOLS TO DETERMINE THE PRESENCE OF PURINERGIC NERVE TRANSMISSION IN SMOOTH MUSCLE

David Satchell

Department of Zoology
University of Melbourne
Parkville, 3052, Australia

INTRODUCTION

It is likely that applied enzymes can access the synaptic region in tissues and can be used to determine whether the substrate is acting as a neurotransmitter provided that the end product has reduced activity.

In guinea-pig, ATP has been suggested as the neurotransmitte mediating relaxation by the non-adrenergic, non-cholinergic (NANC nerves in the taenia coli[1] (Fig.1) and as a cotransmitter mediating the twitch component of the contraction in response to sympathetic nerve stimulation in the vas deferens[2] (Fig.1). Nucleotide pyrophosphatase (NP) converts ATP to AMP. AMP unlike ATP is inactive in the vas deferens and has a much lesser potency than ATP in the taenia coli.

In guinea-pig trachea adenosine has been suggested as the neurotransmitter mediating relaxation by NANC nerves[3]. Adenosine deaminase converts adenosine to inosine which is inactive (Fig.1)

Thus responses to field stimulation of the different nerve types have been determined in the presence and absence of the enzymes described using appropriate controls. Such experiments can be used to examine whether the enzymes can access the synaptic region and specifically interfere with effective neurotransmission by their substrates.

METHODS

Taenia coli strips 1.5 cm long and whole tracheae, were dissected from guinea-pigs (700-900g) and mounted in organ baths at 37°C containing Kreb's solution gassed with 95% O_2 and 5% CO_2. The dissection and mounting of taenia coli and tracheal preparations have been described previously[1,3]. A description of the dissection and recording from guinea-pig vas deferens preparations is described elsewhere in this volume[9].

Tension in taenia coli preparations was registered by means of Grass FT03C transducers and intraluminal pressure in tracheal preparations was measured via Gould Statham P23ID pressure transducers[4]. Both transducers were coupled to Grass Model RPS7C8 polygraphs. Taenia coli preparations were stimulated via ring electrodes and tracheae via coaxial electrodes using a Grass S48 stimulator. Perivascular sympathetic nerves to taenia coli strips were isolated and stimulated via ring electrodes[6]. Hyoscine, dipyridamole, ATP, adenosine, vaso intestinal peptide (VIP), adenosine deaminase (type III), noradrenaline and nucleotide pyrophosphatase were obtained from the Sigma Chemical Company, St Louis. Guanethidine was obtained from Ciba-Geigy.

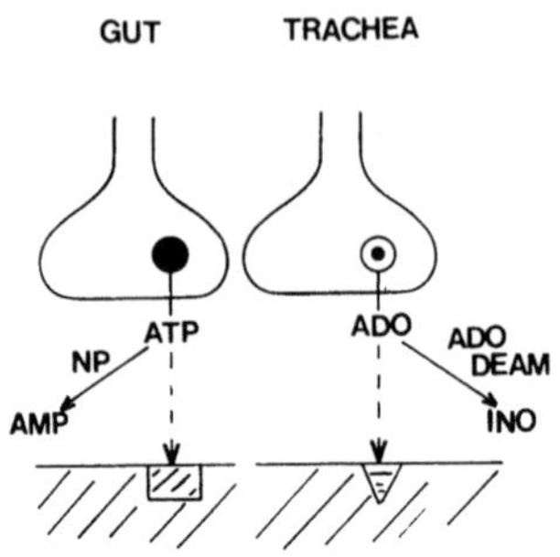

Fig. 1. Schematic representation of the release from non-adrenergic, non-cholinergic, (NANC) nerves in gut and trachea of their putative neurotransmitters ATP and pathways of metabolism of ATP to AMP by nucleotide pyrophosphatase (NP) and adenosine (ADO) to inosine (INO) by adenosine deaminase (ADO DEAM)

RESULTS AND DISCUSSION

It is well established that ATP and noradrenaline cause dose-dependent relaxations of guinea-pig taenia coli preparations[1]. Following treatment of preparations with NP for 15 min relaxations to ATP (1μM) were substantially reduced whereas relaxations in response to noradrenaline (50nM) remained the same (Fig.2). This finding demonstrated that NP not only caused a reduction of relaxations in response to applied ATP but failure of the enzyme to affect responses to noradrenaline supports the idea that its action on the ATP response is specific and is a consequence of its degradation of ATP to AMP prior to receptor occupation.

Relaxations in response to electrical stimulation of taenia coli strips were of two types. Either field stimulation of NANC nerves or stimulation of sympathetic nerves running in the artery supplying the strip. Each type of stimulus was applied alternatel

It should be noted that perivascular sympathetic nerves were stimulated at 30 Hz and the field stimulation of NANC nerves was applied at 1 Hz at which frequency perivascular sympathetic nerves failed to produce a response.

In control untreated preparations both types of electrical stimulation caused relaxations which were similar in amplitude and were largely sustained for the 3 min of stimulus. However, following treatment with NP (1 unit/ml) for 15 min relaxations in response to field stimulation of NANC nerves (F) became smaller, whereas the relaxations in response to perivascular sympathetic nerve stimulation (P) remained unaffected (Fig.2).

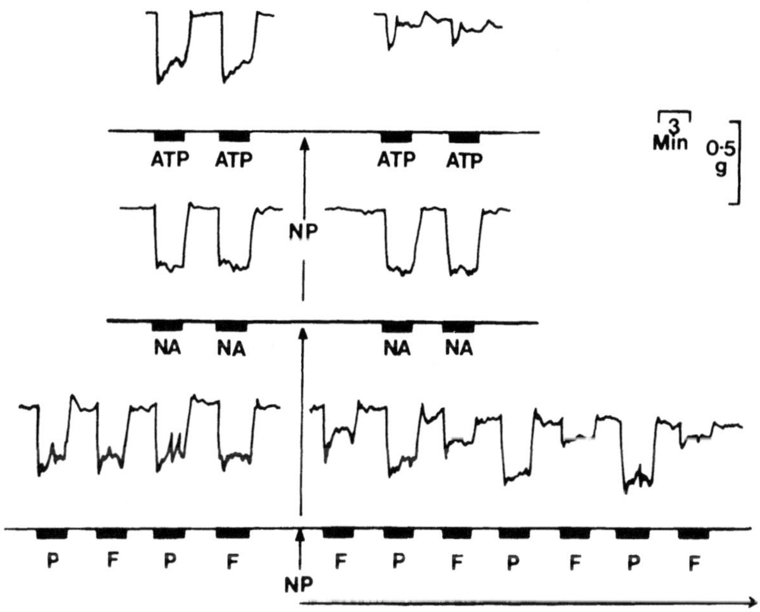

Fig. 2. Relaxations of guinea-pig taenia coli preparations in response to ATP (1μM), noradrenaline (NA 50nM), field stimulation of non-adrenergic, non-cholinergic nerves (F 1 Hz) and stimulation of perivascular sympathetic nerves P 30 Hz) before and after nucleotide pyrophosphatase (NP 1 unit/ml).

In guinea-pig tracheal tube preparations coaxial stimulation for 60 sec caused a biphasic response as a short-lasting increase in pressure followed by a more sustained decrease in pressure (Fig.3). These responses reflect a contraction and relaxation of the trachealis muscle respectively. The short-lasting increase in pressure is hyoscine-sensitive and is thus considered to be a response to field stimulation of cholinergic parasympathetic nerves[7]. The second phase of the response, the more sustained decrease in pressure, is a response to stimulation of NANC nerves in which adenosine is a candidate as the neurotransmitter[3].

In control preparations,(treated with dipyridamole 0.5μM)adenosine (1μM) caused decreases in pressure which reached maximum after 60 sec. VIP (0.5μM) which is also a candidate as the inhibitory neurotransmitter in the NANC nerves of this tissue[8] also caused a decrease in pressure which took 3 min to reach maximum (Fig.3). 15 min after addition of adenosine deaminase (5 units/ml) to the bath decreases in pressure in response to adenosine (1μM) were markedly reduced whereas those in response to VIP were unaffected (Fig 3). Adenosine deaminase (5 units/ml) failed to affect the initial excitatory component of the response to field stimulation (S,10 Hz), but reduced markedly the second inhibitory component (Fig.3).

The finding that adenosine deaminase antagonised the responses to applied adenosine and the inhibitory component of the response to field stimulation is consistent with the suggestion that adenosine is an inhibitory neurotransmitter in NANC nerves in the trachea.

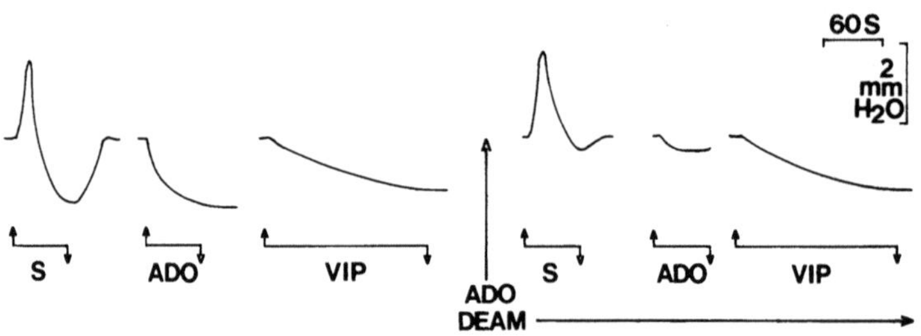

Fig. 3 Changes in intraluminal pressure in guinea-pig trachea in reponse to coaxial stimulation (S 10 Hz), adenosine (ADO 1μM) and VIP (0.5μM) before and after adenosine deaminase (ADO DEAM 5 units\ml). Dipyridamole (0.5μM) and guanethidine (3.5μM) were present in all experiments.

The contraction in response to sympathetic nerve stimulation of the isolated guinea-pig vas deferens has two peaks, an initial twitch component followed by a slower phasic response. ATP has been suggested as the neurotransmitter mediating the former[2] and noradrenaline as the neurotransmitter mediating the latter[2]. Treatment of preparations with NP (1 unit/ml) for 15 min caused a marked reduction in the twitch component of the contraction while the slower phasic component remained unchanged[9] (Fig 4). Contractions in response to ATP were also markedly reduced in the presence of the enzyme which did not, however, affect contractions in response to noradrenaline[9].

In each of the three guinea-pig tissues studied, taenia coli, vas deferens and trachea, NP or adenosine deaminase were able to effectively reduce the response to stimulation of nerves which were reported to release adenine nucleotide or nucleoside as a neurotransmitter.

Apart from providing evidence in favour of neurotransmission by ATP in taenia coli and vas deferens and by adenosine in trachea, the studies suggest that NP and adenosine deaminase can diffuse into the synaptic region and effectively metabolise released transmitter prior to receptor occupation.

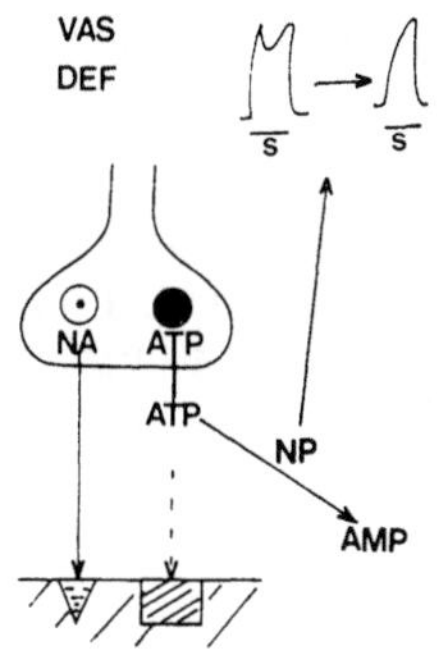

Fig.4. Schematic representation of the release by cotransmission of ATP and noradrenaline (NA) from a sympathetic nerve terminal in the guinea-pig vas deferens. Shown is the pathway by which ATP is metabolised to inactive AMP in the presence of nucleotide pyrophosphatase (NP). Shown in the insert is the effect of NP on contractions of isolated guinea-pig vas deferens to field stimulation(S). Note that in the presence of NP the initial twitch component of the response is abolished. See also reference[9] to poster#39.

ACKNOWLEDGEMENTS

This work carried out with the aid of a grant from the Australian Research Grants Scheme.

REFERENCES

1. G. Burnstock, G. Campbell, D. Satchell, and A. Smythe, `Evidence that adenosine triphosphate or a related nucleotide is the transmitter substance releases by non-adrenergic inhibitory nerves in the gut, Brit. J. Pharmacol. 40:668 (1970).
2. P. Sneddon, and G. Burnstock, Inhibition of excitatory junction potentials in guinea-pig vas deferens by α,β-methylene ATP, European. J. Pharmacol. 100:85 (1984).
3. D.G. Satchell, Adenosine deaminase antagonizes inhibitory responses to adenosine and non-adrenergic, non-cholinergic inhibitory nerve stimulation in isolated preparations of guinea-pig trachea, Brit. J. Pharmacol. 83:323 (1984).
4. J.B. Farmer, and R.A. Coleman, A new preparation of the isolated intact trachea of the guinea-pig, J. Pharm. Pharmacol. 22:46 (1970).
5. D.G. Satchell, Antagonism of the ATP component of sympathetic co-transmission in the rat vas deferens by AMP, European. J. Pharmacol. 132:305 (1986).
6. G. Burnstock, G. Campbell, and M. Rand, The inhibitory innervation of the taenia of the guinea-pig caecum, J. Physiol. 182:504 (1966).
7. R.V. Smith, and D.G. Satchell, Determination of extrinsic pathways of adrenergic nerves to the guinea-pig trachealis muscle using surgical denervation and organ bath pharmacology, Agents and Actions. 19:48 (1986).

8. Y. Matsuzaki, Y. Hamasaki, and S. Said, Vasoactive intestinal peptide : A possible transmitter of non-adrenergic relaxation of guinea-pig airways, Science. 210:1252 (1980).
9. D.G. Satchell, Nucleotide pyrophosphatase antagonizes contractions due to the twitch component of the sympathetic nerve response and to applied ATP in the guinea-pig vas deferens, in : Human Purine and Pyrimidine Metabolism V1 and poster #39 this symposium.

INFLUENCE OF SUCCINYLPURINES ON THE BINDING OF ADENOSINE TO A PARTICULATE FRACTION OF RAT CEREBRAL CORTEX

M. Françoise Vincent and Georges Van den Berghe

Laboratory of Physiological Chemistry, International Institute of Cellular and Molecular Pathology, Brussels, Belgium

INTRODUCTION

Adenylosuccinase deficiency [1] is characterised by the accumulation in cerebrospinal fluid, plasma and urine, of succinylaminoimidazole carboxamide riboside (SAICA riboside) and succinyladenosine (S-Ado). These succinylpurines are formed by dephosphorylation, catalysed by cytosolic 5'-nucleotidase [2], of the two substrates of adenylosuccinase. Adenylosuccinase (adenylosuccinate lyase, EC 4.3.2.2) catalyses two steps in purine synthesis : the conversion of succinylaminoimidazole carboxamide ribotide (SAICAR) into AICAR, and that of adenylosuccinate (S-AMP) into AMP. Its deficiency is transmitted as an autosomal recessive trait and, as apparent from the study of 8 patients in whom the defect has been diagnosed up to now [3], displays marked clinical and biochemical heterogeneity. Severe psychomotor retardation is the principal symptom, although it is strikingly less pronounced in one child. Five patients have epilepsy, 3 display autistic features, and 2 have growth retardation and muscular wasting.

The present work was undertaken as an attempt to elucidate the pathophysiological mechanisms of the neurologic symptoms in adenylosuccinase deficiency. We investigated the hypothesis that, owing to their structural resemblance with adenosine, the succinylpurines might interfere with the binding of adenosine to its cerebral receptors and/or uptake sites. There is increasing evidence that adenosine may have numerous physiological functions. In the central nervous system, these functions may include, among others, inhibition of neurotransmitter release and of nerve cell firing [reviewed in 4,5]. The actions of adenosine are thought to be mediated by adenosine receptors, of which subtypes A1 (with high affinity) and A2 (with low affinity), have been described in most tissues, including brain [reviewed in 6,7]. Adenosine uptake sites may play a role in the termination of the effects of the nucleoside [8].

MATERIALS AND METHODS

Materials. [2-^{3}H]adenosine (24 Ci/mmol) was from The Radiochemical Center, Amersham, U.K. 2'-Deoxycoformycin was from Warner Lambert, (Detroit, MI, U.S.A.). N^6-Cyclohexyladenosine (CHA), N^6-phenylisopropyladenosine (R-PIA), N^6-cyclopentyladenosine (CPA), 2-chloroadenosine, 5'-N-ethylcarboxamide adenosine (NECA), dipyridamole and adenylosuccinate lyase from yeast were from Sigma Chemical Co., St Louis, MO, U.S.A. Adenosine deaminase and alcaline phosphatase were from Boehringer GmbH, Mannheim, Germany. S-Ado was prepared by dephosphorylation of S-AMP : 20 mM of the latter was incubated with 4 mM $MgCl_2$ and alcaline phosphatase (250 U/ml) at 37 °C for 2 h. SAICA riboside was prepared as follows : firstly, SAICAR was synthesized by incubating 10 mM AICAR with 20 mM fumarate, 20 mM Tris-Cl pH 7.4 and adenylosuccinase (1 U/ml) at 37 °C for 14-16 h; thereafter it was dephosphorylated by a 2h incubation in the presence of 4 mM $MgCl_2$ and alcaline phosphatase (250 U/ml). The reaction was stopped by boiling for 2 min. SAICA riboside was separated from AICA riboside and fumarate by chromatography on Whatman 3MM paper, developped in 1 M ammonium acetate pH 3.8/ethanol (3:7.5).

Preparation of particulate fractions of rat cerebral cortex. Adult (250-300 g) fed male Wistar rats were killed by decapitation and their brain rapidly removed and chilled. Cerebral cortex was dissected and immediately homogenized in 10 vol. of 50 mM Tris-Cl buffer pH 7.4, containing 120 mM NaCl, 5 mM KCl, 1.4 mM $CaCl_2$ and 1 mM $MgCl_2$. The homogenate was centrifuged at 22,000 *g* for 10 min. After two washes, the pellet was resuspended in the same buffer, containing adenosine deaminase (4 U/ml), and incubated for 30 min at 25 °C, in order to remove endogenous adenosine. After 2 additional washes, the membrane pellet was resuspended in the same buffer and stored at -80 °C until assayed for binding. Protein was measured according to Lowry et al. [9].

Binding assay. Binding of [2-^{3}H]adenosine (1.25-2.5 nM) was measured in a total volume of 4 ml, in the presence of deoxycoformycin (0.25-1 μM) and of other substances as indicated. Assays were started by the addition of 0.3 mg of membrane protein/ml of incubate. Incubations were carried out in triplicate at 0 °C for 60 min, and were terminated by filtration through a Whatman GF/B filter. Filters were immediately washed thrice with 3 ml of the buffer described above, and their radioactivity was measured.

To verify that, in the conditions of the assay, [2-^{3}H]adenosine was neither degraded nor converted into nucleotides, 10 μl of the incubation medium were spotted, at appropriate time intervals, on cellulose t.l.c. plates on which carrier solutions (50 nmol) of adenosine, inosine, hypoxanthine, xanthine, uric acid, ATP, ADP and AMP had been applied. After development in butan-1-ol/methanol/H_2O/NH_4OH (60:20:20:1), spots were visualised with UV light, cut out and counted.

RESULTS

To demonstrate adenosine receptors, we elected, as others [10,11], to study the binding of [2-^{3}H]adenosine rather than that of labelled, metabolically stable analogues, which bind more specifically to either the A1 or A2 receptor subtypes. This approach requires that adenosine is not metabolized during the assay, but has the advantage that all receptor sites are labelled with a single ligand. We verified that, under the conditions of the binding assay, which included the adenosine deaminase inhibitor deoxycoformycin, [2-^{3}H]adenosine was neither degraded nor incorporated into adenine nucleotides (results not shown). As depicted in Fig. 1, non-specific binding, evidenced by the addition of 0.5 mM non-radioactive adenosine, accounted for approx. 15 % of total adenosine binding. To further validate the technique, we studied the displacing effect of known specific ligands for high affinity A1 receptor sites (2-chloroadenosine, PIA, CHA, CPA), for low affinity A2 receptor sites (NECA), and for uptake sites

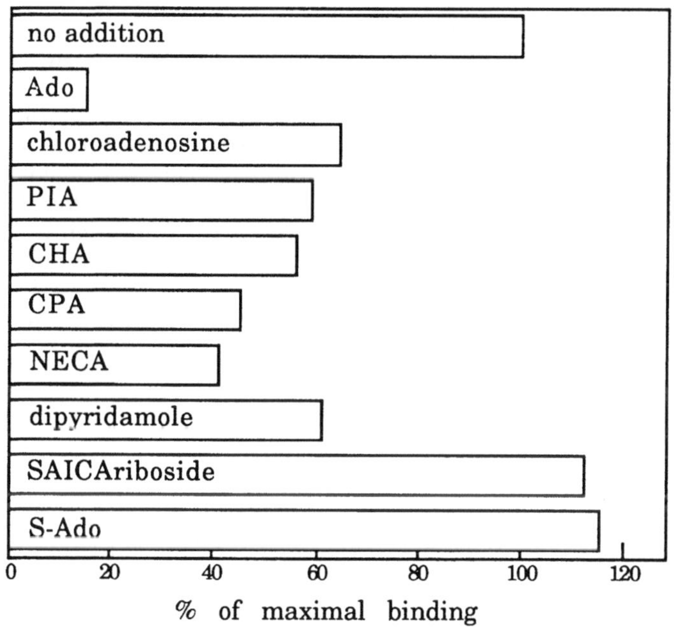

Figure 1. Binding of [2-^{3}H]adenosine to rat brain membranes.
Particulate fractions of rat cerebral cortex were incubated in the presence of 1-2.5 nM [2-^{3}H]adenosine, alone (maximal binding) or with non radioactive compounds, as indicated. Final concentrations were 0.5 mM for Ado, 2.5 μM for 2-chloroadenosine, 25 μM for PIA, CHA and CPA, 200 μM for NECA, 100 μM for dipyridamole, and 200 μM for SAICAriboside and S-Ado.

(dipyridamole). As shown in Fig. 1, all these ligands displaced the binding of [2-^{3}H]adenosine. However, in contrast with the specific analogues, both SAICA riboside and S-Ado, at concentrations in the range of those found in the cerebrospinal fluid of adenylosuccinase-deficient patients (100-200 μM), did not displace [2-^{3}H]adenosine from its binding sites on brain membranes.

DISCUSSION

Several mechanisms have been proposed [3] to explain the symptoms of adenylosuccinase deficiency : *(1)* Impairment of the synthesis of metabolites wich are normally formed distally to the two adenylosuccinase-catalysed reactions, namely the adenine and guanine nucleotides and, hypothetically, AICAR triphosphate (often abbreviated ZTP), a derivative of AICAR. *(2)* Intracellular accumulation of SAICAR and S-AMP, the two substrates of adenylosuccinase. However, neither decreases of purine nucleotide content, nor accumulations of SAICAR and S-AMP could be evidenced in adenylosuccinase-deficient tissues [2], with the exception of a small accumulation of S-AMP in the kidney cortex of two patients. *(3)* Impairment of the purine nucleotide cycle, composed of the enzymes adenylosuccinate synthetase, adenylosuccinase and AMP deaminase. This cycle has been reported to be particularly active in brain [12], but no information is hitherto available with respect to its functioning in the patients. *(4)* A direct effect of extracellular SAICAriboside and S-Ado, the normally undetectable dephosphorylated derivatives of the substrates of adenylosuccinase.

The present study fails to support the hypothesis that interference of SAICAriboside and of S-Ado with the binding of adenosine to its cortical receptor and uptake sites plays a role in the pathogenesis of the neurological symptoms of adenylosuccinase deficiency. Other mechanisms will therefore have to be investigated to explain the cortical dysfunctions in this enzyme defect. These dysfunctions have recently also be evidenced by positron emission tomography, which showed a marked decrease in the uptake of fluorodeoxyglucose in the cerebral cortex, particularly the anterior regions [13].

ACKNOWLEDGEMENTS

We thank Ms T. Timmerman for expert technical assistance. This work was supported by grant 3.4539.87 from the Fund for Medical Scientific Research (Belgium). G.V.D.B. is Director of Research of the Belgian National Fund for Scientific Research.

REFERENCES

1. J. Jaeken and G. Van den Berghe, An infantile autistic syndrome characterised by the presence of succinylpurines in body fluids, *Lancet* 2 : 1058 (1984).
2. G. Van den Berghe and J. Jaeken, Adenylosuccinase deficiency, *Adv. Exp. Med. Biol.* 195 A : 27 (1986).
3. J. Jaeken, S.K. Wadman, M. Duran, F.J. van Sprang, F.A. Beemer, R.A. Holl, P.M. Theunissen, P. De Cock, F. Van den Bergh, M.F. Vincent and G. Van den Berghe, Adenylosuccinase deficiency : an inborn error of purine nucleotide synthesis, *Eur. J. Pediatr.* (1988), in the press.

4. B.B. Fredholm and P. Hedqvist, Modulation of neurotransmission by purine nucleotides and nucleosides, *Biochem. Pharmacol.* 29 : 1635 (1980).
5. P.J. Marangos and J.P. Boulenger, Basic and clinical aspects of adenosinergic neuromodulation, *Neurosci. Biobehav. Rev.* 9 : 421 (1985).
6. C. Londos, J. Wolff and D.M.F. Cooper, Adenosine receptors and adenylate cyclase interactions, *in* : "Regulatory Function of Adenosine", R.M. Berne, T.W. Rall and R. Rubio, eds., Martinus Nijhoff, The Hague (1983).
7. J.W. Daly, Adenosine receptors, *Adv. Cyclic. Nucleot. Prot. Phosphor. Res.* 19 : 29 (1985).
8. P.J. Marangos and J. Duckert, [^{3}H]Dipyridamole binding to guinea pig brain membranes : possible heterogeneity of central adenosine uptake sites, *J. Neurochem*, 48 : 1231 (1987).
9. O.H. Lowry, N.J. Rosebrough, A.L. Farr and R.J. Randall, Protein measurement with the Folin phenol reagent, *J. Biol. Chem.* 193 : 265 (1951).
10. M.E. Newman, J. Patel and H. McIlwain, The binding of [^{3}H]adenosine to synaptosomal and other preparations from the mammalian brain, *Biochem. J.* 194 : 611 (1981).
11. P.H. Snell and C.R. Snell, [^{3}H]adenosine binding sites on 108CC15 neuroblastoma x glioma hybrid cell line and rat brain membranes, *Neurochem. Internat.* 5 : 245 (1983).
12. J.M. Lowenstein, Ammonia production in muscle and other tissues : the purine nucleotide cycle. *Physiol. Rev.* 52 : 382 (1972).
13. A.G. De Volder, J. Jaeken, G. Van den Berghe, A. Bol, C. Michel, M. Cogneau and A.M. Goffinet, Regional brain glucose utilization in adenylosuccinase deficient patients measured by positron emission tomography, *Pediatr. Res.* (1988), in the press.

PROTECTIVE EFFECTS OF CYCLOHEXYLADENOSINE FOLLOWING CEREBRAL ISCHEMIA IN THE GERBIL HIPPOCAMPUS

Jean-Luc Daval, Dag K.J.E. von Lubitz, Jurgen Deckert, and Paul J. Marangos*

Unit on Neurochemistry, BPB, National Institute of Mental Health, 9000 Rockville Pike, Bldg. 10, Room 3C-210, Bethesda, MD 20892 USA and *Gensia Pharmaceuticals, Inc, 11075 Roselle Street, San Diego CA 92121-1207 USA

INTRODUCTION

The brain is very sensitive to an acute episode of ischemia but its response to such an insult is not homogenous. While some areas appear to be quite resistant to an ischemic injury, other regions of the brain are extremely vulnerable.[1] A well known example of selective vulnerability to ischemia is that of the hippocampus. Following a brief period of hypoxia, the pyramidal cells in the CA_1 region of the hippocampus undergo a slowly progressing neuronal damage.[2,3] The precise reason for the high vulnerability of CA_1 neurons is still unclear, but it has been proposed that post-ischemia neuronal hyperexcitability related to synaptically released excitatory amino acids (EAA) is involved in this process.[4,5]

The role of adenosine as a neuromodulator in the central nervous system has been well documented.[6,7] Its action is primarily mediated by 2 types of receptors (A_1 and A_2) linked to the enzyme adenylate cyclase, with either a negative (A_1) or a positive (A_2) coupling.[8] The interaction between the adenosine receptors and adenylate cyclase is mediated via an intermediary nucleotide binding protein (G-protein).[9] In the hippocampus, adenosine is a powerful inhibitor of EAA release and exerts a potent inhibitory action on CA_1 pyramidal cell activity. These depressant effects have been attributed to an interaction with the A_1 receptor subtype.[10-12] Recently, a pronounced down-regulation of adenosine A_1 receptors was reported following transient brain ischemia in the gerbil.[13,14] Therefore, the effects of a post-ischemic administration of cyclohexyladenosine (CHA) were examined on the morphological preservation of the gerbil hippocampus and on the major components mediating the neuromodulatory effect

Correspondence to P.J. Marangos at Gensia

of adenosine in the hippocampal formation: the adenosine A_1 receptors (using [^{3}H]CHA as a radioligand), the G-protein (using [^{3}H]guanylyl-5'-imidodiphosphate, [^{3}H]Gpp(NH)p) and adenylate cyclase (using [^{3}H]forskolin) in a quantitative autoradiographic study.

METHODS

A severe episode of brain ischemia was induced in adult male Mongolian gerbils under light halothane anesthesia (2% halothane, 65% oxygen, 33% nitrous oxide) by a 20 min bilateral occlusion of both common carotid arteries using nylon sutures (Prolene 7.0, Ethicon), whereas non-invasive monitoring of physiological parameters was made during the entire procedure. In control animals, untightened loops were kept in situ for 20 min after surgery. The animals were divided into four groups, with 6 to 8 animals in each group. Five minutes after recirculation, one group received an intraperitoneal (i.p.) administration of CHA (2 mg/kg) dissolved in 10% ethanol (Group I), while another group received only vehicle (Group II). Among control animals, one group received CHA (Group III) and another one was given vehicle (Group IV).

The animals were randomly sacrificed by decapitation either 1 day, 3 days or 6 days after the ischemia insult. None of the animals used in this study showed any seizure activity. Their brains were immediately frozen in isopentane at -50^{0}C and coronal sections (20 um in thickness) were generated using a Bright-Hacker cryostat and thaw-mounted on gelatin-precoated glass slides (2 sections per slide) for the autoradiographic study. For the morphological analysis, adjacent sections were fixed in 10% buffered formalin acetate and stained with 0.1% thionin.

For each binding assay, all animals were examined in a single experiment. The labeling of adenosine A_1 receptors was performed by incubating the slide-mounted sections for 90 min at room temperature in 50 mM Tris-HCl buffer (pH 7.4) with 2 IU/ml adenosine deaminase and 5 nM [^{3}H]CHA (34.4 Ci/mmol, NEN), as previously described.[15] Non-specific binding was measured in the presence of 20 uM unlabeled CHA.

For the forskolin binding assay, the sections were incubated for 10 min at room temperature in 50 mM Tris-HCl buffer (pH 7.4) with 5 mM $MgCl_2$ and 10 nM [^{3}H]forskolin (49.2 Ci/mmol, NEN), while non-specific binding was obtained with 10 uM forskolin.[16]

The measurement of [^{3}H]Gpp(NH)p binding was performed according to Gehlert and Wamsley.[17] After a 30 min preincubation in buffer containing 1 mM EDTA, 0.1% Lubrol PX and 0.002% polyethyleneimine, the sections were incubated for 10 min at room temperature with 200 nM [^{3}H]Gpp(NH)p (11.5 Ci/mmol, Amersham). Non-specific binding was measured in the presence of 100 uM cold GTP.

The sections were finally co-exposed with Amersham [^{3}H] microscales to LKB-Ultrofilm for 2 weeks for [^{3}H]CHA, 4 weeks for [^{3}H]forskolin and 3 days for [^{3}H]Gpp(NH)p. After development, the autoradiographs were analyzed for their optical density.

RESULTS

Morphology

As previously reported[2], no cellular damage could be observed in the hippocampus by light microscopic observation of the sections 1 day after recirculation. Three days after ischemia, a substantial loss of neurons occurred selectively in the CA_1 pyramidal cell layer, while the post-ischemic administration of CHA completely inhibited the cell degeneration (Fig. 1). At 6 days post ischemia, all the pyramidal neurons of the CA_1 subfield were necrotic with also a marked alteration in CA_3 pyramidal cells in 70% of the animals examined. By contrast, the dentate gyrus did not exhibit any morphological changes, indicating a high degree of resistance of this area. In CHA-treated animals, the pyramidal cells were also affected in the CA_1 subfield but significantly less than in the ischemia group with a marked preservation in the CA_3 region (Fig. 1).

Receptor autoradiography

The acute injection of CHA by itself did not change the binding of any ligands when compared to control animals which received only vehicle (data not shown).

One day after ischemia, a significant loss of $[^3H]$CHA binding activity occurred in the pyramidal cells in both CA_1 and CA_3 regions while the reduction in $[^3H]$CHA binding was much less pronounced following CHA treatment (Fig. 2). At this time point, $[^3H]$forskolin binding was also significantly affected as well in the CA_3 as in the CA_1 region, whereas no changes were observed following the administration of CHA. The binding of $[^3H]$Gpp(NH)p remained unaffected, as shown in Fig. 2.

Three days after the ischemia insult, a noticeable reduction in the laminar aspect of the CA_1 subfield was observed in autoradiographs as a consequence of a dramatic decrease in $[^3H]$CHA binding, especially in the pyramidal cell layer (63% loss of binding activity), stratum oriens (43%) and stratum radiatum (38%). Concomitantly, a marked reduction in CHA binding also occurred in the CA_3 subfield. A protective effect could be observed following the treatment with CHA with only a 28% reduction in $[^3H]$Gpp(NH)p binding activities. The forskolin binding was markedly reduced in all CA_1 layers (by 66% in stratum oriens and pyramidal cell layer, by 58% in stratum radiatum), while it was reduced by 43% in the CA_3 pyramidal cell layer (by 21%). The post-ischemia administration of CHA completely prevented the changes in both $[^3H]$forskolin and $[^3H]$Gpp(NH)p binding (Fig. 2).

At 6 days post-ischemia, the effects of a single administration of CHA seemed to be less effective in protecting adenosine A_1 receptors than at 3 days post-ischemia. However, the reduction in $[^3H]$CHA binding was significantly less important following the drug treatment (Fig. 2-Fig. 3). Regarding $[^3H]$forskolin, the binding activity observed in the CA_1 subfield was considerably improved by the pharmacological treatment, whereas the loss in $[^3H]$forskolin binding which

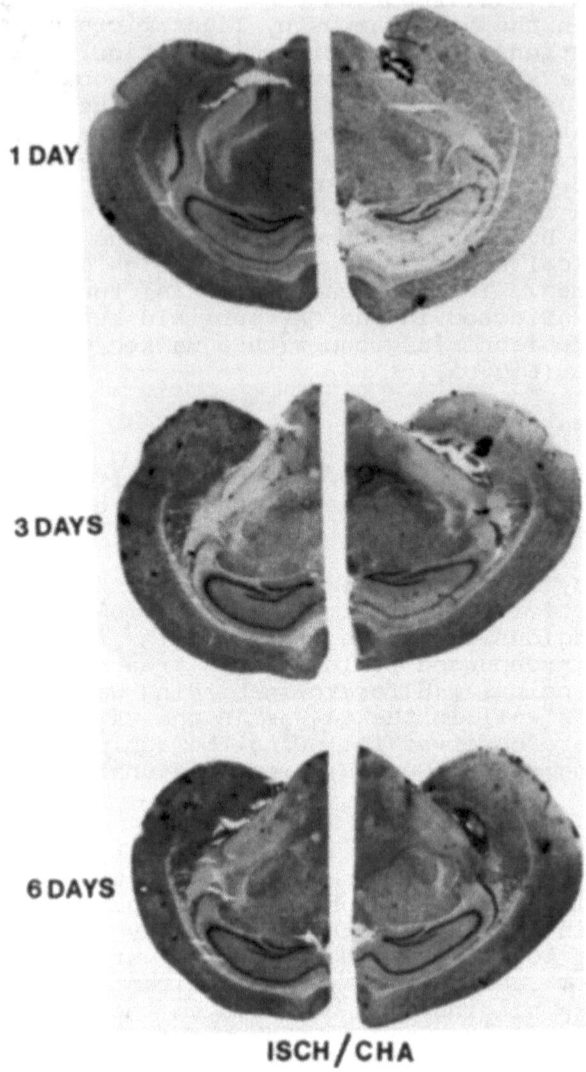

Fig. 1 Effect of CHA on the morphological changes induced by ischemia. Brain sections at the level of anterior hippocampus were fixed and stained with 0.1% thionin. While the hippocampal formation was markedly altered at 3 and 6 days post-ischemia (left-side), a protective effect was observed following the administration of CHA (right side).

occurred in the CA_3 subfield (30%) was totally prevented by CHA. Finally, ischemia reduced by 26% and 19% the binding of [^{3}H]Gpp(NH)p in the CA_1 and CA_3 pyramidal cells, respectively. By contrast, no changes were observed following the administration of CHA (Fig. 2-Fig. 3).

DISCUSSION

Following an episode of brain ischemia in the gerbil, the pyramidal neurons of the CA_1 region of the hippocampus have been reported to undergo progressive cellular damage.[2,3]

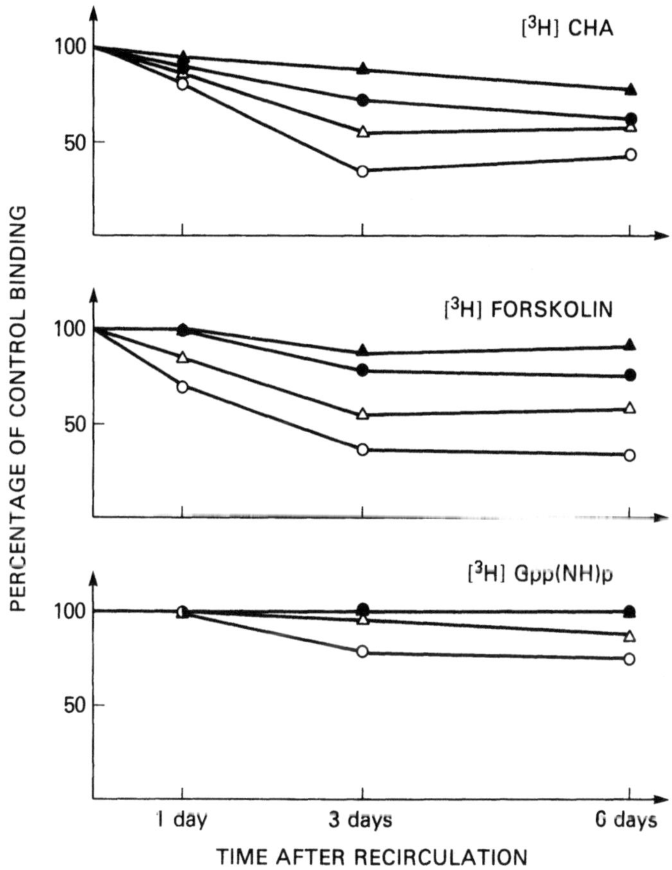

Fig. 2. Time-course of the binding of [^{3}H]CHA, [^{3}H]forskolin and [^{3}H]Gpp(NH)p after ischemia (open symbols) in the pyramidal cells of the CA_1 (circles) and the CA_3 (triangles) subfield of the hippocampus. The effects of an administration of CHA are shown by dark symbols. Data are means obtained from 6 to 8 animals.

Electrophysiologically, the neurons become hyperexcitable for approximately 24 hours and then exhibit a total loss of spontaneous activity.[18] While it has been reported that such cellular damage related to ischemia are not primarily mediated by a failure of energy metabolism,[19] several lines of evidence suggest that synaptically released EAA transmitters initiated the cascade of events that lead to neuronal death.[20,5] Therefore, it has been suggested that drugs which are able to decrease the accumulation of these EAA or block their postsynaptic effects may be effective in preventing ischemic damage. A protective effect of glutamate antagonists in experimentally induced cerebral ischemia has been recently reported.[21,22] However, these molecules are active only at high concentration and cross the blood-brain barrier poorly.[23]

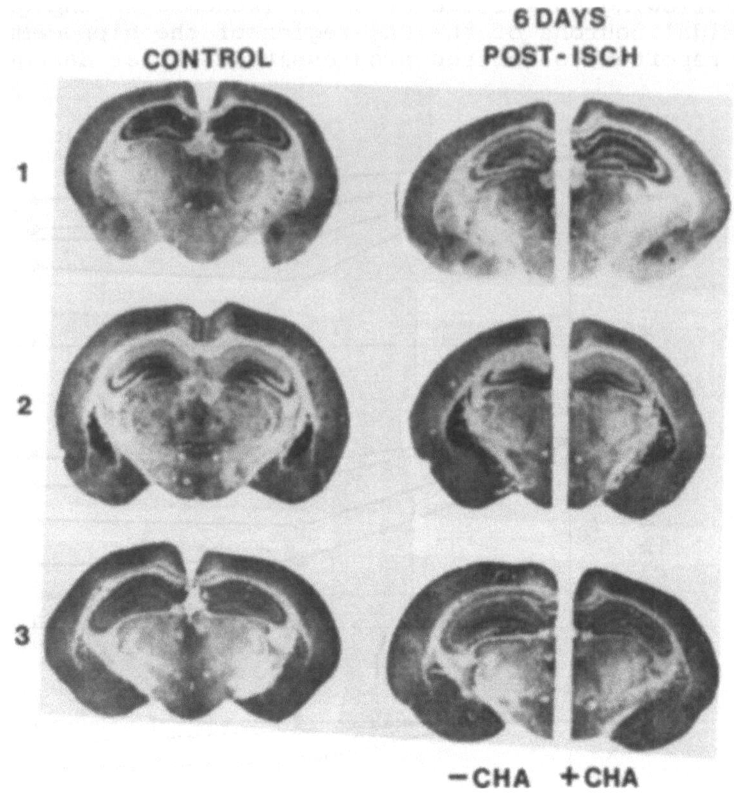

Fig. 3. Protective effect of an i.p. administration of CHA (2 mg/kg) on adenosine A_1 receptors (1), forskolin (2) and Gpp(NH)p (3) binding sites 6 days after a 20 min episode of brain ischemia in the gerbil. For each ligand, the non-specific binding was very close to background (data not shown).

von Lubitz et al.[24] have previously shown that an intraventricular injection of the stable adenosine analogue CHA, which interacts almost selectively with the adenosine A_1-subtype receptor, may considerably improve survival following a 30 min period of cerebral ischemia in the gerbil. In the present study, an i.p. administration of CHA (2 mg/kg) significantly reduced the morphological alteration occurring in the hippocampal formation after transient cerebral ischemia in the gerbil and was also able to prevent the ischemia-induced reduction in the binding of $[^3H]$CHA, $[^3H]$Gpp(NH)p and $[^3H]$forskolin. As observed in additional experiments, the decrease in the binding activity for these compounds mainly reflects a loss of binding sites rather than a change in affinity (data not shown). These data indicate that CHA protects against the ischemia-induced reduction in adenosine A_1 receptors, guanine nucleotide and forskolin binding sites, the 3 essential components mediating the neuromodulatory action of adenosine.

The nature of the protective effect of CHA is still unknown. However, the elevated extracellular concentration of glutamate is one of the major destructive elements characterizing ischemia and there is substantial evidence that adenosine and its derivatives regulate glutamatergic neurotransmission, an action mediated via A_1 receptors.[5,10-12] Therefore, CHA may enter the brain (more easily after ischemia) and reduce neuronal hyperactivity by interacting with adenosine receptors. This hypothesis is further supported by the distribution of both glutamate and adenosine A_1 receptors that are highly concentrated in the vulnerable regions.[25,15] The synaptic action of adenosine agonists also includes inhibition of Ca^{++} influx into the neurons[6] and such an action may contribute to the protective effect of CHA.

In conclusion, the present study provides new insights concerning the relevance of adenosine derivatives in protecting against ischemia-related brain injury. Such data are in good agreement with the recent observation that 2-chloroadenosine, another adenosine analogue, also attenuates ischemic injury in rat brain.[26]

REFERENCES

1. J. B. Brierley, in "Greenfield's Neuropathology," W. Blackwood and J. A. N. Corsellis, eds. Edward Arnold, London (1976) pp. 41-85.

2. T. Kirino, Brain Res. 239:57-69 (1982).

3. W. A. Pulsinelli, J. B. Brierley and F. Plum, Ann. Neurol. 11:491-498 (1982).

4. R. P. Simon, J. H. Swan, T. Griffiths and B. S. Meldrum, Science 226:850-852 (1984).

5. S. M. Rothman and J. W. Olney, Ann. Neurol. 19:105-111 (1986).

6. J. W. Phillis and P. H. Wu, in "Physiology and Pharmacology of Adenosine Derivatives," J. W. Daly, Y. Kuroda, J. W. Phillis and M. V. Shimizu, eds., Raven Press, New York (1983), pp. 219-236.

7. T. V. Dunwiddie, in "International Review of Neurobiology," Vol. 27, J. R. Smythies and R. J. Bradley, eds., Academic Press, New York (1985) pp. 63-140.

8. D. Van Calker, M. Muller and B. Hamprecht, J. Neurochem. 33:999-1005 (1979).

9. P. F. Worley, J.M. Baraban and S. H. Snyder, Ann. Neurol. 21:217-223 (1987).

10. G. R. Siggins and P. Schubert, Neurosci. Lett. 23:55-60 (1981).

11. K. Lee, P. Schubert and U. Heinemann, Brain Res. 321:160-164 (1984).

12. J. Fastbom and B. B. Fredholm, Acta Physiol. Scand. 125:121-123 (1985).

13. K. S. Lee, W. Tetzlaff and G. W. Kreutzberg, Brain Res. 380:155-158 (1986).

14. H. Onodera, G. Sato and K. Kogure, Brain Res. 415:309-322 (1987).

15. R. R. Goodman and S. H. Snyder, J. Neuroscience 2:1230-1241 (1982).

16. D. R. Gehlert, T. M. Dawson, H. I. Yamamura and J. K. Wamsley, Brain Res. 361:351-360 (1985).

17. D. R. Gehlert and J. K. Wamsley, Eur. J. Pharmacol. 129:169-174 (1986).

18. R. Suzuki, T. Yamaguchi, C. L. Li and I. Klatzo, Acta Neuropathol. 60:217-222 (1983).

19. B. S. Meldrum, T. Griffiths and M. C. Evans, in "Protection of Tissues Against Hypoxia," A. Wauquier, M. Borgers and W. K. Amery, eds., Elsevier Biomedical Press, Amsterdam (1982) pp. 275-286.

20. S. Rothman, J. Neuroscience 4:1884-1891 (1984).

21. E. Ozyurt, D. I. Graham, G. N. Woodruff and J. McCulloch, J. Cereb. Blood Flow Metab. 8:138-143 (1988).

22. J. H. Swan, M. C. Evans and B. S. Meldrum, J. Cereb. Blood Flow Metab. 8:64-78 (1988).

23. B. Meldrum, Clin. Sci. 68:113-122 (1985).

24. D. K. J. E. von Lubitz, J. M. Dambrosia and D. J. Redmond, J. Neuroscience (1988), in press.

25. D. T. Monaghan, V. R. Holets, D. W. Toy and C. W. Cotman, Nature (London) 306:176-179 (1983).

26. M. C. Evans, J. H. Swan and B. S. Meldrum, Neurosci. Lett. 83:287-292 (1987).

DIFFERENTIAL EXPRESSION OF ECTO-NUCLEOTIDE METABOLIC ENZYMES DURING IMMUNOGLOBULIN GENE REARRANGEMENTS IN HUMAN PRE-B-CELLS

Jerzy Barankiewicz, May Hui, Amos Cohen and Hans-Michael Dosch

Division of Immunology/Rheumatology
Research Institute
The Hospital for Sick Children
555 University Ave.
Toronto, Ontario
Canada M5G 1X8

INTRODUCTION

Nucleotide metabolism has long been regarded as an exclusively intracellular process where high cytoplasmic concentrations of nucleotide prevent the passage of extracellular nucleotides into the cell and where all enzymes characteristic of nucleotide metabolic pathways have been demonstrated within the intracellular milieu (1). The presence, in body fluids, of considerable levels of extracellular ATP as well as the expression of functional adenosine receptors on many cells are therefore not well understood in their role.

More recent studies with a variety of intact cells unequivocally demonstrated that some but not other cell types express nucleotide metabolic enzymes on the external face of their plasma membranes (2). In detailed studies of human lymphocytes, it was found that members of the B lymphocyte lineage expressed high extracellular nucleotide degrading activity whereas intact T lymphoid cells were entirely unable to degrade extracellular nucleotides (3). On the other hand, only T-, but not B lymphocytes express functional receptors for adenosine, the product of the catabolic ectoenzyme pathway expressed on the B cell surface. These findings lead to speculations that B cell surface ecto-nucleotidase activity might provide a physiological source of adenosine within lymphoid tissue, perhaps relevant as a pathway of B⇒T cell interactions at a metabolic level. (3,4).

EXPERIMENTAL STRATEGY: A CAUTION

Because the high levels of cytoplasmic purine metabolic enzymes, studies of purine ecto-enzymatic activities must critically rely upon sustained integrity of the cells under study. In the present series of experiments care was taken to maintain viability in the various fresh and cultured lymphoid cell preparations. This was achieved through the use of complete culture media rather than merely buffered salt solutions, gentle washings through cushions of calf serum and repeated Ficoll-Hypaque gradient centrifugation to remove the up to ten percent non-viable cells from lymphoblast lines. In addition, cell integrity was frequently monitored, using the membrane impermeant RNA/DNA dye, propidium iodide, and the actively excluded trypan blue. In addition, we routinely measure, in the extracellular space, two intracellular enzymes (pyruvate kinase and adenylate deaminase) as a marker of leaky cell membranes. When judged by these criteria, data presented where thus obtained with cell suspension of near 100% viability.

ECTO NUCLEOTIDE METABOLIC ENZYMES ON HUMAN LYMPHOCYTES

Guinea Pig and Rat leukocytes have previously found to express ecto-ATPase, ecto-AMPase and ecto-paranitrophenol-phosphatase (5-7). Intact human lymphocytes have similarly been found to express a number of purine ecto-enzymes and in cell separation experiments these activities can be exclusively assigned to the B cell lineage (Table 1). In some contrast to rodent lymphocytes, human lymphocyte ecto-enzyme expressions does not reflect the state of cell activation. For example, exposure of human lymphocytes to lectins such as Con A and PHA fails to induce enzyme activity in T cells, and in B cells lectins inhibit ecto-AMPase (5'NT) activity leaving ecto-AMPase and ecto-ATPase activity unaffected (data not shown).

Table 1 Differential Expression of Purine Ecto-Enzymes in Purified Human Lymphoid Cells from Tonsil

Ecto-Enzyme	**Lymphocyte Lineage** **B Cells**	**T Cells**
	μmoles Product/10^6 Cells/60 min	
ectoATPase	840	0
ectoADPase (ecto 5'NT)	860	0
ectoAMPase	67	0
ectoAMP deaminase	0	0
ecto Adenosine deaminase	0	0
ecto purine nucleoside phosphorylase	0	0
nonspecific phosphatases	0	0

Purified tonsillar B and T lymphocytes were incubated with 500 μm 2,8-^{3}H-ATP, 2,8-^{3}H-ADP, G-^{3}H-AMP, 8-^{3}H- adenosine, 8-^{3}H, adenosine or 10 μM-nitrophenyl phosphine, respectively. Products of reactions were determined according to (3).

Ecto-nucleotidase activity on human B lymphocytes shows low substrate specificity with efficient degradation of both ribonucleotides and deoxyribonucleotides (3). While tissue levels of most phosphonucleosides are at best low, steady state levels of ATP are high (~20μM). Given the enzyme activities described (Table 1), equivalent adenosine levels can be expected in lymphoid tissues containing significant proportions of B lineage cells. If correct, then this could delineate a novel metabolic axis of B⇒T cell interaction, where the generated adenosine triggers adenosine receptors on T cells. This pathway would allow signaling directed from B- towards T cells, a signal vectored in opposite direction to that of the many T cell derived B cell growth and maturation factors. Such a regulatory circuit could thus affect the ability of T cells to promote or reduce B cell development, in a feed back mechanism.

Based on our previous observation of varying ecto-nucleotide metabolic enzyme activities in relatively immature B-lymphocytes (3), we have now asked if ecto-enzyme expression could more closely be related to distinct stages and events in B cell ontogeny. Towards this end we correlated immunoglobulin gene rearrangement and expression patterns and ecto-enzyme expression in a recently generated panel of EBV transformed, fetal bone marrow derived B-progenitor, pre-B, and immature (IgM secreting) B lymphocyte clones.

As shown in Fig. 1., fetal bone marrow derived B-progenitor clones were characterized by low ectoenzyme levels (Fig.1a). These levels were, however, still significantly higher than those observed in T lineage cells (See Table 1). These early B-progenitor cells express several B cell properties (e.g. surface CD20), but showed no immunoglobulin related mRNA species and had not rearranged v(H) or C(H) genes (Fig. 1b).

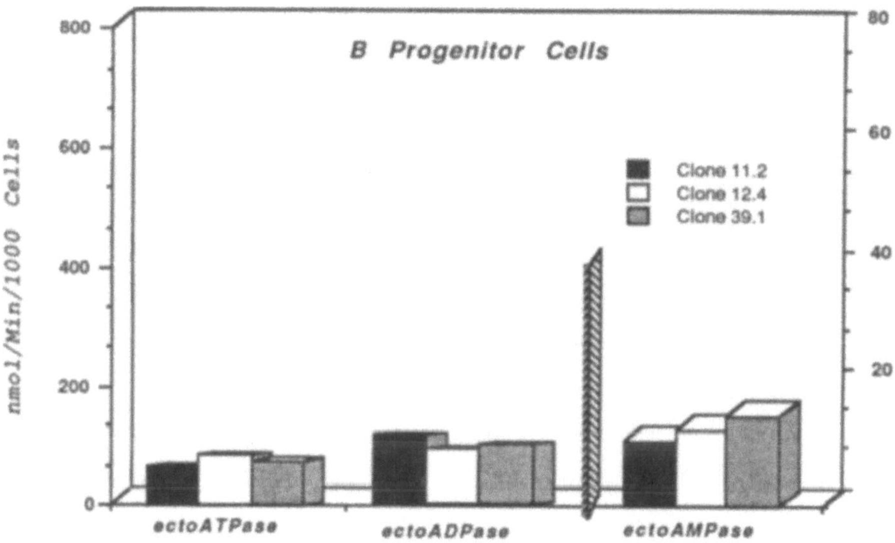

Figure 1a

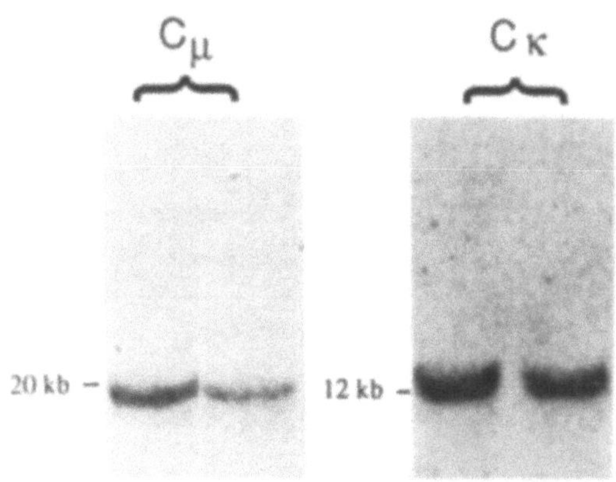

Figure 1b. *Southern blot of genomic and B-progenitor cell DNA, probed with genomic DNA probes for heavy (Cμ) and light chain (Cκ) genes.*

Classical pre-B-cells are shown in Fig. 2. With the rearrangement of IgM heavy chain gene (at least 1 allele) and independent of the expression of μ-mRNA (Fig. 2b), clones in this group of cells show a parallel increase in the activity of both ecto-ATPase and ecto-ADPase This rise in ecto-enzyme activity is dramatic and covers at least one order of magnitude Fig. 2a). Ecto-AMPase levels in these cells were the highest found among any of the B cells studied.

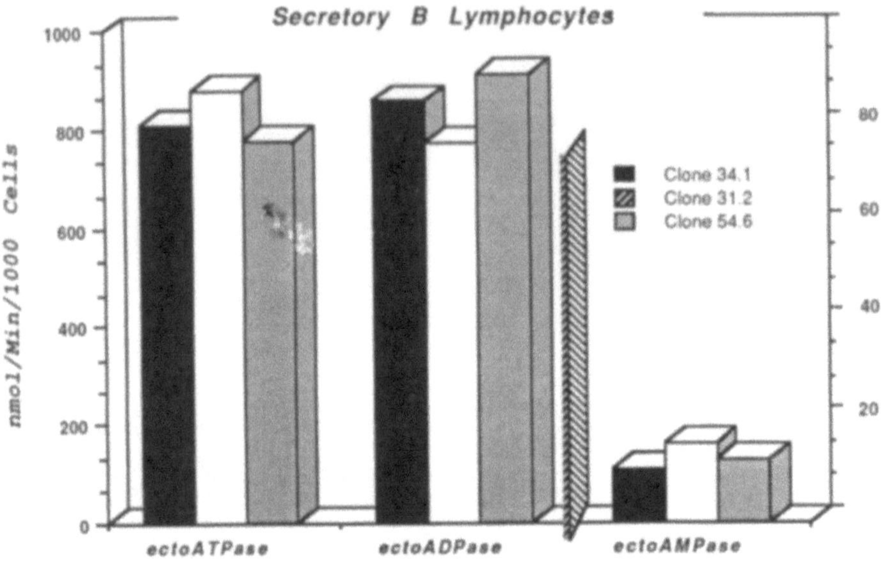

Figure 2a

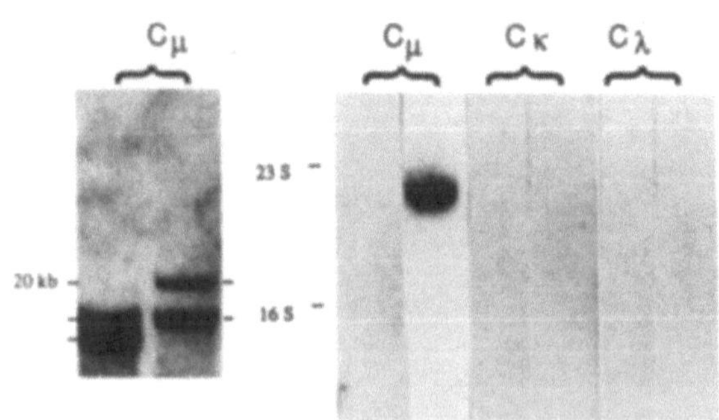

Figure 2b. *Southern blot of pre-B cell DNA demonstrating single or dual allel Cμ rearrangements (clones 8.1 & 18.2 respectively, left 2 gels). Northern blots of the same cells, demonstrating Cμ expression in one, but absent light chain mRNA expression in both clones.*

The differentiation event characteristic of the pre-B⇒B-Cell transition is the rearrangement and expression of light chain genes (Fig. 3b). With the completion of this step in development, 5'NT activities (Fig. 3a) revert to the low levels found in the very immature B-progenitor cells as well as in the more mature, secretory B lymphoid cells (Table 1). In contrast, levels of ecto-ATPase and ecto-ADPase showed further elevation (Fig. 3a), reaching the levels typical for mature B lymphoid cells in peripheral tissues.(see Table 1)

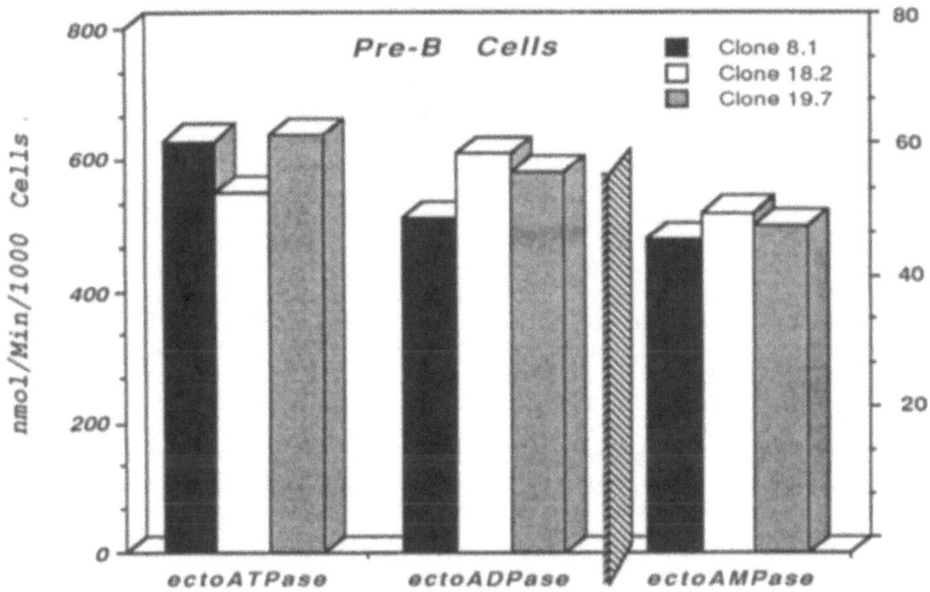

Figure 3a

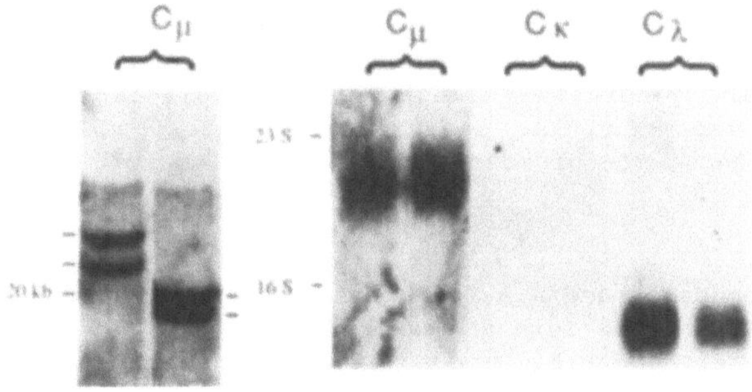

Figure 3b. Southern and Northern blots of B lymphocyte DNA demonstrating Cμ rearrangements and expression of λ-mRNA in IgM secreting clones 34.1 & 31.2 respectively).

These observations establish the expression of ectonucleotide metabolic enzymes as attractive B-cell differentiation markers, and they provide correlates between identifiable cross roads in B cell ontogeny and phosphonucleoside catabolism. The most drastic change in enzyme activity can be localized to a distinct stage in B cell development, characterized by the single or dual allel rearrangement (but not necessarily expression: see clone 8.1, Fig. 2b) of heavy chain vDJ⇒Cμ. With the highest 5'NT activity expressed at this stage of B cell differentiation, the full degradative pathway is at peak capacity with respect to adenosine production in these "classical" pre-B cells.

HYPOTHESIS

Triggering of T lymphocyte adenosine receptors generally results in the development and/or enhancement of regulatory (suppressive) function (13,14 and unpublished observations). We propose a model where B-cell ectonucleotide catabolic enzymes, in the presence of normal (i.e. high) levels of extracellular ATP, provide for a continuous negative restraint of B cell function mediated by adenosine activated regulatory T lymphocytes. This negative restraint could involve either the presence of T suppressor or the absence of T helper/inducer cell activity.

Although many regulatory T cell products/functions have been found to promote or retard B cell differentiation and growth, it is presently unclear if and how such T⇒B cell interactions may be controlled. B(-and in particular pre-B-) cell generated adenosine may provide a feedback mechanism which allows the controlled expansion/containment of B cell growth in a given tissue compartment as a function of local B cell mass. The adenosine mediated restraint of growth promoting T cell function would disappear with acute B cell depletion of a given lymphoid tissue, e.g. during an immune response. B cell depletion would reduce the production of adenosine, thus reduce the restraint mediated by adenosine triggered T cells and permit an accelerated accumulation of immature B lineage cells. In this model, adenosine levels in lymphoid tissue would provide a read-out system for T lymphocytes, to recognize and maintain level B cell mass in the face of immune response-directed, dynamic fluctuations of B cell immigration, emigration and proliferation.

CONCLUSION

Apart from triggering T-cell adenosine receptors, adenosine may play additional physiological roles as intracellular regulator of purine and pyrimidine biosynthesis de novo or through S-adenosylhomcysteine hydrolase pathways (10, 11). A central role of extracellular adenosine production is consistent with the observation that intracellular adenosine generation in human lymphocytes is at best extremely small. Intracellular ATP degradation proceeds exclusively *via* AMP delineation represents an avid substrate for both adenosine deaminase and adenosine kinase. Adenosine generated through the extracellular pathway may therefore play an important and exclusive regulatory role and it provides perhaps the sole source of adenosine in human lymphoid tissue.

References

1. J.F. Henderson and A.R.P. Paterson: Nucleotide metabolism Academic Press, New York and London pp. 1-271, (1973).
2. J.F. Manery and E.E. Dryden: Ectoenzymes concerned with nucleotide metabolism. Physiological and Regulatory Functions of Adenosine and Adenine Nucleotides. Eds. H.P. Baer and G.I. Drummond, Raven Press, N.Y., pp. 323-339, (1979)

3. J. Barankiewicz, H.M. Dosch, A. Cohen,Extracellular Nucleotide Metabolism in Human B and T lymphocytes. J.Biol. Chem. 263:7094-7098, (1988).
4. J. Barankiewicz, H.M. Dosch and A. Cohen:Source of adenosine production in human B and T lymphocytes: In Cellular Basis of Immune Modulation (J.G. Kaplan, and D.R. Green eds) A.R.Liss.Inc., (in press) (1988)
5. J.W. DePierre and M.L. Karnovsky, J.Biol.Chem., 249:7111-7120 (1974).
6. J.W. DePierre and M.L. Karnovsky, J.Biol.Chem., 249:7111-7129 (1974).
7. J.W. DePierre and M.L. Karnovsky, In:In vitro methods Cell Mediated Tumor Immunology,(B.R. Bloom and J.R. David, eds.) Academic Press, New York, pp. 353-358.
8. G. Pommier, G. Ripert, E. Azoulay and R. Diepieds, Biochem. Bioplys. Acta, 389:485-494(1975).
9. J.Barankiewicz, A. Cohen, J.Biol.Chem., 259:15178-15181, (1984)
10. F. Snyder, M.S. Hershfield, J.E. Seegmiller, Cancer Res., 38:2357-2362 (1977).
11. M.S. Hershfield, N.M. Kredich, Science, 202:757-760 (1978)
12. L.A. Schultz, Kammer G.M., Radolph, S.A. FASEB J.2:244-250, (1988)
13. T. Henriksson, Immol. Letters 7:171 (1983)
14. A.L.Schwartz, F.C. Stenard, S.H. Palrman, Clin. Immunol. Immunopathol. 9:499 (1978).

OPTIMISED CONDITIONS FOR THE ROUTINE HPLC SEPARATION OF NUCLEOTIDES IN CELL EXTRACTS

David Perrett, Karl E. Herbert, G. Morris* and H. Anne Simmonds*

Department of Medicine, St Bartholomew's Hospital, London EC1A.7BE

and Purine Laboratory*, Guys Hospital, London SE1 U.K.

INTRODUCTION

High Performance Liquid Chromatography (HPLC) is now the accepted method for the quantitaive and qualitative study of nucleotides, nucleosides and bases in cell and tissue extracts (1). For nucleotides a variety of chromatographic approaches including reversed-phase and reversed-phase with a number of different ion-pairing agents have been described (2) but to date the use of microparticulate anion exchange procedures remains the most popular. However the accurate analysis of intracellular nucleotide pools requires careful attention to sample collection, sample extraction, chromatographic operation, peak identification and quantitation in addition to correct choice of suitable chromatographic mode (3). This study addresses the optimisation of two of these goals namely extraction and chromatography. and draws upon experience of some 14 years with the HPLC of nucleotides during which time many thousands of analyses have been made at our two centres.

EXPERIMENTAL AND DISCUSSION

All studies have been performed around previously described HPLC (4) and red cell extraction techniques (5).

Extraction of nucleotides from cells

The efficient extraction of nucleotides from biological samples prior to HPLC requires the optimisation of various parameters including maximal extraction of the total intracellular nucleotide pool, minimum nucleotide interconversion, maximal removal of protein and excess extractant and high recoveries of added nucleotides. We have optimised these parameters for the extraction of nucleotides from human red blood cells.

Although a number of precipitant/extractant methods have been described the two most commonly used with mammalian cells are the acid extractants perchloric acid (PCA) and trichloroacetic acid (TCA). Initial studies showed that significant losses of added nucleotides (200nmol/ml PRBCs) could occur by adsorption onto the potassium perchlorate percipitate formed when PCA was used. (Table 1). Only slight losses were found when using TCA.

Table 1 Comparison of nucleotides recovery from RBCs with TCA and PCA

Extractant	Mean recovery (% of control) n = 6		
	AMP	ADP	ATP
12% TCA	101 ±4.0	92 ±5.1	93 ±7.4
5% PCA	97 ± 5.2	83 ±6.3	74 ±5.3

Similar findings of ATP losses have been reported previously by workers (6) using the luciferase assay for ATP. All our subsequent studies have therefore employed TCA.

The total amounts of nucleotides extracted from cells, as determined by the total integrator output from the HPLC when corrected for volume changes by use of cXMP as the internal standard, was found to vary with both the strength of the TCA and the ratio of the volume of cells to volume of extractant (Table 2). The results suggest that it was not only necessary to precipitate proteins and release the nucleotides but also a minumum final volume was necessary to solublise the nucleotides. The highest recovery was achieved with a 1:3 extraction ratio and 15% TCA.

Table 2 Total Amounts of Nucleotides Extracted from RBC's (% of maximum value)*

Ratio Vol cells:Vol TCA	% w/v TCA in water: 5	10	15	20	25
1 : 1	protein	69	68	92	79
1 : 2	66	93	98	93	90
1 : 3	77	94	100	93	94
1 : 4	88	93	94	89	87
1 : 5	80	83	76	91	82

* Mean of three experimental determinations

However too high a concentration of TCA or too large a volume of acid was found to change the ATP/ADP ratio significantly. At lower volume ratios more ADP was apparently released than ATP so reducing the ratio whereas at higher acid concentration hydrolysis of ATP was probably occurring. The optimum ratio was found to be 1:2 when the ATP/ADP varied from 5.00, 5.94, 9.40, 6.86 and 4.74 for 5, 10, 15, 20 and 25% TCA solutions. For all the extraction conditions investigated we determined the efficiency of protein removal by measurement of protein remaining the the final extract using the bicinchonnic acid procedure (7). Except for 1:1 ratio at 5% TCA protein removal, an important requirement for maximal column life, was greater than 99.95% and finding in agreement with similar studies on plasma extraction (8).

In later studies optimum extract conditions which combined satisfactory extraction without undue dilution of the sample were employed. These conditions were 1 vol of fresh packed red blood cells with the buffy coat and top 10% of red cells removed extracted by adding slowly to 12% w/v ice cold TCA while whirli-mixing. For accurate pipetting of red cells postive displacement pipettes are prefered. The use of cyclic nucleotides e.g cXMP added to give a final concentration of 50umol/l are excellent internal standards since they are relatively stable and elute centrally on the anion exchange HPLC system.

Excess TCA should be removed as soon as possible to prevent acid hydrolysis of ATP etc and it can be removed by either water saturated diethyl ether or the Freon/amine procedure of Khym (9). Both methods increased the pH and removed the TCA peak, which elutes between GMP and NADP on our HPLC system. However 5 x 3ml extractions with ether were needed compared to 3 x 2ml of the Freon/amine mixture to achieve equal neutralisation.

Chromatographic Variables

For routine analysis of nucleotides we prefer to use amino-propyl silica, a weak anion-exchange packing, with gradient elution although other chromatographic modes are employed for particular applications. The HPLC system has been previously described in detail (4) but at present both 4.6 x 100mm (Barts) and 4.6 x 250mm (Guys) columns of 5um APS-Hypersil are employed. The principle requirement outlined in the initial study was the importance of pH control of both buffers in effecting good column selectivity in all parts of the chromatogram. In general initial pH conditions for both buffer components of the gradient should be approximately pH 2.8 -3.2. Some variation in these values may be necessary depending on the exact column employed in order to acheive the standard separation for a newly packed column (Fig 1). Because of the variations in manufacturing quality it is adviseable to purchase loose packing material in bulk if the most consistant chromatography is required yet even then some variation between newly packed columns was observed.

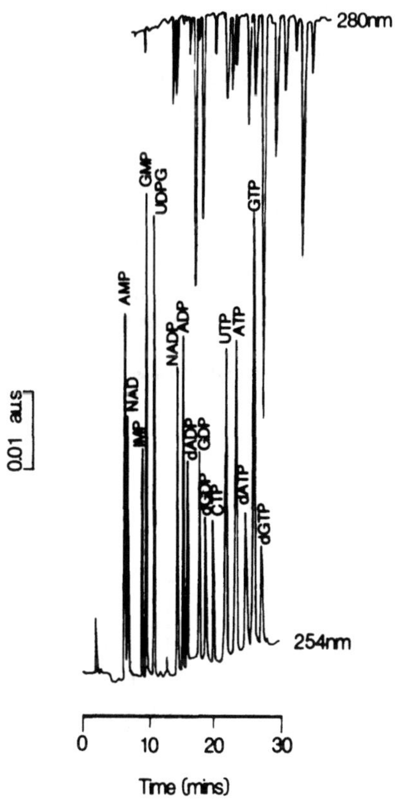

FIG 1 Standard anion-exchange (4.6 x 250mm) HPLC of a mixture of nucleotides (1nmol injected).

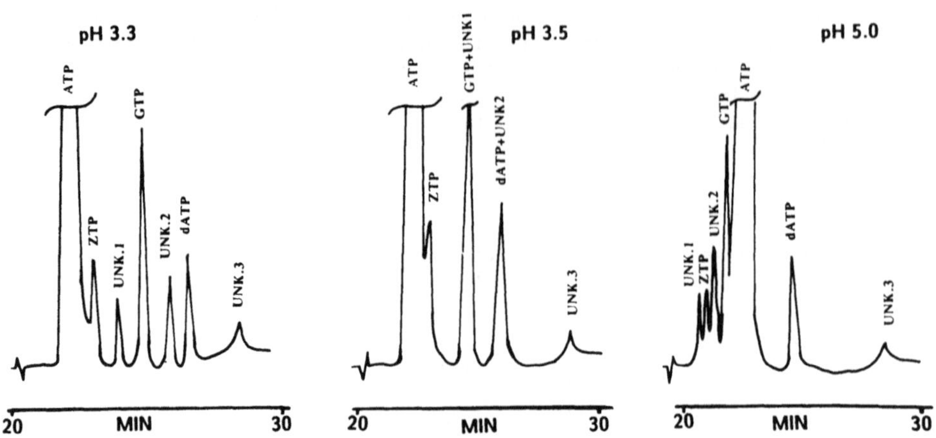

FIG 2 Effect of changes in the pH of the second buffer on the resolution of the triphosphate nucleotides. N.B. Only later part of the chromatogram is shown

The retention times of AMP, NAD, UDPG are the most sensitive to the pH of the first buffer. At higher pH values NAD runs with AMP but UDPG elutes before ADP. A particularly troublesome problem encountered was the change in retention time of NADP with column ageing. NADP initially elutes prior to ADP but with column useage they co-elute and finally separate with the NADP running after ADP. Incorporation of upto 10% MeOH into either buffer did not stop this change of retention time Changes in the pH of the second buffer signifcantly change the elution order of the triphosphates. Manipulation of the pH of the second buffer was important for the accurate quantitation of some of minor triphosphates such as dATP (Fig 2). It is therefore necessary to carefully balance buffer pH in order to optimise any separation of interest.

With use the ion-exchange capacity of the packing falls although this change as little effect on the chromatographic selectivity of the column, except for NADP mentioned above. It does however cause a gradual reduction in retention times (Fig3). This change is particularly troublesome to automatic integration systems. It is therfore necessary to reduce the molarity of the final eluent composition either by reducing the % composition of the gradient or by using a weaker buffer.

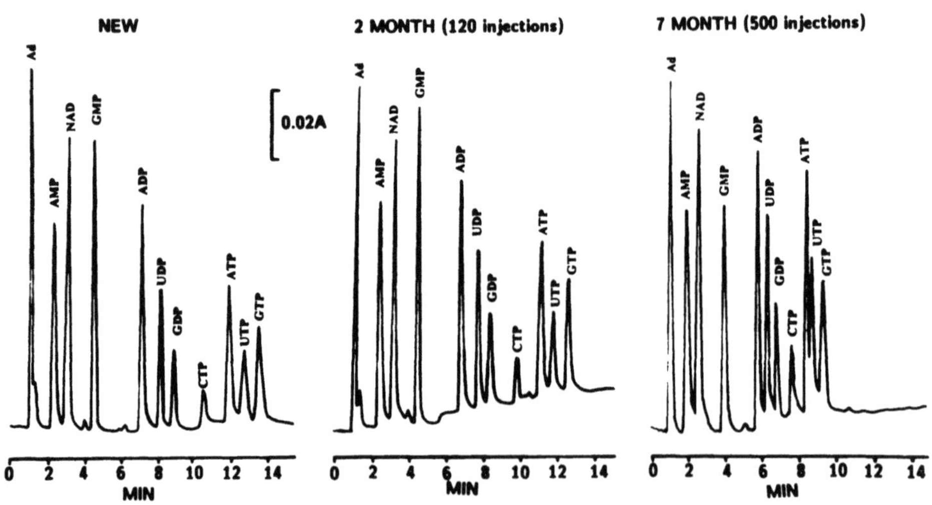

FIG 3 The effect of use on the retention times of a 10cm anion column

A number of potential interferences that are not nucleotides but are retained by the column have been encountered. Table 3 lists those impurities and suggests where they might run on a chromatogram. Their exact elution times relative to depends on pH etc.

Table 3 Potential Interferences with the Anion-exchange HPLC of Nucleotides

Compound	Comment
Ascorbic Acid	Elutes near CMP
Ficoll/Hypaque	Two peaks one near GMP one near GDP
Orotidine	Elutes near AMP
Excess nucleosides and bases	Large frontal peak with smaller peaks on the tail
TCA	Broad peak eluting after GMP
EDTA	Multiple peaks near IMP
Water impurities	Contributes to baseline rise
Impurities in phosphate	Contributes to baseline rise and gives broad peak before CTP

CONCLUSIONS

HPLC has been a major tool in improving of our knowledge of purine metabolism in man and other species. This study has attempted to address some of the problems associated with gaining maximum qualitative and quantitative accuracy using anion-exchange HPLC for nucleotides. Although optimised extraction procedures for the isolation of nucleotides from red blood cells have been detailed it must be stressed that these conditions are not universal and individuals should optimise their own methods along similar lines if working with other cells or tissues. Even red cells from other species do not necessarily behave in the same way as those from man (3,5).

Changes in the performance of the column packing materials must be improved by the manufacturers in order to increase system reliability particularly for those laboratory involved in routine screening procedures. Until then it is essential that the analyst understands the chromatographic variables discussed here and learns to optimise any particular system and column. Using the adaptations reported here, reliable comparisons of nucleotide levels can be obtained in a single patient over periods of years. This is important for the monitoring of therapy. It is also essential for the detection and identification of abnormal nucleotides or altered levels of normal nucleotides, in inherited disorders.

REFERENCES

1. Krustolvic, A.M. (ed.) Handbook of Chromatography of Nucleic acids and related compounds Vol I C.R.C. Press. Boca Raton (1987)
2. Perrett, D. Nucleotides, nucleosides and bases chapter 9 in HPLC of small molecules; a practical approach pp 221-259 ed. C.K. Lim IRL Press, Oxford, (1986)
3. Perrett, D. The isolation and extraction of tissues prior to HPLC in Handbook of Chromatography of Nucleic acids and related compounds Vol I Part B pp3-29 ed A.M.Krustolvic C.R.C. Press. Boca Raton USA (1987)
4. Perrett, D. Rapid Anion-Exchange Chromatography of Nucleotides in Physiological Fluids. Chromatographia 16, 211-214 (1982)
5. Dean, B.M., Perrett, D. and Sensi, M., Changes in Nucleotide concentrations in the Erythrocytes of man, rabbit and rat during short-term storage. Biochem. Biophys. Res. Comm. 80, 147-154 (1978)
6. Wiener, S., Weiner, R., Urivetzky, M. and Meilman, E., Coprecipitation of ATP with Potassium Perchlorate: The effect on the firefly enzyme assay of ATP in tissue and blood. Anal. Biochem. 59, 489-500, (1974)
7. Smith, P.K., Krohn, R.I., Hermanson, G.T., Mallia, A.K., Gartner, F.H. Provenzano, M.D., Fujimoto, E.K., Goeke, N.M., Olson, B.J. and Klenk, D.C. Measurement of Protein using Bicinchonnic Acid. Anal. Biochem., 150, 76-85, (1985)
8. Blanchard, J., Evaluation of the relative efficacy of various techniques for deproteinizing plasma samples prior to high performance liquid chromatographic analysis, J. Chromatogr., 226, 455-460, (1981)
9. Khym, J.X., An analytical system for rapid separation of tissue nucleotides at low pressures on conventional anion exchangers. Clin. Chem., 21, 1245-1252, (1975)

QUALITATIVE AND QUANTITATIVE STUDIES ON THE NUCLEOTIDES OF INTESTINAL MUCOSA

David Perrett, Alastair Watson* and David Rolston*

Departments of Medicine and Gastroenterology*
St. Bartholomew's Hospital
London EC1A 7BE. U.K.

INTRODUCTION

The conversion of hypoxanthine and/or xanthine to urate via xanthine oxidase (xanthine: oxygen oxidoreductase (E.C. 1.1.3.22.) may when ischaemic tissues are re-perfused in the presence of oxygen generate radical oxygen species. The generalised nature of this phenomena is now questioned since only very low levels of xanthine oxidase occur in the majority of tissues (1) from most species particularly man (2) and recent studies have indicated that the rate of conversion of the dehydrogenase form of the enzyme to the necessary oxidase appears to be relatively slow (3). Moderately high levels of xanthine oxidase are only present in the liver and the intestine (1).

Such ischaemia and the resultant re-perfusion injury may, however, therefore still be of importance in the damage caused to the intestinal mucosa by surgical procedures(4) including the difficulties associated with the transplant of viable intestine. Recently it has been proposed that ischaemia re-perfusion injury may also be involved in the adverse effects of non-steroidal anti-inflammatory drugs (NSAIDS) on the human gastrointestinal tract (5).

We are interested in changes to intestinal nucleotides that may occur during active transport processes as well as ischaemic situations. There is however little qualitative or quantitative information concerning the nucleotide content of intestine available and none of the available data was produced using chromatographic techniques. This studies reports some preliminary observations on the nucleotide content of rat intestine using high performance liquid chromatography (HPLC) and whether the nucleotide content is readily susceptible to ischaemic changes.

METHODS

Following an overnight fast male wistar rats (weight 180-240g) were anesthesised with pentobarbitone (60mg/kg/i.p.) and the abdominal cavity opened. The appropriate segments of gut were washed with a few ml of warm isotonic saline to remove food residues or faeces. Segments of gut were eased free of other organs and rapidly freeze-clamped with tongs that had previously been cooled to liquid nitrogen temperatures.

Approximately 15mm of gut were trapped by the tongs. The frozen segment was crudely separated from the rest of the tissue and all unfrozen tissue carefully removed. Freeze-clamped pieces of other organs were similarly prepared.

We have determined the nucleotide profiles of rat jejunum and for comparison liver and heart by HPLC and quantified the individual nucleotides. Figure 1 shows the nucleotide profile of a typical extract of jejunum. Intestine clearly contains significant quantities of nucleotides other than the expected adenosine nucleotides. Quantitative data for the major nucleotides in jejunum is compared to those of liver and heart in Table 1. The dry weight/wet weight ratio of the three tissues varied slightly and was 0.210±.012, 0.313±.015 and 0.228±.021 (mean+1 s.d. n=12) for jejunum, liver and heart respectively.

TABLE 1. Nucleotide content of various rat organs

	Jejunum (n=10)	Liver (n=6)	Heart (n=5)
		nmol/g wet wt	
ATP	1936±401	2930±565	3210±450
ADP	616±100	960±77	690±81
AMP	252±22	222±50	120±24
GTP	444±72	400±127	115±21
GDP	213±19	90±15	42±11
GMP	49±24	34±14	trace
UTP	377±32	260±33	107±45
CTP	67±21	99±23	trace
NAD	184±31	809±106	699±42

As expected warm ischaemia at room temperature caused a rapid fall in ATP content of rat jejunum with a rise in both ADP and AMP concentration However due to rapidity of the changes we have not yet been able to determine the rate of loss of ATP with any accuracy.

In preliminary studies we have determined the effect of short term incubation of rat jejunum in vivo to NSAIDs. Table 2. shows the effect of indomethacin and aspirin on the major the nucleotide of rat jejunum Under the test conditions employed neither drug significantly affected the total levels of nucleotides in intestinal mucosa.

TABLE 2 Effect of NSAIDs on rat jejunum nucleotides

	AMP	NAD	ADP	ATP	UTP	GTP
			nmol/g wet wt			
Control loops (n=6)	105±30	194±29	611±82	1496±150	364±49	267±47
Indomethacin	167	281	939	1780	350	311
	125	260	212	1616	446	273
	195	430	805	2150	450	402
Aspirin	93	143	402	1256	327	182
	514	146	786	657	111	111
	282	278	909	1610	462	308

Nucleotides were immediately extracted from a weighed piece of frozen tissue (approximately 100mg) by homogenising in a micro Ultra-turrax homogeniser with 1ml of 20% w/v trichloroacetic acid (TCA) containing cXMP (50umol/l) as an internal standard. Excess TCA was removed with 4 washes of water-saturated diethyl ether (5vol). Extracts were stored at -20C until analysis. A small portion of tissue was retained in order to determine its water content by measuring the change in weight following freeze-drying for 24h.

The effect of NSAIDs was determined 'in vivo' using a closed loop technique. Two 5cm long loops of jejunum were cannulated washed with warm isotonic saline saline, One loop was filled via a syringe with 2 ml of either indomethacin, aspirin and adjacent loops were filled with saline as controls. The drugs were dissolved in saline at 2mmol/l. After 30min incubation the loops were freeze-clamped as described above.

Nucleotides were determined by HPLC as previously described (6). In outline the chromatograph consisted of a 100 x 4.6mm stainless steel column (HETP, Macclesfield, U.K.) containing a weak anion exchange packing, 25ul loop injector (Specac, U.K.) a Cecil CE212 U.V. monitor operating at 254nm and an ACS 300/01 pump (ACS Macclesfield U.K.) The output from the monitor was to a Trivector Trio integration system. Gradients were generated using a Pye Unicam LC-XP gradient programmer. Columns were packed in the laboratory with 5um APS-Hypersil (Shandon, Runcorn, U.K.) and columns with efficiencies of 3500-4500 plates per 10cm were used. The column was eluted at 1ml/min at room temperature. A linear gradient from 0.04M KH_2PO_4 pH 2.9 to 0.5M KH_2PO_4 plus 0.8M KCl pH 2.9 in 14 minutes was used to elute the nucleotides. The $KH_2PO_4$was HPLC grade (Fisons) but other reagents were A.R. grade. Nucleotide standards, indomethacin and aspirin were obtained from Sigma.

RESULTS

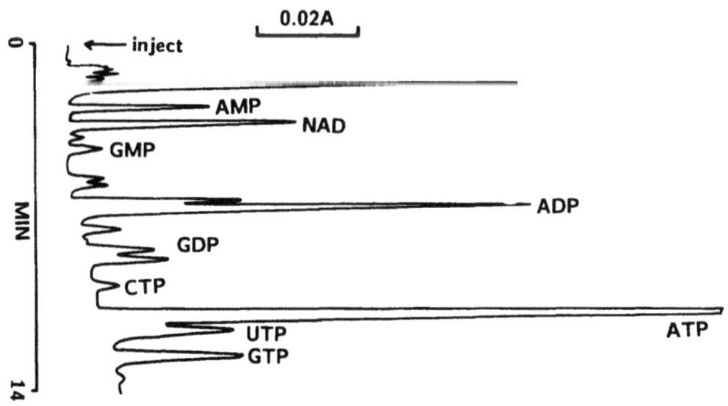

FIG 1 Nucleotide profile of a typical portion of rat jejunum determined by anion exchange HPLC

DISCUSSION

Gastroenterology opinion as usually considered the intestinal mucosa's nucleotide pool as mainly providing a rich source of energy i.e. ATP to drive active intestinal transport processes in a manner equivalent to the specialised muscles. From the complexity of nucleotides found the results of the present study suggests that the intestine is a major site of nucleotide metabolism with both nucleotide concentrations and nucleotide patterns equivalent to liver rather than heart.

Previous workers have often only reported levels of adenosine nucleotides for the intestine and even then many results on substantially ischaemic tissues have been reported. Three groups have reported data for ATP, ADP and AMP on reasonable preparations i.e. freeze-clamping in situ but using non-chromatographic means of analysis. Bronk and Leese (7) reported ATP, ADP and AMP levels in rat jejunum of 3000±100, 2800±150 and 2700±150 nmol/dry wet wt and Faelli et al. (8) 7600±600, 4600±900, and 1200±200 nmol/g dry wt respectively. In most tissues high ATP/ADP ratios are considered to characterise both a good sampling technique and an initially viable tissue preparation (9) In both the above studies the ATP/ADP was approximately 1 compared to 3.1 in the present study. Lamers and Hulsmann (10) reported a higher and more equivalent ATP/ADP ratio of 5.1 and levels of 1090±40, 190±10 and 27±2 nmol/g wet wt for ATP,ADP and AMP respectively. However there is disagreement between the total adenosine nucleotide levels reported here and those reported in the three studies using non-chromatographic methods. Some differences in levels may result from variations in water content although in the present study such variations were small.

The intestine of both the guinea pig (11) and the rat (12) under normal dietary regimes have been shown to possess an inactive de novo pathway for purine biosynthesis and to be dependent on the salvage pathways for nucleotide synthesis However LeLeiko et al. (12) have shown that the de novo pathways in the rat are activated when purines are ommitted from the diet. The levels of both the major and minor nucleotides found suggest that if de novo synthesis is minimal then the salvage pathways must be particularly active and further work is required to define these pathways.

Although rapid degradation of nucleotides was caused by warm ischaemia no conistant changes were induced by either of the two NSAIDs tested. There was neither a reduction in the total adenosine nucleotide levels in the intestine nor a change in the energy charge. Such findings may be due to the doses employed or the short length of the incubation period but if correct question the exact mechanism of the proposed radical damage to the intestine

REFERENCES

1. Parks, D.A. and Granger, D.N., Xanthine Oxidase: biochemistry, distribution and physiology. Acta Physiol. Scand. suppl 548 87-99 (1986)
2. Watts, R.W.E., Watts, J.E.M. and Seegmiller, J.E., Xanthine oxidase activity in human tissues and its inhibition by allopurinol (4-hydroxypyrazolo[3,4-d]pyrimidine). J.Lab. Clin. Med. 66, 688-697 (1965)

3. Engerson, T.D., McKelvey, T.G., Rhyne, D.B., Boggio, E.B., Snyder, S.J., Jones, H.P. Conversion of Xanthine Dehydrogenase to Oxidase in Ischemic Rat Tissues. J. Clin. Invest, 79, 1564-1570 (1987)
4. Granger, D.N., Rutili, G. and McCord, J.M., The role of superoxide radicals in intestinal ischemia. Gastroenterolgy, 81, 22-29 (1981)
5. Rainsford, K.D. Prevention of Indomethacin-induced gastro-intestinal ulceration in rats by glucose-citrate formulations: Role of ATP in Mucosal defences (Abstract) Brit Soc. of Rheumatology (1987)
6. Perrett, D. Rapid Anion-Exchange Chromatography of Nucleotides in Physiological Fluids. Chromatographia 16, 211-214 (1982)
7. Bronk, J.R. and Leese, H.J. Changes in the adenine nucleotide content of preparations of the rat small intestine in vitro J. Physiol. 235, 183-196 (1973)
8. Faelli, A., Esposito, G. and Capraro, V. Energy-rich phosphates and transintestinal transport in rat intestine incubated in vitro and different temperatures. Biochim. Biophys. Acta 455, 759-766. (1976)
9. Perrett, D. The isolation and extraction of tissues prior to HPLC in Handbook of Chromatography of Nucleic acids and related compounds Vol I Part B 3-29 ed A.M.Krustolvic C.R.C. Press. Boca Raton (1987)
10. Lamers, J.M.J. and Hulsmann, W.C., The effect of fructose on the stores of energy-rich phosphate in rat jejunum in vivo. Biochim. Biophys. Acta 313, 1-8 (1973)
11. Mackinnon, A.M., and Deller, D.J., Purine Nucleotide Biosynthesis in Gastrointestinal Mucosa. Biochim. Biophys. Acta 319, 1-4 (1973)

12. LeLeiko, N.S., Bronstein, A.D., Baliga, B.S. and Munro, H.N., De Novo Purine Nucleotide synthesis in the rat small and large intestine: Effect of Dietary Protein and Purines. J. Pediat. Gastro. and Nutr. 2, 313-319 (1983)

RELATIONSHIP BETWEEN EXTRACELLULAR AND INTRACELLULAR NUCLEOTIDE METABOLISM IN HUMAN LYMPHOCYTES

Jerzy Barankiewicz[1], Hans-Michael Dosch, Roy Cheung and Amos Cohen

Division of Immunology/Rheumatology, Research Institute, The Hospital for Sick Children, Toronto, Ontario, M5G 1X8, Canada

[1]Present address: Gensia Pharmaceuticals, Inc., 11075 Roselle Street, San Diego, CA 92121, USA

Ectoenzymes present on the surface of B lymphocytes, but not T lymphocytes, can degrade extracellular ATP to produce adenosine (1). Since ATP is released from different cell types into the extracellular space to attain micromolar concentrations in human plasma (2), ectonucleotidases present on B cells can provide a significant physiological source of adenosine in lymphoid tissues. Extracellular ATP and adenosine, on the other hand, can affect lymphocyte function (3-5). Although the biological role of ectonucleotidases is not clear, they are capable of regulating the relative amounts of extracellular ATP and adenosine.

It has been reported that micromolar concentrations of extracellular ATP can inhibit antibody dependent cytotoxicity as well as natural killer cell activity both in humans and rodents (3). Extracellular ATP also stimulates DNA synthesis in bone marrow and thymus cells, but inhibits it in spleen, lymph nodes or peripheral blood lymphocytes (4,5). Acting via plasma membrane ATP receptors, extracellular ATP may induce changes in membrane permeability and modify cation fluxes. The mode of action of extracellular adenosine and ATP is not clear yet but, it has recently been demonstrated that thymocytes contain P_2 ATP receptor (6) as well as A_2 adenosine receptors (7).

While ATP cannot cross the cell membrane, adenosine, the product of extracellular ATP degradation, is readily transported into cells, rapidly deaminated to inosine, or phosphorylated to AMP and ATP. Acting intracellularly, adenosine can also cause cytostatic effects, which are promoted by inhibition of adenosine deaminase. Adenosine can inhibit purine and pyrimidine biosynthesis de novo (8) as well as S-adenosylhomocysteine hydrolase activity (9).

In this paper we discuss the effect of extracellularly added ATP and its degradation product, adenosine, on intracellular purine metabolism in B and T human lymphocytes.

Extracellular ATP Degradation is Providing Precursors for Intracellular Nucleotide Synthesis

The products of extracellular ATP degradation can be incorporated into intracellular nucleotides and can thus influence cellular nucleotide metabolism. This aspect of extracellular/intracellular metabolic cooperation can be an important source of purines especially to resting lymphocytes with no de novo purine biosynthetic activity and total dependence on purine salvage for replenishment of purine nucleotide pools. Adenosine generated, following exposure of B cells to 2 mM extracellular ATP, entered the cells and was both deaminated to inosine and hypoxanthine and phosphorylated to ATP via AMP and ADP (Fig. 1). Although the phosphorylation of adenosine to nucleotides (mainly ATP) is about 3 times less than its deamination, it can be clearly concluded that the degradation of extracellular ATP contributes significantly to the synthesis of intracellular ATP.

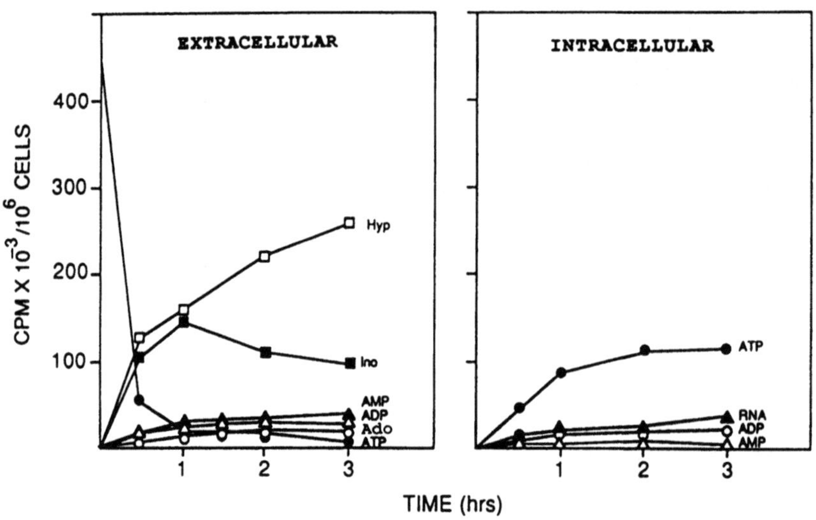

Figure 1. Relationship between extracellular ATP degradation and intercellular nucleotide synthesis in human tonsillar B cells. Cells (1×10^6) were incubated with 1 uCi of [2.8 - ^{3}H] ATP (2 mM initial concentration). After incubation nucleotides nucleosides and bases were analyzed separately in the medium end in the cell extracts as described in (1).

Effects of Extracellular ATP and Adenosine of Intracellular Purine Metabolism

To study the effects of extracellular ATP and adenosine on intracellular purine metabolism, B and T lymphocytes were incubated with radioactive purine metabolic precursors (adenine, hypoxanthine, guanosine and formate) and different concentrations of unlabeled ATP or adenosine (Table 1).

Table 1. Effect of extracellular ATP and adenosine on the intracellular metabolism of adenine, hypoxanthine, guanosine and formate in human B and T lymphocytes.

Tonsil B (non-E rosetting cells) or T (E-rosetting) lymphocytes (1 x 10^6) were incubated in 0.1 ml with different concentrations of nonlabeled ATP or adenosine and 1 uCi of radioactive adenine, hypoxanthine, guanosine, or formate. Incorporation into purine ribonucleotides was measured using TLC chromatography according to (1).

	B Lymphocytes		T Lymphocytes	
	Extracellular			
uM	ATP	Ado	ATP	Ado
	Total incorporation of radioactivity into intracellular purine nucleotides (cpm/10^6 cells)			
	Adenine			
0	53020	47950	15950	17155
100	31040	23300	13410	3415
500	27951	19752	12350	2950
1000	20150	18350	12255	2930
	Hypoxanthine			
0	8520	11400	7950	7480
100	2145	3140	6910	1405
500	2040	1775	5545	1245
1000	1750	1510	4388	900
	Guanosine			
0	14475	15945	11335	8485
100	3955	2485	11455	3140
500	3490	1815	12685	1210
1000	2655	1220	10441	988
	Formate			
0	5282	4915	2771	2533
100	2661	3177	2652	1352
500	2527	2115	2658	1271
1000	2195	2003	2489	987

In purified B lymphocytes (Table 1) extracellular ATP as well as adenosine inhibit adenine, hypoxanthine, guanosine and formate metabolism with comparable dose dependence. These data are consistent with the view that the effect of added ATP is mediated by the formation of adenosine, its extracellular degradation product. On the other hand, the marginal effect of extracellular ATP on purified T lymphocytes is consistent with the absence of an extracellular catabolic pathway in T lineage cells.

To determine whether adenosine itself or a product of its deamination are the active compounds in the inhibition of purine metabolism, we examined the salvage of radioactive adenine into cells incubated with different concentrationsof either 2-chloroadenosine (a poorly deaminated adenosine analog) or with inosine (the product of adenosine deamination).

Table 2. Effect of 2-chloroadenosine and inosine on adenine salvage in human B and T lymphocytes

Tonsil B (non-E rosetting cells) or T (E rosetting cells) (1×10^6) were incubated with different concentrations of nonlabeled 2-chloroadenosine or inosine and 0.5 uCi radioactive adenine. Incorporation of radioactivity into total purine nucleotides was measured by TLC according to (1).

	B Lymphocytes		T Lymphocytes	
	Extracellular			
uM	2-chloroAdo	Ino	2-chloroAdo	Ino
	Total incorporation of radioactivity into purine intracellular nucleotides (cpm/10^6 cells)			
0	27690	26340	24825	25950
100	13340	25245	13825	25895
500	12650	26095	12730	26360
1000	5640	24945	7200	26502

These results (Table 2) show that 2-chloroadenosine significantly affect the adenine salvage pathway whereas inosine, the product of adenosine deamination, has no significant effect.

In conclusion, the results presented here provide evidence for metabolic cooperation between extracellular and intracellular nucleotide metabolism, as well as metabolic cooperation between B and T lymphocytes. Degradation of extracellular ATP by B cells produce extracellular adenosine which can be easily utilized by both B and T lymphocytes.

Adenosine produced extracellularly affected several reactions of purine metabolism, acting probably by both adenosine receptors and intracellularly.

REFERENCES

1. J. Barankiewicz, H.M. Dosch, and A. Cohen <u>J. Biol. Chem</u>. 263:7094-7088 (1988).

2. J.L. Gordon <u>Biochem. J</u>. 233:309-319 (1968).

3. A. Schmidt, J.R. Ortaldo and R.B. Herberman <u>J. Immunol</u>. 132:146-149 (1984).

4. S. Gregory and M. Kern <u>Biochem. Biophys. Res. Comm</u>. 83:1111-1116 (1978).

5. S. Ikehara, R.N. Pahwa, D.G. Lanzer, R.A. Good and M. J. Modak J. Immunol. 127:1834-1837 (1981).

6. C. El-Moatassim, J. Dornand and J.L. Mani Biochim. Biophys. Acta 927:437-444 (1987).

7. L.A. Schultz, G.M. Kammer, and S.A. Rudolph FASEB J. 2:244-250 (1988).

8. F.F. Snyder, M.S. Hershfield, and J.E. Seegmiller Cancer Res. 38:2357-2362 (1977).

9. M.S. Hershfield and N.M. Kredich Science 202:757-760 (1978).

EFFECT OF SOME PURINE METABOLITES CONTAINED IN OYSTER ON PLATELET AGGREGATION

T.Ohta, M.Nakatsuka, J.Nomura, T.Tanaka, K.Satoh and Y.Shibata

Department of Biochemistry
Aichi Medical University
Aichi, Japan

INTRODUCTION

Oyster extract is a rich source of taurine and metals such as zinc. Taurine is an amino acid containing sulfonic acid group in its molecule, and is biosynthesized from methionine via cysteine. The biological function of taurine has well been studied, and its biological activity in preventing cardiovascular dysfunction such as hypertension,stroke,and myocardial infarction. Authors previously reported that the oyster extract suppressed platelet aggregation caused by ADP and epinephrine in vivo and in vitro.[1),2)]

Shibata et al. reported that long term oral taurine administration increased 5-hydroxytryptamine(5-HT;or serotonin) level in liver[3)].

The platelet aggregation has been thought to be somewhat enhanced by 5-HT[4)], one of Tryptophan metabolites. It seems likely that various factors fluctuates Trp metabolism especially under morbid condition[5)].For instance, under vitamin B_6 deficient state, urinary xanthurenic acid level increase. However the interrelation between Trp metabolites, taurine, oyster extract and platelet aggregation has remained unknown.

On the other hand, low molecular fraction of oyster extract, which contains nucleic acid related substances, suppressed ADP induced platelet aggregation. Effect of cyclic nucleotide,especially cyclic AMP, on platelet aggregation has been well studied[6)],however the effect of other nucleic acid related substances has not so discussed.

This is a preliminary work to examine the effect of these purine derivatives and Trp metabolites on platelet aggregation.

Materials and Methods

Chemicals used for platelet aggregation

Tryptophan metabolites examined for platelet aggregation test were kynurenine sulfate, 3-hydroxykynurenine, anthranilic acid, 3-hydroxyanthranilic acid, 5-hydroxyindole acetic acid (5-HIAA), which were purchased from Sigma Chemical Co.St.Louis,Mo,and 5-hydroxytryptamine (5-HT) purchased from Aldrich Chemical Co., Milwaukee, Wis.
L-Tryptophan was purchased from Ajinimoto Co., Tokyo. All the chemicals used in the present work were guaranteed reagent grade.

Animals

Experimental animal used for platelet aggregation was male Wistar rats of ca. 150 gram.

To examine platelet aggregation of human blood, fresh blood were collected from two healthy male volunteers of 29 and 36 years of age without any medication.

Platelet aggregation test

To examine the rats' platelet aggregation,blood was collected from hearts by punctuation of six rats without starvation. Fresh blood thus collected was gently mixed with 3.8% sodium citrate, then centrifuged at 1000rpm for 10 min to obtain platelet rich plasma (PRP). Platelet poor plasma (PPP) prepared by centrifugation of the remaining precipitates of the blood was used as the reference of platelet aggregation test. The platelet aggregation test was evaluated by measuring turbidity of the incubation mixture using ADP.

To examine the effect of addition of Trp metabolites on the ADP induced platelet aggregation, Trp metabolites were added in the incubation medium at the final concentration as shown in Table 2.

Determination of urinary tryptophan metabolites

To survey the influence of tryptophan and its metabolites on platelet aggregation, L-tryptophan was injected to rats intraperitoneally at the dose of 10mg/100g body weight.
In some experiments chelating compounds injected together with T-Trp[7].
The amount of excreted tryptophan metabolites into urine was measured by Brown et al's method[8].

Preparation and Analysis of nucleic acid related substances in oyster

Cold 10% perchloric acid was added to oyster extract. After neutralization with 5N-KOH, it was centrifuged. Its supernatant was absorbed to the active carbon column and eluted by 2.8% ammonia 50% ethanol solution to obtain the nucleic acid related substances.

The ion exchange chromatography(Dowex 1X8)was carried out by the stepwise elution method to fractionate the nucleic acid related substances in oyster extract. Formic acid and formic acid natrium were used for the elution buffer. These fractions were desalted by the active carbon, lyophilized and examined for the platelet aggregation test.High-performance liquid chromatography(Shimadzu Ltd.) was carried out to analyze the nucleic acid related substances in oyster extract.

Some authentic samples of the nucleic acids and their related substances were also examined for the platelet aggregation test.

RESULTS

Glucose Tolerance Test(GTT) in Alloxan Diabetic Rats(Fig-2)

Raw oyster was extracted with hot water and fractionated.(Fig-1)
20mg of low molecular fraction of oyster extract/kg/day were orally administered in diabetic rats for 21 days.(ALMO 20)
Alloxan diabetic rats(AC)showed the high level of blood glucose caused by 50g GTT and it continued for 48 hours. Rats fed oyster extract(ALMO 20), on the other hand,showed significantly lower level of blood glucose than alloxan diabetic rats.

Raw Oyster

Extraction with hot water at 80°C

Centrifuge

ppt — sup

Chromatography on Sephadex G-50

High molecular Fraction — Low molecular Fraction

Rechromatography on Sephadex G-10

Frn.1 — Frn.2 — Frn.3 ($NaBH_4$ treated)

Freeze Drying

Fig 1. Scheme for the fractionation of hot water soluble oyster extract.

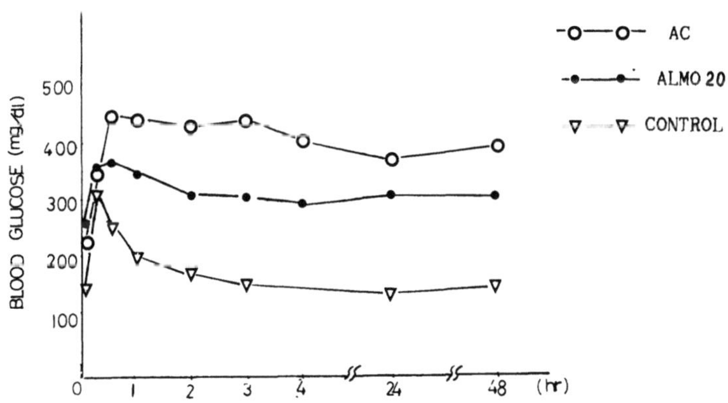

*Control rats were fed only normal diet.

Fig 2. Effect of low molecular fraction of oyster extract on alloxan diabetic rats in 50g GTT.

Level of Urinary Tryptophan Metabolites of Rats under Different Conditions

Quantitative analysis of Trp metabolites secreted into urine was done for Wistar rats kept under different nutritional conditions.(as shown in Table 1) Rats to which taurine was administered orally for 21 days showed no significant difference in xanthurenic acid level between control diet fed rats. Meanwhile xanthurenic acid secretion rose about 1.4 times in the urine of rats administered methionine and serine simultaneously. Rats fed oyster extract, on the other hand, showed significantly lower level of urinary xanthurenic acid than control diet fed rats.

Table 1. Excretion amount of xanthurenic acid in urine after oral Trp administration.

	μg/day in urine
Taurine fed group	2689.1
Ser+Met fed group	3666.7
Oyster extract fed group	1851.1
Control group	2681.4

*Each group except control group was given the feed described the table added to normal diet. Control group was fed only normal diet.

Influence of in vitro Addition of Tryptophan Metabolites on ADP-induced Platelet Aggregation

Different kinds of Trp metabolites were added in the incubation medium to measure ADP-induced platelet aggregation. Data are shown in Table 2.

Externally added 5-HT of incubation medium did not enhance the platelet aggregation but rather suppressed slightly. This result suggests that the externally added 5-HT should not be dealt equally in its role on the platelet aggregation.

3-Hydroxykynurenine strikingly suppressed the ADP-induced platelet aggregation. In the meantime 5-Hydroxytryptophan enhanced the platelet aggregation to some extent.

Table 2. Influence of Trp metabolite on ADP-induced platelet aggregation in vitro

Trp metabolites	Final concentration (μg/ml)	Platelet aggregation
5-HIAA	83.3	↓↑
5-HT	37.5	↓
Anthranilic acid	74.2	↓↑
3-Hydroxanthranilic acid	83.3	↓↑
Kyunurenic acid	83.3	↓
Xanthrenic acid	85.8	↓
3-hydroxy-L-Kynurenine	83.3	↓↓
5-Hydroxytryptophan	85.8	↓

↑↑ : Strongly accelerate
↑ : Weakly accelerate
↓↑ : no effect
↓↓ : Strongly inhibit
↓ : Weakly inhibit

Influence of in vitro Addition of Nucleic Acid Related Substances in Oyster Extract on ADP-induced Platelet Aggregation

9 fractions were mainly fractionated by the ion exchange chromatography(Fig. 3). HPLC analysis of their fractions showed the main component contained in each fraction, i.e. Adenosine in Frn.2, Inosine in Frn.3, AMP in Frn.7, IMP in Frn.8, GMP in Frn 9. Each fraction (lyophilized sample) and the authentic sample of nucleic acids were solely added in the incubation medium to measure ADP-induced platelet aggregation. Data are shown in Fig. 4 and 5.

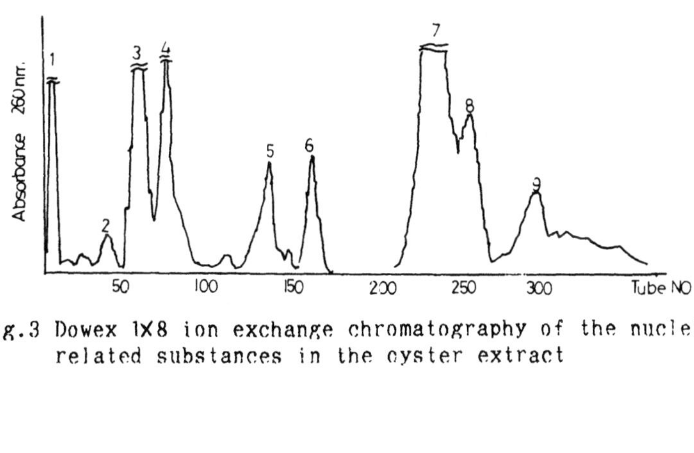

Fig.3 Dowex 1X8 ion exchange chromatography of the nucleic related substances in the oyster extract

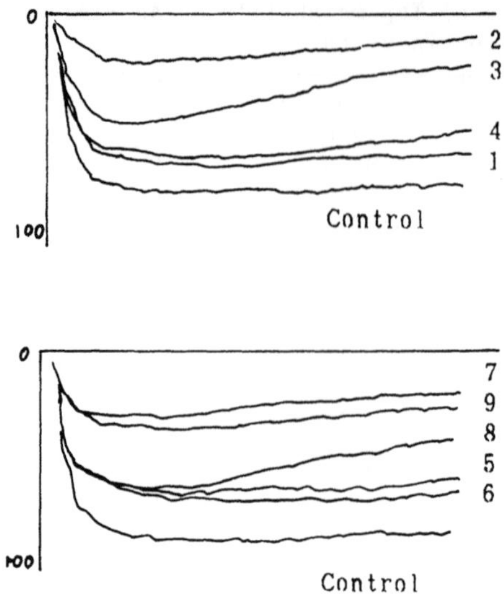

*Final conc.83.3μg/ml

Fig.4 Influence of every fraction (Dowex 1x8) on ADP-induced platelet aggregation

Fig.4 shows the influence of every fraction (Dowex 1X8) on ADP-induced platelet aggregation. the ordinate shows the turbidity,i.e. 0% means that platelet aggregation is suppressed perfectly. The abscissa shows the time (minutes)after adding the inducer(ADP). Frn.2, Frn.3, Frn.7, Frn.8 and Frn.9 at the concentration of 83.3μg/ml of the incubation medium suppressed the ADP-induced platelet aggregation as in Fig. 4.

The authentic samples of nucleic acid related substances (AMP,GMP, UMP,CMP,IMP,inosine,guanine,guanosine cytidine,cytosine,urasil,adenine, adenosine,hypoxantine)were also examined. Among these substances,adenosine adenine,AMP(mainly purine derivatives)suppressed the platelet aggregation. as shown in Fig.5.

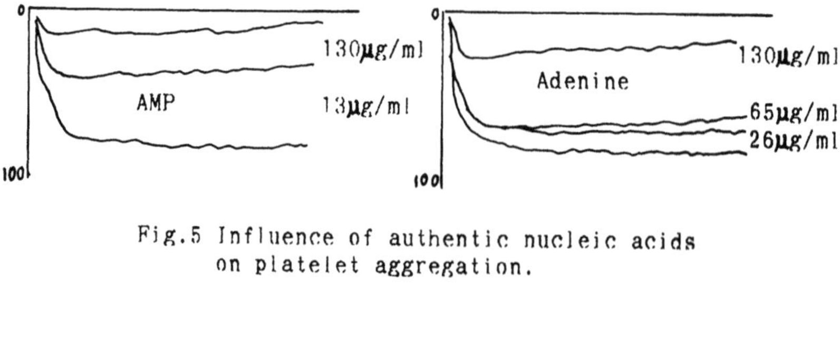

Fig.5 Influence of authentic nucleic acids on platelet aggregation.

DISCUSSION

As described before, the oyster extract suppressed the ADP-induced platelet aggregation of alloxan diabetic rats. The oyster extract also recovered 50g GTT of alloxan diabetic rats. Under the vitamin B_6 deficient state, platelet aggregation is sometimes accelerated like under the diabetes mellitus and vitamin B_6 has an important role in Trp metabolism. In the Trp administered rats, the oyster extract suppressed the excretion of xanthurenic acid in urine. One of Trp metabolites(3-hydroxy-L-kynurenie) strongly suppressed the platelet aggregation in vitro.

Other components of oyster(AMP, adenosine, inosine), they have purine bases, also suppressed the platelet aggregation. However, pyrimidine base group did not suppressed so strongly.

From these facts, we speculate that the oyster extract has some effect on Trp. metabolism. We need to examine if Trp metabolites in serum really participate in the platelet aggregation in vivo, and if so to what extent does it contribute to the suppression of platelet aggregation in vivo.

In the case of the nucleic acid related substances in oyster extract, th purine base seems to need in the suppression of ADP-induced platelet aggregation. We also need to examine the influence of purine and pyrimidine base group on platelet aggregation in vivo.

REFERENCES

1)T.Ohta. M.Ookubo. S.Okumura. et al. Effect of oyster extract on human platelets aggregation. Proceedings the 2nd symposium on trace nutrients research Tokyo, 1985. 169

2)T.Ohta. M.Ookubo.Y.Shibata. Effect of oyster extract on rat platelet aggregation. Proceedings the 3rd symposium on trace nutrients research,1986.61

3)ibid 75
Y.Shibata. R.Tsubouchi.

4)a)F.DE Clerck, B.Xhonneux and R. Van DE Wiele. Agents and Actions, Vol 17.2:220~228. 1985.
b)J. Bevan and S.Heptinstall Nannyn-Schmiedeberg'a Arch Pharmacol 334:341~345. 1986.

5)T.Igari. M.Tsuchizawa. Tohoku J. exp Med. 1987.153. 79~86

6)Malmstem,C. Granstorm, E and Samuelsson,B Cyclic AMP inhibits synthesis of prostaglandin endoperoxide (PGG_2) in human platelets. Biochim. Biophys. Res. Commun. 1976.68.569~576

7)Y.Shibata. et al.Tryptophan metabolism and Mineral Proceedings The 2nd Symposium on Trace Nutrients Research.1985.

8)a)Lepkovsky, S.Roboz, E & Haggen-Smit,A.J. :Xanthurenic acid and its role in tryptophan metabolism of pyridoxine deficient rats. J.Biol. Chem 149:195~201,1943
b)Brown, R.R. & Price,J.M. Quantitative studies on metabolites of Trp in the urine of the dog, cat, rat, and human. J. Biol. Chem.291:985~997,1956

ENZYMES OF PURINE NUCLEOTIDE METABOLISM IN MATURE CULTURES OF RAT BRAIN NEURONS

S. Brosh, O. Sperling, Y. Bromberg and Y. Sidi

Tel-Aviv University Sackler School of Medicine, Dept. of Clinical Biochemistry and Dept. of Medicine D, Beilinson Medical Center, Petah-Tikva and Dept. of Chemical Pathology Ramat Aviv, Israel

INTRODUCTION

Several inherited disorders of purine metabolism are assosiated with severe neurological dysfunction: complete hypoxanthine-guanine phosphoribosyltransferase (HGPRT) deficiency is the cause of the Lesch-Nyhan syndrome (1), purine nucleoside phosphorylase (PNP) deficiency is associated with retarded development, hypotonia and nerve deafness (2), and adenylosuccinate lyase deficiency is associated with autism and mental retardation (3). These disorders demonstrate the crucial importance of purine metabolism for normal neuronal function. Nevertheless, many aspects of purine metabolism in neuronal cells are unclarified as yet. Recently a short term rat neuronal cell culture has been characterized (4), which enables us to study intensively the linkage between purine metabolism and neuronal function. This culture system allows ordered differentiation to occur, as manifested by improved ability of these cells to release and store dopamine along with other neurotransmitters (4). We initiated the study of purine metabolism in these cells, at full functional maturity, by measuring the activity (Vmax) of several key enzymes in purine nucleotide metabolism.

METHODS

Rat neuronal cultures were established from brains of rat fetuses (18 days of gestation; 4). All assays were done on 10-14 day-old cultures. De novo purine synthesis was gauged by the rate of incorporation of (14C)formate into intracellular and extracellular purines. Enzymes were assayed radiochemically (5).

RESULTS

No activity could be detected for IMP dehydrogenase, xanthine oxidase and uricase. Table 1 summarizes the measured enzyme activities.

The rate of 14-C formate incorporation into purines was linear over 3 hour period.

Table 1. Activity of enzymes of purine metabolism in 14 day-old cultured neuronal cells and in brains from 14 day-old rats.

	Activity (nmoles/mg protein /h)	
Enzyme	Cultured neurons	Rat brains
AMP deaminase	443∓189 (7)*	634∓228 (3)
AMP 5'-nucleotidase	269∓143 (14)	142∓ 83 (3)
Adenosine kinase	316∓100 (10)	251∓ 88 (3)
Adenosine deaminase	700∓374 (6)	408∓184 (3)
IMP 5'-nucleotidase	72∓ 18 (6)	141∓ 68 (3)
Purine nucleoside phosphorylase	2358∓356 (8)	1390∓325 (4)
Hypoxanthine-guanine PRT	19∓ 13 (25)	183∓ 64 (3)
Adenine PRT	113∓ 86 (13)	125∓ 35 (3)
Guanase	38∓ 20 (8)	N.D**

* The numbers in parentheses indicate number of experiments, each in duplicates.

** Not done.

DISCUSSION

The results indicate that cultured neurons are capable of purine biosynthesis by both de novo and salvage pathways and of purine degradation to hypoxanthine (from inosine) and to xanthine (from guanine).

The rate of neuronal de novo purine synthesis is similar to that of primary cultures of rat skeletal muscle myocytes, assayed in our laboratory (9).

Of the salvage enzymes, adenosine kinase appears to have the highest specific activity. Adenosine deaminase, which competes on adenosine with adenosine kinase, has a higher activity (2.2 folds) than that of the kinase, but its affinity for adenosine is markedly lower. A similar condition has been recently described in cardiac and skeletal muscle cultures (5,6), in which tissues, only a small fraction of adenosine, at physiological concentrations, is deaminated to inosine. The specific activity of HGPRT in neurons is very low, but in absence of xanthine oxidase, the HGPRT may be physiologically functional in the salvage of hypoxanthine to IMP. Recently, some doubt was raised concerning the quantitative role of HGPRT in various tissues (5). The salvage of guanine by this enzyme in the neurons is also questionable, in view of the considerable activity of guanase, which may compete with this reaction. Indeed, rat brain slices were found to have a very limited ability of guanine salvage (8). Undoubtedly, more studies are needed to elucidate the physiological significance of hypoxanthine and guanine salvage in the neurons. The specific activity of APRT, another salvage enzyme, was found to be considerably higher than that of HGPRT, but in view of the questionable availability of adenine, the physiological role of this enzyme is not fully understood. Thus, of all the salvage enzymes, adenosine kinase appears to be the most important.

Concerning the catabolism of adenine nucleotides, it appears that the degradation of adenine nucleotides by deamination to IMP has a slightly better capacity than the dephosphorylation to adenosine. This is evident from the ratio between the activities of AMP deaminase and AMP nucleotidase, the enzymes which can catabolise AMP, and from the known Km of these enzymes for AMP. It should be mentioned, however, that the ratio between the AMP deaminase and nucleotidase is only 1.6, whereas this ratio in cultured cardiomyocytes and skeletal muscle cells is 7 and 150, respectively (5,6).

The comparison between the activity of the enzymes in mature neuron cultures and in brains of 14 day old rats (Table 1), disclosed that HGPRT is the only enzyme which activity differs markedly (almost 10 folds) between the two tissues. Previously, HGPRT activity has been found to increase markedly with maturation of the brain tissue, as studied in rat embryos and after birth (10). The low HGPRT activity in the mature cultured neurons may reflect either lack of 'maturation° of HGPRT in the cultured neurons, or that the increase in HGPRT activity with maturation of rat brains in-vivo reflects changes in cell types, which do not occur in the cultured neurons. This latter system contains about 90% neuronal cells.

ACKNOWLEDGEMENT: Supported by the USA-Israel Binational Science Foundation.

REFERENCES

1. Seegmiller JE, Rosenbloom FM, Kelley WN. 1967. An enzyme defect associated with a sex-linked human neuological disorder and excessive purine synthesis. Science 155: 1682-1685.

2. Simmonds HA, Fairbanks LD. Morris FS, Morgan G, Watson AK, Timms P, Singh B. 1987. Central nervous system dysfunction and erythrocyte guanosine triphosphate depletion in purine nucleoside phosphorylase deficiency. Arch Dis Child 62: 385-391.

3. Jaeken J, Van den Berghe G. 1984. An infantile autistic syndrome characterized by the presence of succinylpurines in body fluids. Lancet 1: 1058-1061.

4. Zurgil N, Zisapel N. 1981. Induced neurotransmitter release from primary cultures of rat brain neurons. Life Sci 29:2265-2271.

5. Zoref-Shani E, Kessler-Icekson G, Sperling O. 1988. Pathways of adenine nucleotide catabolism in primary rat cardiomyocyte cultures. J Mol Cell Cardiol 20: 23-33.

6. Zoref-Shani E. Shainberg A. Sperling O. 1982. Characterization of purine nucleotide metabolism in primary rat muscle cultures . Biochim Biophys Acta 716: 324-330.

7. Sperling O, Liberman UA, Frank M, De Vries A. 1971. Xanthinuria. an additional case with demonstration of xanthine oxidase deficiency. Am J Clin Path 55: 351-354.

8. Wong PCL, Henderson JF. 1972. Purine ribonucleotide biosynthesis, interconversion and catabolism in mouse brain in vitro. Biochem J 129: 1085-1094.

9. Zoref-Shani E, Kessler-Icekson G, Wasserman L, Sperling O. 1984. Characterization of purine nucleotide metabolism in primary rat cardiomyocyte cultures. Biochim Biophys Acta 804: 161-168.

10. Allsop J. Watts RWE. 1980. Activities of amidophosphoribosyltransferase and the purine phosphoribosyltransferases and the phosphoribosylpyrophosphate content of rat central nervous system at different stages of development. J Neurolog Sci 46: 221-232.

NUCLEOTIDE PYROPHOSPHATASE ANTAGONIZES CONTRACTIONS DUE TO THE TWITCH COMPONENT OF THE SYMPATHETIC NERVE RESPONSE AND TO APPLIED ATP IN THE GUINEA-PIG VAS DEFERENS

David Satchell

Department of Zoology
University of Melbourne
Parkville, 3052, Australia

INTRODUCTION

Stimulation of sympathetic nerves to isolated preparations of guinea-pig vas deferens causes an initial twitch contraction followed by a slower phasic contraction. Explanations as to the mechanism of neurotransmission mediating the twitch component have proved elusive. One theory is that sympathetic neurotransmission in this tissue involves the process of cotransmission whereby noradrenaline is released and mediates the slow contraction and at the same time ATP is co-released from the same nerve terminal and mediates the twitch[1].

Nucleotide pyrophosphatase (NP) converts ATP to AMP. In guinea-pig vas deferens preparations ATP causes a rapid contraction whereas AMP is inactive[2]. Thus if NP can gain access to the synaptic region in vas deferens, ATP released as a neurotransmitter might well be metabolised largely to an inactive form prior to receptor occupation. Under such conditions NP would be expected to interfere with the twitch response to sympathetic nerve stimulation. Thus NP has been applied to isolated preparations of guinea-pig vas deferens in organ baths to test the theory that ATP is a cotransmitter released by the sympathetic nerves in the tissue.

METHODS

The vasa deferentia were dissected from male guinea-pigs (700-900g) and opened flat by a lengthwise cut on the opposite side to the main deferential blood vessels. Preparations were mounted in 5 ml organ baths in Kreb's solution gassed with 95% O_2, 5% CO_2 at 37°C. Preparations were attached to Grass FT03. transducers coupled to a Grass polygraph at a resting tension of 0.25g. Stimuli were delivered via platinum ring electrodes at supra-maximal voltage every 4 min and agonists were added at 8 min intervals. Drugs used were ATP, noradrenaline, nucleotide pyrophosphatase (Sigma) and phentolamine mesylate (CIBA).

RESULTS

ATP (25μM) elicited a rapid contraction of the vas deferens. However after treatment with NP (1 unit/ml) this response was progressively reduced (Fig.1A).

Field stimulation of the vas deferens caused a biphasic response. Following treatment with NP (1 unit/ml) the initial twitch component of the response was progressively reduced while the slower phasic component remained unaffected (Fig.1B). Noradrenaline (10μM) caused a slowly developing contraction of the vas deferens preparation. Treatment of preparations with NP (1 unit/ml) failed to affect contractions in response to noradrenaline (10μM) (Fig.1C).

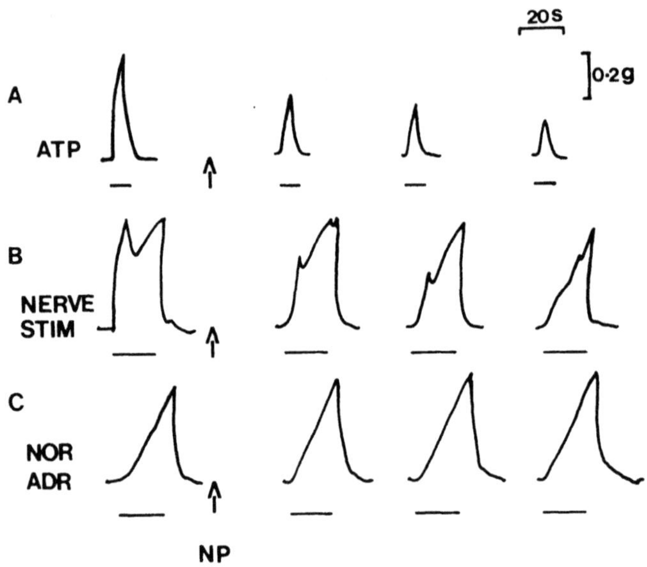

Fig. 1. Contractions of guinea-pig vas deferens before and after nucleotide pyrophosphatase (NP 1 unit/ml). (A) ATP (25 μM); (B) Intramural nerve stimulation (8 Hz, 0.2 ms supramaximal voltage for 20 S); (C) noradrenaline (NOR ADR 10 μM).

Treatment of preparations with phentolamine (10μM) failed to affect contractions in response to ATP (25μM) (Fig.2A), although, responses to noradrenaline (10μM) were almost abolished (Fig.2C). Phentolamine (10μM) abolished the second phase of biphasic response to field stimulation leaving the initial twitch component unchanged (Fig.2B). However, further treatment of the preparation with NP (1 unit/ml) largely abolished the twitch component as well, leaving only a small transient residual contraction (Fig.2B).

DISCUSSION

The mechanism underlying the initial twitch component of the response of the vas deferens to sympathetic nerve stimulation has remained unresolved for many years. A suggestion that ATP is the transmitter mediating this component has opened the way for further experimentation to answer this question. The finding that NP specifically reduced the twitch component and responses to ATP while leaving the slow component of the response and the noradrenaline response unaffected provides support for the ATP-noradrenaline cotransmitter concept.

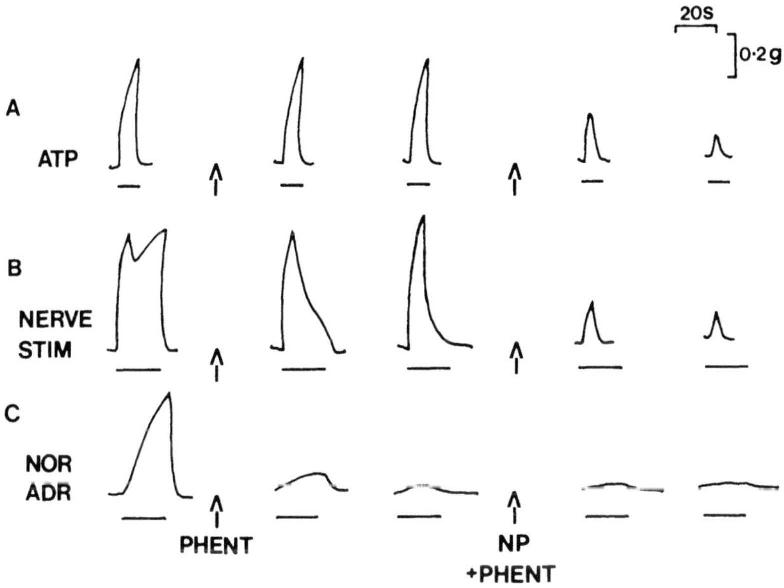

Fig. 2. Contractions of guinea-pig vas deferens before and after phentolamine (PHENT, 10μM), after addition of nucleotide pyrophosphatase (NP 1 unit/ml) together with phentolamin added previously; (A) ATP (25 μM); (B) Intramural nerve stimulation (8 Hz, 0.2 ms supramaximal voltage for 20 S); (C) noradrenaline (NOR ADR 10 μM).

These results are in agreement with earlier experiments which were also designed to interfere with the production of the response by ATP. The ATP receptor antagonist arylazido-aminopropionyl-ATP specifically reduced the twitch component of the response to sympathetic nerve stimulation in guinea-pig vas deferens[3], as did AMP probably as a result of ATP P_2 receptor antagonism[2]. Desensitization of P_2 receptors by the α,β-methylene isostere of ATP also caused a specific reduction of the twitch component in this tissue leaving the slower phasic response unchanged[1].

It will be of interest to further extend these studies to the nature of the response to sympathetic nerve stimulation in blood vessels.

ACKNOWLEDGEMENTS

This work was carried out with the aid of a grant from the Australian Research Grants Scheme.

REFERENCES

1. P. Sneddon, and G. Burnstock, Inhibition of excitatory junction potentials in guinea-pig vas deferens by α,β-methylene ATP, European. J. Pharmacol. 100:85 (1984).
2. D.G. Satchell, Antagonism of the ATP component of sympathetic co-transmission in the rat vas deferens by AMP, European. J. Pharmacol. 132:305 (1986).
3. J.S. Fedan, G.K. Hogaboom, J.P. O'Donnell, J. Colby, and D.P.Westfall, Contribution by purines to the neurogenic response to the vas deferens of the guinea-pig, European. J. Pharmacol. 69:41 (1981).

REGULATION OF PURINE NUCLEOTIDE METABOLISM IN HYPOXIC LIVER AND INTESTINE OF RATS: RADICAL SCAVENGING EFFECTS OF ALLOPURINOL AND OXYPURINOL

Gerhard Gerber, Werner Siems, Andreas Werner, Reinhard Stoesser[+], Zénó Foeldes-Papp, Jan Kowalewski, Wolfram Schneider, and Tilman Grune

Institut für Biochemie, Bereich Medizin (Charité) and Sektion Chemie[+]
Humboldt-Universität zu Berlin
Hessische Straße 3 - 4
Berlin 1040, DDR

INTRODUCTION

Ischemia-reoxygenation induces alterations in metabolism and structure of cells which are of tremendous significance for organ injuries related to a number of health problems such as myocardial infarction, stroke, intestinal inflammations and organ transplantation. Availability of oxygen during reperfusion following ischemia is assigned to be the reason for a burst of activated oxygen generated by xanthine oxidase (McCord and Roy, 1982). The rate of activated oxygen formation in the rat liver is elevated already during hypoxia, in the course of which the rates of nucleotide degradation as well as hydrogen superoxide production depend on the extent of oxygen deficiency (Siems et al., 1983; Räder et al., 1985). It can be attenuated by inhibition of the xanthine oxidase with allopurinol (Siems et al., 1983). In this manner the detrimental consequences of hypoxia and ischemia may be lessened.

However, further explanations of the cytoprotective action of allopurinol were debated: (i) facilitation of the electron transport during reperfusion when the lipid components of the mitochondrial respiratory chain can be expected to be disordered (Peterson et al., 1986); (ii) scavenging properties of allopurinol and oxypurinol toward radicals (Moorhouse et al., 1987; Das et al., 1987).

This contribution is aimed at the following problems:

(i) To what extent shares the purine nucleotide degradation via xanthine oxidase and urate oxidase producing activated oxygen to the rate of total oxygen consumption in the hypoxic rat liver?

(ii) What are the changes in the concentrations of nucleobase compounds during and following anoxia in hepatocytes and ischemia in small intestine?

(iii) What is the mechanism at the molecular level of the protective action of allopurinol and oxypurinol?

MATERIAL AND METHODS

The rat livers were prepared and perfused as previously described (Räder et al., 1985); hepatocytes were prepared following the procedure of Vincent et al. (1982). More than 85 percent of these cells were not stained with trypan blue.

The experiments on intestinal ischemia were performed on male Wistar H-strain rats of 350 g body weight. Allopurinol was injected intraperitoneally 24 hours and 30 minutes before the experiment (50 mg/kg body weight). The ani-

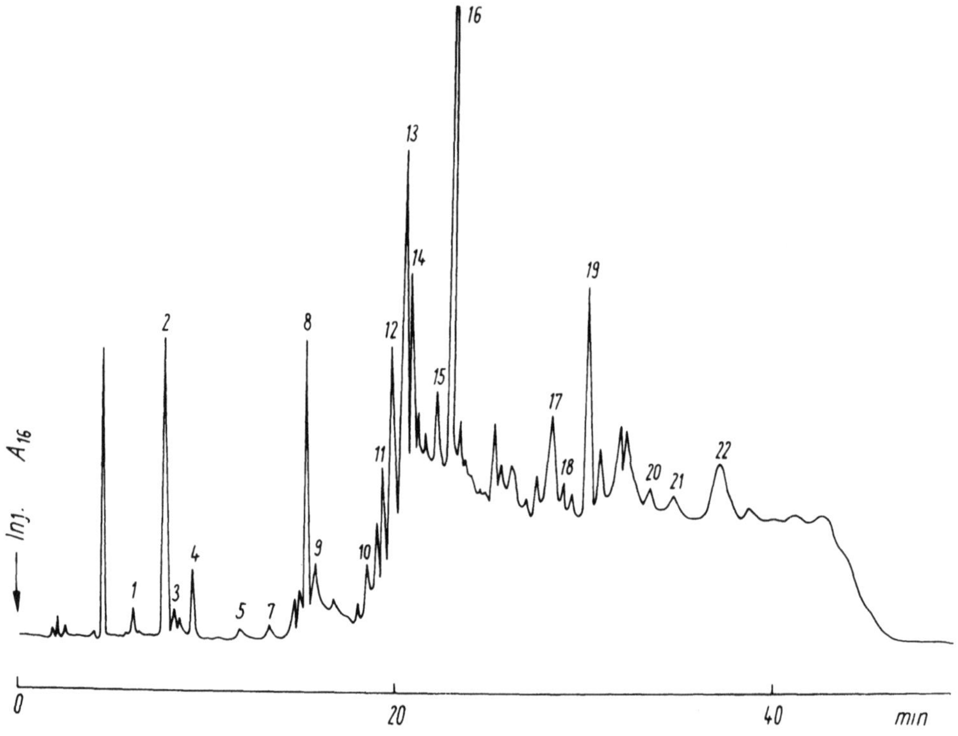

Fig. 1. Separation of nucleobase compounds in an extract from rat small intestine after 60 min of ischemia by ion-pair reversed-phase HPLC. Peaks: 1 = uracil; 2 = hypoxanthine; 3 = uridine; 4 = xanthine; 5 = oxypurinol; 6 = allopurinol; 7 = uric acid; 8 = inosine; 9 = guanosine; 10 = adenine; 11 = adenosine; 12 = NAD; 13 = IMP; 14 = GMP; 15 = UMP; 16 = AMP; 17 = GDP; 18 = UDP; 19 = ADP; 20 = GTP; 21 = UTP; 22 = ATP

Table 1. Concentrations of Purine Compounds in Rat Hepatocytes at Anoxia and Reoxygenation (mMol/l cells; mean ± S.D.)

Metabolite	Initial Value	30 min Anoxia	10 min Reoxygenation
ATP	2.16 ± 0.17	0.21 ± 0.12	2.20 ± 0.53
ADP	0.43 ± 0.04	0.25 ± 0.08	0.26 ± 0.11
AMP	0.27 ± 0.03	0.34 ± 0.06	0.22 ± 0.04
IMP	0.31 ± 0.05	1.45 ± 0.09	1.09 ± 0.04
Ado	0.12 ± 0.02	0.62 ± 0.42	0.32 ± 0.02
Ino	0.06 ± 0.03	0.47 ± 0.02	0.14 ± 0.04
Guo	0.03 ± 0.02	0.13 ± 0.03	0.11 ± 0.01
Hyp	<0.01	0.60 ± 0.16	0.07 ± 0.03
GTP	0.03 ± 0.02	0.09 ± 0.02	0.21 ± 0.01
GDP	0.06 ± 0.03	0.10 ± 0.09	0.17 ± 0.05
Sum of AdN	2.86 ± 0.24	0.82 ± 0.26	2.67 ± 0.67
ATP:ADP	5.02 ± 0.39	0.86 ± 0.49	8.55 ± 2.05
ATP:AMP	7.96 ± 0.62	0.63 ± 0.36	10.04 ± 2.41
GTP:GDP	5.09 ± 0.27	0.88 ± 0.15	1.23 ± 0.07

mals were anaesthetized with diethyl ether and the small intestine was prepared. After tissue sampling the superior mesenteric arterial vessel was ligated by a microvascular clamp. The clamp was released after one hour and the small intestine was reperfused. All samples were immediately frozen in liquid nitrogen.

Nucleotides, nucleosides and nucleobases were determined in the perchloric acid extract by a reversed-phase ion-pair HPLC as previously reported (Werner et al., 1988). In order to achieve improved separation of nucleosides and free bases the linear gradient was varied in the following way: a concave gradient from Buffer A (10 mM ammonium phosphate, pH 5.5, 2 mM tetrabutylammonium phosphate) to 80 percent Buffer B (which contains Buffer A supplemented with 20 percent acetonitrile) was applied within 12 minutes and continued by the isocratic mode.

In Fig. 1 a chromatogram is depicted showing sufficient separation of 22 compounds in the extract from small intestine which was subjected to 60 minutes of ischemia. An excellent separation of free bases and nucleosides was accomplished. Allopurinol was not detectable obviously due to its conversion to oxypurinol (Peak 5).

The radical scavenging properties of allopurinol and oxypurinol were studied by electron paramagnetic resonance measurements using two model systems with different radical generating sources. The EPR spectra were recorded on a Varian Spectrometer E 4.

RESULTS AND DISCUSSION

Data on the concentrations of purine compounds in rat hepatocytes and their changes during anoxia and reoxygena-

Table 2. Contribution of Xanthine Oxidase (XOD) and Urate Oxidase (UO) Mediated Hydrogen Superoxide Production to the Entire Oxygen Consumption in Severe Hypoxic Rat Liver (µMol/g w.w. x h)

	O_2 Consumption	H_2O_2 Production		
		XOD	UO	Total
Normoxia	26.4	0.2	0.4	0.6
Hypoxia	5.5	1.1	1.8	2.9

tion are summed up in Table 1. In accordance with observations on hepatocytes and intact rat liver (Marubayashi et al., 1980; Vincent et al., 1982; Adkison et al., 1986, Siems et al., 1988) there is a fall in ATP, GTP as well as in the ratios ATP:ADP, ATP:AMP and GTP:GDP during 30 min of anoxia; simultaneously a distinct accumulation of IMP, adenosine, inosine, guanosine and hypoxanthine happens. During 10 min. reoxygenation most of these changes are compensated. The exceptionally high IMP concentration may be explained by the impairment of AMP resynthesis via adenylosuccinate due to a limited supply of aspartate.

The contribution of the oxygen consumed by the xanthine oxidase and urate oxidase mediated purine degradation to urate and allantoin in comparison with the entire rate of respi-

Table 3. Concentrations of Purine Compounds in the Rat Small Intestine at Ischemia and Reperfusion
A. Control Group
B. Animals Pretreated with Allopurinol
(nMol/mg protein; mean ± S.D.)

Metabolite	Initial Value	60 min Ischemia	20 min Reperfusion
A)			
ATP	10.4 ± 1.9	3.7 ± 1.8	3.7 ± 2.7
ADP	3.2 ± 1.5	6.0 ± 1.8	6.3 ± 2.2
AMP	2.6 ± 0.6	3.7 ± 3.0	6.6 ± 1.9
GTP	9.3 ± 2.3	5.3 ± 2.8	2.7 ± 2.2
GDP	2.3 ± 0.9	2.9 ± 1.5	3.9 ± 1.4
Hyp	0.04 ± 0.01	0.29 ± 0.09	0.21 ± 0.08
Uric acid	3.8 ± 1.8	3.8 ± 3.3	4.0 ± 1.8
B)			
ATP	11.2 ± 4.6	2.2 ± 1.6	8.1 ± 4.1
ADP	5.5 ± 2.0	4.0 ± 1.4	3.1 ± 1.4
AMP	5.1 ± 2.2	4.6 ± 4.2	2.2 ± 1.4
GTP	9.0 ± 3.4	2.6 ± 1.9	7.6 ± 1.9
GDP	7.0 ± 2.8	3.2 ± 1.7	3.7 ± 1.0
Hyp	0.2 ± 0.1	1.7 ± 1.1	0.7 ± 0.6
Uric acid	3.1 ± 1.9	3.2 ± 3.5	1.0 ± 0.8

ration was investigated in the severely hypoxic rat liver (pO_2 = 5 kPa). For the calculation of the cytosolic $O_2^{\cdot -}$ and hydrogen superoxide production by xanthine oxidoreductase the following items were considered: (i) shares of O-type and D-type of the enzyme, (ii) inhibition of the D-type activity by free NADH estimated from the lactate:pyruvate ratio. The contribution of the O-type increased from 20 % to 27 % within one hour of hypoxia. The D-type was hardly inhibited by the elevated free NADH concentration (<10 %). The rate of hydrogen superoxide generation by the cytosolic xanthine oxidase increased almost sixfold (Table 1) and that by the peroxisomal urate oxidase 4.5-fold. Thus, in severe hypoxia both enzymes drain for hydrogen superoxide formation about half of the oxygen consumed.

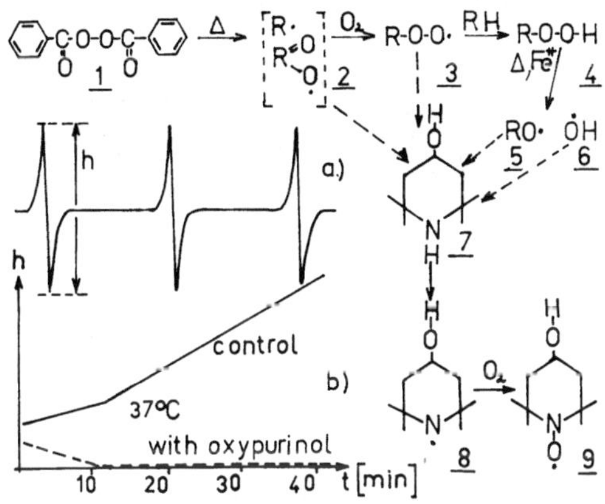

Fig. 2. EPR detection of radicals generated by the thermal decomposition of dibenzoylperoxide using (9) as monitor (a). The effect of oxypurinol is shown in (b).

A synopsis of the alterations in the nucleotide pattern of the ischemic rat small intestine is given in Table 3. The concentrations of ATP and GTP scale down to about one-third within one hour of ischemia and only in the presence of allopurinol recover to about three-forth of the initial values. In the control animals those values remain low. There is a more distinct accumulation of hypoxanthine in the presence of allopurinol.

Elevations of the glutathione disulfide concentrations and those of thiobarbituric acid-reactive substances indicate an oxidative load during the reperfusion of the intestine. Pretreatment with allopurinol diminished these changes (Siems et al., 1988).

The molecular mechanisms of the radical quenching properties of allopurinol and oxypurinol were explored and the results are depicted in figures 2 and 3. Dibenzoylperoxide (1) in aqueous solution (Fig. 2) decomposes at 37° C in a thermic reaction yielding the radical species (2), (3), (5), and (6). By hydrogen abstraction from the indicator molecule (7) the intermediate (8) is obtained which after the reaction with biatomic oxygen yields the nitroxyl radical (9)

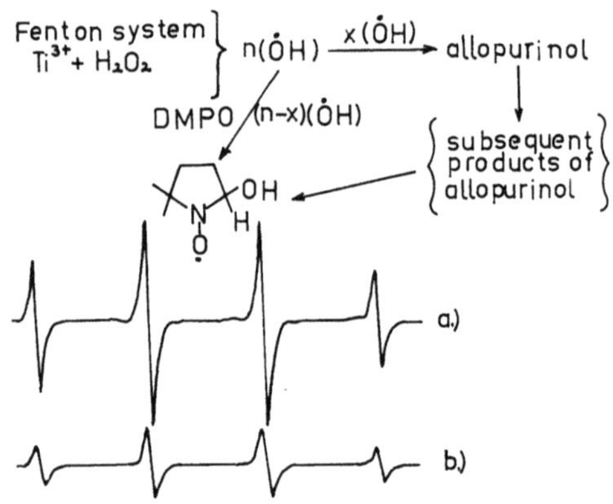

Fig. 3. OH• generation in a Fenton type reaction and trapping by 5,5-dimethyl-1-pyrroline-N-oxide (a). The effect of allopurinol is shown in (b).

(Stoesser et al., 1988). This serves as a monitor for the short-living radical products (2), (3), (5), and (6) as illustrated by the EPR spectrum in the insert a. The evolution of the indicator radical concentration with time is presented in insert b of Fig. 2. The scavenging effect of oxypurinol on the short-living radicals is obvious extending recent observations on OH• radical scavenging properties of oxypurinol (Moorhouse et al., 1987).

Up to this point it is not elucidated whether allopurinol/oxypurinol per se or radical induced products of them are the effective agents. In the model system presented in Fig. 3 OH• radicals were generated by a Fenton-type reaction and monitored as the spin-adduct of DMPO (part a). In the presence of allopurinol the amplitudes decreased in the course of several minutes (part b). If allopurinol was added after starting the reaction the DMPO-OH spectrum retained. Thus, it is concluded that a fraction of OH• radicals induces reactions of allopurinol yielding long-living products. These products react with DMPO-OH with loss of its radical properties. The novel observation summarized in Fig. 3 may contribute to a thorough understanding of the multifacetted cytoprotective action of allopurinol and oxypurinol.

SUMMARY

The relationship between nucleotide catabolism and generation of activated oxygen species was investigated in liver, hepatocytes and small intestine of rats. In severe hypoxia nucleotide degradation via xanthine oxidase and urate oxidase requires about half of the oxygen consumed. Data on the changes of nucleobase compounds in rat hepatocytes and small intestine during ischemia and reoxygenation and the effects of allopurinol and oxypurinol thereon are presented. From EPR measurements it is concluded that OH• radicals induce reactions of allopurinol yielding long-living products which are able to react with DMPO-OH with loss of its radical properties.

REFERENCES

Adkison, D., Höllwarth, M. E., Benoit, J. N., Parks, D. A., McCord, J. M., and Granger, D. N., 1986, Role of free radicals in ischemia-reperfusion injury to the liver, Acta Physiol. Scand., Suppl. 548:101-107.

Das, D. K., Engelmann, R. M., Clement, R., Otani, H., Prasad, M. R., and Rao, P. S., 1987, Role of xanthine oxidase inhibitor as free radical scavenger: A novel mechanism of action of allopurinol and oxypurinol in myocardial salvage. Biochem. Biophys., Res. Commun. 148: 314-319.

Marubayashi, S., Takenaka, M., Doki, K., Ezaki, H. and Kawasaki, T., 1980; Adenine nucleotide metabolism during hepatic ischemia and subsequent blood reflow periods and its relation to organ viability. Transplantation 30:294-296.

McCord, J. M., and Roy, R. S., 1982; The pathophysiology of superoxide: roles in inflammation and ischemia, Can. J. Physiol. Pharmacol. 60:1346-1352.

Moorhouse, P. C., Grootveld, M., Halliwell, B., Quinlan, J. G., and Gutteridge, J. M. C., 1987, Allopurinol and oxypurinol are hydroxyl radical scavengers. FEBS-Letters 213:23-28.

Peterson, D. A., Kelly, B., and Gerrard, J. M., 1986, Allopurinol can act as an electron transfer agent. Is this relevant during reperfusion injury? Biochem. Biophys. Res. Commun. 137:76-79.

Räder, L., Siems, W., Müller, M., and Gerber, G., 1985, Formation of activated oxygen in the hypoxic rat liver, 1985, Cell Biochem. Funct. 3: 289-296.

Siems, W., Mielke, B., Müller, M., Heumann, C., Räder, L., and Gerber, G., 1983, Status of glutathione in the rat liver. Enhanced formation of oxygen radicals at low oxygen tension. Biomed. Biochim. Acta 42:1079-1089.

Siems, W., Kowalewski, J., Jakstadt, M., Schreiter, C., Schimke, I., and Gerber, G., 1988, Effects of allopurinol on the glutathione status and thiobarbituric acid reactive substances (TBA-RS) in ischaemic rat intestine, 1988, Pharmazie, 43:438-439.

Stoesser, R., Thurner, J. U., Warncke, W., Tomaschewski, G., and Taplick, T., 1988, Reaktionen aliphatischer Diazoverbindungen; photo- und radikalinduzierte Reaktionen unter Beteiligung von Sauerstoff, 1988, Z. Chem., accepted for publication.

Vincent, M.-F., Van den Berghe, G., and Hers, H.-G., 1982, The pathway of adenine nucleotide catabolism and its control in isolated rat hepatocytes subjected to anoxia, Biochem. J. 202:117-123.

Werner, A., Siems, W., Gerber, G., Schmidt, H., Gruner, S., and Becker, H., 1988, Application of reversed-phase and ion-pair HPLC for investigation of nucleotide metabolism in erythrocytes, reticulocytes and tumour cells. Chromatographia 25:237-241.

LOCALIZATION OF XANTHINE DEHYDROGENASE IN LIVER

Y.Hattori, T.Nishino*, C.Usami*, K.Tsushima*, M.Ichikawa**, and A.Ichikawa**.

Dept. of Neurosurgery, Dept. of Biochem.*, Dept. of Anatomy**, Yokohama City Univ. Sch. of Med., Kanazawa-ku, Yokohama 236, Japan

INTRODUCTION

Xanthine dehydrogenase has been reported to be present in the endothelial cells in beef or rat liver[1] and also in peroxisomes of the hepatic cells[2]. However, cellular and intracellular localization in liver is still obscure because the enzyme was recovered mostly in 100,000 x g supernatant solution of rat or chicken liver homogenate[3-5]. Immunohistochemical method has gained importance in the detection and localization of polypeptides and proteins, including numerous enzymes, because of their high specificity. In chickens, which are uricotelic animals, the specific activity of xanthine dehydrogenase in the liver was increased[6-14] upon increase of the content of protein in the diet. Thus it is interesting to know the cellular or subcellular localization of this enzyme after changing the content of dietary protein. In the present study we reexamined localization of xanthine dehydrogenase in the tissue of chicken liver by immunohistochemical method and also reexamined the enzyme activity in isolated hepatocytes.

MATERIALS AND METHODS

Preparation of antibody against xanthine dehydrogenase

Chicken liver xanthine dehydrogenase was purified by the method of Nishino[4]. A sterile 0.1% solution of the purified chicken liver xanthine dehydrogenase, emulsified in an equal volume of Freund's complete adjuvant, was injected intracutaneously into male white rabbits (New Zealand White rabbits, weighing 2-2.5kg). Each rabbit received 1 mg xanthine dehydrogenase at 4 different sites. After 5 weeks from first injection, 0.1mg (in 1ml) of xanthine dehydrogenase was injected intravenously once a week as a booster shot. After 3 times of booster injections (9 weeks after first injection), the obtained antisera (up to 1:64 dilution) and purified enzyme gave a distinct single precipitin line in Ouchterlony double diffusion reactions[15] against purified chicken liver xanthine dehydrogenase or crude extract.

Immunohistochemical experiment

Animals. Chickens (male, Dekalb strain) were obtained from a local dealer and maintained for three weeks on a commercial diet of 20% protein before synthetic diets were fed. Chickens were divided into three groups. After starvation for 24 hrs these groups were fed on different synthetic diets ad libitum for 3 days. 75%, 20%, and 5% of protein diets were given to these groups.

Fixation of the tissue. The liver of chickens (weighing about 150g) was perfused through the left ventricle with a rinsing solution (Heparin 0.25g, Procain HCl 5.0g in 1.000ml, pH 7.14, 346mOs/kg) for 1 min and followed by infusion of a fixing solution (periodate-lysine-paraformaldehyde fixing solution: 0.01M $NaIO_4$, 0.075M lysine, 0.075M Phosphate buffered 4% paraformaldehyde pH 5.78, 1298mOs/kg) for 5 min. The hepatic tissue was excised and cut into small blocks. They were stocked in the same fixing solution for 6 hrs followed by washing with running water overnight, dehydrated with ethanol, and embedded in paraffin.

Immunostaining. Five micrometer-thick sections of the liver were incubated in anti-xanthine dehydrogenase antisera at room temperature (25°C) for 24 hrs, washed in five change of PBS for 15 min, and then reacted with the peroxidase-labeled anti-rabbit gamma-globulin for 30 min. After washing with PBS, the sections were incubated in a mixture of 20 mg 3,3'-diamino benzidine and 0.003% hydrogen peroxide in 100ml of 0.05M Tris buffer(pH 7.6) at room temperature for 10 min, washed in distilled water, and counterstained with hematoxylin.

Isolation of hepatocytes

The procedure of isolation of hepatocytes described by Mapes and Krebs[16] was adopted with some modifications. Chickens (4 weeks, weighing 150-300g) were starved for 16 hrs prior to sacrifice. Chicken was anesthetized with intraperitoneal injection of Nembutal (5mg/100g). The anterior thorax was opened by two longitudinal incisions. The chest wall was cut off and a loose ligature was placed around the inferior vena cava close to the heart. The silicone rubber cannula was pushed into the inferior vena cava from right atrium. The ligature around the cannula was secured. Then perfusion was started and portal vein was cut. The initial volume of the perfusion medium (Ca^{++} free Krebs-Henseleit buffer) was 150ml. The first 50ml of the medium passed through the liver were discarded to remove the blood. At this point 30mg of collagenase suspended in small volume of the perfusion medium were added. The fluid in the abdominal cavity was collected and returned to the reservoir. The perfusion was continued for 20-25 min. After the liver became soft and slightly swollen at the end point, the liver was transferred to a petri dish and HEPES buffer(pH 7.4) was added to the dish. The liver cells were liberated from the connective-vascular tissue by careful raking with comb(about 3 mm between teeth). The connective tissue was discarded and the cell suspension was filtered through a nylon mesh. The filtrate was cooled in an ice bath and centrifuged at 50 x g for 2 min at 4°C and the cells were resuspended in the ice cold HEPES buffer(pH 7.4). This cell suspension was washed 4 times with HEPES buffer(pH 7.4).

Preparation of crude extracts. Isolated hepatocytes and liver were homogenized with a Polytron Tissue Homogenizer for 20 sec. The homogenates were centrifuged at 100,000 x g for 60min at 4°C. The supernatants were used for determination of enzyme activity and protein concentration.

Assay of xanthine dehydrogenase. Xanthine dehydrogenase activity was assayed at 25°C in the mixture described previously[4] by following the change of difference between absorbance at 295 nm and 310 nm using a dual wavelength spectrophotometer(Hitachi 557 equipped with temperature controler). One unit of activity was defined as the formation of 1 micro-mol of uric acid per min.

Protein determination. Protein concentration of the extract was determined by the method of Lowry et al.[17] with crystalline bovine serum albumin as a standard.

RESULTS AND DISCUSSION

Characterization of antibody

Ouchterlony double diffusion analysis gave a single precipitin line when the antisera was allowed to react with a purified xanthine dehydrogenase or crude extracts obtained from chicken liver.

Localization of xanthine dehydrogenase in chicken liver

In light microscopy of the liver of chickens, fed 20% and 75% protein diets, distinct positive reaction of xanthine dehydrogenase was detected in hepatocytes, Kupffer cells and endothelial cells with immunoperoxidase technique. Staining pattern in hepatocytes was rather diffuse, suggesting the existence of the enzyme in cytosol. It should be noted that staining intensity of the samples of chicken fed 20% and 75% protein diet was higher than in chickens fed a 5% diet. Similar results were also recognized in Kupffer and endothelial cells. Since the activity of xanthine dehydrogenase in the hepatic tissue is known to be changed dramatically depending on the protein contents in diet, the present result may be due to the changes in xanthine dehydrogenase content in these cells examined.

Determination of xanthine dehydrogenase activity in isolated hepatocytes

Inspection by light microscopy of the cells obtained by the procedure described in Materials and Methods showed that most of the obtained cells had similar shape and size, suggesting that the obtained cells are mostly hepatocytes. Viability of cells examined by dye (trypan blue) exclusion test[18] was found to be more than 80%. Xanthine dehydrogenase activity of the obtained cells was determined in the cell homogenate and 100,000 x g supernatant solution of the cells. It was found that xanthine dehydrogenase activity was detected in the supernatant solution of cell homogenate in comparable amount to that in liver homogenate. As shown in table 1, the specific activity of supernatant solution is about 80% of that of liver supernatant solution. It would be possible that some amount of enzyme might be lost during the procedure of cell isolation in such a way as damaging the cell membrane to cause a selective leakage of cytosolic protein from the cells. It should be noted that most of the activity of xanthine dehydrogenase was recovered in a supernatant solution, suggesting the localization of this enzyme in cytosol fraction of hepatocytes, although we have not yet examined the existence of this enzyme in peroxisome[2]. It has been reported that the enzyme in beef is localized in endothelial cells in most of the tissues including liver[1]. We could detect the staining not only in endothelial cells, but also hepatocytes and Kupffer cells.

ACKNOWLEDGEMENTS

This work was supported by a Grant-in-Aid for scientific research from Japanese Ministry of Education, Science and Culture (62480135) and by a Research Grant for Intractable Diseases from Japanese Ministry of Health and Welfare.

Table 1. Activity of xanthine dehydrogenase

	Activity of xanthine dehydrogenase in 100,000 x g supernatant solution of homogenate of isolated cells or perfused liver. Data were obtained from separate experiments. Specific activity (U/mg protein)
Isolated cells	6.06×10^{-3}
	7.06×10^{-3}
Perfused liver	7.69×10^{-3}
	8.84×10^{-3}

REFERENCES

1. E.D.Jarasch, C.Grund, G.Bruder, H.W.Heid, T.W.Keenan and W.W.Franke, Localization of xanthine oxidase in mammary-gland epithelium and capillary endothelium, Cell 25:67 (1981)
2. S.Angermuuller, G.Bruder, A.Voulkl, H.Wesch, H.D.Fahimi, Localization of xanthine oxidase in crystalline core of peroxisomes. A cytochemical and biochemical study, Eur. J. of Cell Biol. 45:137 (1987)
3. T. Ikegami and T. Nishino, The presence of desulfo xanthine dehydrogenase in the purified and crude enzyme preparations from rat liver, Arch. Biochem. Biophys. 247:254 (1986)
4. T.Nishino, Purification of hepatic xanthine dehydrogenase from chicken fed a high protein diet, Biochim. Biophys. Acta 341:93 (1974)
5. C.F.Strittmatter, Studies on avian xanthine dehydrogenase, J. Biol. Chem. 240:2557 (1965)
6. C.Remy and W.W.Westerfeld, The effect of diet on xanthine dehydrogenase in chicken tissues, J. Biol.Chem. 193:659 (1951)
7. G.Litwack and H.Fisher, Role of essential amino acids in the early formation of avian liver xanthine dehydrogenase activity, Am. J. Physiol. 191:355 (1957)
8. W.W.Westerfeld, D.A.Richert and A.C.Hermans, Glowth and liver xanthine dehydrogenase in chicks and poults fed casein or soy protein diets, J. Nutr. 76:475 (1962)
9. E.D.Corte and F.Stripe, Regulation of xanthine dehydrogenase in chick liver, Biochem. J. 102:520 (1967)
10. R.W.Scholz and W.R.Featherston, Effect of alterations in protein intake on liver xanthine dehydrogenase in the chicken, J. Nutr. 95:271 (1968)
11. W.R.Featherston and R.W.Scholz, Changes of liver xanthine dehydrogenase and uric acid excretion in chicks during adaptation to high protein diet, J. Nutr. 95:393 (1968)
12. R.W.Scholz and W.R.Featherston, Dietary-induced depletion and repletion of avian liver xanthine dehydrogenase, J. Nutr. 98:193 (1969)
13. R.Itoh and K.Tsushima, Comparison of adaptations to diet of enzymes involved in uric acid production from IMP in chickens and rat, J. Biochem. 75:715 (1974)
14. R.Itoh, T.Nishino, C.Usami and K.Tsushima, An immunochemical study of the changes in chicken liver xanthine dehydrogenase activity during dietary adaptation, J. Biochem. 84:19 (1978)

15. O.Ouchterlony and L.-A.Nilsson, pp. 32.1-32.17, in: "Handbook of Experimental Immunology, Vol.1," D.M.Weir ed., Blackwell Sci. Pub., Oxford (1986)
16. J.P.Mapes and H.A.Krebs, Rate-limiting facter in urate synthesis and gluconeogenesis in avian liver, Biochem. J. 172:193 (1978)
17. O.H.Lowry, N.J.Rosebrough, A.L.Farr and R.J.Bray, Protein measurement with the folin phenol reagent, J. Biol. Chem, 237:3714 (1951)
18. M.K.Patterson, Jr., Measurement of Growth and Viability of cells in culture, in: "Methods in enzymology," W.B.Jakoby and I.H.Pastaned., Academic Press, New York, 108,141 (1979)

DETERMINATION OF ALLANTOIN IN MAMMALIAN TISSUES

Anna Di Stefano, Maria Pizzichini, Germano Resconi* and Enrico Marinello

Inst. of Biological Chemistry, University of Siena
* Inst, of Mathematics, University of Brescia, Italy

INTRODUCTION

Allantoin is the final product of purine degradation in most mammals and is considered "the key compound" of purine catabolism. The other important product of purine catabolism is uric acid, particularly in man.

A specific assay for allantoin in biological extracts is not yet available; most methods reported up to now are based on the Rimini-Schryver reaction[1,2,3]. This procedure consists in three steps: a) alkaline hydrolysis of allantoin to allantoic acid; b) acid hydrolysis of allantoic acid with formation of glyoxylic acid and two moles of urea; estimation of glyoxylic acid as a dinitrophenylhydrazine derivative. Rimini-Schryver reaction is not specific because other tissue components of low molecolar weight (lactic and uric acid, glycine, glucose) interfere with glyoxylic determination.

In order to overcome this interference, we have attempted: 1) to estimate an other product of acid hydrolysis of allantoic acid,i.e., urea; 2) to separate allantoin from interfering substances in tissue prior to execution of the Rimini-Schryver reaction.

We have also developed a specific procedure for purification of allantoin and uric acid, which permits the calculation of their labelling with isotopes for kinetic studies in different tissues.

MATERIALS AND METHODS

Male albino Wistar rats, weighing 250 g and on a standard diet were used. All the animals were fasted for 12 hours before decapitation. Immediately after killing, the rat liver was rapidly removed, washed in a saline solution and homogenized at 5 °C with 5 % trichloracetic acid (1 g + 5 ml 5 % TCA). After centrifugation and neutralization, the supernatant was utilized for determination of allantoin using the three following procedures:

A) <u>Conventional procedure</u>

Determination of allantoin using the Rimini-Schryver reaction as modified by Dietriech and Borries[4] which involves: treatment with 0.1 N alkali (7 min at 100 °C) and formation of allantoic acid; subsequent treatment with 0.025 N HCl and production of urea and glyoxylic acid; final determination of the latter compound as 2,4-dinitrophenylhydrazine derivative.

B) Enzymatic determination of urea (derived from allantoin) with urease.

The supernatant was treated with urease for 30 min at 37 °C to destroy endogenous urea, then submitted to alkali and acid hydrolysis as in Rimini-Schryver reaction; finally urea was determined, instead of the glyoxylic acid. The urea thus derived from allantoin was converted - by further treatment with urease - to NH_3, which was determined by the colorimetric procedure of Berthelot as modified by Chaney and Marbach[5](scheme 1).

C) Treatment with mercuric acetate

The liver supernatant was submitted to acid hydrolysis (1 h at 100 °C in H_2SO_4 1 N); after neutralization, allantoin, uric acid and free purine bases derived from nucleotides, were precipitated by addition of mercuric acetate solution (1 g of mercuric acetate and 10 g of sodium acetate in 100 ml , v/v).

After standing overnight, the flocculent precipitate was collected by centrifugation and washed three times with distilled water. The precipitate was dried *in vacuo* over P_2O_5, then ground up with water and decomposed with hydrogen sulfide[6].

After removal of mercuric sulfide, the solution was aerated and the allantoin determined as reported in procedure A.

Purification of allantoin and uric acid

One aliquot of the sample obtained after treatment with mercuric acetate was applied to a column (20 X 0.4 cm) of Dowex (AG 50W-X8, 200-400 mesh, cation exchange resin), equilibrated with distilled water (flow rate 12 ml/h; fractions every 10 min). Allantoin and uric acid were eluted with distilled water, purine bases with HCl 6 N.

The fractions containing allantoin (3th-6th) were concentrated and brought to pH 2, again submitted to chromatography on a Dowex (1-X8, 200-400 mesh, anion exchange resin) column (20 X 0.4 cm) equilibrated with distilled water; the allantoin was eluted with distilled water; uric acid with 0.02 N HCl and unidentified compounds with 0.06 N HCl.

SCHEME 1

DETERMINATION OF ALLANTOIN WITH UREASE

Rat liver (1 g + 5 ml 5% TCA)

|

Homogenized at 4 °C

|

Centrifuged at 8,000 x g for 15 min

|

Supernatant neutralized with 10 N NaOH

|

Urease 30 min at 37 °C

|

0.1 N NaOH, 7 min at 100 °C

|

18 h of lyophilization for elimination of NH_3

|

brought to pH 1.5 with 10 N HCl, 5 min at 100 °C

|

Neutralized with 10 N NaOH

|

Determination of urea (derived from allantoin) with urease

Table 1 Mean levels of normal rat liver allantoin determined by different procedures. The reported values are the means ± SD of 5 cases.

PROCEDURE	ALLANTOIN CONTENT (µg / g)
Rimini-Schryver	58 ± 3
Rimini-Schryver after precipitation with mercuric acetate	27 ± 1
Enzymatic method with urease	16 ± 9

The fractions containing allantoin (4th- 6th) were then treated with charcoal (Norit A) (1 ml + 10 µl of 50 % Norit A) and the allantoin content was determined as described in procedure A.

Uric acid was determined according to Praetorius and Poulsen[7].

RESULTS

The concentration of allantoin in TCA extracts determined according to Rimini-Schryver reaction was 58 ± 3 µg/g (tab. 1).

The allantoin content, evaluated by the same method, after precipitation with mercuric acetate, was about 50 %, 27 ± 1 µg/g (tab. 1); linearity with tissue concentration was observed.

When a standard solution of allantoin was added to the extract, the mean recovery was 96 ± 3 % . The content of allantoin corrected for the uric acid interference was 23 ± 1 µg / g.

The allantoin was also determined enzymatically with urease. The use of this enzyme is specific , but it requires large amounts of the sample and is effective only after elimination of endogenous urea and NH_3 (scheme 1).

The mean values resulting from this procedure were lower than those obtained after treatment with mercuric acetate (tab. 1); in addition standard deviation of values was high (16 ± 9 µg/g), and recovery of standard allantoin varied (70 ± 15%); this procedure gave often a non linear response when applied to different tissue concentrations, therefore being of limited use.

The purification of allantoin and uric acid is illustrated in the following figures.

Fig. 1 shows the 1st chromatography of liver extracts samples after mercuric acetate precipitation.

Allantoin was still contamined by uric acid and unidentified compounds, which show a maximal absorbance at 260 nm.

The 2nd chromatographic run, reported in Fig.2, separated allantoin from uric acid and other compounds, but some interference was still evident, which was eliminated by treatment with Norit A.

DISCUSSION

From our results it is evident that the allantoin content in rat liver differs according to procedure adopted. The following conclusions can be drawn from the our experiments:

1) Rimini-Schryver reaction, when directly applied to liver extracts, gives excessively high values, which are lower after precipitation with mercuric acetate.

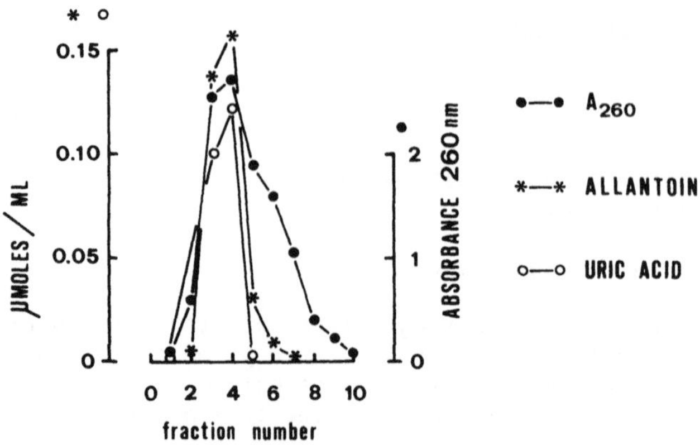

FIG. 1 Cation-exchange chromatography of liver exctracts samples after mercuric acetate precipitation on a Dowex (AG 50W-X8). For experimental conditions see Materials and Methods. The allantoin peak was contaminated by uric acid and unidentified compounds (A_{260})

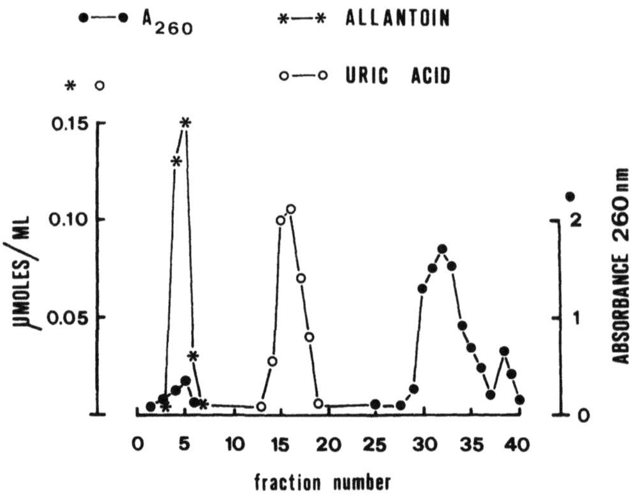

FIG. 2 Anion-exchange chromatography on a Dowex (1-X8) of fractions containing allantoin and uric acid, concentrated and brought to pH 2. For experimental conditions see Materials and Methods. The allantoin was separated from uric acid and unidentified compounds (A_{260})

The difference is obviously due to the presence of interfering compounds, as already reported[1].

2) The enzymatic method based on urease indicates that the "true" values of liver allantoin are lower, but neither the accuracy, precision nor the reproducibility are adequate.

3) We conclude that the Rimini-Schryver reaction is still a suitable procedure for a rapid and accurate determination of allantoin in biological tissues, provided that a preliminary, specific treatment of the extract with mercuric acetate is carried out, in order to handle a more purified preparation.

Further purification by ion exchange chromatography guarantees excellent recovery both of allantoin and of uric acid and permits the determination of their labelling with isotopes for kinetics studies in different tissues.

The chromatographic procedure optimized in this study can be used for the purification of allantoin and uric acid in other tissues.

REFERENCES

1. E. Young and C.F. Conway, On the estimation of allantoin by the Rimini-Schryver reaction, J. Biol. Chem., 142: 839 (1942).
2. R. Borchers, Allantoin determination, Anal. Biochem., 79: 612 (1977)
3. M. Grootveld and B. Halliwell, Measurement of allantoin and uric acid in human body fluids, Biochem. J., 243: 803, (1987)
4. L. S. Dietriech and E. Borries, On the determination of xanthine oxidase activity in animal tissues, J. Biol. Chem., 208: 287 (1954).
5. A. L. Chaney and C.P. Marbach, Modified reagents for determination of urea and ammonia, Clin. Chem., 8 (2): 130 (1962).
6. E. D. Korn, Excretory products of purine and pyrimidine metabolism, in " Methods in Enzymology", S. P. Colowick and N. O. Kaplan, ed, Academic Press Inc., Publishers, New York 4: 631 (1957)
7. E. Praetorius and H. Poulsen, Enzymatic determination of uric acid with detailed directions, Scandinavian J. Clin. Lab. Inv., 5: 273 (1953)

CHARACTERIZATION OF NUCLEOSIDE TRANSPORT DURING LEUKEMIC CELL DIFFERENTIATION

Tetsuya Takimoto, Toshiyuki Kitoh, Akihiko Tanizawa, Yuichi Akiyama, Yukio Kiriyama, Masaru Kubota, and Haruki Mikawa

Kyoto university, Department of Pediatrics, Kyoto Japan

INTRODUCTION

Several nucleoside and nucleobase analogs have been used as therapeutic agents against leukemia cells. Since such agents must be transported into cells to exert its toxicity, it is clinically important to elucidate the mechanisms of nucleoside and nucleobase transport across the cell membrane.

On the other hand, leukemia cells are considered to be a consequence of maturation arrest at different stages of differentiation and leukemic transformation. And it is generally accepted that leukemia cells with mature phenotype are more resistant to anticancer drugs. Therefore, the relationship between cell differentiation and drug resistance is another issue of great interest.

Previously, we reported that human T-lymphoblastoid MOLT-4 cells became resistant to the cytotoxic effects of 1-β-D-arabinofuranosylcytosine (Ara-C) after 12-O-tetradecanoylphorbol-13-acetate (TPA)-induced differentiation and this acquired resistance to Ara-C was attributed to decreased membrane transport (1).

In the present study, we further examined and characterized the change in nucleoside transport during cellular differentiation.

METHODS

Cell Culture

The human T-lymphoblastoid cell line MOLT-4 and promyelocytic cell line HL-60 were cultivated in RPMI 1640 medium supplemented with 10% heat inactivated fetal bovine serum, 2mM L-glutamine, penicillin (100U/ml), and streptomycin (100μg/ml). Cultures were maintained at 37°C in a humidified atomosphere of 95% air/5% CO_2. The cells regularly checked for Mycoplasma contamination.

Differentiation Induction

MOLT-4 cells were suspended at a density of $2x10^5$/ml with 10^{-8}M TPA and incubated for 4 days. For determination of cellula differentiation, we evaluated the activity of purine degradativ enzymes, adenosine deaminase, purine nucleoside phosphorylase, and 5'-ectonucleotidase, since it has been shown that different stages during T-cell maturation are characterized and defined b specific patterns of these enzymes (2). The details of the results was demonstrated elsewhere (1). In the case of HL-60, cells ($1x10^5$/ml) were exposed to 1μM retinoic acid (RA) for 4 days. After that, nitroblue tetrazolium dye reduction was deter mined as an indicator for differentiation. More than 90% of RA-treated cells showed a positive reaction with the dye.

Ara-C Sensitivity

Cells ($2x10^5$/ml) were cultivated in media containing various concentration of Ara-C. After 48hr of cultivation, the concentration required to inhibit growth by 50% (ID_{50}) was determined by trypan blue dye exclusion.

Determination of Ara-C Transport

To measure the rate of Ara-C transport, we preincubated th cells ($4x10^6$/ml) at 37°C for 10 min. Transport of Ara-C was started by rapid addition of [5-^{3}H]Ara-C (0.6Ci/ml) to a final concentration of 1μM. At various intervals, 0.5ml of samples of the cell suspention were removed and placed immediately in micr centrifuge tubes containing 1ml silicon oil-paraffin (specific gravity 1.03g/ml). The tubes were centrifuged at 9000xg for 10 sec. The radioactivity in the cell pellets was counted in a liq uid scintillation after solubilization with NCS.

RESULTS

Drug Sensitivity

The sensitivity to Ara-C after 48 hr of exposure was shown in Table 1. The ID_{50} values of Ara-C were raised about 2400-fold for TPA-treated MOLT-4 and about 220-fold for RA-treated HL-60 cells compared to those of the control cells.

Ara-C Transport

A representative time course of Ara-C transport in MOLT-4 and HL-60 cells before and after differentiation induction was shown in Fig.1. After cells were treated with differentiation inducers, the transport of Ara-C was drastically decreased. Ara-C accumulation in TPA-treated MOLT-4 cells was only about 15% of the control value at 120 sec. It was also true in RA-treated HL-60 cells, although to a lesser degree.
The similar results were obtained in respect to the transport of the naturally occurring nucleoside, thymidine (data not shown).

Intracellularly, Ara-C is rapidly converted to its monophosphate, Ara-CMP, by deoxycytidine kinase (dCK) and is subsequently phosphorylated to Ara-CTP. Therefore, to examine the effect of these intracellular metabolism on the transport event we evaluated the transport of Ara-C in a mutant of HL-60 cells

Table 1. Ara-C Toxicity before and after Differentiation Induction

		ID_{50} (M)
MOLT-4	Control	6.6×10^{-9}
	+TPA	1.6×10^{-5}
HL-60	Control	3.8×10^{-8}
	+ RA	1.7×10^{-6}

deficient in dCK. These mutants were selected in our laboratory on the basis of being resistant to Ara-C and contained less than 1% dCK activity of wild type cells (3). Fig.2 demonstrates that Ara-C transport in the mutant lacking dCK was almost the same as wild cells.

Effects of Transport Inhibitors

Recently several transport inhibitors such as 4-nitrobenzyl-6-thioinosine (NBTI) and dipyridamole have been used to estimate the characteristics of the nucleoside transport (4). As shown in Fig.3, a preincubation with 10µM NBTI resulted in only about 40% inhibition of Ara-C transport in HL-60 cells,

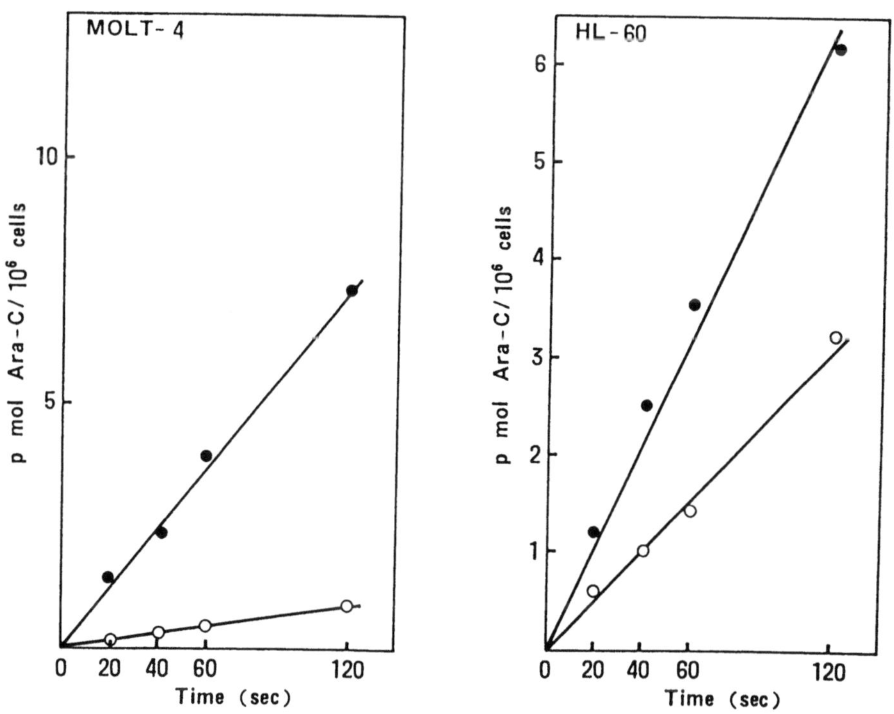

Fig. 1. Accumulation of [5-^{3}H]Ara-C by (left) TPA-treated (O) and untreated (●) MOLT-4 cells, (right) RA-treated(O) and untreated (●) HL-60 cells.

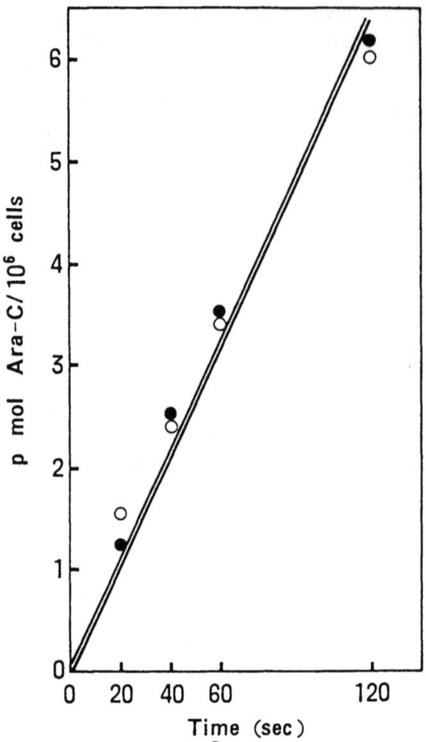

Fig. 2. Accumulation of [5-^{3}H]Ara-C by wild type (●) and a mutant deficient in deoxycytidine kinase (O).

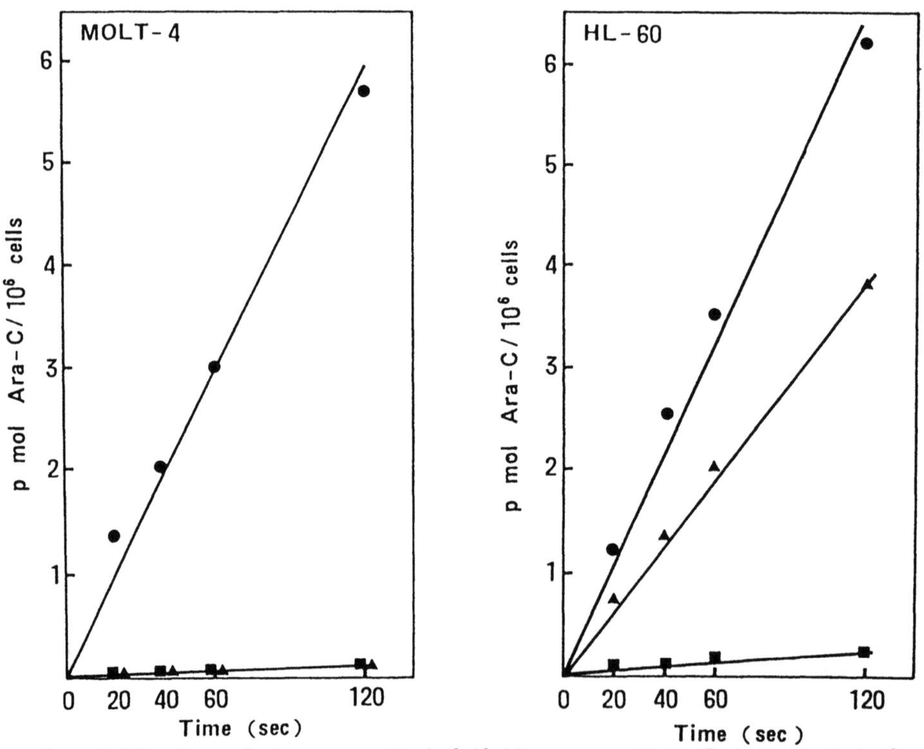

Fig. 3. Effects of transport inhibitors on Ara-C transport in (left) MOLT-4 cells, and (right) HL-60 cells; no treatment (●), NBTI (▲), and dipyridamole (■).

suggesting the presence of nucleoside transport sites resistant to NBTI. In contrast, MOLT-4 cells showed no detectable Ara-C transport up to 120 sec when coincubated with NBTI. 10μM dipyridamole inhibited more than 90% of Ara-C transport in both cell lines.

Such NBTI-resistant transport sites were also retained in HL-60 cells after RA-induced differentiation (Fig.4).

Ara-C Transport in HL-60 Cells in their confluent Phase

Cell differentiation is accompanied by altered growth characteristics;i.e., the retarded growth rate and the changes in cell cycle distribution (1). Thus, we examined the drug transport at different grow stages.

Ara-C transport of HL-60 cells decreased by about 50% in their confluent phase compared to that in logarithmic phase (Fig.5), although the viability of the cells did not change significantly (>95%). In addition, our preliminary examination on cell cycle distribution by cytofluorometry showed a significantly different pattern between cells in logarithmic and confluent phase. Namely, the proportion of cells in the G_1 phase was 56% in logarithmic, and 73% in confluent phse.

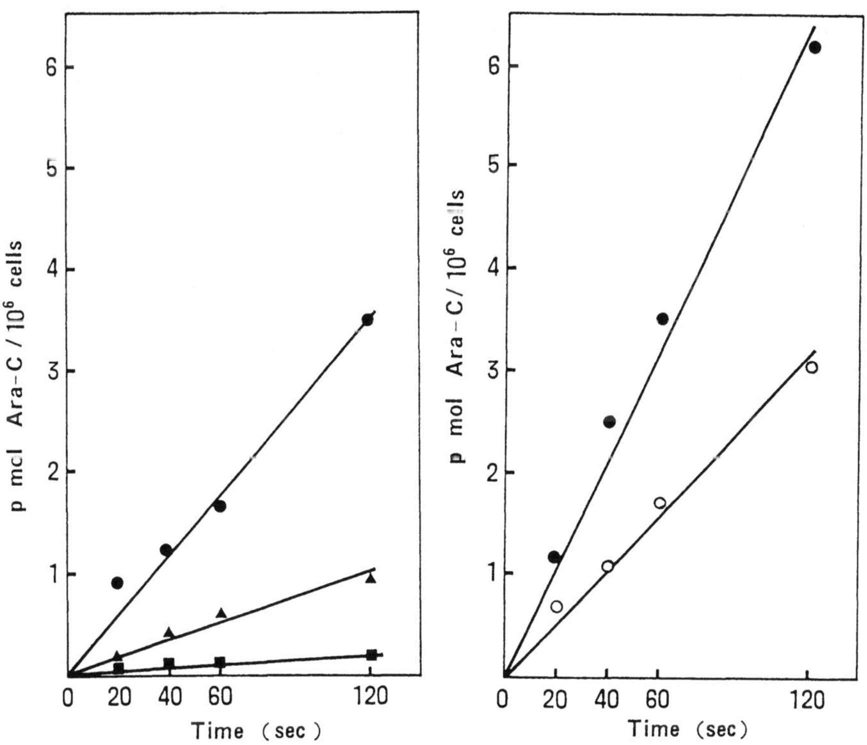

Fig. 4. (left) Effects of transport inhibitors on Ara-C transport in HL-60 cells after RA-induced differentiation; no treatment (●), NBTI (▲), and dipyridamole (■).

Fig. 5. (right) Accumulation of [5-^{3}H]Ara-C by HL-60 cells in their logarithmic (●) and confluent (O) growth phase.

DISCUSSION

Mammalian cells have been postulated to possess a single nucleoside transporter that transports all natural ribo- and deoxyribonucleosides. This transport system can be divided intc two types, namely, NBTI-sensitive and NBTI-resistant, and the former expresses a high-affinity NBTI binding sites (5).

Since various synthetic nucleosides, such as Ara-C, were also transported into cells by this nucleoside specific transport system, the investigation of the characteristics of nuclec side transport is important to clarify the mechanisms of drug resistance.

In the present study, we demonstrated that the transport of Ara-C in human leukemia MOLT-4 and HL-60 cells were decrease after differentiation induction by different inducers and the concomitant increase of the ID_{50} values of Ara-C was observed. HL-60 cells, but not MOLT-4, had NBTI-resistant nucleoside transport sites, and such was retained after cellular differentiation.

Up to now, several studies have shown that the ability in the membrane transport falls during cellular differentiation. Gordon et al. reported the decreased uridine transport in Frier erythroleukemia cells after dimethylsulfoxide exposure (6). More recently, Chen et al. demonstrated that the induction of myeloid differentiation in HL-60 cells by the treatment with 0.8% N,N-dimethylformamide is accompanied by profound decrease in nucleoside transport, which was confirmed by a 19-fold decrease in the number of specific binding site per cell for NBTI (7).

On the other hand, it is noteworthy that cell differentiation is accompanied by the retarded growth rate. Actually, the doubling times of TPA-treated MOLT-4 cells increased 2.6-fold compared to controls. Such alteration in growth characteristics is inevitably accompanied by changes in cell cycle distributior Therefore, the finding that HL-60 cells in their confluent phas revealed a reduction of Ara-C transport to a similar extent as shown with RA-treated cells suggests this decrease of nucleosic transport is, in part, considered to be a function of altered cell cycle distribution after differentiation induction. Namely, an accumulation of cells in the G_1 phase with a concomitant decrease in the S and G_2+M phase.
This is further supported by the observation that the number of NBTI binding sites per cell increased as cells progressed from G_1 through S phase in HeLa cells (8).

Our present results suggests one possible biochemical basi for the decreased sensitivity to various nucleoside analogs in more differentiated leukemia.

REFFERENCES

1. T. Takimoto, M. Kubota, S. Tsuruta, T. Kitoh, A. Tanizawa, Y. Akiyama, Y. Kiriyama, and H. Mikawa, Changes in sensitivity to anticancer drugs during TPA-induced differentiation in a human T-lymphoblastoid cell line (MOLT-4), Leukemia (in press)

2. D.D.F. Ma, T. Sylwestrovics, G. Janossy, and A.V. Hoffbrand, The role of purine metabolic enzymes and terminal deoxynucleotidyltransferase in intrathymic T-cell differentiation, Immunol. Today 4:65 (1983).
3. M. Kubota, T. Takimoto, A. Tanizawa, Y. Akiyama, and H. Mikawa, Differential modulation of 1-β-D-arabinofuranosylcytosine metabolism by hydroxyurea in human leukemic cell lines, Biochem.Pharmac. 37:1745 (1988).
4. M. Kubota, T. Takimoto, T. Kitoh, A. Tanizawa, Y. Kiriyama, Y. Akiyama, and H. Mikawa, Prevention of 1-beta-D arabinofuranosylcytosine toxicity by 4-nitrobenzyl-6-thioinosine or dipyridamole in human leukemia cell lines, Anticancer Res. 8:339 (1988).
5. P.G.W. Plagemann, and R.M. Wohlhueter, Nucleoside transport in cultured mammalian cells; Multiple forms with different sensitivity to inhibition by nitrobenzylthioinosine or hypoxanthine, Biochem.Biophys.Acta 773:39 (1984).
6. P.B. Gordon, and M.S. Rubin, Membrane transport during erythroid differentiation, J.Membrane Biol. 64:11 (1982).
7. S.-F. Chen, J.S. Cleaveland, A.B. Hollmann, M.C. Wiemann, R.E. Parks,Jr., and J.D. Stoeckler, Changes in nucleoside transport of HL-60 human promyelocytic cells during N,N-dimethylformamide induced differentiation, Cancer Res. 46: 3449 (1986).
8. C.E. Cass, E. Dahlig, E.Y. Lau, T.P. Lynch, and A.R.P. Paterson, Fluctuations in nucleoside uptake and binding of the inhibitor of nucleoside transport, nitrobenzylthioinosine, during the replication cycle of HeLa cells, Cancer Res. 39:1245 (1979).

EXPRESSION OF A NOVEL HIGH AFFINITY PURINE BASE TRANSPORT SYSTEM IN MUTANT MOUSE S49 CELLS DOES NOT REQUIRE A FUNCTIONAL NUCLEOSIDE TRANSPORTER

Joanne Beck and Buddy Ullman

Department of Biochemistry
The Oregon Health Sciences University
Portland, Oregon 97201

ABSTRACT

A novel type of somatic mutation that causes the expression of a high affinity purine base permease has been inserted into murine S49 lymphoma cells that are deficient in nucleoside transport. Two classes of mutants expressing this nucleobase permease were generated. The first, as exemplified by the AE_1HADPAB2 cell line, possesses an augmented capacity to transport low concentrations of the three purine bases, hypoxanthine, guanine, and adenine. The second class of mutants, as typified by the AE_1HADPAB5 clone, possesses an augmented capability to translocate low levels of hypoxanthine and guanine, but not adenine. Neither the AE_1HADPAB2 nor the AE_1HADPAB5 cells can transport nucleosides suggesting that the expression of the high affinity base transporter did not revert the mutation in the nucleoside transport system. Fusion of the AE_1HADPAB2 and AE_1HADPAB5 cell lines with wild type cells indicated that the expression of the high affinity base transporter behaved in a dominant fashion, while the nucleoside transport deficiency was a recessive trait. These data suggest that the high affinity purine base transporter of mutant cells and the nucleoside transport function of wild type cells are products of different genes and that expression of the former probably requires the unmasking or alteration of a specific genetic locus that is silent or different in wild type cells.

INTRODUCTION

Nucleoside transport into animal cells is a facilitated diffusion process. Kinetic and genetic evidence support the existence of a single nucleoside transporter which mediates the permeation of all purine and pyrimidine nucleosides across the mammalian plasma membrane (1). Mutant S49 T lymphoblasts have been generated and characterized which are genetically deficient in nucleoside transport capability (2). Most nucleoside transport-deficient S49 cell lines also have a markedly impaired capacity to transport purine nucleobases (3). The genetic deficiency in base transport can be mimicked pharmacologically by either nitrobenzylthioinsoine (NBMPR) or dipyridamole (DPA) (3). These data suggest that the nucleoside transport system also mediates the entry of purine bases into wild type S49 cells.

Recently rare mutants have been isolated from wild type cells that, unlike wild type S49 cells, are capable of proliferation in hypoxanthine-azaserine-DPA containing medium (4). These cells express a high affinity transport system for hypoxanthine, guanine, and adenine, which can be differentiated from the wild type nucleoside-nucleobase carrier by a variety of biochemical criteria. In order to demonstrate that the expression of the high affinity base transporter does not require nucleoside transport activity, this mutation was inserted into nucleoside transport-deficient cells (5). We now observe that the biochemical phenotype of the expressed mutant nucleobase transporter can vary considerably with respect to substrate specificity and kinetic parameters among mutant cell lines. The phenotypes of these cells illustrate a novel type of somatic mutation in animal cells.

MATERIALS AND METHODS

Chemicals and reagents. All chemicals and reagents were of the highest purity available commercially, and their sources have been described (5).

Cell culture conditions. The wild type S49 cell line is a T lymphocytic lymphoma originally derived from a Balb/c mouse. All S49 cell lines were routinely cultured in Dulbecco's modified Eagle medium containing 10% heat inactivated horse serum in a humidified 10% CO_2 atmosphere.

Cell lines. The nucleoside transport-deficient AE_1 cell line was isolated from a mutagenized wild type S49 cell population by virtue of its resistance to adenosine toxicity. Its nucleoside transport-deficient phenotype has been described in great detail (2, 3, 6). The JPA2 cell line which expresses the high affinity nucleobase permease was selected from wild type cells by its ability to salvage low concentrations of hypoxanthine in the presence of an inhibitor of purine biosynthesis (4). JPA2 cells transport nucleosides normally.

To insert the mutation causing the expression of the high affinity purine base permease into nucleoside transport-deficient cells, mutagenized AE_1 cells were plated in semi-solid medium containing 20uM hypoxanthine-11uM azaserine-10uM DPA (7). AE_1 cells were mutagenized by a 3 hr exposure to 2ug/ml N-methyl-N-nitroso-N'-nitroguanidine, after which the mutant phenotypes were allowed to express for 7 generations before selection. Two clones, AE_1HADPAB2 and AE_1HADPAB5, were isolated for detailed characterization.

To generate tetraploid hybrids, a parental diploid cell line was generated, 24.6.1.-TgC, that lacked both cyclic AMP-dependent protein kinase and hypoxanthine-guanine phosphoribosyltransferase activities. The deficiency of the former is a dominant mutation, whereas the lack of the latter enzyme is a recessive trait. 24.6.1.-TgC cells behaved similarly to wild type cells with respect to nucleoside and nucleobase transport phenotype in that they expressed the wild type nucleoside transporter but not the high affinity purine base carrier system.

Growth rate determinations. The abilities of all S49 cells to salvage hypoxanthine in the absence or presence of inhibitors of nucleoside transport were ascertained as previously described (4,5). Hypoxanthine salvage was determined in the presence of azaserine, an inhibitor of phosphoribosylformylglycineamide synthetase (8), an enzyme of the de novo purine pathway. The growth sensitivities of wild type and mutant cells toward cytotoxic nucleosides and nucleobases were compared by previously described methods (2, 6).

Transport measurements. Rates of nucleobase and nucleoside transport over very short time intervals were measured by a modification of the rapid sampling kinetic technique originally described by Aronow et al. (9). Cells were harvested by centrifugation and resuspended at a density of $5x10^7$ cells/ml in Dulbecco's modified Eagle medium lacking horse serum. 100ul of transport medium containing either tritiated nucleobase or nucleoside at 10uCi/ml and varying specific activities were layered over 100ul of an inert oil mixture in a polypropylene Eppendorf centrifuge tube. Transport measurements were initiated by the instananeous addition of the cells and terminated by sedimentation of the cells through the oil layer at 10,000xg in a Microfuge B centrifuge. The upper aqueous layer was removed by aspiration, the oil layer washed twice with 1.0 ml of water or phosphate buffered saline, the oil removed, and the cell pellets lysed in 100ul of 1% Triton X-100. The radiolabel associated with the cells was quantitated by liquid scintillation.

Selection of intraspecies tetraploid hybrids. To determine whether the mutations in the AE_1HADPAB2 and AE_1HADPAB5 cell lines were dominant or recessive, the relevant intraspecies tetraploid hybrids were generated. Parental diploid cells were fused with the 24.6.1.-TgC cells by the polyethylene glycol technique described previously by this laboratory (9). Tetraploid cells were isolated in selective medium containing 1.0mM dibutyryl cyclic AMP, 1mM hypoxanthine, and 11uM azaserine.

RESULTS

Growth phenotypes. The AE_1HADPAB2 and AE_1HADPAB5 clones were isolated from the nucleoside transport-deficient S49 cell line, AE_1, by their ability to grow in medium containing 20uM hypoxanthine-11uM azaserine-10uM DPA. Figure 1 demonstrates that the AE_1HADPAB2 and

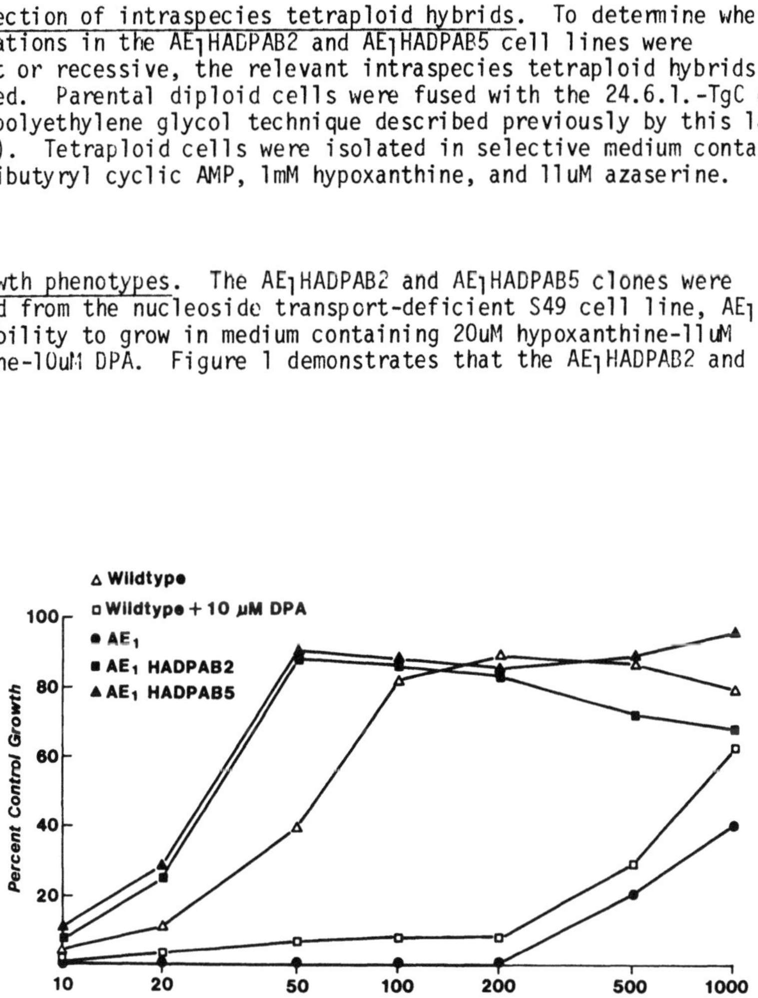

Fig. 1 Abilities of Cells to Salvage Hypoxanthine. The abilities of wild type, AE_1, AE_1HADPAB2, and AE_1HADPAB5 cells to salvage increasing concentrations of hypoxanthine was performed in the presence of 11uM azaserine as described in Materials and Methods (7). The effects of 10uM DPA on the ability of wild type S49 cells to salvage hypoxanthine is also shown.

AE_1HADPAB5 cell lines were far more capable of salvaging low concentrations of hypoxanthine than parental AE_1 cells. Whereas both the AE_1HADPAB2 and AE_1HADPAB5 cell lines grew normally in 50uM hypoxanthine-11uM azaserine, the AE_1 cells could not salvage sufficient hypoxanthine to grow even minimally unless the extracellular hypoxanthine concentration was greater than 200uM. As expected, hypoxanthine salvage in these cells, unlike that in wild type cells (6), was insensitive to inhibition by either 10uM DPA or 10uM NBMPR (data not shown), since the parental line does not possess the DPA-sensitive, NBMPR-sensitive nucleoside transport system. Moreover, both the AE_1HADPAB2 and AE_1HADPAB5 cell lines retained the nucleoside transport-deficient growth phenotype of the parental AE_1 cell line in that they grew normally in the presence of cytotoxic nucleosides, including 100nM 5-fluorouridine, 100nM 5-fluorodeoxyuridine, and 100uM deoxyguanosine, and died in medium containing 0.5mM hypoxanthine-0.4uM amethopterin-16uM thymidine (HAT medium), conditions which require thymidine transport for survival.

Transport phenotypes. A cell line, JPA2, that expresses a high affinity transport system for hypoxanthine, guanine, and adenine (4) was isolated from a wild type cell population under selective conditions similar to those used in the selection of the AE_1HADPAB2 and AE_1HADPAB5 cells. Therefore, the abilities of AE_1HADPAB2 and AE_1HADPAB5 cells to transport 5uM concentrations of these three purine bases were compared to that of wild type and parental AE_1 cells. The data in Figure 2 indicate that the AE_1HADPAB2 and AE_1HADPAB5 cell

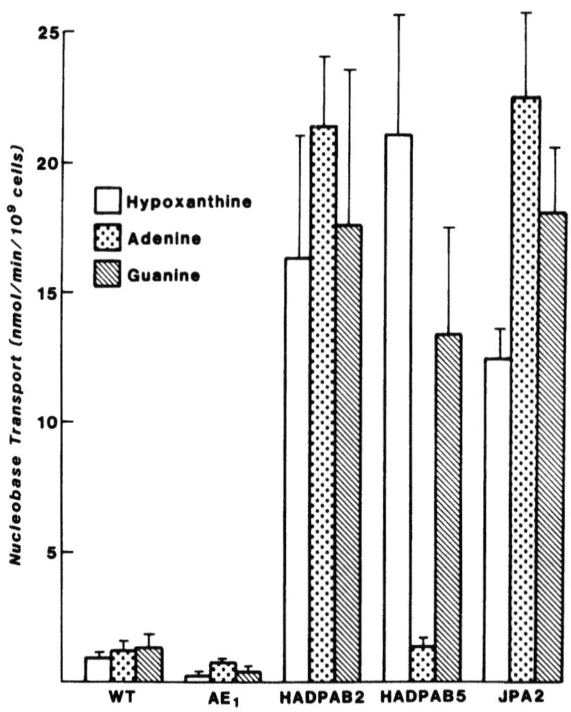

Fig. 2 Purine Transport Studies. The abilities of wild type, AE_1, AE_1HADPAB2, AE_1HADPAB5, and JPA2 (4) cells to transport 5uM concentrations of purine bases was carried out by the rapid sampling technique described in Materials and Methods (7).

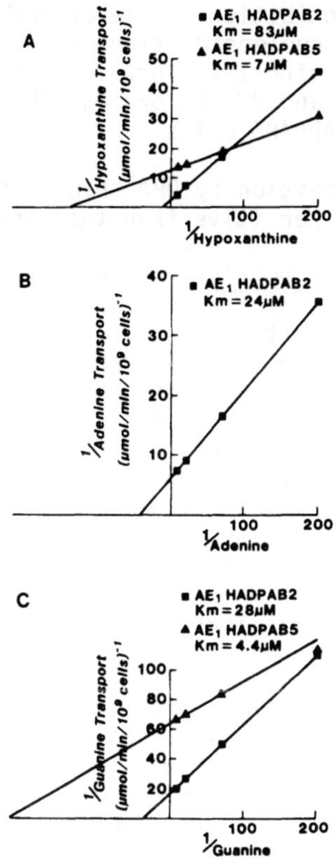

Fig. 3 Lineweaver-Burk Plots of High Affinity Purine Base Transporter Systems in Mutant Cells. The abilities of AE_1HADPAB2 andAE_1HADPAB5 cells to transport radiolabelled purine bases was carried out over a range of concentrations and plotted (7).

lines have a markedly augmented capacity to transport 5uM concentrations of both hypoxanthine and guanine. Surprisingly however, AE_1HADPAB5 cells did not transport 5uM adenine at a rate greater than either parental or wild type cells. The AE_1HADPAB2 and AE_1HADPAB5 cells retained their nucleoside transport deficiency, since they failed to take up [^{3}H]guanosine, [^{3}H]inosine, [^{3}H]uridine, and [^{3}H]cytidine from the culture medium.

Lineweaver-Burke analyses. The augmented capacities of mutant cells to transport low concentrations of purine bases were analyzed by Lineweaver-Burke analyses. The data in Figure 3 indicated that the kinetic parameters of the base transporters in AE_1HADPAB2 and AE_1HADPAB5 cells differed from one another. Apparent K_m values for base transport by the AE_1HADPAB2 cell line were 83uM hypoxanthine, 24uM adenine, and 28uM guanine and for the AE_1HADPAB5 cell line apparent K_m values were 7uM hypoxanthine and 4.4uM guanine. Previously determined apparent K_m values for the transport of hypoxanthine and adenine into wild type S49 cells were 0.7-1.0mM and 0.3mM, respectively (3, 4). The V_{max} values for the transport of hypoxanthine and guanine into AE_1HADPAB2 cells were 2-3 fold greater than into AE_1HADPAB5 cells.

Sensitivities to inhibitors of nucleoside transport. The sensitivities of the high affinity purine base transporters to inhibition by NBMPR and DPA differed in the AE_1HADPAB2 and AE_1HADPAB5 cell lines (data not shown). The ability of AE_1HADPAB2 cells to transport purine bases was refractory to inhibition by either 10uM DPA or 10uM NBMPR, conditions which abrogate nucleoside transport into wild type S49 cells, while hypoxanthine and guanine transport into AE_1HADPAB5 cells were inhibited 50% by 10uM DPA and 20-30% by 10uM NBMPR. Thus, the transport of hypoxanthine into AE_1HADPAB5 cells

was more sensitive to inhibition by DPA than hypoxanthine transport into AE_1HADPAB2 cells but less sensitive than that into wild type cells (6).

Intraspecies tetraploid hybrids. To establish differences for the genetic bases of expression of the nucleoside transport deficiency and the augmented nucleobase transport capacity, intraspecies cell-cell hybrids were generated between the nucleoside transport-deficient cells possessing a high affinity nucleobase permease and cells possessing wild type transport capabilities for both nucleosides and nucleobases. The tetraploid WTxHADPAB2 and WTxHADPAB5 hybrid cells were both capable of salvaging much lower concentrations of hypoxanthine than a WTxWT hybrid. Indeed, the WTxWT cells grew very slowly in 1mM hypoxanthine-11uM azaserine, indicating that hypoxanthine salvage was rate limiting for growth. Moreover, 5uM DPA inhibited the limited hypoxanthine salvage capability of WTxWT cells but did not interfere with hypoxanthine utlization by WTxHADPAB2 and WTxHADPAB5 cells. In comparative growth experiments with cytotoxic nucleosides, all three tetraploid lines were killed by either 100nM 5-fluorouridine or 100nM 5-fluorodeoxyuridine and were capable of surviving in HAT medium (data not shown). Thus, by these growth criteria, the expression of the high affinity nucleobase transporter and the nucleoside transport deficiency in the AE_1HADPAB2 and AE_1HADPAB5 cells behaved as dominant and recessive characteristics, respectively. Direct transport measurements using radiolabelled nucleobases and nucleosides confirmed these observations. WTxHADPAB2 and WTxHADPAB5 were both capable of transporting the nucleosides adenosine and inosine, and possessed a notably elevated capacity to transport low concentrations of hypoxanthine (data not shown). Only the WTxHADPAB2 tetraploid cell line, as expected from the diploid parental phenotypes, transported adenine at a rate faster than the WTxWT hybrid. Thus, the growth and transport phenotypes of the tetraploid hybrid cells established that the mutations to nucleoside transport deficiency and to augmented nucleobase transport capacity behaved differently; the first is recessive, while the second is a dominant trait.

DISCUSSION

Considerable controversy exists with respect to the multiplicity and substrate specificities of purine base transporters in mammalian cells. Genetic studies in S49 cells have demonstrated a role of the nucleoside transport system in mediating the permeation of purine bases across the plasma membrane (3, 4). Nucleoside transport-deficient S49 cell lines have been generated which transport hypoxanthine and adenine at lower rates than wild type parental cells (3). As a consequence of this genetic deficiency in hypoxanthine transport, nucleoside transport-deficient cells cannot salvage hypoxanthine efficiently in the presence of azaserine, the glutamine antimetabolite, which blocks purine biosynthesis. The effects of a genetic deficiency in nucleoside transport on base transport and hypoxanthine salvage can be mimicked pharmacologically by either of two potent inhibitors of nucleoside transport, DPA and NBMPR (3). Exploiting this pharmacologic inhibition a selective system was designed which permitted the isolation of cells expressing a high affinity purine base carrier (4, 5).

To determine whether the expression of the unique high affinity base transporter required the expression of the wild type nucleoside transport system, the mutation to high affinity purine base transport was introduced into nucleoside transport-deficient cells. The ability to insert this mutation into nucleoside transport-deficient cells without restoring nucleoside transport capacity strongly suggests that the expression of the high affinity transport system is unrelated to the expression of a functional nucleoside transporter. Biochemical criteria substantiate this hypothesis, since the two transporters can be distinguished by multiple biochemical parameters, including substrate specificity, affinities for common substrates, and sensitivities toward specific inhibitors of nucleoside transport.

The genetic basis for the expression of the high affinity purine base permease is not completely understood. Two models consistent with the data can be proposed. First, the expression of the low K_m base transporter could be caused by the activation of a gene locus that is 'silent' in wild type cells. Alternatively, the base transport capacity could have been derived by genetic alteration of an already expressed gene, such as one which codes for a preexisting membrane protein. That N-methyl-N-nitrosoN'-nitroguanidine, but not 5-azacytidine, increases the frequency of cells in a population that express this high affinity transporter, is consistent with but does not prove the latter hypothesis (4). Further support for the 'mutational' hypothesis stems from the biochemical phenotypes of the various cells lines expressing high affinity base transport. The two cell lines described herein expressed high affinity transporters differing in substrate specificity, inhibitor sensitivity, and kinetic parameters. Subsequently, the high affinity base transporters of other independently selected mutant cell lines have been evaluated, and the kinetic properties and substrate specificities of these transporters vary considerably. Thus, the activation of a single silent gene locus, which should give rise to a single phenotype among cells expressing high affinity purine base transport, is difficult to reconcile with the expression of multiple mutant high affinity purine base transport phenotypes.

ACKNOWLEDGMENTS

This work was funded by Public Health Service grant DK38809 from the National Institutes of Health. B.U. is a recipient of a Research Career Development Award from the National Institutes of Health.

REFERENCES

1. Ullman B. Mutational Analysis of Nucleoside and Nucleobase Transport, Chemical Rubber Company Critical Reviews, in press.

2. Cohen, A., B. Ullman, and D.W. Martin Jr. Characterization of a mutant mouse lymphoma cell with deficient transport of purine and pyrimidine nucleosides. J. Biol. Chem. 254:112 (1979).

3. Aronow, B. and B. Ullman. Role of the nucleoside transport function in the transport and salvage of purine nucleobases. J. Biol. Chem. 261:1986 (2014).

4. Aronow, B., D. Toll, J. Patrick, P. Hollingsworth, K. McCartan, and B. Ullman. Expression of a novel high-affinity purine nucleobase transport function in mutant mammalian T-lymphoblasts. Mol. Cell. Biol. 6:2957 (1986).

5. Ullman, B., Patrick, J., and McCartan, K. Expression of the High Affinity Purine Base Transporter in Mutant Mouse S49 Cells Does Not Require a Functional Wild Type Nucleoside-Nucleobase Transporter, Mol. Cell. Biol. 7:97 (1987).

6. Aronow, B. and B. Ullman. Thymidine incorporation in nucleoside transport-deficient lymphoma cells. J. Biol. Chem. 260:16274 (1985).

7. Ullman, B., Patrick, J., and McCartan, K. Expression of the High Affinity Purine Base Transporter in Mutant Mouse S49 Cells Does Not Require a Functional Wild Type Nucleoside-Nucleobase Transporter, Molec. Cell. Biol. 7:97 (1987).

8. Melnick, I. and J.M. Buchanan. Biosynthesis of the purines. XIV. Conversion of (alpha-N-formyl)glycinamid ribotide to (alpha-N-formyl)-glycinamidine ribotide; purification and requirements of the enzyme system. J. Biol. Chem. 225:157 (1962).

9. Aronow, B., K. Allen, J. Patrick, and B. Ullman. Altered nucleoside transporters in mammalian cells selected for resistance to the physiological effects of inhibitors of nucleoside transport. J. Biol. Chem. 260:6226 (1985).

CHARACTERIZATION OF A MUTANT CLONE OF LEISHMANIA DONOVANI DEFICIENT IN FOLATE TRANSPORT

Joanne Beck and Buddy Ullman

Department of Biochemistry
The Oregon Health Sciences University
Portland, Oregon 97201

INTRODUCTION

Current chemotherapeutic manipulations for the treatment of leishmaniasis are far from ideal. The design of a biochemically rationale chemotherapeutic strategy for the treatment of leishmaniasis or any other parasitic disease depends on a fundamental understanding of the processes by which parasites incorporate and become resistant to drugs. Methotrexate (MTX) is a prototype drug which has been exploited to study molecular mechanisms of drug resistance in mammalian cell systems. MTX is a potent inhibitor of the enzyme dihydrofolate reductase (DHFR) and causes a depletion of cellular tetrahydrofolate pools. Resistance to MTX cytotoxicity has been reported to occur by a number of mechanisms in mammalian cells, including impaired MTX transport, defective MTX binding to the intracellular DHFR, and increased DHFR levels associated with the amplification of the DHFR gene.

The amplification of DNA sequences has been also shown to be a mechanism by which *Leishmania* can become resistant to MTX. MTX-resistant *Leishmania major* have been isolated which overproduce the bifunctional DHFR-thymidylate synthasecomplex (1) as a result of gene amplification (2, 3). Recently, however, MTX-resistant *Leishmania major* have been characterized with phenotypes that have been ascribed to a transport defect (4, 5).

We have used somatic cell genetic approaches to generate, isolate, and characterize a MTX-resistant clone of *Leishmania donovani*, which fails to transport folate or MTX. The characterization of this unique mutation in a species of parasitic protozoa has important implications concerning how these organisms utilize pterins and folates and become refractory to drugs.

MATERIALS AND METHODS

Cell culture. *Leishmania donovani* (Sudanese strain) were grown in continuous culture as the promastigote form (insect vector form) at 26° in a humidified 5% CO_2 atmosphere in a completely defined DME-L growth medium.

Isolation of the methotrexate-resistant cell line. The MTXA5 clone was isolated in semi-solid growth medium containing 1mM methotrexate.

Growth sensitivities to cytotoxic agents. The growth sensitivities of wild type parental and MTXA5 mutant promastigotes to methotrexate and other agents were ascertained in Costar multiwell plates (24 wells) as described previously (6).

Ability of cells to grow in pterin supplemented growth medium. The ability of various exogenous sources of pterin containing compounds to serve as growth effectors for wild type and MTXA5 cells were performed in pterin-supplemented folate-deficient Dulbecco's modified Eagle medium. Experiments were initiated with cells that had been incubated 3 days in folate-deficient medium to ensure depletion of stored cellular folate pools. Generally, Leishmania donovani were capable of 2-3 cell divisions (doubling time 24 hr) after transfer to folate-deficient medium.

Rapid sampling transport assays. The rates of transport of [^{3}H]methotrexate and [^{3}H]folate were measured by a rapid sampling kinetic assay similar to that previously described by Aronow et al. (7) for measuring nucleoside transport.

RESULTS

Growth phenotype of MTXA5 cells. The MTXA5 cell line was isolated from a mutagenized wild type (DI700) population by virtue of its resistance to 1 mM MTX. In comparative growth rate experiments, the MTXA5 cells were completely insensitive to growth inhibition by 4 mM MTX, whereas the effective dose of MTX which inhibited wild type cell growth by 50 % (E.D.$_{50}$ value) was approximately 50 uM (Figure 1). Thymine or thymidine could circumvent the pharmacologic block of MTX indicating that the site of MTX action was the leishmanial DHFR-TS complex (8).

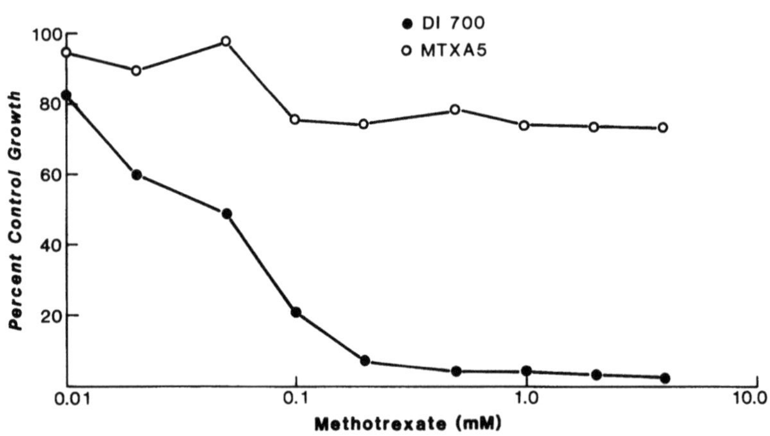

Figure 1. Growth sensitivity of DI-700 and MTXA5 cells to inhibition by MTX. The abilities of wild type (DI700) and MTXA5 cells to proliferate in DME-L growth medium containing various concentrations of methotrexate were compared. The results depicted are those of a single typical experiment, which has been repeated in part or in total on 5-10 other separate occasions.

Transport of MTX and folate into wild type and mutant Leishmania. One mechanism by which parasitic protozoa can become resistant to cytotoxic agents is by their failure to take up the drug efficiently. For instance, Leishmania donovani isolated by virtue of their resistance to either one of the cytotoxic nucleoside analogs, formycin B or tubercidin, possess stable genetic alterations in one of their two specific nucleoside transport systems (6, 7). Therefore, the abilities of wild type and MTXA5 cells to transport [^{3}H]MTX from folate-deficient culture medium were compared. The results of these experiments indicated that very little transport of either 0.5 uM [^{3}H]MTX (Figure 2) or 0.55 uM [^{3}H]folate (Figure 3) into MTXA5 cells suspended in folate-deficient medium could be detected for intervals up to 16 min, whereas transport into wild type parental cells was easily measurable, linear with time for a few minutes, and possessed a saturable component (8). Therefore, the translocation assays suggested that the genetic defect in the MTXA5 cell line was in a common folate-MTX transport system.

Competition assays revealed that excess MTX, folate, tetrahydrofolate, dihydrofolate, folinic acid, and p-aminobenzyoylglutamate, but not pterins or pteroic acid, could block the influx of both [^{3}H]folate and [^{3}H]MTX into DI-700 cells. The transport of both MTX and folate were analyzed by Lineweaver-Burk analysis (data not shown). The results indicated that the apparent Km values of the cells for folate and MTX transport were 0.2uM and 0.4uM, respectively.

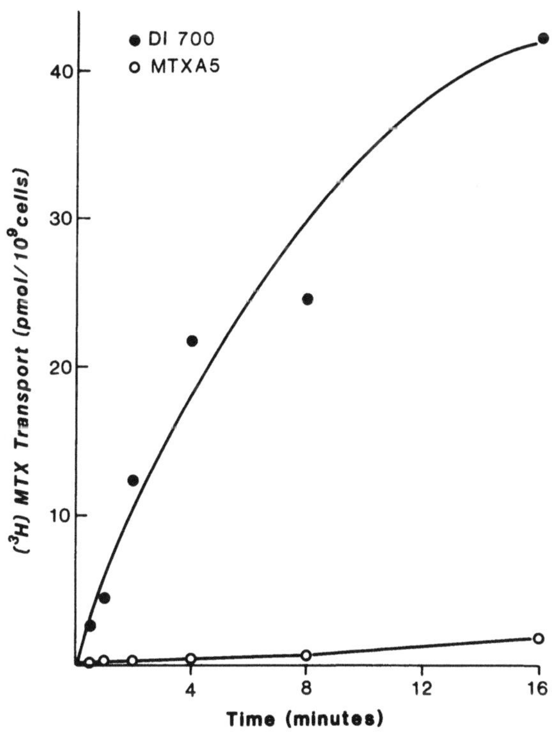

Figure 2. Transport of [^{3}H]methotrexate by wild type and mutant Leishmania donovani. The abilities of DI700 and MTXA5 cells to transport 0.50uM [^{3}H]methotrexate (20Ci/mmol) were ascertained by the rapid sampling kinetic assay described in Materials and Methods.

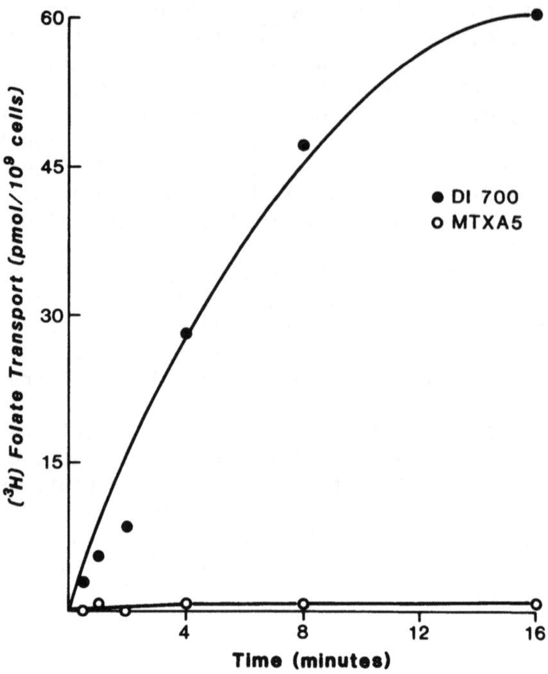

Figure 3. Transport of [^{3}H]folate by DI700 and MTXA5 cells. The abilities of DI-700 and MTXA5 cells to transport 0.55uM [^{3}H]folate (18Ci/mmol) were measured as described in Materials and Methods.

Affinity labeling. To initiate a biochemical analysis of the folate/ methotrexate transport function, an affinity probe of MTX was synthesized, which has been shown to induce an irreversible inactiviation of the mammalian L1210 MTX transport system (9). By allowing [^{3}H]MTX or [^{3}H]folate to react with 1-ethyl-3(3-dimethylaminopropyl)carbodiimide to form a relatively unstable complex, a protein in wild type Leishmania could be affinity labeled that migrated on SDS polyacrylamide gels with a molecular weight of 46,000 daltons. No affinity labeled protein band was observed when MTXA5 cells were treated with the derivatized MTX or folate.

Growth of wildtype and MTXA5 cells in folate and biopterin. That MTXA5 cells did not transport 0.55 uM radiolabelled [3',5',7,9-^{3}H]folate efficiently suggested that Leishmania donovani might not require a source of exogenous folate for growth. However, wild type and mutant cells resuspended in folate-deficient growth medium, could only undergo 2-3 cell doublings prior to permanent growth arrest. Interestingly, however, the concentrations of folate required to stimulate the maximal growth of wild type and MTXA5 cells were identical, approximately 100-300 nM. Moreover, biopterin and neopterin promoted the growth of wild type Leishmania in folate-deficient medium, while MTXA5 cells, which grew well in 10uM folate, failed to thrive in 10uM biopterin. 6-hydroxymethylpterin, however, served as a growth factor for both leishmanial strains. These data suggest that MTXA5 cells, unlike wild type Leishmania, lack the ability to convert certain pterins to folate.

DISCUSSION

The biochemically defined mutation in the MTX-resistant MTXA5 Leishmania donovani strain described in this manuscript appears to be in system required for the transport of both folate and MTX. These mutant organisms failed to take up either [^{3}H]MTX or [^{3}H]folate present in the culture medium in both long term incorporation studies and short term transport studies. Affinity labeling of wild type cells with derivatized MTX identified a candidate for the folate/MTX transporter, a protein with a molecular weight around 46kd. This biochemical genetic analysis argues for the existence of a common folate-MTX carrier in Leishmania donovani. The genetic identification of a common folate-MTX transporter in Leishmania is substantitated by the kinetic and genetic studies in Leishmania major (5). Whether the mutation to transport deficiency in the MTXA5 cells is in a structural gene encoding the folate-MYX transporter, in a regulatory gene, or in a gene governing membrane composition or molecular architecture cannot be determined from these somatic cell genetic studies.

The biochemical and growth phenotype of the mutant MTXA5 strain has some important implications concerning the mechanism(s) by which Leishmania utilize pterins and folates. First, these organisms have an absolute requirement for pterins since they cannot grow unless their growth medium is supplemented with either folate or biopterin. Second, Leishmania have an absolute requirement for pterins other than folate, since thymine and thymidine can circumvent MTX toxicity but do not permit wild type Leishmania to grow in folate-deficient medium. Third, biopterin can replace folate as a growth factor for wild type Leishmania. This raises the intriguing speculation that Leishmania, like prokaryotes but dissimilar to mammalian cells, have the ability to synthesize tetrahydrofolate from biopterin.

ACKNOWLEDGMENTS

These studies were supported by research grant No. AI23682 from the National Institute of Allergy and Infectious Diseases. The author is a recipient of a Research Career Development Award from the National Institutes of Health.

REFERENCES

1. Coderre, J.A., Beverley, S.M., Schimke, R., and Santi, D.V. Overproduction of a bifunctional thymidylate synthetase-dihydrofolate reductase and DNA amplification in methotrexate-resistant Leishmania tropica. Proc. Natl. Acad. Sci. U.S.A. 80:2132 (1983).

2. Beverley, S.M., Coderre, J.A., Santi, D.V., and Schimke, R.T. Unstable DNA amplifications in methotrexate-resistant Leishmania consist of extrachromosomal circles which relocalize during stabilization. Cell 38:431 (1984).

3. Washtien, W.L., Grumont, R.A, and Santi, D.V. DNA amplification in antifolate-resistant Leishmania. J. Biol. Chem. 260:7809 (1985).

4. Dewes, H., Ostergaard, H.L., and Simpson, L. Impaired drug uptake in methotrexate resistant Crithidia fasciculata without changes in dihydrofolate reductase activity or gene amplification. Mol. Biochem. Parasitol. 19:149 (1986).

5. Ellenberger, T. and Beverley, S.M. Reductions in methotrexate and folate influx in methotrexate-resistant lines of *Leishmania major* are independent of R or H region amplification. *J. Biol. Chem.* 262:13501 (1987).

6. Iovannisci, D.M., Kaur, K., Young, L., and Ullman, B. Biochemical genetic analysis of nucleoside transport in *Leishmania donovani*. Mol. Cell Biol. 4:1013 (1984)

7. Aronow, B., Kaur, K., McCartan, K., and Ullman, B. Two high affinity nucleoside transporters in *Leishmania donovani*. *Mol. Biochem. Parasitol.* 22:29 (1987).

8. Kaur, K., Coons, T., Emmett, K., and Ullman, B. Methotrexate-resistant *Leishmania donovani* Geneticallys Deficient in the Folate-Methotrexate Transporter, J. Biol. Chem. 263:7020 (1988).

9. Henderson, G.B., Zevely, E.M., and Huennekens, F.M. Irreversible inactivation of the methotrexate transport system of L1210 cells by carbodiimide-activated substrates. *J. Biol. Chem.* 255:4829 (1980).

INFLUENCE OF THE TEMPERATURE ON HYPOXANTHINE TRANSPORT THROUGH HUMAN ERYTHROCYTE MEMBRANES[1]

C. Salerno, E. Capuozzo, A. Giacomello, and C. Crifò

Laboratory of Biochemical and Clinical Analysis, University of Chieti, and Department of Biochemical Sciences, University of Rome, Italy

INTRODUCTION

Because of lacking of many of the enzymes necessary for *de novo* purine synthesis (1), mature human erythrocytes require preformed purines to maintain the intracellular nucleotide pool. The salvage pathways for exogeneous purine bases in their simplest form can be viewed as two-component systems (2), comprised of transmembrane equilibration of preformed nucleobases coupled with enzymatic phosphoribosylation within the cells. Literature data are consistent, on the whole, with the independent operation of the two steps. For example, the uptake of purine bases can be measured in the absence of metabolic conversions in cells lacking of the specific phosphoribosyltransferases (3-4). Moreover, phosphoribosylation kinetics can be assessed under conditions where the intracellular pool of free purine bases has attained a steady state (5-6). In other cases (7-8), the overall time course of purine base incorporation was found to be described by a two-step process.

Transmembrane equilibration of hypoxanthine can be easily studied in human erythrocytes, since phosphoribosylation of the purine base is very inefficient in these cells in media not containing or containing nearly physiological concentrations of phosphate (6,9-10). Most of the experimental results have been obtained at relatively low temperature, owing to the rapid incorporation of hypoxanthine at 37°C. It has been reported (7) that radiolabeled hypoxanthine accumulated in the cells up to 3-fold at medium concentrations below 1 μM. By contrast, at relatively high hypoxanthine concentrations, the extracellular level of the purine base at equilibrium greatly exceeded the intracellular one (10). These effects have been attributed either to functioning of additional high-affinity, low-capacity, active transport systems (7) or to saturable binding to cellular components (7,10).

In this paper, we present evidence that hypoxanthine uptake by human erythrocyte strongly depends on the temperature of the cellular suspension. Zero-trans influx of labeled hypoxanthine at 37°C behaves as a monoexponential process independently of the initial concentration of the purine base. Under these conditions, the intracellular concentration of the labeled base at equilibrium always equals that in the medium.

---

[1] This work was partially supported by a grant from Ministero della Pubblica Istruzione.

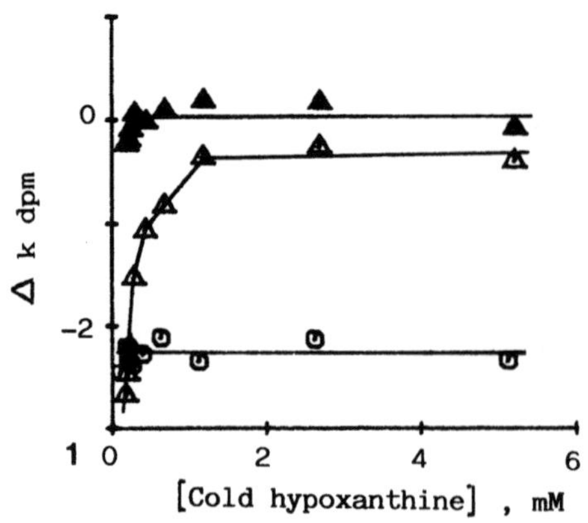

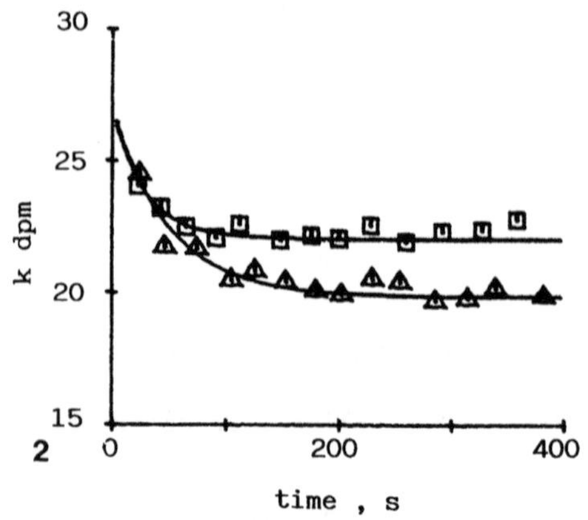

Fig. 1. (left) Total amount of 14-C hypoxanthine taken up at equilibrium by native (△, o) or protease-treated (▲) erythrocytes as a function of initial concentration of cold hypoxanthine in the suspending medium. The temperature of the incubation mixture was 0°C (△, ▲) or 37°C (o). Initial 14-C hypoxanthine concentration was held constant at 1 μM.

Fig. 2. (right) Time course of 14-C hypoxanthine uptake by erythrocytes at 0°C. Initial 14-C hypoxanthine concentration was 8 μM. Initial concentration of cold hypoxanthine was 0 (△) and 50 μM (◘).

EXPERIMENTAL PROCEDURES

Human erythrocytes were prepared from blood freshly drawn in heparin from healthy adult donors and washed twice in isotonic glucose-NaCl solution (5 mM glucose and 155 mM NaCl). Plasma and buffy coat were removed by aspiration.

Protease-treated erythrocytes were prepared by incubating the packed cells at 37°C for 60 min in the presence of 0.1 mg/ml proteinase K. At the end of the incubation period, the cells were washed in isotonic glûcose-NaCl solution and immediately used for incorporation experiments.

The packed cells were suspended in equal volume of isotonic glucose-NaCl solution containing 14-C labeled and/or cold hypoxanthine and incubated in a thermostated bath. At regular intervals, samples of the incubation mixture were removed. Erythrocytes were separated from extracellular medium by centrifugation as previously described (10).

RESULTS AND DISCUSSION

In phosphate-free isotonic media, 14-C hypoxanthine is taken up by erythrocytes, but less than 5% is metabolized to IMP (10). The kinetics of hypoxanthine uptake are strongly dependent on the temperature and on the initial hypoxanthine concentration in the suspending medium.

At 0-5°C, but not at higher temperatures, the uptake of radiolabeled purine base is inhibited by the presence of cold hypoxanthine in the incubation mixture (fig. 1). The inhibition results in a decrease of the total amount of labeled purine base incorporated in the cell at equilibrium, rather than in a significant variation of the relaxation time of the overall

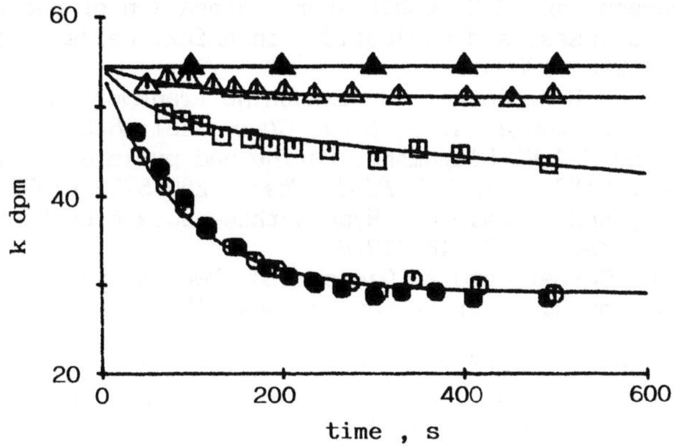

Fig. 3. Time course of 14-C hypoxanthine uptake by erythrocytes at different temperatures. Initial 14-C hypoxanthine concentration was 2 mM. The temperature of the incubation medium was 0°C (△), 10°C (■), 25°C (o). Closed symbols refer to experiments performed in the presence of protease-treated erythrocytes.

process (fig. 2).

This inhibitory effect can be also observed by varying the initial concentration of labeled hypoxanthine in the medium at 0-5°C in the absence of the cold nucleobase. At relatively low 14-C hypoxanthine concentrations (about 1 μM), the uptake results in concentrative accumulation of the labeled base in the cells. At higher 14-C hypoxanthine concentrations (1-10 mM), the labeled compound remains in large extent in the extracellular medium. By contrast, at 37°C, the uptake seems to be independent of the initial concentration of 14-C hypoxanthine in the mixture and always results in uniform distribution of radioactivity in the two compartments.

Other differences can be observed by studying the effect of partial proteolysis of the cell membrane on the uptake of the purine base. At 0-5°C, protease-treated erythrocytes are practically unable to take up hypoxanthine. At 25-37°C, the time course of hypoxanthine incorporation in protease-treated erythrocytes appears to be very similar to that in native cells (fig. 3).

We do not have unequivocal explanation for these phenomena. They could reflect the presence of two trasport systems with different affinity for hypoxanthine, but other factors might be responsible, since we were unable to observe 14-C hypoxanthine binding to the membrane fraction in erythrocyte lysates or a michaelian behaviour of the uptake kinetics. At low temperatures, perhaps because of changes in membrane fluidity (11), hypoxanthine could be trapped in the erythrocyte membrane shell, which could be regulated by protease-sensible transmembrane components (12-13). Another hypothesis is that enzymes involved in reutilization and catabolism of hypoxanthine could have different sensibility to temperature changes and, thus, could contribute to the variation of the steady state level of the purine base in the cell.

REFERENCES

1. A.W. Murray, The biological significance of purine salvage, *Ann. Rev. Biochem.* 40:811 (1971)

2. P.G.W. Plagemann, and R.M. Wohlhueter, Permeation of nucleosides, nucleic acid bases, and nucleotides in animal cells, *Curr. Top. Membr. Transp.* 14:225 (1980)
3. B.L. Alford, and E.M. Barnes, Hypoxanthine transport by cultured Chinese hamster lung fibroblasts, *J. Biol. Chem.* 251:4823 (1976)
4. J.M. Zylka, and P.G.W. Plagemann, Purine and pyrimidine transport by cultured Novikoff cells, *J. Biol. Chem.* 250:5756 (1975)
5. A. Giacomello, and C. Salerno, Hypoxanthine uptake by human erythrocytes, *FEBS letters* 107:146 (1979)
6. C. Salerno, G. Gerber, and A. Giacomello, Regulatory aspects of hypoxanthine uptake by human erythrocytes, *Adv. Exp. Med. Biol.* 195B:75 (1986)
7. P.G.W. Plagemann, C. Woffendin, M.B. Puziss, and R.M. Wohlhueter, Purine and pyrimidine transport and permeation in human erythrocytes, *Biochim. Biophys. Acta* 905:17 (1987)
8. U.V. Lassen, Hypoxanthine transport in human erythrocytes, *Biochim. Biophys. Acta* 135:146 (1967)
9. A. Hershko, A. Razin, and J. Mager, Regulation of the synthesis of phosphoribosylpyrophosphate in intact red blood cells and in cell-free preparations, *Biochim. Biophys. Acta* 184:64 (1969)
10. E. Capuozzo, M.C. Gigante, C. Salerno, and C. Crifò, Hypoxanthine transport through human erythrocyte membranes, *Adv. Exp. Med. Biol.* 195B:71 (1986)
11. J. Yguerabide, and E.E. Yguerabide, Role of membrane fluidity in the expression of biological functions, *in:* "The enzymes of biological membranes, Vol.1", A.N. Martonosi, ed., Plenum Press, New York (1985)
12. M.P. Sheetz, and J. Casaly, 2,3-Diphosphoglycerate and ATP dissociate erythocyte membrane skeletons, *J. Biol. Chem.* 255:9955 (1980)
13. D.L. Siegel, S.R. Goodman, and D. Branton, The effect of endogenous proteases on the spectrin binding proteins of human erythrocytes, *Biochim. Biophys. Acta* 598:517 (1980)

ADENINE UPTAKE BY HUMAN ERYTHROCYTES

C. Salerno, A. Werner, W. Siems, G. Gerber,
A. Giacomello and C. Crifò

Laboratory of Biochemical and Clinical Analysis, University of Chieti, and Department of Biochemical Sciences, University of Rome, Italy; Institute of Biochemistry, Humboldt University, Berlin, GDR

INTRODUCTION

In view of the absence of the *de novo* pathway (1), the mechanism by which mature human eryhtrocytes turn over purine nucleotides is not completely clear. In these cells, AMP could be synthesized from adenine and phosphoribosylpyrophosphate (PRPP) through the action of the specific phosphoribosyltransferase (APRT), or from adenosine via adenosine kinase and ATP.

As far as the role of PRPP in adenine nucleotide synthesis is concerned, it has been found that PRPP is not necessary to maintain the nucleotide pool, since unexpectantly elevated levels of ATP, ADP, and AMP are present in erythrocytes from subjects lacking for PRPP synthetase (2). This observation seems to be in line with the findings that erythrocytes from both homozygotes and heterozygotes for APRT deficiency have normal ATP and PRPP concentrations (3,4). A confusing aspect of homozygotes for the latter inherited disease is the presence of abnormally large amounts of adenine and its oxidative products in the urine (5), which suggests, by contrast, an important role of APRT in the salvage of adenine in healty subjects.

When 14-C labeled adenine was given intravenously to rabbits, most of the radioactivity was found in adenine mononucleotides of different tissues, including red cells, and only 20% of adenine and its catabolites were found in the urine (6). When the purine base was injected in healty human subjects, free adenine and its oxidative products were excreted in the urine to approximately the same extent as had been found with rabbits (7).

Other evidence of an active adenine salvage pathway in human erythrocytes comes from studies performed to improve cell survival during storage. It has been reported (8) that long-stored human erythrocytes, depleted of most of their phosphate metabolites, including ATP, could resynthesize ATP to the concentration of the fresh cell by incubation with inosine plus adenine, but not with inosine alone. Moreover, it has been shown (9) that addition of a small amount of adenine at the outset of storage of human blood could extend the outdating time to 5-6 weeks and that the improved erythrocyte survival was associated with higher levels of ATP. The addition of adenine to blood before storage is a practice common in western countries.

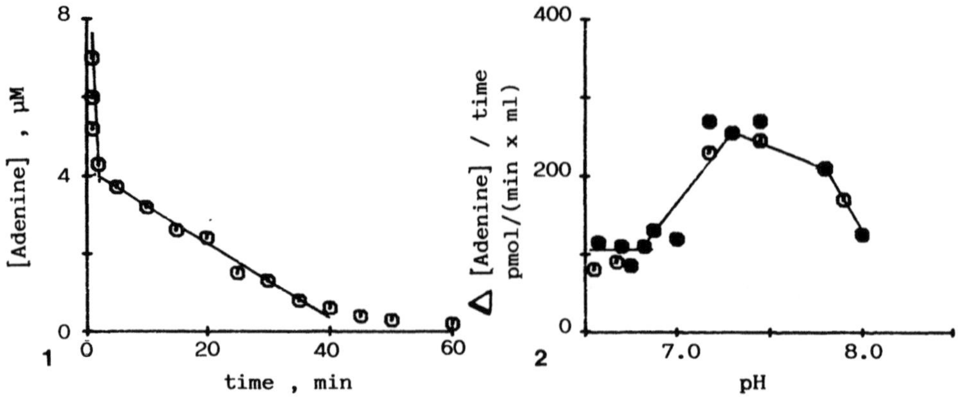

Fig. 1. (left) Time course of 14-C adenine uptake by human erythrocytes. Erythrocytes were suspended in isotonic Tris-NaCl-glucose buffer, pH 7.4, containing 8 µM 14-C adenine and incubated at 37°C under nitrogen atmosphere. At regular intervals, samples of the incubation mixture were removed and the separated medium was employed for liquid scintillation counting.

Fig. 2. (right) Slope of the slower phase of 14-C adenine uptake by human erythrocytes as a function of pH of incubation medium. The cells were equilibrated in nitrogen atmosphere (o) or in oxygen atmosphere (o). Other conditions were as described in fig. 1.

In this paper we report data concerning adenine uptake by human erythrocytes suspended in isotonic medium. Evidence is presented that exogeneous adenine is utilized by the cells for nucleotide synthesis even in the presence of very low PRPP concentrations.

EXPERIMENTAL PROCEDURES

Human erythrocytes were obtained from blood freshly drawn in heparin. The cells were washed with isotonic Tris-NaCl-glucose buffer and equilibrated with appropriate gas atmosphere as previously described (10). Xanthine oxidase-treated erythrocytes were prepared by incubating the cells for 10-60 min at 37°C in an equal volume of isotonic saline, pH 7.4, containing 40-100 mU/ml xanthine oxidase (XOD). The enzyme was then removed from the erythrocyte suspension by exhaustive washing in Tris-NaCl-glucose buffer and the cells were immediately used to study the uptake of purine bases.

Packed erythrocytes were suspended in isotonic media containing 14-C labeled and/or cold purine bases to give 40-50% hematocrit and incubated at 37°C. At regular intervals, samples of the incubation mixture were removed. The medium, separated from the cells by centrifugation (10), and the cellular extract, obtained from the packed cells by osmotic lysis, were employed for liquid scintillation counting and high-performance liquid chromatography (11).

RESULTS AND DISCUSSION

14-C Adenine (1-10 µM) is taken up by human erythrocytes in phosphate-free isotonic media through a biphasic process (fig. 1). At first, the radioactive purine is rapidly incorporated by the cells. When the intracel-

Table I. Uptake of 14-C adenine by human erythrocytes

Experimental conditions°	Slope of the slower phase of 14-C adenine uptake (dpm/s)
native cells	-2.81
native cells + 340 µM cold adenine	-0.69
native cells + 2 mM cold hypoxanthine	-2.82
XOD-treated cells*	-6.15

°The cells were suspended in isotonic medium, pH 7.4, containing 7 µM 14-C adenine and incubated at 37°C.
*The cells were preincubated for 60 min in the presence of 40 mU/ml XOD.

lular concentration of the labeled base becomes approximately equal to the external one, the uptake continues at lower velocity leading to a total consumption of extracellular 14-C adenine. ATP, ADP, AMP, and adenosine account for more than 70% of the radioactive material present in the cells at the end of the slower phase. When cold adenine is used instead of radiolabeled adenine, the kinetics of uptake are very similar and the cold base is incorporated to the same extent as has been found with the 14-C base. Cold adenine in the suspending medium markedly inhibits the slower phase of the uptake of the labeled base, suggesting that adenine incorporation and nucleotide synthesis are related to saturable processes (table I).

An outstanding result is that, under the experimental conditions employed, erythrocytes incorporate much more adenine than hypoxanthine into the nucleotide pool (12). Phosphoribosylation of both purine bases is markedly stimulated by inorganic phosphate (13). Phosphate-induced nucleotide synthesis is particularly evident at unphysiologically low pH and in anaerobiotic conditions. By contrast, adenine nucleotide synthesis in phosphate-free media is faster at physiological pH and independent of the oxygen partial tension (fig. 2).

The pathway involved in the phosphoribosylation of adenine in phosphate-free media is not clear. It has been recently proposed (14) that IMP, inosine, and hypoxanthine have a share in a futile cycle, which includes the coupled reactions catalyzed by 5′-nucleotidase, nucleoside phosphorylase, and hypoxanthine-guanine phosphoribosyltransferase and accounts for the very low intracellular PRPP concentration (10). Adenine could react with PRPP (which is continuously formed by the specific kinase) and, by contrast with hypoxanthine, could accumulate into the nucleotide pool owing to the inefficiency of nucleoside phosphorylase towards adenosine.

Some of our results are not in line with this hypothesis. As shown in table I, adenine uptake is not inhibited by exogeneous hypoxanthine (up to 2 mM), which would compete with adenine for PRPP. Xanthine oxidase-treated cells, which are expected to contain more PRPP, incorporate adenine more rapidly, but take up roughly the same amount of hypoxanthine as native erythrocytes.

Another proposal is that the phosphoribosylation of exogeneous adenin does not involve *de novo* PRPP synthesis, but requires simultaneous pyrophosphorolysis of endogeneous nucleotides. We obtained evidence of such a mechanism by using 14-C IMP enriched erythrocytes (15-16). Adenine nucleotide synthesis by native erythrocytes in phosphate-free artificial media could be accounted for, at least in part, by IMP pyrophosphorolysis to hyp xanthine, which accumulates in the extracellular compartment (1).

Whatever the mechanism of the observed phenomenon, our data indicate that human erythrocytes incorporate exogeneous adenine into the nucleotide pool even in the presence of an intracellular PRPP concentration very low and insufficient to allow an appreciable uptake of hypoxanthine as IMP.

REFERENCES

1. A.W. Murray, The biological significance of purine salvage, *Ann. Rev. Biochem.* 40:811 (1971)
2. W.N. Valentine, H.M. Anderson, D.E. Paglia, E.R. Jaffè, P.N. Konrad, an S.R. Harris, Studies on human erythrocyte metabolism. Nonspherocytic hemolytic anemia, high red cell ATP, and ribosephosphate pyrophosphokinase deficiency, *Blood* 39:674 (1972)
3. I.H. Fox, J.C. Meade, and W.N. Kelley, Adenine phosphoribosyltransferas deficiency in Man. Report of a second family, *Am. J. Med.* 5:614 (1973)
4. B.M. Dean, D. Perret, H.A. Simmonds, A. Sahota, and K.J. Van Acker, Adenine and adenosine metabolism in intact erythrocytes deficient in adenosine monophosphate pyrophosphate phosphoribosyltransferase. A study of two families, *Clin. Sci. Mol. Med.* 55:407 (1978)
5. K.J. Van Acker, H.A. Simmonds, C. Potter, and J.S. Cameron, Complete deficiency of adenine phosphoribosyltransferase. Report of a family, *N. Engl. J. Med.* 297:127 (1977)
6. G.R. Bartlett, Metabolism by the rabbit of intravenously administred adenine, *Transfusion* 17:351 (1977)
7. G.R. Bartlett, Metabolism by man of intravenously administred adenine, *Transfusion* 17:367 (1977)
8. A.W. Shafer, and G.R. Bartlett, Phosphorylated carbohydrate intermediates of the human erythrocyte during storage in acid citrate dextrose, *J. Clin. Invest.* 41:690 (1962)
9. E.R. Simon, Red cell preservation: further studies with adenine, *Blood* 20:485 (1962)
10. C. Salerno, G. Gerber, and A. Giacomello, Regulatory aspects of hypoxanthine uptake by human erythrocytes, *Adv. Exp. Med. Biol.* 195B:75 (1986)
11. P.D. Schweinsberg, and T.L. Loo, Simultaneous analysis of ATP, ADP, AMP and other purines in human erythrocytes by high-performance liquid chromatography, *J. Chromatogr.* 181:103 (1980)
12. E. Capuozzo, M.C. Gigante, C. Salerno, and C. Crifò, Hypoxanthine transport through human erythrocyte membranes, *Adv. Exp. Med. Biol.* 195B: 71 (1986)
13. A. Giacomello, and C. Salerno, Hypoxanthine uptake by human erythrocytes, *FEBS letters* 107:146 (1979)
14. P.A. Berman, D.A. Black, L. Human, and E.H. Harley, An oxypurine cycle in human erythrocytes regulated by pH, inorganic phosphate and molecular oxygen, *Klin. Wochenschr.* 65(Suppl X):28 (1987)
15. C.Salerno, and A. Giacomello, Hypoxanthine-guanine exchange by intact human erythrocytes, *Biochemistry* 24:1306 (1985)
16. A. Giacomello, and C. Salerno, Adenine-induced hypoxanthine release from IMP-enriched human erythrocytes, *Biochim. Biophys. Acta* 756:403 (1983)

CONTRIBUTORS

INDEX

MIX
Papier aus verantwortungsvollen Quellen
Paper from responsible sources
FSC® C105338

If you have any concerns about our products,
you can contact us on
ProductSafety@springernature.com

In case Publisher is established outside the EU,
the EU authorized representative is:
Springer Nature Customer Service Center GmbH
Europaplatz 3, 69115 Heidelberg, Germany

Printed by Libri Plureos GmbH
in Hamburg, Germany